SUSTAINABLE ENERGY SOLUTIONS IN AGRICULTURE

Sustainable Energy Developments

Series Editor

Jochen Bundschuh
University of Southern Queensland (USQ), Toowoomba, Australia
Royal Institute of Technology (KTH), Stockholm, Sweden

ISSN: 2164-0645

Volume 8

Sustainable Energy Solutions in Agriculture

Editors

Jochen Bundschuh

*University of Southern Queensland, Faculty of Health, Engineering and Sciences &
National Centre for Engineering in Agriculture, Toowoomba, Queensland, Australia &
Royal Institute of Technology, Stockholm, Sweden*

Guangnan Chen

*University of Southern Queensland, Faculty of Health, Engineering and Sciences &
National Centre for Engineering in Agriculture, Toowoomba, Queensland, Australia*

CRC Press
Taylor & Francis Group
Boca Raton   London   New York   Leiden

CRC Press is an imprint of the
Taylor & Francis Group, an **informa** business

A BALKEMA BOOK

CRC Press/Balkema is an imprint of the Taylor & Francis Group, an informa business

First issued in paperback 2017

© 2014 Taylor & Francis Group, London, UK

Typeset by MPS Limited, Chennai, India

Library of Congress Cataloging-in-Publication Data

Sustainable energy solutions in agriculture / editors, Jochen Bundschuh, University of
    Southern Queensland, Faculty of Health, Engineering and Sciences & National Centre for
    Engineering in Agriculture, Toowoomba, Queensland, Australia; Guangnan Chen,
    University of Southern Queensland, Faculty of Health, Engineering and Sciences &
    National Centre for Engineering in Agriculture, Toowoomba, Queensland, Australia.
    pages cm. — (Sustainable energy developments, ISSN 2164-0645 ; volume 8)
Includes bibliographical references and index.
    ISBN 978-1-138-00118-3 (hardback : alk. paper) 1. Agriculture—Energy consumption.
2. Agriculture—Energy conservation. I. Bundschuh, Jochen. II. Chen, Guangnan, 1962–
    S494.5.E5S87   2014
    338.1—dc23

                                                         2013050612

Published by: CRC Press/Balkema
                 P.O. Box 11320, 2301 EH Leiden, The Netherlands
                 e-mail: Pub.NL@taylorandfrancis.com
                 www.crcpress.com – www.taylorandfrancis.com

ISBN 13: 978-1-138-07774-4 (pbk)
ISBN 13: 978-1-138-00118-3 (hbk)

About the book series

Renewable energy sources and sustainable policies, including the promotion of energy efficiency and energy conservation, offer substantial long-term benefits to industrialized, developing and transitional countries. They provide access to clean and domestically available energy and lead to a decreased dependence on fossil fuel imports, and a reduction in greenhouse gas emissions.

Replacing fossil fuels with renewable resources affords a solution to the increased scarcity and price of fossil fuels. Additionally it helps to reduce anthropogenic emission of greenhouse gases and their impacts on climate change. In the energy sector, fossil fuels can be replaced by renewable energy sources. In the chemistry sector, petroleum chemistry can be replaced by sustainable or green chemistry. In agriculture, sustainable methods can be used that enable soils to act as carbon dioxide sinks. In the construction sector, sustainable building practice and green construction can be used, replacing for example steel-enforced concrete by textile-reinforced concrete. Research and development and capital investments in all these sectors will not only contribute to climate protection but will also stimulate economic growth and create millions of new jobs.

This book series will serve as a multi-disciplinary resource. It links the use of renewable energy and renewable raw materials, such as sustainably grown plants, with the needs of human society. The series addresses the rapidly growing worldwide interest in sustainable solutions. These solutions foster development and economic growth while providing a secure supply of energy. They make our society less dependent on petroleum by substituting alternative compounds for fossil-fuel-based goods. All these contribute to minimize our impacts on climate change. The series covers all fields of renewable energy sources and materials. It addresses possible applications not only from a technical point of view, but also from economic, financial, social and political viewpoints. Legislative and regulatory aspects, key issues for implementing sustainable measures, are of particular interest.

This book series aims to become a state-of-the-art resource for a broad group of readers including a diversity of stakeholders and professionals. Readers will include members of governmental and non-governmental organizations, international funding agencies, universities, public energy institutions, the renewable industry sector, the green chemistry sector, organic farmers and farming industry, public health and other relevant institutions, and the broader public. It is designed to increase awareness and understanding of renewable energy sources and the use of sustainable materials. It also aims to accelerate their development and deployment worldwide, bringing their use into the mainstream over the next few decades while systematically replacing fossil and nuclear fuels.

The objective of this book series is to focus on practical solutions in the implementation of sustainable energy and climate protection projects. Not moving forward with these efforts could have serious social and economic impacts. This book series will help to consolidate international findings on sustainable solutions. It includes books authored and edited by world-renowned scientists and engineers and by leading authorities in economics and politics. It will provide a valuable reference work to help surmount our existing global challenges.

Jochen Bundschuh
(Series Editor)

Editorial board

Table of contents

Section 1: Introduction

Section 2: Energy efficiency and management

Section 4: Access to energy

Contributors

Jukka Ahokas: Department of Agricultural Sciences, University of Helsinki, Helsinki, Finland, E-mail: jukka.ahokas@helsinki.fi

Syed Ameer Basha: Mechanical Engineering Department, Jubail University College, Saudia Arabia, E-mail: sdameer2001@yahoo.co.in

Nicole van Beeck: NL Agency, Energy and Climate Change Division, Utrecht, The Netherlands, E-mail: nicole.vanbeeck@agentschapnl.nl

Andrzej Białowiec: Institute of Agricultural Engineering, Faculty of Life Sciences and Technology, Wroclaw University of Environmental and Life Sciences, Wroclaw, Poland, E-mail: bialowiec@gmail.com

Susan I. Blackburn: CSIRO Energy Transformed Flagship, Division of Marine and Atmospheric Research, Hobart, TAS, Australia, E-mail: susan.blackburn@csiro.au

Malcolm R. Brown: CSIRO Energy Transformed Flagship, Division of Marine and Atmospheric Research, Hobart, TAS, Australia, E-mail: malcolmraybrown@gmail.com

Jochen Bundschuh: Faculty of Health, Engineering and Sciences & National Centre for Engineering in Agriculture, University of Southern Queensland, Toowoomba, Queensland, Australia, E-mail: jochen.bundschuh@usq.edu.au

Guangnan Chen: Faculty of Health, Engineering and Sciences & National Centre for Engineering in Agriculture, University of Southern Queensland, Toowoomba, Queensland, Australia, E-mail: guangnan.chen@usq.edu.au

Raymond Louis Desjardins: Agriculture & Agri-Food Canada, Ottawa, Canada, E-mail: ray.desjardins@agr.gc.ca

James Athur Dyer: Agro-environmental Consultant, Cambridge, Ontario, Canada, E-mail: james-dyer@sympatico.ca

Alessandro Flammini: Climate, Energy and Tenure Division, Food and Agriculture Organization of the United Nations (FAO), Rome, Italy, E-mail: alessandro.flammini@fao.org

Jüri Frorip: Department of Energy Engineering, Institute of Technology, Estonian University of Life Sciences, Tartu, Estonia, E-mail: juri@monte.ee

Xinqun Gui: Power Systems Group, Deere and Company, Waterloo, Iowa, USA, E-mail: guixinqun@johndeere.com

Paul Harris: Retired from School of Agriculture, Food and Wine, The University of Adelaide, Adelaide, South Australia, Australia, E-mail: harrisfm@aapt.net.au

Tamara Jackson: Graham Centre, School of Agricultural and Wine Sciences, Charles Sturt University, Wagga Wagga, Australia, E-mail: tajackson@csu.edu.au

Munir A. Hanjra: International Water Management Institute (IWMI), Southern Africa Regional Office, Pretoria, South Africa & Institute for Land, Water and Society, Charles Sturt University, Wagga Wagga, Australia, Email: mahanjra@gmail.com

Eugen Kokin: Department of Energy Engineering, Institute of Technology, Estonian University of Life Sciences, Tartu, Estonia, E-mail: eugen.kokin@emu.ee

Kees W. Kwant: NL Agency, Energy and Climate Change Division, Utrecht, The Netherlands, E-mail: kees.kwant@agentschapnl.nl

Xavier Lemaire: Energy Institute, University College London, London, UK, E-mail: x.lemaire@ucl.ac.uk

Brian Glenn McConkey: Agriculture & Agri-Food Canada, Swift Current, Canada, E-mail: brian.mcconkey@agr.gc.ca

Hannu Mikkola: Department of Agricultural Sciences, University of Helsinki, Helsinki, Finland, E-mail: hannu.mikkola@helsinki.fi

Albert Moerkerken: NL Agency, Energy and Climate Change Division, Utrecht, The Netherlands, E-mail: albert.moerkerken@agentschapnl.nl

Shahbaz Mushtaq: Australian Centre for Sustainable Catchments, University of Southern Queensland, Toowoomba, Queensland, Australia, E-mail: shahbaz.mushtaq@usq.edu.au

Hans Oechsner, State Institute for Farm Machinery and Farm Structures, University of Hohenheim, Stuttgart, E-mail: oechsner@unihohenheim.de

Janusz Piechocki: Department of Electric and Power Engineering, Faculty of Technical Sciences, University of Warmia and Masury, Olsztyn, Poland, E-mail: jpt@uwm.edu.pl

Väino Poikalainen: Department of Food Science and Technology, Institute of Veterinary Medicine and Animal Sciences, Estonian University of Life Sciences, Tartu, Estonia, E-mail: vaino.poikalainen@emu.ee

Jaan Praks: Department of Food Science and Technology, Institute of Veterinary Medicine and Animal Sciences, Estonian University of Life Sciences, Tartu, Estonia, E-mail: jaan.praks@emu.ee

Mari Rajaniemi: Department of Agricultural Sciences, University of Helsinki, Helsinki, Finland, E-mail: mari.rajaniemi@helsinki.fi

Winfried Schäfer: Animal Production Research, MTT Agrifood Research Finland, Jokioinen, Finland, E-mail: winfried.schafer@mtt.fi

Ralph E.H. Sims: Centre of Energy Research, Massey University, Palmerston North, New Zealand, E-mail: r.e.sims@massey.ac.nz

Bert Stuij: NL Agency, Energy and Climate Change Division, Sittard, The Netherlands, E-mail: bert.stuij@agentschapnl.nl. Also associated with: Energy Delta Institute, Groningen, The Netherlands, E-mail: stuij@energydelta.nl

Jeff N. Tullberg: National Centre for Engineering in Agriculture, University of Southern Queensland, Toowoomba & School of Agriculture and Food Sciences, University of Queensland, Gatton & CTF Solutions, Brisbane, Queensland, Australia, E-mail: Jtullb@bigpond.net.au

Imbi Veermäe: Department of Animal Health and Environment, Institute of Veterinary Medicine and Animal Sciences, Estonian University of Life Sciences, Tartu, Estonia, E-mail: imbi.veermae@emu.ee

Lijun Wang: Biological Engineering, School of Agricultural and Environmental Sciences, North Carolina Agricultural and Technical State University, Greensboro, NC 27411, USA, E-mail: lwang@ncat.edu

Dariusz Wiśniewski: Department of Electric and Power Engineering, Faculty of Technical Sciences, University of Warmia and Masury, Olsztyn, Poland, E-mail: dwisniewski@o2.pl

Foreword by Bill Stout

To be sustainable, agriculture must have energy in the right form at the right time. This includes energy embedded in fertilizer and agricultural chemicals, liquid and gaseous fuels, electricity, energy for grain drying, food processing and delivery to consumers. Agriculture is both a consumer of energy and a producer.

This book is focused on effective and efficient use of energy and also renewable energy technologies to meet future food and agricultural needs. Providing food security and a balanced and nutritious diet for forecasted nine billion people by the year 2050 will be a big challenge and will require both more efficient utilization of energy and also increased energy supply. Costs of emerging and novel technologies using both conventional and renewable energy sources are discussed while taking into account environmental and social impacts.

The goal of this book is to provide professional engineers, students, agricultural scientists, policy makers, and the general public knowledge of agriculture's energy needs and details of major renewable energy technologies as well as strategies that reduce energy consumption. This book contains 15 chapters that document agriculture's energy needs in an era where fossil fuels are under review because of greenhouse gases (GHG) produced and the resulting climate change. The fossil fuel supply picture is rapidly changing as new technology is being applied to both increase production and to control GHG problems associated with their combustion.

To a great extent, renewable forms of energy avoid most GHG issues. New technologies are discussed to reduce the cost of renewable energy which must compete with the price of fossil fuels. All these technologies and issues are dealt with in the book. Most of the chapters have extensive reference lists to enable further study by readers.

The range of topics in the 15 chapters and the geographic diversity and experience of the authors is impressive and adds greatly to the value of the book. Emphasis is placed on energy efficiency as well as lowering the cost of renewable energy so that it will be competitive with fossil fuels which so far have not been required to include the environmental impact in their price.

Energy efficiency is emphasized specifically in the crop production system including on-farm uses, more efficient Diesel engines that produce fewer emissions, food processing operations, and livestock production. Renewable energy technologies such as producing fuels from biomass, biogas technology, and gasification are dealt with in various chapters.

The social impact of energy for and from agriculture in developing countries is also discussed. The need for increasing energy availability for rural areas is highlighted.

Along with energy, water is a crucial need for sustainable agriculture. Efficient use of water in irrigated cropping systems is discussed in detail.

Readers will profit from the vast experience of the authors and their insight into the complex issues of energy efficiency and renewable sources necessary to provide sustainable agriculture to feed a world of nine billion people by the year 2050.

Professor Bill Stout, P.E.
22 September 2013

Editors' Foreword

Agriculture and the related primary industry is an increasingly energy demanding sector to cover the ever growing food demand of our planets population. Energy is needed to different extent in all the stages of the agrifood chain involved: production and operation of agricultural machinery, production of agrochemicals where particularily the production of fertilizers is very energy intensive, on-farm crop production where irrigation is the key energy consumer, fisheries, livestock production, post-harvest operations such as food processing which also requires water, and transport of agricultural products and food end-products.

At present, this energy is predominantly provided from fossil fuel sources. The increasing scarcity and correspondingly increasing costs of fossil fuels and freshwater, and demand of reducing greenhouse gas emissions, demands a modernisation of the agricultural sector towards more energy and water efficiency and replacement of fossil fuel use by energy from renewable energy sources. Such a climate-smart agriculture that produces energy-smart food as promoted by the FAO (Food and Agriculture Organization) in their Energy-Smart Food (ESF) Program is an approach that contributes to both, food security and climate protection which are global key challenges for the next decades.

This book seeks to provide knowledge and information on the energy use along the agrifood chain and linkages between energy and agrifood systems and showcases where and how energy-efficient systems and renewable energy resources, can be used for reducing economic costs by depending less on fossil fuels and at the same time contributing towards more energy, water, and food security and climate protection.

This book is divided in four sections: Section I contains the introductory Chapter 1 focusing on the energy-water-food nexus in the agricultural food chain water and future needs for a shift to a more energy-smart food in order to secure future food supply and at the same time contribute to climate protection. Section II contains 8 chapters dealing with differen aspects of energy efficiency and management whereas Section III provides detailed insights in the production of bio-energy from different sources and Section IV taggles with the topic of access to energy services in agriculture. Chapter 2 provides details on energy demand along the food chain and options to improve on-farm energy efficiency to improve energy security as well as food security. Chapter 3 discusses present and potentialmore sustainable energy technologies for different crop production systems. Chapter 4 provides a detailed analysis of CO_2 emissions from agriculture showcasing the example of Canada. Chapter 5 shows different technology options how to improve energy efficiency in agriculture and the food processing industry whereas Chapter 6 provides details on technologies, practices and policies for a shift from the present agricultural practices towards an energy-smart food based agriculture. Chapter 7 showcases the nexus between energy, water and food for irrigated crop systems. The last both chapters in this section deal with sustainable energy options in intensive livestock production (Chapter 8) and on emission reduction of diesel-powered on-farm machinery including also options of increased biodiesel use (Chapter 9). Section III describes in Chapter 10 in detail the species, properties, production, and harvesting of microalgae and their conversion to biofuels. Chapter 11 discusses the performance and greenhouse gas emissions of biodiesel and compares them with diesel produced from fossil fuels. The technologies for biogas production and thermal gasification of biomass waste are discussed in Chapters 12 and 13, respectively. Chapter 14 showcases why we need a more bio-based economy and how this shift can be done using a cascading approach for sustainable development of biomass that is shown for the case of The Netherlands. Finally in Section IV: "Access to Energy Services" Chapter 15 provides information on options how to improve the access to energy services in rural areas.

We believe that this book will provide the readers with a thorough understanding of technological options to improve energy efficiency and to replace fossil fuels through energy from renewable energy resources in agriculture and associated industries, in order to reduce the dependence of the fossil fuels, hence reducing cost and at the same time greenhouse gas emissions.

We wish that the book will be used as incentive for what and how new technologies provide the innovative, effective and more sustainable solutions for agriculture by using sustainable, environmental friendly, renewable energy sources and energy efficient technologies for agrictural production to satisfy the continually increasing demand for food and fibre in an economically sustainable way, and to overcome major challenges such as the reduced availability of water, energy and agricultural land, energy demand in agriculture using modern technologies, while contributing to global climate change mitigation.

We hope that this book will help all readers, in the professional, academics and non-specialists, as well as key institutions that are working in agriculture, sustainable energy development and climate change mitigation projects. It will be useful for leading decision and policy makers, agricultural and energy sector representatives and administrators, policy makers from the governments, business leaders, business houses in energy sector, independent power producers, and energy engineers/scientists from industrialized and developing countries as well. It is expected that this book will become a standard, used by educational institutions, and Research and Development establishments involved in the respective issues.

Jochen Bundschuh
Guangnan Chen
(editors)
January 2014

About the editors

Jochen Bundschuh (1960, Germany), finished his PhD on numerical modeling of heat transport in aquifers in Tübingen in 1990. He is working in geothermics, subsurface and surface hydrology and integrated water resources management, and connected disciplines. From 1993 to 1999 he served as an expert for the German Agency of Technical Cooperation (GTZ) and as a long-term professor for the DAAD (German Academic Exchange Service) in Argentine. From 2001 to 2008 he worked within the framework of the German governmental cooperation (Integrated Expert Program of CIM; GTZ/BA) as adviser in mission to Costa Rica at the Instituto Costarricense de Electricidad (ICE). Here, he assisted the country in evaluation and development of its huge low-enthalpy geothermal resources for power generation. Since 2005, he is an affiliate professor of the Royal Institute of Technology, Stockholm, Sweden. In 2006, he was elected Vice-President of the International Society of Groundwater for Sustainable Development ISGSD. From 2009–2011 he was visiting professor at the Department of Earth Sciences at the National Cheng Kung University, Tainan, Taiwan. By the end of 2011 he was appointed as professor in hydrogeology at the University of Southern Queensland, Toowoomba, Australia where he leads a working group of 26 researchers working on the wide field of water resources and low/middle enthalpy geothermal resources, water and wastewater treatment and sustainable and renewable energy resources (http://www.ncea.org.au/groundwater). In November 2012, Prof. Bundschuh was appointed as president of the newly established Australian Chapter of the International Medical Geology Association (IMGA).

Dr. Bundschuh is author of the books "Low-Enthalpy Geothermal Resources for Power Generation" (2008) (Balkema/Taylor & Francis/CRC Press) and "Introduction to the Numerical Modeling of Groundwater and Geothermal Systems: Fundamentals of Mass, Energy and Solute Transport in Poroelastic Rocks". He is editor of the books "Geothermal Energy Resources for Developing Countries" (2002), "Natural Arsenic in Groundwater" (2005), and the two-volume monograph "Central America: Geology, Resources and Hazards" (2007), "Groundwater for Sustainable Development" (2008), "Natural Arsenic in Groundwater of Latin America (2008). Dr. Bundschuh is editor of the book series "Multiphysics Modeling", "Arsenic in the Environment", and "Sustainable Energy Developments" (all Balkema/CRC Press/Taylor & Francis).

Dr. Guangnan Chen graduated from the University of Sydney, Australia, with a PhD degree in 1994. Before joining the University of Southern Queensland as an academic in early 2002, he worked for two years as a post-doctoral fellow and more than five years as a Senior Research Consultant in a private consulting company based in New Zealand. Dr. Chen has extensive experience in conducting both fundamental and applied research. His current research focuses on the sustainable agriculture and energy use. The researches aim to develop a common framework and tools to assess energy uses and greenhouse gas emissions in different agricultural sectors. These projects are funded by various government agencies and farmer organsations. In addition, Dr Chen has also conducted significant research to compare the life cycle energy consumption of alternative farming systems, including the impact of machinery operation, conservation farming practice, irrigation, and applications of new technologies and alternative and renewable energy. Dr. Chen has so far published 80 papers in international journals and conferences, including 7 invited book chapters. He serves as a member of editorial board for the International Journal of Agricultural & Biological Engineering (IJABE), and was the Guest Editor of a special issue on agricultural engineering, Australian Journal of Multi-Disciplinary Engineering in both 2009 and 2011. He is currently a member of Board of Technical Section IV (Energy in Agriculture), CIGR (Commission Internationale du Génie Rural), one of the world's top professional bodies in agricultural and biosystems engineering.

Acknowledgements

The editors thank the reviewers of the individual chapters for their valuable comments that significantly contributed to the quality of the book: Galip Akay (UK), Muhammad Amer (USA), Julie A. Bailey (USA), Thomas Banhazi (Australia), Michael A. Borowitzka (Australia), Harriëtte L. Bos (The Netherlands), Nolan Caliao (Australia), Tim Chamen (UK), Sibel Dikmen (Turkey), Theodor Friedrich (Cuba), Girish Ganjyal (USA), Mathias Gustavsson (Sweden), Pavel Kic (Czech Republic), Eugen Kokin (Estonia), Fanbin Kong (USA), Song Charng Kong (USA), Tek Maraseni (Australia), Bernardo Martin-Gorriz (Spain), Aurelian Mbzibain (UK), Sven Nimmermark (Sweden), Hans Oechsner (Germany), Matti Pastell (Finland), Renaud Sanscartier (Canada), Peer M. Schenk (Australia), Robert Schock (USA), Bagas Wardono (Saudi Arabia), Jale Yanik (Turkey), Qin Zhang (USA). The editors and authors thank also the technical people of the Taylor & Francis Group for their cooperation and the excellent typesetting of the manuscript.

Section 1
Introduction

CHAPTER 1

Towards a sustainable energy technologies based agriculture

Jochen Bundschuh, Guangnan Chen & Shahbaz Mushtaq

> *"Modern society continues to rely largely on fossil fuels to preserve economic growth and today's standard of living. However, for the first time, physical limits of the Earth are met in our encounter with finite resources of oil and natural gas and its impact of greenhouse gas emissions onto the global climate. Never before has accurate accounting of our energy dependency been more pertinent to developing public policies for a sustainable development of our society, both in the industrial world and the emerging economies."*
>
> Minutes, Debate of Senate (Eerste Kamer), 2009 (in Dutch)[1]

1.1 INTRODUCTION

Agriculture, including associated primary industry such as production of machinery, fertilizers, feed concentrates for livestock, agrochemicals, water and agroprocessing, is an energy intensive activity. To manage the escalating global food demand, agriculture is increasingly becoming energy intensive, and importantly, the agricultural frontiers are expanding into areas that are not ideally suited for farming. This further translates to an increase in demand for energy, which is not proportional to the food production increase achieved. In many cases, energy cost may represent up to 20–50% of the total agricultural production inputs cost, including the cost of manufacturing and transporting inputs such as fertilizers. Nutrient-poor soils not only require large quantities of fertilizers but may also require large volumes of irrigation water, which, in some instances, must be pumped from ever-greater depths. The last is caused by increasing limitations on the availability of surface water due to seasonal fluctuations and its continuous quality degradation as a result of anthropogenic contamination, which in many areas makes groundwater the principal source for irrigation and other agricultural purposes. The importance of groundwater for agriculture will undoubtedly further increase in the future with the need to sustain food supplies to an increasing world population. Thermal desalination of seawater or saline or brackish groundwater, which in many areas will be the only available option for regional or national food production, is even more energy demanding.

The sustainable provision of increasingly large amounts of energy and freshwater demanded for agricultural production is crucial in a world facing both population and economic growth. This must be achieved while also avoiding or mitigating greenhouse gas emissions that would occur if the additional energy demand were to be covered by fossil fuels and conventional agricultural technologies. Such a scenario of increasing competition for water and energy resources urgently requires the development and implementation of innovative integrated agricultural approaches. Greatest efforts are required in the developing world to meet their food requirements, with their high population growth rates, rapidly-expanding emerging economies and increases in living standards. This can be clearly demonstrated using the parameter "electricity demand", the world-average of which is predicted in the World Energy Outlook 2012 (IEA, 2012) to increase by 70% in the 2010–2035 period (2.2% per year on average), with an annual average grow rate of 3.3%

---

[1]Source: Minutes of the debate of the Senate (Eerste Kamer) of the Dutch Parliament, March 31st, 2009, http://www.eerstekamer.nl/stenogram/stenogram_254/f=x.pdf.

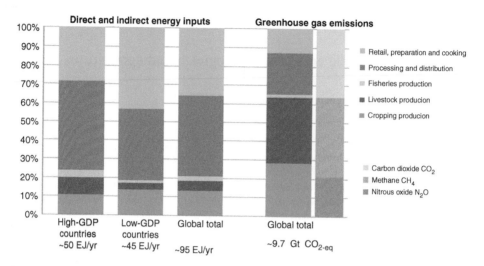

Figure 1.1.   Global sectorial shares of the energy consumption of the total food sector and related GHG emissions and distribution to high-GDP and low-GDP countries (source: FAO, 2011).

in developing countries (non-affiliated with OECD: Organization for Economic Cooperation and Development), compared with a growth rate of only 0.9% in industrialized countries (OECD countries) (IEA 2012, New Policies Scenario). For comparison, the Current Policies Scenario and the 450 Scenario result in average annual world electricity demand increases of 2.6% and 1.7%, respectively (IEA, 2012).

According to estimates of the FAO (Food and Agriculture Organization of the United Nations), global food production will need to be increased by 70% to feed the world population, which is forecasted to reach nine billion by the year 2050. Since the global primary energy demand is forecasted to increase by one third in the 2010–2035 period (IEO, 2012) and because the annual global freshwater withdrawal is expected to grow by about 10–12% per decade, corresponding to an increase of 38% from 1995 to 2025 (UNESCO, 1999), the securing of energy and water supply is the key challenge facing modern society. Increasing stress on limited freshwater resources – the largest part used for agriculture – will intensify competition for water for farming and industrial purposes, as well as for consumption in cities. This may require massive production of freshwater from alternative sources, for example from seawater and brackish or saline groundwater.

The FAO (2011) has estimated that about 30% of the global total energy consumption corresponds to the food sector. Figure 1.1 shows energy consumption and GHG emission of different parts of the food sector for developing (low-GDP) countries and developed (high-GDP) countries. Only about 20% of the total food energy is demanded by the primary farm production (i.e. cropping and livestock production) but these activities produce about 65% of the total food sector's GHG emissions (Fig. 1.1).

The energy demands of the principal energy consuming technologies for crop and livestock production are shown in Figure 1.2 for the example of New Zealand. For some crops, on-farm direct inputs such as chemical and fertilizers can account for up to 70–80% of total energy used in their production (Chen *et al.*, 2013).

Fossil fuel energy resources, which are still the globally dominating energy source in the agricultural sector, are becoming more and more limited. In the case of oil, peak production is already exceeded and production is forecasted to decline by 2030 to half of its 2010 value (Fig. 1.3: EWG, 2007). Decreased availability translates into increasing prices of oil (Fig. 1.4) and other fossil fuels and it can be expected that one day even scarcity prices, i.e. a premium determined by the supply and demand situation must be paid. The same limitation is true for freshwater, the

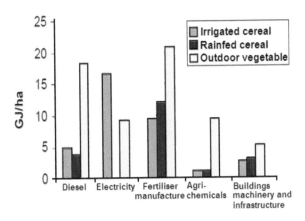

Figure 1.2.   Direct and indirect energy inputs into different parts of the crop and livestock production for different agricultural New Zealand enterprises (source: Barber, 2004).

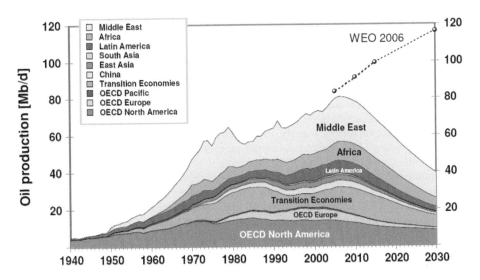

Figure 1.3.   Global oil production *versus* World Energy Outlook (WEO) 2006 (source: EWG, 2007).

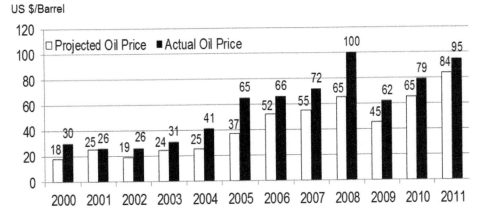

Figure 1.4.   Projected and actual price for oil (source: Fell, 2012).

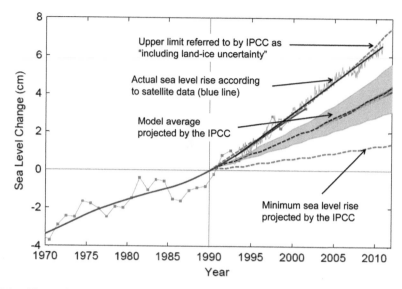

Figure 1.5.   Observed sea level rise versus scenarios projected by IPCC projections (source: Rahmstorf *et al.*, 2007; updated by Rahmstorf *et al.* with data from the year 2010).

demand for which is increasing but the availability of which is decreasing due to anthropogenic contamination (UNEP, 2008; UNESCO, 1999).

In addition to the increasing scarcity, fossil fuels have the drawback that they are the principal contributors to global warming. As earlier mentioned, agriculture contributes significantly to global GHG emissions, causing global warming the consequences of which advance much faster and more dramatically than formerly thought (Fell, 2012). They comprise melting of the arctic polar ice cap (The Cryosphere Today, 2012) leading to increasing sea levels (Fig. 1.5) (Rahmstorf *et al.*, 2007; updated by Rahmstorf *et al.* with data from the year 2010) contributing to increased risk of flooding of coastal regions worldwide. Global warming is also a cause of an already observed increase in the frequency and severity of weather phenomena, such as major droughts, floods, extremely heavy rainfalls (and related mudslides, etc.), typhoons/hurricanes and forest/bush fires causing severe damage to property and infrastructures, economic loss and increased risk to populations. In addition, global warming results in the thawing of permafrost, which releases large volumes of the greenhouse gas methane.

The above discussion suggests that the future food supply is intrinsically linked with energy, water and climate issues and must therefore be managed in an integrated way. The availability of freshwater and energy are intrinsically linked to human social and economic development. The use of fossil fuels and conventional agricultural technologies to meet the increasing energy demand in agriculture would result in a CO_2 emission increase proportional to the energy demand increase.

Food security requires energy and water security. Hence, securing energy and water supply – fundamental to feeding our global population – while mitigating climate change is the key challenge facing modern society. Implementation of energy- and water-efficient and renewable energy technologies, together with energy conservation methods and greenhouse gas sequestration from atmospheric CO_2 and other greenhouse gases into soil and biomass are therefor essential. Global food security may be compromised if innovative and cost effective technologies and low-cost solutions are not developed. Governments and industries around the world must look into cutting-edge energy-efficient, low-emission technologies and management practices for the agricultural sector.

There is a widespread belief among decision makers that these options are economically unviable. However, the opposite is true: the use of new technologies will contribute to the creation of new jobs and economic growth if these energy-efficient and low-emission technologies are developed and produced in large numbers. This has been successfully demonstrated in Germany where, since 1998, employment in the renewable energy sector increased by more than a factor of 10, from about 66,600 (1998) to 377,800 (2012) with the aim to reach 500,000 in the year 2020, providing the second-highest share of employment in Germany after the automotive industry (Agentur für Erneuerbare Energien, 2013).

In many cases it can be already observed that high and ever increasing fossil fuel prices have increasingly negative financial effects on the agricultural sector and favor the introduction of energy-efficient low-emission technologies in agriculture. Additional benefits include the avoidance of external costs such as those related to human health costs and environmental costs due to consumption of large volumes of fossil fuels, while the use of energy-efficient and low-emission technologies minimizes costs for both the agricultural sector and the consumer of the agricultural products.

Energy-efficient and low-emission technologies in agriculture can grow much faster than generally assumed, similar to the introduction of other technologies such as plasma TVs and mobile phones, which only took few years, have shown (Fell, 2012). A number of sustainable technologies suitable for applications in agriculture already exist in large numbers and at commercial scales, but this does not take away the need to continuously develop new technology. However, the rate at which these technologies are developed and adopted is dependent on favorable political, policy and financial conditions. This requires close cooperation between policy makers and the financial sector; mechanisms, incentive and compensation models and options can be used for the agricultural sector analogous to those described and discussed in detail in the book by Fell (2012).

Adoption of sustainable energy technologies for agriculture requires policies which actively facilitate market entry and wide market penetration as well use as wide application by the farmers. It is essential that the manufacturers recognize markets in the agricultural sector. Therefore, state regulations are required, to allow the diversion of private sector money into sustainable energy technologies investments (both, the manufacturers and the users in the agricultural sector). As investments in sustainable energy technologies start to yield returns, increased scales of production are likely to lead to reduction in the cost of production and wider market penetration. Analogous to the period of 15–20 years estimated by Fell (2012) for the development of self-sustaining renewable energy technologies or measures to become economically mature and self-propelling, given effective state regulation and active political support, we can conclude the same timeframe for the advent of green agricultural technologies.

In many countries the agricultural sector is subsidized by government; this also includes subsidizing non-sustainable energy technologies, such as subsidies or tax exemptions for diesel used on farm. The aim must be to remove such perverse benefits and to instead provide economic incentives for investing in sustainable energy technologies to boost their relative market penetration. Only then, the private sector – which counts on much more capital than governments – will invest in developing and scaling up production of these technologies, leading to cost reduction and wide market penetration as it could be observed for solar panels, for example, which no longer need state subsidization.

Finally, mass production of energy-efficient and low-emission technologies in agriculture is likely to result in a reduction in purchase prices, while more and more people will realize that these technologies will lead to energy savings; adoption of the new technologies will lead to affordable energy costs as the costs of energy from conventional sources (fossil fuels), especially due to the increasing scarcity and the costs of their external damages will further increase. Therefore, it is evident that energy-efficient and low-emission technologies in agriculture will soon become the more economic option.

An overview of technological and other options is given in Section 1.2.

Table 1.1.   Growth rates demand projects, per cent p.a. (source: Alexandratos and Bruinsma, 2012).

	1970–2007	1980–2007	1990–2007	2005/ 2007–2030	2030–2050	2005/ 2007–2050
Demand (all commodities – all uses), total						
World	2.2	2.2	2.3	1.4	0.8	1.1
Developing countries	3.6	3.6	3.5	1.7	0.9	1.3
Sub-Saharan Africa	*3.1*	*3.4*	*3.5*	*2.6*	*2.1*	*2.4*
Near East/North Africa	*3.3*	*2.8*	*2.8*	*1.7*	*1.1*	*1.5*
Latin America and the Caribbean	*2.8*	*2.6*	*2.6*	*1.7*	*0.6*	*1.2*
South Asia	*3*	*3*	*2.7*	*2*	*1.3*	*1.7*
East Asia	*4.3*	*4.4*	*4.4*	*1.4*	*0.5*	*1*
Developed countries	0.5	0.3	0.4	0.6	0.2	0.5

1.1.1   *Challenges*

Despite a predicted slowdown in the rate of global demographic growth of -0.75% per year over the next 40 years to 2050 (Alexandratos and Bruinsma, 2012), FAO projections indicate that by 2050 a 70% increase (or 1.1% averaged increase per year) in food production over 2005–2007 levels will be necessary to meet the expanding demand for food (Table 1.1). In the USA, it has been estimated that the operation of current food production systems, including agricultural production, food processing, packaging, and distribution, accounted for approximately 19% of the national fossil fuel energy use (Pimentel, 2006). In another study, Pimentel and Giampietro (1994) found that in the USA about 1500 L of oil equivalents are expended annually to feed each American. It is believed that in many developed countries, fossil fuel consumption by food systems often rivals that of transport systems.

Population growth, increases in per capita consumption and changes in diets leading to the consumption of more livestock products are the suggested drivers of such expected changes (Alexandratos and Bruinsma, 2012; FAO, 2012). These production gains are largely expected to come from increases in the productivity of crops, livestock and fisheries (FAO, 2009). However, and more importantly, as populations expand and economies grow, the global demand for energy and water is also expected to increase as discussed earlier.

In low-carbon-water starved future, realizing such production improvements will not be a simple task for several reasons:

- Land and water resources are now much more stressed than in the past and are becoming scarcer, both in quantitative terms (per capita) and qualitative ones, following soil degradation, salinization of irrigated areas and competition from uses other than for food production (Figs. 1.6 and 1.7).
- There has been a decline in the development of arable land in developing countries, particularly since the 1980s (Fig. 1.6). However, according to Fischer *et al.* (2011), at the global level there is a significant amount of land with potential for rainfed production, although with various degrees of suitability. However, not all land is suitable for cultivation. After allowing for forest, non-agricultural uses such as human settlement, and marginal and poor quality soil, it is estimated that there are about 1.4 billion ha of prime/good land that could be brought into cultivation (Alexandratos and Bruinsma, 2012).
- Water is another critical resource. Historically, irrigated agriculture has contributed significantly to food security. World irrigated areas have increased more than twofold (by 300 million ha) since 1960 (Alexandratos and Bruinsma, 2012); however, irrigated agriculture is under

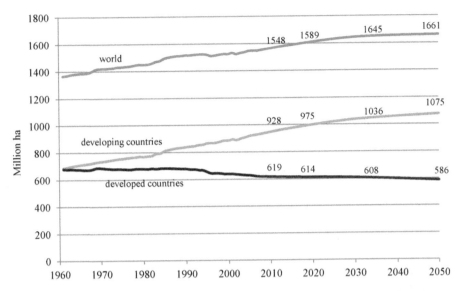

Figure 1.6.   Arable land and land under permanent crops: past and future (source: Alexandratos and Bruinsma, 2012).

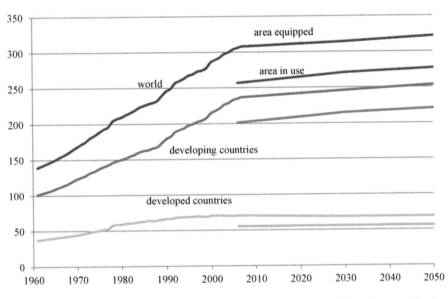

Figure 1.7.   Arable irrigated land: equipped and in use (million ha) (source: Alexandratos and Bruinsma, 2012).

immense pressure due to increased competition with non-agricultural sectors while the potential impacts of climate change may change rainfall distributions (Mushtaq *et al.*, 2012; Torriani *et al.*, 2007) and further exacerbate competition for increasingly scarce water resources. The potential for further expansion of irrigation, therefore, is limited without investment in water saving irrigation to improve water use efficiency and water reuse technologies (Mushtaq *et al.*, 2103). There are significant quantities of renewable water resources globally; but they are extremely scarce in regions such as the Near East/North Africa, or northern China, where they are most needed (Alexandratos and Bruinsma, 2012). The irrigation expansion project suggested by Alexandratos and Bruinsma (2012) indicates that the area equipped[2] for irrigation could be expanded by 20 million ha (or 6.6%) over the period from 2005/07 to 2050, nearly all of it in the developing countries (Fig. 1.7). There is further potential to increase the production of irrigated areas, which could effectively expand by 34 million ha with an increase in multiple cropping on both existing and newly irrigated areas. However, again, sustainable technologies need to play a major role in achieving such growth.

- Energy: Fossil fuels have been the primary energy source for our world for more than a century. However, because fossil fuels are a limited resource, improvements in farming energy efficiency are essential. Continuous high fuel price, the increasing demand for "green food" and significant reductions in greenhouse gas emissions also make the exploration of new alternative and renewable energy sources essential.

1.2   SUSTAINABLE ENERGY OPTIONS IN AGRICULTURE

> *"Tackling the challenges of food security, economic development and energy security in a context of ongoing population growth will require a renewed and re-imagined focus on agricultural development," . . . "Agriculture can and should become the backbone of tomorrow's green economy," . . . "It's time to stop treating food, water and energy as separate issues and tackle the challenge of intelligently balancing the needs of these three sectors, building on synergies, finding opportunities to reduce waste and identifying ways that water can be shared and reused, rather than competed for," . . . "Climate-smart farming systems that make efficient use of resources like water, land, and energy must become the basis of tomorrow's agricultural economy."*

Alexander Mueller
FAO Assistant Director-General for Natural Resources
Bonn 2011 Nexus Conference

In practice, most of the technologies and other options needed for providing sustainable energy solutions in agriculture already exist. The suggested measures have been described many times, and importantly, components of such systems, such as solar panels, have been available in mass markets. Practical on-farm demonstrations (Chen *et al.*, 2009) have also been undertaken to assess the viability of their use in agriculture and clarify the potential technical issues in their applications.

1.2.1   *Energy efficiency and energy conservation*

Improving energy efficiency and energy conservation in agriculture are essential to reduce energy demand and therefore reduce costs. Improving energy efficiency, and thus reducing reliance on fossil fuels, will further reduce greenhouse gas emissions. In addition, it must be taken into consideration that a reduced energy demand will also proportionally reduce investment costs for farm expansions or for shifting from fossil fuels to on-site renewable energy sources.

Everywhere in agriculture where energy is used, its demand can be reduced. For example, it has been shown that fossil energy use in the current food system could be significantly reduced by appropriate technology changes. In the USA, it is estimated (Pimentel, *et al.*, 2008) that the total energy in corn production could be reduced by more than 50% with the following changes of practices: (i) using smaller machinery and less fuel; (ii) replacing commercial nitrogen

---

[2]Data reported in FAOSTAT on arable irrigated land refer to 'area equipped for irrigation'.

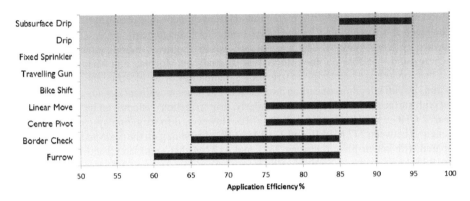

Figure 1.8.   Application efficiencies for various irrigation systems (source: Mushtaq and Maraseni, 2011).

applications with legume cover crops and livestock manure; and (iii) adopting alternative tillage and conservation techniques.

Pellizzi *et al.* (1988) showed that with improved management and operation, energy saving of around 12–15% of present consumption can be realistically obtained for tractors, 30% for soil tillage, and 10% for harvesting machinery. Brown and Elliot (2005) found that the largest on-farm energy savings are available in motorized systems, especially irrigation pumping. Pathak and Bining (1985) showed that for irrigation, fuel savings of over 50% were feasible through improvements in irrigation equipment and water management practices. In the USA, energy efficiency audits on irrigation systems have on average identified savings of at least 10% of the energy bill – and in many instances up to 40%. Very often, the irrigators who owned these inefficient systems were unaware of any problems.

Irrigated agriculture is a vital part of world agriculture, particularly in semi-arid and arid crop-ping areas. Potentially twice as productive as rain-fed agriculture (Entry *et al.*, 2002), it not only makes a significant contribution to global food production but it also contributes significantly to national economies. With global population growth, global climate change and growing com-petition for scarce water resources with different sectors including the environment, irrigated agriculture is under considerable pressure to adopt best practice to ensure efficiency in terms of water use and productivity.

Conventional irrigation practices are generally characterized by low water use efficiencies and there is potential in irrigated production systems for significant water savings, resulting in either increased productivity or increased water availability for alternative uses (Clemmens, 1998; Green *et al.*, 1996; Mushtaq *et al.*, 2013; Robinson, 2004). However, there may be adverse economic and environmental consequences if water savings achieved result in increases in energy consumption and GHG emissions by agriculture. Agriculture currently relies heavily on the use of fossil fuels. Given 'peak oil' predictions, adoption of irrigation technologies are likely to be challenged by higher energy costs (Foran, 1998).

1.2.1.1   *Enhancing irrigation and energy efficiency of the irrigated systems*
Pressurized systems have the potential to increase water use efficiency. In terms of maximum possible efficiencies, drip technology outperforms sprinkler systems, and both deliver greater maximum efficiencies than surface irrigation (Fig. 1.8). However, system design and management can have a big impact on water use efficiency. For example, a flood irrigation system using border check, a sprinkler irrigation system using center-pivot or linear move, and a drip system can have the same level of application efficiency depending on the system design and level of management for each system. However, as the percentile bands for sprinkler and drip systems are higher than

for flood systems, the benefit of greater financial investment in a pressurized system is that it will be much less susceptible to water losses from poor management.

Surface irrigation systems, which are usually based on gravity-based system, do not require any energy and therefore do not emit GHGs. A study by Mushtaq and Maraseni (2011) estimates that, on average, center-pivot irrigation systems run by an electric pump increases GHG emissions by 906 kilograms of carbon dioxide equivalent per megaliter (kg CO_{2-e}/ML) when compared with surface irrigation systems. Similarly, drip irrigation systems run by an electric pump increased GHG emissions by 568 kgCO_{2-e}/ML. However, drip irrigation systems required 28% less energy (between 777 MJ/ML and 3262 MJ/ML), depending on the scale and farming system, compared with center-pivot (between 2321 and 4127 MJ/ML) and lateral-move (between 2884 and 4195 MJ/ML) systems. Similarly, drip irrigation produced around 25% less GHGs compared with center-pivot and lateral-move.

Economic efficiency is also a major consideration. While the level of water savings that can be achieved are substantial and the conversions can be proved to be economically feasible, the energy costs associated with these conversions are considerable. This has implications for production costs as well as infrastructure requirements, depending on the spatial distribution of energy demand. Irrigators will also bear increased costs of pumping, particularly in surface water regions where irrigation water would previously have been applied in a relatively energy 'free' way, and therefore with low or no costs for application. Costs could vary from A\$ 120-1000 per hectare with drip and sprinkler irrigations systems, depending on, among other factors, crop water use, irrigation method selected and fuel source (Mushtaq and Maraseni, 2011; Mushtaq *et al.*, 2013). This is an option that would need to be carefully considered by farmers and policy makers as any significant increase in the operating costs mean that the overall financial position of farmers is compromised.

1.2.1.2 *Cooling and heating*

There is a considerable requirement for heating and cooling in agriculture. For example, greenhouse heating may be essential to the year-round production of vegetables, fruit and flowers. Temperature controlled storage and refrigeration systems also consume large amounts of electricity and thereby contribute greatly to the running costs of businesses which have considerable cooling requirements, particularly in the horticulture and vegetable industry. Improvements to technical elements and operation of modern refrigeration systems have the potential to reduce energy consumption by 15–40%. This will become more important as a price is placed on greenhouse gas emission and as energy prices rise.

Improved thermal insulation of buildings to reduce the costs of heating and cooling would result in reduced demand for energy, while the on-farm production of energy from renewable sources (e.g. solar), can produce more energy than is needed. The energy efficiency of all electrical devices used in agriculture can also be continuously increased. For example, energy consumption in the lighting sector can be reduced by shifting to energy-saving fluorescent lamps, LEDs or OLEDs. There are many agricultural production processes, the energy efficiency of which can further be increased.

1.2.2 *Use of biomass and biomass waste for carbon-neutral production of biofuel, electricity and bio-coal fertilizers*

Biomass can be produced by cultivating suitable crops and used for production of different types of biofuels. Bio-ethanol can be used to replace gasoline, and biogas to substitute natural gas. Biodiesel and pure plant oils can be produced from oil-rich plants. This biodiesel can substitute for fossil-fuel based diesel used in many agricultural activities and requires only minor changes to the diesel engines. These biofuels can also be used for electricity production, which is a very economic option, in particular in off-grid applications. Compared with biodiesel, pure plant oil can be easily produced without great technical effort. Older diesel engines, which are often used in developing and transitioning countries to supply electricity, can often run without conversion with pure plant oils as fuel. However, modern diesel engines require technical conversions to be fueled with pure plant oils (Fell, 2012).

Therefore, biofuels can reduce the need for agricultural installations to purchase expensive fossil fuels, which must often be transported over long distances, and hence significantly reduce production costs.

Of many crops, only a small part is extracted. The rest is often disposed without any use. An example is given by Fell (2012) from Brazil where annually 100 megatons of sugar is produced from about 1 gigaton of sugar cane. About 90% of the sugar cane crop is burned without any further use. Such biomass waste can be used either for biofuel production and, if needed, electricity. Alternatively, after converting it to bio-coal (e.g. by hydrothermal carbonization) it can be used as fertilizer hence reducing the need to purchase mineral fertilizers. All in all, such measures increase economic benefits and contribute to climate protection.

1.2.3 *Decentralized renewable energy systems (solar, wind, geothermal)*

Many processes and applications in agriculture require energy either in the form of heat, mechanical energy or electricity, which can be provided by solar, wind and/or geothermal energy, depending on the local sources and the specific agricultural application. For example, wind and solar energy can economically produce electricity to power off-grid machinery such as pumps for irrigation; wind energy can also be used as mechanical energy for pumping; solar heat can be used directly for space heating/cooling and warm water production while solar and geothermal heat can economically power thermal water desalination and the treatment of agricultural effluents. Electricity produced from wind and solar energy sources can also be used for water desalination using membrane technologies but at higher cost than thermal methods. Geothermal heat with temperature differences of a few degrees centigrade to the ambient temperature can be used through heat pumps for space heating/cooling. Depending on the temperature of the available resource, geothermal heat has many applications in agriculture such as for dehydration of agricultural products, and heating for greenhouses, soils and aquaculture. Biomass produced on-site can also provide an energy source as biofuel for machinery, as heat or as electricity produced from it.

1.2.4 *Economic benefit of green food*

That the production of "green food" is also of economic benefit for the farmers can be shown in the case of Germany, where the sales of green food have increased by about 300% in the period from 2000–2011 (BÖLW, 2011; Fell, 2012). Organic food may play an important role in climate protection while contributing also to the population's health and are therefore particularly important in schools, hospitals and retirement homes. Healthy-nourished officials are more efficient than those having consumed unhealthy food for years (Fell, 2012).

1.3 CONCLUSIONS

Agriculture is typically highly reliant on fossil fuels and energy is a significant input cost to production. Production of food and other agricultural products accounts for 70% of global freshwater withdrawals. In addition, in the 2010–2035 period, world primary energy demand is forecasted to rise by one-third and electricity demand by 70% increasing the cost of energy and agricultural production. Some of these demands will be met from bioenergy which will in turn intensify competition for resources, particularly water for food and fiber production and therefore the need to maximize the efficient use of these resources becomes increasingly important.

As fossil fuel costs continue to increase, so does the focus on energy efficiency to help minimize the impacts of rising energy costs on profitability and competitiveness. This includes a growing number of renewable energy sources that could be considered as alternatives to fossil fuels.

Renewable energy sources may include solar, wind, hydropower, biomass, biogas and geothermal power. Where the opportunities are appropriate, integrating renewable energies into the farming operations is likely to save energy, costs, and greenhouse gas emissions. Examples of specific applications include solar crop drying, solar space and water heating, solar irrigation and using biomass for heating purpose and electricity generation. Other applications include off-grid electric fences, lighting, irrigation, livestock water supply, wastewater treatment pond aeration, communication and remote equipment operation and others.

Overall, the long-term future for renewable energy is definitely positive, since the prices of fossil fuels will continue to rise as the resources are depleted while the prices of renewable energy will continue to decrease.

There are already a good number of successful examples of application of alternative energy sources in the agriculture industry. However, it has also been found that practical on-farm demonstrations are essential for the widespread adoption of these technologies and research will be needed to further assess the viability of their use in agriculture and clarify the technical issues raised. Further research is also still required to identify suitable pathways and policy frameworks to encourage future market uptake.

REFERENCES

Agentur für Erneuerbare Energien: Entwicklung der Arbeitsplätze im Bereich Erneuerbare Energien. 2013, http://www.unendlich-viel-energie.de/uploads/media/AEE_Entwicklung_EE-Arbeitsplaetze_98-12_Mar13.pdf (accessed September 2013).

Alexandratos, N. & Bruinsma, J.: World agriculture towards 2030/2050: the 2012 revision. ESA Working paper No. 12-03, Food and Agriculture Organization of the United Nations (FAO), Rome, Italy, 2012.

Antonietti, M., Murach, D. & Titirici, M.M.: Opportunities for technological transformations. From climate change to climate management? In: H.-J. Schellnhuber (ed): *Global sustainability: A nobel cause.* Cambridge University Press, Cambridge, UK, 2010, pp. 319–330.

Barber, A.: Seven case study farms: total energy and carbon indicators for New Zealand arable and vegetable production. Agrilink New Zealand Limited, Auckland, New Zealand, 2004.

BÖLW (Bund ökologische Lebensmittelwirtschaft e.V.: Zahlen, Daten, Fakten: Die Bio-Branche 2011. Berlin, Germany, 2011, http://www.boelw.de/uploads/media/pdf/Dokumentation/Zahlen__Daten__Fakten/ZDF2011.pdf (accessed September 2013).

Brown, E. & Elliot, R.N.: Potential energy efficiency savings in the agriculture sector. The American Council for an Energy-Efficient Economy, Washington, DC, 2005, http://www.aceee.org/pubs/ie053.htm (accessed September 2013).

Chen, G., Baillie, C. & Kupke, P.: Evaluating on-farm energy performance in agriculture. *Austral. J. Multi-Discipl. Eng.* 7:1 (2009), pp. 55–61.

Chen, G., Baillie, C, Eady, S. & Grant, T.: Developing life cycle inventory for life cycle assessment of Australian cotton. *Australian Life Cycle Assessment Conference*, 14–18 July 2013, Sydney, 2013.

Clemmens, A.J.: Achieving high irrigation efficiency with modern surface. *IA Expo Technical Conference*, Irrigation Association, November 1998, Falls Church, VA, 1998, pp. 161–168.

Entry, J.A., Sojka, R.E. & Shewmaker, G.E.: Management of irrigated agriculture to increase organic carbon storage in soils. *Soil Sci. Soc Am. J.* 66:6 (2002), pp. 1957–1964.

EWG (Energy Watch Group): Crude oil. The supply outlook. *EWG-Series* No. 3, 2007, http://www.energywatchgroup.org/fileadmin/global/pdf/EWG_Oilreport_10-2007.pdf (accessed September 2013).

FAO (Food and Agriculture Organization of the United Nations): How to feed the world in 2050. Food and Agriculture Organization of the United Nations, Rome, Italy, 2009, www.fao.org/fileadmin/templates/wsfs/docs/expert_paper/How_to_Feed_the_World_in_2050.pdf (accessed September 2013).

FAO (Food and Agriculture Organization of the United Nations): Energy-smart food for people and climate. Issue Paper, Food and Agriculture Organization of the United Nations, Rome, Italy, 2011, http://www.fao.org/docrep/014/i2454e/i2454e00.pdf (accessed September 2013).

FAO (Food and Agriculture Organization of the United Nations): Energy-smart food at FAO: An overview. FAO, Rome, Italy, 2012, http://www.fao.org/docrep/015/an913e/an913e.pdf (accessed September 2013).

Fell H.-J.: *Global cooling: Strategies for climate protection.* CRC Press, Boca Raton, FL, 2012.

Fischer, G., van Velthuizen, H. & Nachtergaele, F.: GAEZ v3.0 – Global Agro-ecological Zones Model documentation. IIASA, Luxemburg, 2011, http://www.iiasa.ac.at/Research/LUC/GAEZv3.0/gaez2010-Flyer_1final1.pdf (accessed September 2013).

Foran, B.: Looking for opportunities and avoiding obvious potholes: some future influences on agriculture to 2050. In: D.L. Michalk & J.E. Pratley (eds): *Proceedings of the 9th Australian Agronomy Conference*, 20–23 July 1998, Charles Sturt University, Wagga Wagga, NSW, Australia, 1998.

Green, G., Sunding, D. & Zilberman, D.: Explaining irrigation technology choices: a microparameter approach. *Am. J. Agr. Econ.* 78:4 (1996), pp. 1064–1072.

IEA (International Energy Agency): World Energy Outlook 2012. Paris, France, 2012, http://www.eia.gov/forecasts/ieo/pdf/0484(2013).pdf (accessed September 2013).

Mushtaq, S. & Maraseni, N.T.: Technological change in the Australian irrigation industry: Implications for future resource management and policy development. *Waterlines Report Series* No [53.], August 2011, National Water Commission, Canberra, ACT, Australia, 2011, http://archive.nwc.gov.au/__data/assets/pdf_file/0013/10921/Waterlines_53_PDF_Fellowship-_Technological_change_in_the_irrigation_industry.pdf (accessed September 2013).

Mushtaq, S., Chen, C., Hafeez, M., Maroulis, J. & Gabriel, H.: The economics value of improved agrometeorological information to irrigators amid climate variability. *Int. J. Climatol.* 32:4 (2012), pp. 567–581.

Mushtaq, S. Maraseni, T.N. & Reardon-Smith, K.: Climate change and water security: estimating the greenhouse gas costs of achieving water security through investments in modern irrigation technology. *Agr. Syst.* 117 (2013), pp. 78–89.

Pathak, B.S. & Bining, A.S.: Energy use pattern and potential for energy saving in rice-wheat cultivation. *Energy Agri.* 4 (1085), pp. 271–278.

Pellizzi, G; Cavalchini, A.G. & Lazzari, M.: *Energy savings in agricultural machinery and mechanization*. Elsevier Science Publishing Co. New York, 1988.

Pimentel, D.: Impacts of organic farming on the efficiency of energy use in agriculture. The Organic Center, Cornell University, 2006, http://www.organicvalley.coop/fileadmin/pdf/ENERGY_SSR.pdf (accessed September 2013).

Pimentel, D. & Giampietro, M.: Food, land, population and the U.S. economy. Carrying Capacity Network, 11/21/1994, http://www.dieoff.com/page55.htm (accessed September 2013).

Pimentel, D., Williamson, S., Alexander, C., Gonzalez-Pagan, O., Kontak, C. & Mulkey, S.: Reducing energy inputs in the US food system. *Human Ecology* 36 (2008), pp. 459–471.

Rahmstorf, S., Cazenave, A., Church, J.A., Hansen, J.E., Keeling, R.F., Parker, D.E. & Somerville, R.C.J.: Recent climate observations compared to projections. *Science* 316:5825 (2007), p. 709, http://www.pik-potsdam.de/~stefan/Publications/Nature/rahmstorf_etal_science_2007.pdf, updated with 2010 data, http://www.pik-potsdam.de/~stefan/update_science2007.html (accessed September 2013).

Robinson, D.W.: *Economic analysis of deficit irrigation on broadacre crops to improve on-farm water use efficiency*. Master of Economics Thesis, University of New England, Armidale, NSW, Australia, 2004.

The Cryosphere Today: A webspace devoted to the current state of our cryosphere. The University of Illinois at Urbana Champaign, 2012, http://arctic.atmos.uiuc.edu/cryosphere/ (accessed September 2013).

Torriani, D., Calanca, P., Lips, M., Ammann, H., Beniston, M.& Fuhrer, J.: Regional assessment of climate change impacts on maize productivity and associated production risk in Switzerland. *Reg. Environ. Change* 7:4 (2007), pp. 209–221.

UNEP: Vital Water Graphics. United Nations Environmental Programme, Nairobi, Kenya, 2008.

UNESCO: Summary of the Monograph 'World Water Resources at the beginning of the 21st Century', prepared in the framework of IHP UNESCO, Paris, France, 1999.

Section 2
Energy efficiency and management

CHAPTER 2

Global energy resources, supply and demand, energy security and on-farm energy efficiency

Ralph E.H. Sims

2.1 INTRODUCTION

By 2050, a 70% increase in food output, in relation to production levels attained in 2005–2007, will be needed to meet rising demands for higher protein diets from a growing world population (FAO, 2009a). To feed the projected world population of over 9 billion people by 2050 will require further intensification of crop and animal production. This is largely expected to come from increased productivity of crops, livestock and fish resources, equating to ∼1000 megatons (Mt) per year of extra production of cereals and ∼200 Mt/year of meat and fish. However, these targets could well be constrained by future dependence on fossil fuel supplies, especially if global greenhouse gas (GHG) emission abatement targets are to be successfully met.

There are sufficient reserves of most types of energy resources, including uranium, coal, natural gas and oil, to last at least several decades at current rates of use in conversion technologies with relatively high energy-efficient designs. How best to use these resources in an environmentally acceptable manner while providing for the needs of growing populations and developing economies will be a considerable challenge. The known recoverable reserves and potential resources of conventional and unconventional supplies of oil, natural gas and coal are significant, as are the stocks of carbon that would be released to the atmosphere on combustion of these fossil fuels (Fig. 2.1). The cumulative emissions of carbon dioxide released to the atmosphere from the combustion of fossil fuels for heat supply, electricity generation and transport since the mid-19th century from the start of the industrial revolution, is over 1100 Gt CO_2 (gigatons or billion tons of carbon dioxide) (Sims *et al.*, 2007). This is relatively small when compared with the volumes that could be emitted during this century if society continues to depend upon fossil fuels as in the past (Fig. 2.1). IPCC analysis, based upon integration assessment models, has determined that such emissions will take the world on a pathway to a mean global temperature rise of around 4°C or above with consequential sea level rise, more extreme weather events, loss of biodiversity and adverse impacts on food production and human health (IPCC, 2007).

In essence, the world will not reach peak production of fossil fuels or start to run out for many years. We may have passed the era of "cheap" liquid fuels because the costs of exploration, extraction and distribution of conventional crude oil have tended to increase with remoteness (such as deep-water, off-shore fields and in the Artic) and the production and refining costs of heavy oils, tar sands and coal-to-liquids are higher than for light crude. Conversely, the recently expanded natural gas supplies in the USA, and elsewhere, resulting from the relatively new extraction process from shale formations known as "fracking" have reduced the gas price considerably down to around 4 US$/GJ. Some oil is now also produced using this technique along with horizontal drilling such as in northern Dakota.

Hence, the major current issues arising from fossil fuel use in many countries are not related to scarcity of supplies or the delivered costs of energy carriers. They involve providing access to energy services for around 3 billion people currently relying on fuelwood, animal dung and other sources of traditional biomass for their cooking and heating needs, and reducing the negative environmental impacts from continued fossil fuel use, both for climate change and human health.

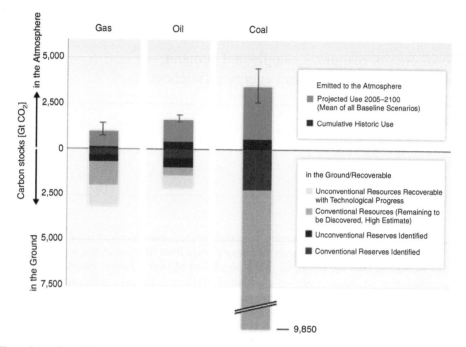

Figure 2.1.   Cumulative amounts of carbon dioxide released to the atmosphere since 1750 from combustion of natural gas, oil and coal; projections for future emissions this century under business-as-usual scenarios (showing standard deviations for the range of 24 baseline scenarios assessed); and assessments of carbon stocks of fossil fuels remaining underground and deemed to be recoverable (Moomaw *et al.*, 2011).

Recent discussions of these and other energy-related issues can be found in the report "Global Energy Assessment – Towards a Sustainable Future" (GEA, 2012). Discussions concerning the nexus of food/energy/water/climate linked with future sustainable development can be found in various reports including "Energy-Smart Food for People and Climate (FAO, 2011a), "Climate Change – a Scientific Assessment for the GEF" (GEF, 2012), "Water for Energy" (WEC, 2010); and "Water for Energy" chapter 17 of World Energy Outlook, 2012 (IEA, 2012). This chapter constrains the discussion to energy issues and use in the primary food production industry.

2.1.1   *Energy access*

At the small farm holder level, energy access can play an important role in productivity and also increasing the available labor supply for food production. Energy poverty requires that a substantial amount of time is spent collecting fuel wood from long distances to meet basic household energy needs. Improved access to energy resources would result in time and labor savings from collecting fuel wood which could then be better utilized for increased food production. However, energy access, equity and sustainable development are compromised by higher and rapidly fluctuating prices for oil and gas. These factors may increase incentives to deploy zero- and low-carbon energy technologies, but conversely, could also encourage the market uptake of cheaper coal and unconventional hydrocarbons and technologies with consequent increases in CO_2 emissions (Sims *et al.*, 2007).

The relationship between expansion of energy access and sustainable agri-food production systems in developing countries can be cost effective. A fundamental barrier to sustainable development in many developing countries is the poor availability of efficient modern energy services in many rural regions. Their provision is a basic necessity for meeting the millennium development goals (MDGs) and to help improve both the health and livelihoods of millions of people. The provision of energy services can aid food production, storage and security. At present, almost 3 billion people have limited access to modern energy services for heating and cooking and 1.4 billion have zero or limited access to electricity.

A balance exists between energy access, efficient use of the energy resources available, and their affordability for potential consumers. In some countries, improved energy access has been achieved by governments expanding the deployment of renewable energy technologies in rural areas by introducing appropriate support measures (REN21, 2012). This can result in economic, social and environmental co-benefits at the local level while also providing energy directly to the agri-food system. Renewable energy can also provide basic services for non-agricultural activities in rural areas including domestic lighting, cooking, entertainment, information and communication, and motive power. Producing local biofuels at the small scale can enable agricultural machinery to be utilized and the road transport of food products to local market. For example, using straight vegetable oils directly in diesel engines to generate electricity or to run farming equipment is technically feasible in warmer climates (though there is a risk of engine damage from the higher viscosity oils so their conversion to biodiesel (methyl or ethyl esters) is recommended).

Co-benefits from renewable energy deployment include employment opportunities, reduced drudgery, improved livelihoods, a better gender balance for work responsibilities, local development, up-skilling of local tradespeople, improved health due to reduced air pollution, social cohesion, and an enhanced sense of community spirit (IPCC, 2011). It is therefore a preferable option for governments to support the deployment of renewables than to take the alternative approach of subsidizing the retail prices paid by energy users for indigenous or imported fossil fuels and petroleum products.

"Renewable energy can enhance access to reliable, affordable and clean modern energy services; it is particularly well-suited for remote rural populations, and in many instances can provide the lowest cost option for energy access" (Mitchell *et al.*, 2011). The trade-off between the possible high cost of some renewable energy technologies, such as small wind turbines, mini-hydro schemes, solar PV systems, anaerobic digesters and small bioenergy CHP plants, and realizing their co-benefits depends on local resource availability and circumstances. Solar, biogas and other improved designs of traditional biomass cooking stoves, including those using ethanol gels, dimethyl ether (DME), or liquid petroleum gas (LPG), may be less laborious, more efficient in terms of energy output, produce lower GHG emissions, and reduce health impacts from local air pollution. The use of more efficient cooking stoves can reduce the demand for traditional fuelwood by more than half compared with the use of open fires. However, the affordability of such new technologies needs to be carefully considered based on low annual incomes since their investment costs can be higher (Geoghegan *et al.*, 2008; UNDP, 2009). Cultural issues also have to be acknowledged. For example, cooking in the heat of the day using a solar oven may be rejected, especially by women with young children, compared with cooking in the cooler evenings using traditional biomass.

Around one quarter of the 2.7 billion people who rely on traditional biomass for cooking (plus the 0.3 billion who rely on coal), now use improved cooking stove designs in 166 million households, two-thirds being in China (UNDP, 2009). The dissemination of improved designs of domestic cook stoves has succeeded mainly when micro-finance is made available to support the capital investment needed (IPCC, 2011). Similarly, programs to introduce renewable energy technologies have mainly succeeded when micro-financing arrangements have been made available by national and local governments, aid agencies or the private sector to overcome the relatively high capital investment costs needed when installing solar, small wind, or mini-hydro systems on small farms.

In the many developing countries where subsistence farming/fishing is the most common type of food production, limited or zero access to modern energy services is a major constraint (UNEP, 2011a). Where energy can be provided in an economic manner, resulting increases in food productivity can result, together with reduced food losses from better storage leading to improved livelihoods. For example the co-benefits from using enclosed solar dryers as opposed to traditional open air sun drying for post-harvest processing, could reduce losses and produce a surplus for sale or feed an extra farm family member. The provision of access to modern energy services by applying appropriate technologies needs to be balanced by affordability. Ideally, any newly introduced system should make human labor more effective, minimize operating costs, avoid the need for future investments, improve food storage and processing activities, and enhance current crop production as a result.

The goal of energy access for all will require making basic and affordable energy services available using a range of energy resources and innovative conversion technologies while reducing adverse effects on human health, minimizing GHG emissions, and improving other local and regional environmental impacts. The method used to achieve optimum integration of heating, cooling, electricity and transport fuel provision with more efficient food production systems will vary with the region, local growth rate of energy demand, existing infrastructure and by identifying and valuing all the co-benefits. To accomplish this will require the global energy industry and society as a whole to collaborate with governments on an unprecedented scale.

2.1.2   *Environmental impacts*

The entire agri-food supply chain, including primary agricultural production, contributes approximately 22% of total annual anthropogenic GHG emissions (FAO, 2011a), plus an additional 15% contributed from land use changes particularly associated with deforestation to gain more agricultural land (IPCC, 2007). Additional risks from potential climate change impacts on food supply security means that the resilience of the agri-food sector requires careful evaluation (see Chapter 6). Analysis of probable climate change impacts on agricultural productivity up to 2050 has shown that negative effects are likely, leading to a reduction in food availability and human well-being, particularly in all developing regions (Fischer, 2009; Spielman and Pandya-Lorch, 2010).

Addressing environmental impacts usually depends on the introduction of regulations and tax incentives rather than relying on market mechanisms (Sims *et al.*, 2007). Large-scale energy-conversion plants with a life of 30–100 years give a slow rate of turnover of around 1–3% per year. Thus, decisions taken today that support the deployment of carbon-emitting technologies, especially in countries seeking supply security to provide sustainable development paths, could have profound effects on GHG emissions for the next several decades. Smaller-scale, distributed energy plants using local energy resources and low- or zero-carbon emitting technologies, can give added reliability, be built more quickly, and be highly efficient by utilizing both heat and power outputs locally (including for cooling using adsorption and other developing technologies (IEA, 2007a)).

2.1.3   *Food price and energy nexus*

The "green revolution" of the 1960s and 1970s solved the global food shortage problem at the time, not only through conventional plant breeding but also by tripling the application of inorganic fertilizers, expanding the land area under irrigation, and increasing direct and indirect fossil fuel energy inputs to provide additional services along the agri-food chain including mechanization, agri-chemicals, transport, refrigeration, etc. Today, the annual incremental yield increases of major cereal crops are in decline and some fossil fuels are becoming relatively scarce and hence more costly. A link between food prices and crude oil prices is therefore evident (Fig. 2.2) as is the correlation of the broader energy commodity prices with food prices (Fig. 2.3).

Partly due to fluctuating world energy prices reaching a peak, a related increase in global food prices was evident in 2008 which hit developing countries the hardest. In the poorest urban households where food can account for 50–80% of total expenditure, this increase resulted in societal unrest in some regions. In an average household in OECD countries, the food bill is typically 7–15% of total expenditure, though food price rises are still of concern.

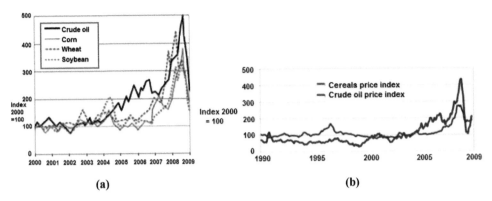

(a)   **(b)**

Figure 2.2.   Comparative trends of global crop commodity and crude oil price indices (a) from 2000–2009
with 2000 as base year (Heinberg and Bomford, 2009) and (b) from 1990 to 2009 with 2004 as
base year (Kim, 2010).

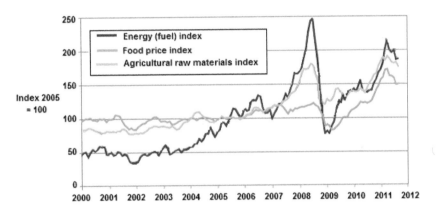

Figure 2.3.   Trends of commodity price indices for fuel (energy), food, and agricultural raw materials from
2000 to 2011 compared with base year 2005 (Index Mundi, 2012).

The global financial recession in 2008 had a significant impact on these commodity trends
which reduced prices significantly before rising again. The BRICS group of large emerging
economies (Brazil, Russia, India, China and South Africa) rebounded relatively quickly in
2010/11 (after one year in 2009 of slow performance) giving rise to increasing commodity prices,
even though OECD growth remained sluggish. As a result, economies that are large commodity
exporters improved their balance-of-payments but for the least-developed countries, high and
volatile commodity prices raised concerns about the availability and security of both food and
energy supplies.

Global climate change policy priorities have been affected as a result of these concerns. Where
climate policies have been linked to broader economic and energy security issues as in Australia,
South Korea, Brazil and even in Europe, it has proved relatively easy to still achieve some
political action for GHG mitigation. Conversely, where economic and security concerns have
tended to dominate, as in USA and developing countries, climate change policies have been
largely neglected (ADB, 2009; GEA, 2012; IEA, 2007b).

2.2   GLOBAL ENERGY TRENDS

"Since the advent of the industrial revolution, societies have relied on increasing supplies of energy to meet their aspirations for goods and services. Current energy systems, unless radically changed, will most likely fail to meet the aspirations of affordable, safe, secure, and environmentally sound energy services for all in the future." (GEA, 2012).

The ever-growing global demand for energy is projected to continue for at least the next few decades under business-as-usual. Energy-related GHG emissions therefore continue to increase annually despite:

- greater deployment of low- and zero-carbon technologies, particularly those utilizing renewable energy;
- the implementation of various policy support mechanisms by many states and countries;
- the advent of carbon trading in some regions; and
- a substantial increase in world energy commodity prices.

Energy-related emissions account for around 70% of total GHG emissions. These include carbon dioxide, methane and some nitrous oxide as well as short-lived climate forcers such as black carbon. To continue to extract and combust the world's rich endowment of coal, oil and natural gas, as well as peat, unconventional gas and oil sources at current or increasing rates, and so release more of the stored carbon into the atmosphere, is no longer environmentally sustainable, unless carbon dioxide capture and storage (CCS) technologies currently being developed can be successfully and widely deployed.

At the 15th Conference of Parties of the UN Framework Convention on Climate Change (UNFCCC) in Copenhagen, 2009, it was agreed by all countries present to limit the future global mean temperature rise to no more than 2°C above pre-industrial levels. Agreements made at the 16th and 17th Conference of Parties of the UNFCCC (COP 16 in Cancun, 2010 and COP 17 in Durban, 2011) recognized the need, and set a goal for deep cuts in global GHG emissions. To achieve this will require a rapid transition to a low carbon society with greater reliance on energy efficiency measures, renewable energy systems, nuclear power and the combustion of fossil fuels linked with carbon dioxide capture and storage (CCS).

Strategic investments may not necessarily encourage the greater uptake of lower carbon-emitting technologies leading to insecurity of energy supply issues and the missed opportunity to gain future co-benefits (Sims *et al.*, 2007). The various concerns about the future security of conventional oil, gas and electricity supplies could aid the transition to more low-carbon technologies. However, these same security concerns could also encourage the greater uptake of unconventional oil and gaseous fuels as well as increase the demand for coal and lignite in countries with abundant supplies in order to help meet national energy-supply security strategies.

Under the *New Policies Scenario*[1] of the International Energy Agency, fossil fuels will continue to dominate the growing demand for heat, electricity and transport fuels, along with an increased share for renewable energy systems (Fig. 2.4).

Under this scenario, GHG emissions will continue to increase, reaching over 50 Gt $CO_{2\text{-eq}}$ (CO_2 equivalent) in 2035 with energy-related emissions contributing over 70% of the global total (Table 2.1). The IEA's more stringent *450 Policy Scenario*, which puts emissions on a path to avoid the global mean temperature increasing above 2°C, needs to be implemented if the world is to reduce this continuing growth in GHG emissions.

The wide range of energy sources and carriers that provide energy services need to offer long-term security of supply, be affordable and have minimal impact on the environment. However,

---

[1]This scenario assumes implementation of all the existing commitments and plans announced up until mid-2011 by countries to tackle climate change, energy insecurity, and atmospheric pollution, including the Cancun Agreements, G-20 decisions, and Asia Pacific Economic Community (APEC) country initiatives.

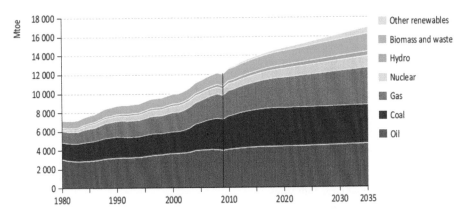

Figure 2.4.   World primary energy demand from 1980 until 2009 with projections out to 2035 under the
New Policies Scenario (IEA, 2011).

Table 2.1.   Projected world anthropogenic greenhouse gas emissions by scenario compared with actual
emissions in 2009 [Gt $CO_{2\text{-eq}}$] (IEA, 2011).

Greenhouse gases	Emissions 2009	New Policies Scenario 2035	450 ppm Scenario 2035
CO_2 energy	28.8	36.4	21.6
CO_2 industry	1.4	1.1	0.8
Methane	7.7	7.1	5.1
Nitrous oxide	3.2	3.2	2.7
F-gases	0.7	0.9	0.5
CO_2 land use change	5.2	1.9	1.9
Total	47.1	50.6	32.6

Note: F-gases include hexaflurocarbons (HFC's), perflurocarbons (PFC's) and sulfur hexafluoride (SF6)
from several sectors, mainly industry.

these goals often compete in terms of government priorities. An assessment of reserves and
resources was made in the IPCC 4th Assessment Report, Mitigation (Sims *et al.*, 2007).

- Conventional oil reserves may or may not have peaked, but it is uncertain exactly when or what
  will be the extent of the current transition to alternative liquid fuels such as coal-to-liquids, gas-
  to-liquids, oil shales, tar sands, heavy oils, oil from fracking, and biofuels. It remains uncertain
  what shares of the market these alternatives might take and what the resultant changes in global
  GHG emissions will be as a result.
- Conventional natural gas reserves are more abundant in energy terms than conventional oil,
  but they are also distributed less evenly across regions. Unconventional gas resources such as
  shale gas are also abundant and have recently been economically developed in the USA and
  elsewhere at a surprising rate. Future economic development of these resources, together with
  biogas from anaerobic digestion as being deployed in Europe (Chum *et al.*, 2011), is uncertain,
  but promising.
- Coal is very abundant but varying in quality and unevenly distributed leading to large volumes
  being traded. It can be converted to heat, electricity, liquids and gases. More intense utilization
  will require viable CCS technologies to be in place if GHG emissions from its use are to be
  constrained.
- There is a trend towards using energy carriers with increased efficiency and convenience,
  thereby moving away from solid fuels to liquid and gaseous fuels and eventually to more
  electricity generation both for heating and transport applications.

- Nuclear energy, currently at about 15% of total electricity generation, could make an increasing contribution to low-carbon electricity as well as heat supply in the future. The major barriers are long-term fuel resource constraints without recycling, economics, safety, radioactive waste management, security, weapons proliferation, and adverse public opinion, particularly after the Fukishima tsunami disaster in 2011. This caused several countries, particularly Germany and Japan, to review their existing nuclear plants and to reconsider future plans for expansion of the industry.
- Renewable energy resources (particularly solar, wind, biomass, low-grade ground source heat and small hydro, but with the exceptions of large hydro and high-grade geothermal) are widely dispersed compared with fossil fuels, which tend to be concentrated at individual locations (such as coal mines and oil fields) and therefore require distribution. Renewable energy can therefore either be used in a distributed manner or concentrated to meet the higher energy demands of cities and large industries.
- Non-hydro renewable energy-supply technologies, are currently small overall contributors to global heat and electricity supplies, but their shares are rapidly increasing (IPCC, 2011). Costs, as well as social and environmental barriers, are restricting this growth. Therefore, increased rates of deployment may need further supportive government policies and measures including subsidies, mandated shares of total energy, priority grid access, RD&D and public education, (REN21, 2012).
- The use of traditional biomass for domestic heating and cooking still accounts for more than 10% of global energy supplies but dependence on it could be reduced by more modern biomass systems (including the use of ethanol gels and di-methyl ether (DME) for cooking fuels) and other renewable energy systems, as well as by fossil-based domestic fuels such as kerosene and liquefied petroleum gas (LPG).

The transition from a world of cheap, surplus fossil fuel resources to one of new "clean-energy" supplies and conversion technologies, faces regulatory and acceptance barriers in order to achieve rapid implementation. Market competition alone may not lead to reduced GHG emissions. The energy systems of many nations are evolving from their historic dependence on fossil fuels in response to their increasing reliance on global energy markets, possible market failure of the supply chain, and the threat of climate change, thereby necessitating the wiser use of energy in all sectors.

Improved energy-efficient technologies can aid food and energy supply security by reducing future energy-supply demands as well as any associated GHG emissions (FAO, 2011a and Section 2.5 of this chapter). However, as shown by business-as-usual baseline scenarios, the present adoption path for energy efficiency, together with low- and zero-carbon supply technologies, will not reduce emissions significantly or at the desired rate.

A more rapid transition towards low-carbon intensity energy supply systems needs to be managed to minimize economic, social and technological risks and to co-opt those incumbent stakeholders who retain strong interests in maintaining the status quo. In rural areas of developing countries, distributed electricity systems could help avoid the need for transmission infrastructure costs and elsewhere offset the high investment costs of upgrading existing distribution networks that are close to full capacity (GEA, 2012, Chapter 15). The electricity, building and industry sectors are beginning to become more proactive and help governments make the transition happen as are local governments (IEA, 2009). Sustainable energy systems emerging as a result of government, business and private interactions should not be selected on cost and GHG mitigation potential alone but also on their other co-benefits that include improved local air pollution, health, employment and sustainable development.

2.2.1   *Bridging the emissions gap*

The United Nations Environment Programme (UNEP) recently undertook a holistic view of the entire spectrum of GHG emissions and reduction opportunities in all sectors in order to estimate

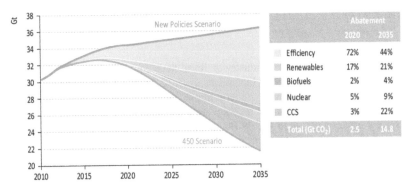

Figure 2.5.   Abatement potential of annual global energy-related CO_2 emissions in the IEA *450 Policy Scenario* compared with the *New Policies Scenario* through to 2035 (IEA, 2011).

the emissions gap between business-as-usual emissions (with no national pledges or policies implemented) and reduced emissions consistent with a 66% chance to stay below the 2°C target. The UNEP (2011b) report argued that 14 to 20 Gt $CO_{2\text{-}eq}$ of GHG annual emissions could be avoided without major technological breakthroughs by reducing emissions in the major sectors. Looking more closely at the energy sector, UNEP (2011b) and IEA (2011) both estimated an emission gap of over 4 Gt $CO_{2\text{-}eq}$ per year by 2020, increasing to over 20 Gt $CO_{2\text{-}eq}$ per year by 2035. It was calculated that improved energy efficiency measures and renewables could provide up to 90% of the required reduction to bridge this gap (Fig. 2.5), with the other 10% coming from biofuels, nuclear and CCS.

Energy efficiency options, where available, tend to provide least cost mitigation opportunities. However, the clear message from IEA (2011) is that removal of government subsidies that encourage the wasteful consumption of fossil fuels, and that in 2010 totaled over US$ 4 billion, will first be needed if energy efficiency and renewables are to gain significant traction.

Beyond the energy sector, a range of mitigation options are available for abatement of agricultural-related GHG emissions, emissions arising from deforestation and forest degradation (REDD+), particularly in developing countries, and short-lived black carbon and organic aerosols. As well as from agriculture and forestry management, key mitigation opportunities include improved energy efficiency, renewable heat and power generation, low-carbon transport options, urban systems design and infrastructure, (IEA, 2011; UNEP, 2011b). More efficient and innovative energy supply technologies are best combined with improved end-use efficiency technologies to give a closer matching of energy supply with demand in order to reduce both losses and GHG emissions. In this regard, the Global Energy Assessment (GEA, 2012) stated:

"It is possible for humanity to transform its energy system into one that provides everyone with access to clean, affordable, and secure energy supply, while capping climate warming under 2°C, and containing the environmental and other ancillary risks of energy systems. Nevertheless, such a change will require a major transformation from today's energy systems to technologically available, alternative energy systems resting on new pillars.

Global investments in combined energy efficiency and supplies will need to increase from about 1.7 to 2.2 trillion dollars per year as compared to the present levels of about 1.3 trillion dollars per year[2] (about 2% of current world GDP)."

---

[2]The UNEP green economy report (UNEP, 2011b) estimated the annual financing needs to green the global economy were in the range of USD 1.05–2.59 trillion or about 2% of the global GDP or 10% of the global total annual investments. A significant part of these investments would go into reducing the carbon footprint of major sectors.

2.3   OTHER MAJOR RELATED ISSUES

2.3.1   *Economic viability*

The heavy reliance of the agri-food industry on fossil fuels could jeopardize the future economic viability of some food supply chain businesses given the present price volatility and possible future supply disruptions. This raises concerns about the future security and affordability of food, as well as the continuing profitability for some agri-food industry enterprises.

If fossil fuel prices continue to widely fluctuate, oil remains around 100 US$/barrel or above, and carbon charges are added to cover the externality costs of CO_2 and other GHG emissions released during combustion, then the present dependency of the agri-food system on fossil fuels will, in time, make key inputs such as tractor and boat fuel, agri-chemical and fertilizer manufacture, processing, packaging and transport, all the more costly. This, in turn, could challenge the view that since farm land and fishing stocks are limited, future increases in food production to meet the growing population will have to mainly come from continued crop yield increases, particularly as a result of higher external energy inputs in less intensive systems. (Increased productivity from genetically modified crops may also contribute but this possibility is not discussed here). Therefore, if the desire is to move towards a low-carbon, less fossil fuel dependent, agri-food sector, then further agricultural intensification (in terms of level of direct and indirect energy inputs per unit of food produced), together with any expansion in beyond-farm gate activities, should ideally be disconnected from additional fossil fuel demands.

2.3.2   *Competing land uses*

Higher consumption of milk and meat products over the next few decades is projected due to growth in GDP and personal incomes in several emerging and developing economies, particularly China, India and other Asian countries (FAO, 2011a). These animal products will increase the demand for cereals for livestock feed and land use for grazing. In addition, if the demand for commodities such as corn (maize) used as a feedstock for transport biofuels continues to grow, as expected in the USA where ethanol now provides around 10% of liquid transport fuels by volume, then this will also put additional pressure on the demand for cereal production. In some regions, urban development and desertification are also placing significant pressures on agricultural land and contributing to increased competition for land use.

Fluctuating energy prices, future energy security and concerns at rising GHG emissions are challenges for future development of the agri-food sector and the aims to reduce its environmental impacts and support sustainable development. Due to the increasing competition for land and water, rising energy and hence fertilizer prices, and the anticipated impacts of climate change, a new paradigm is needed whereby farmers, fishers, food processors and distributors will need to learn to "save and grow" (FAO, 2011b).

2.3.3   *Dangerous climate change*

The United Nation's Framework Convention on Climate Change, adopted in May 1992, aims to *"achieve the stabilization of greenhouse gas concentrations in the atmosphere at a level that would prevent dangerous anthropogenic interference with the climate system. Such a level should be achieved within a timeframe sufficient to allow ecosystems to adapt naturally to climate change, to ensure that food production is not threatened and to enable economic development to proceed in a sustainable manner"* (UNFCCC, 1992).

IPCC (2007) suggested that *"defining what is dangerous interference with the climate system is a complex task that can only be partially supported by science, as it inherently involves normative judgments. Different approaches to defining danger, and an interpretation of Article 2, are likely to rely on scientific, ethical, cultural, political and/or legal judgments"*. Based on the available knowledge at that time, a 2°C increase was determined to be 'an upper limit' beyond which the risks of grave damage to ecosystems, and of non-linear responses, were expected to increase rapidly. Even 2°C is considered too high by some scientists arguing the world should aim at stabilization of warming at <1.5°C (Hansen, 2009).

The world has already warmed by about 0.8°C compared to the pre-industrial level of 14°C (Huber and Knutti, 2012). Continued warming could cross the 2°C threshold by as early as the 2030s (Smith *et al.*, 2011) and reach 4–6°C before the turn of the century (IEA, 2011). The risks of resulting extreme weather events have recently been reported by IPCC (2012) showing that many countries, particularly developing countries, face severe challenges in coping with climate-related disasters. However, regional variations in experiencing extreme events largely remain uncertain.

2.3.4 *Existing efforts are inadequate*

Global CO_2 emissions rose by 45% over the period 1990–2010 with a record increase of about 6% in 2010 (Peters *et al.*, 2012). With the USA not ratifying the Kyoto Protocol agreement and Canada, Japan and New Zealand withdrawing from it in 2011 and 2012, less than one fifth of total global GHG emissions will be regulated by remaining signatories in its second commitment period. The present growth in global GHG emissions exceeds their marginal reduction by the Kyoto signatories. The voluntary GHG reduction pledges made at COP 15 (Copenhagen, 2009), plus additional pledges made at later COPs are also insufficient and would lead to a projected warming of around 3.2°C which would give a rapid shift in species (Chen *et al.*, 2011). IEA (2011) concluded that the door of opportunity to keep global temperatures below 2°C is fast closing unless urgent mitigation actions are implemented before 2017. Even with all these international agreements in place, the risk of exceeding 2°C remains very high.

An integrated approach to energy for sustainable development is needed wherein energy policies are coordinated with policies in other sectors such as food, industry, buildings, urbanization, transport, health, environment, climate, and security, to make them mutually supportive (GEA, 2012). Using the wide range of available low- and zero-carbon technologies, the total mitigation potential by 2030 for the energy sector alone, at carbon prices below 20 US$/t $CO_{2\text{-eq}}$ ranges between 2.0 and 4.2 Gt $CO_{2\text{-eq}}$/year. Developing countries could provide around half of this potential. All other sectors, including agriculture, can also make a contribution in all regions (Fig. 2.6).

In the longer term, bridging the emissions gap (Section 2.2.1) will require some CO_2 to be removed from the atmosphere before the end of this century. From the agricultural perspective, this

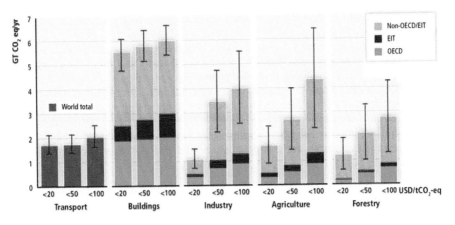

Figure 2.6.   Annual economic greenhouse gas emission reduction potentials by sector and region in 2030 at carbon prices of <20, <50 and <100 US$/t $CO_{2\text{-eq}}$ compared to respective baselines as assumed in sectoral assessments (IPCC, 2007). Note: The energy supply and waste sectors are not shown here. EIT = economies in transition.

objective could be assisted by reducing atmospheric CO_2 concentrations through either linking biomass combustion with CCS and regrowing the biomass to absorb more CO_2, or increasing the carbon content of soils, possibly using biochar (Lehmann and Joseph, 2009). The pyrolysis of biomass (heating in the exclusion of air/oxygen), produces the useful energy-containing gases, carbon monoxide and hydrogen, together with a stable, carbon rich, solid char co-product. If this biochar is added to soils, then a system is created that has abatement potential. This could also have long-term, positive effects on crop productivity by up to a 15% increase, but depending on the soil type, the type of biochar and its biomass source (Jeffery *et al.*, 2011).

2.4   GLOBAL ENERGY SUPPLY FOR AGRICULTURE

Globally[3], primary production directly consumes around 2 EJ/year[4] of total final energy for fisheries and aquaculture, mainly associated with boat propulsion, pond aeration and water pumping (FAO, 2009b) and around 6 EJ/year for direct energy demand on-farms, excluding human and animal power (Fig. 2.7). Currently around 1 EJ/year of this total final end-use demand is supplied by renewable energy (excluding traditional biomass). By 2035 it is projected that the agricultural sector end-use energy demand will grow steadily to reach ~9 EJ/year with renewables doubling its share to ~2 EJ/year (Sims *et al.*, 2011). The agri-food supply chain sector includes the 8 EJ/year of direct energy consumption for primary production (agriculture and fisheries) as well as shares of the transport, buildings and industry end-use demands for a range of activities including fertilizer manufacturing, food processing, distribution, storage and refrigeration, cooking, etc. (FAO, 2011a). Only primary production activities are discussed here.

On-farm energy demand in OECD countries is mainly for water pumping, livestock housing, cultivation, harvesting, heating protected crops in greenhouses, crop drying, and storage (OECD, 2008). In addition, indirect energy demand for the manufacturing and delivery of boats, tractors, machinery and fertilizer is approximately 9 EJ/year (GoS, 2011). As an illustration of how heavily dependent agriculture is on the energy sector, the synthesis of nitrogenous fertilizers alone consumes approximately 5% of annual natural gas demand (~5 EJ/year). An additional ~0.4 EJ/year of indirect energy is embedded in aquaculture feedstuffs (Smil, 2008).

Fossil fuel consumption for on-farm production in high-GDP countries is around 20.4 GJ/ha, almost double that for farms in low-GDP countries[5] (~11.1 GJ/ha) (FAO, 2011a). Fossil fuel energy inputs have served to reduce human labor inputs (typically being around 152 MJ of fossil fuel inputs for every man-hour of labor input in high-GDP countries versus 4 MJ/man hour in low-GDP countries) and hence have reduced the drudgery involved. They can also lead to a lower level of energy intensity when crop yields are increased, partly as a result of additional energy inputs (Table 2.2).

In high-GDP countries, energy inputs used for processing, transport and food preparation along the global food supply chain usually total around three to four times the energy used in primary production (Smil, 2008). In low-GDP countries, energy inputs (excluding human labor and animal power) can be up to 10 times but are complex, difficult to assess and vary widely between different regions (Fig. 2.8). In the USA, the current energy input/food energy output ratio is around 7 to 1 (Heller and Keoleian, 2000). It is lower in Africa where the dominance of energy used for cooking in subsistence farming is evident, with lower shares of energy used for production, processing and transport as might be expected given that much of the food produced

---

[3]When global agricultural data are quoted they can sometimes mask the differences between OECD/high GDP and non-OECD/low GDP agri-food systems since the latter often have poor data availability which is therefore not always accurately represented in the total data set.

[4]Data is uncertain with estimates in the literature varying by over 50%.

[5]The term "high-GDP" loosely groups together the top 50 or so countries measured in terms of their gross domestic product on a purchasing power parity basis divided by their population. Hence, "low-GDP" comprises the remaining 176 or so nations (http://www.indexmundi.com/g/r.aspx?v=67).

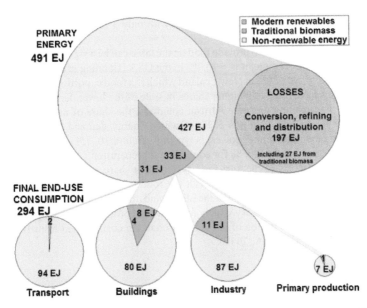

Figure 2.7.   Global final end-use consumption energy in 2008 from renewables, non-renewables (oil, coal, natural gas and nuclear) and traditional biomass primary energy sources by major sector. Note: Data from IEA (2010) converted to direct equivalent accounting method. Figure based on Sims *et al.* (2011).

Table 2.2.   Crop yields and energy intensities for corn production in the USA in 1945 compared with 2007 (Smil, 2008).

	1945	2007
Energy inputs [GJ/ha]	6	18
Corn yield [t/ha]	2.2	9.0
Energy intensity [GJ/t]	2.7	2.2

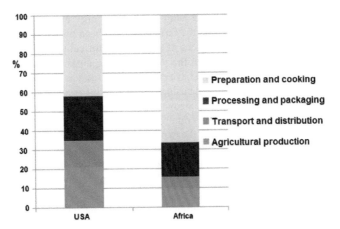

Figure 2.8.   Indicative shares of energy inputs in the agri-food supply chain comparing extreme examples of high-GDP (USA) and low-GDP (Africa) regions (Heller and Keoleian, 2000).

is typically consumed by the subsistence farmers with only a small portion delivered and sold fresh at local markets.

The total energy input needed to provide food on the table can be a significant share of a nation's total end-use energy. For example, it is ∼15.7% in the USA (Canning *et al.*, 2010), ∼20% in UK (GoS, 2011) and as high as ∼30% in New Zealand which is a food exporting country (CAE, 1996). Low-GDP countries have relatively higher shares in spite of the lower demands for transport and food processing. For example in some African countries, the share of national energy used for the agri-food chain can be as high as 55%, of which energy demand for primary production (excluding human labor and animal power) is typically around 10–15% of the total, 15–20% for transport and processing, and 65–75% for cooking and preparation.

Total energy-related costs in the agri-food system as a share of the total purchase price per unit of food to the consumer vary widely with the food product, but the proportion is usually relatively high in high-GDP countries. On the other hand, the total energy-related costs as a share of on-farm food production costs vary widely for each food product. For example, in the USA the energy-related costs as a proportion of total crop production costs ranged from 10% for soybean to 31% for corn (maize) (DEFRA, 2010).

2.5   ENERGY EFFICIENCY IN AGRICULTURE

For primary production systems, the aim should be to produce more or similar amounts of food per unit of land and water but using fewer energy inputs to do so. Several detailed assessments concur that increased intensification and technical progress require greater inputs of energy but that the energy should be utilized efficiently to maximize the services desired per unit of energy input.

Schneider and Smith (2009) compared agricultural energy intensities in low- and high-GDP countries. Steadily increasing fertilizer consumption and machinery use in China and India in particular have led to rising energy intensities since the 1960s. In high-GDP countries, declining energy intensities were observed since the mid-1980s, partly because average, annual, incremental crop yields continued to increase. Taking these opposing trends into account, overall global energy intensities started to decline slightly after the 1980s, though this trend varied widely between countries. Raising the national agricultural energy efficiency level of below-average countries to reach the national average level could be achieved by employing a range of energy efficiency improvements. This has the potential to reduce annual GHG emissions by up to 0.5 Gt CO_{2-eq} (Schneider and Smith, 2009).

Reducing the energy intensity[6] throughout the entire agri-food chain depends upon a combination of the development and deployment of improved farming/fishing management practices, behavioral changes, and new technologies with improved energy efficiency design specifications. Energy efficiency measures behind the farm gate have been promoted for several decades in high-GDP countries with varying degrees of success. Historically, direct energy input costs have been a relatively small component of the total operating costs for many farm businesses, hence incentives to reduce energy demand have not been as strong as they have now become. As energy costs increase and more agricultural businesses set targets to reduce their carbon footprints, renewed interest in improving energy efficiency to gain win-win benefits is becoming apparent.

For low-GDP and newly industrialized countries, efforts are being made to minimize energy intensities as energy demands increase in the expanding agricultural business sector (Schneider and Smith, 2009). This can involve changes to existing farming and processing practices for zero or minimum costs, though often a change in behavior by farm owners, managers and staff will also be required. Significant capital investment in modern efficient equipment can be a constraint to low-GDP countries wishing to emulate high-GDP countries that have already adopted improved energy efficient technologies. These are outlined in the following sections and include precision

---

[6]Using less energy to get the same result, such as reducing energy inputs per unit of food produced (MJ/kg).

Table 2.3.   Examples of direct or indirect energy efficiency improvements in agriculture and fisheries through technical and social interventions.

Direct energy	Indirect energy
Fuel efficient tractor engines/better maintenance. More precise irrigation water applications. Precision farming for accurate fertilizer application. Adopting minimum or no-tillage practices. Better control of building environments. Improved heat management of greenhouses. Better propeller designs of fishing vessels.	Less input-demanding crop varieties and animal breeds. Agro-ecological farming practices and nutrient recycling. Reducing water demand and losses. Improved fertilizer and machinery manufacture. GPS identification of fish stock locations to reduce trawling distances.

farming, irrigation water use monitoring, more efficient boat propeller designs, the use of GPS[7] to aid transport logistics, light emitting diodes, heat exchangers, variable speed electric motors etc. A balance is usually needed between deploying energy efficient technologies, assessing their affordability over a complete life-cycle, projecting future energy costs, and improving energy access.

There are many successful examples of reducing the energy intensity of food production as measured by reductions in energy inputs per kg of food produced (MJ/kg), energy inputs per hectare of land area (MJ/ha), and energy input/output ratios. Energy reduction strategies across the diverse range of agri-food management options are complex and can result in trade-offs having to be made. Two key points in this regard relating to primary production management practices should be emphasized:

- Any method used to reduce energy inputs that also lowers crop productivity (such as simply cutting back on fertilizer applications rather than optimizing the amount applied) is rarely beneficial and should be avoided.
- Intensive input production systems do not necessarily have higher energy intensities, especially when they result in increased crop yields (Table 2.2). Conversely, low-input farming systems can have relatively high energy intensities where lower yields result.

For small-scale and subsistence farming systems, there may be a case for increasing direct and indirect energy inputs over time in order to improve productivity and water use efficiency. In this case efficient use of energy could possibly best result from agro-ecological integrated farming practices that has the potential to achieve both high yields and also benefit livelihoods (Bogdanski *et al.*, 2010).

Energy conservation and efficiency measures in agriculture can be achieved in several ways (Table 2.3). These can either result from direct savings due to technology improvements, behavioral changes, or from indirect savings as a co-benefit from the use of agro-ecological farming practices. For both large and small farming systems, any potential food losses should also be avoided since they normally result in considerable waste of the energy embedded in the food production system and at the same time increase the growing competition for land and water use.

Since the annual direct end-use energy demand of the primary production sector is only a small percentage of total consumer energy in most countries[8], energy efficiency measures will not make a significant contribution to reducing national energy demands. However, energy saving measures

---

[7] Global positioning satellite systems can track truck routes, optimise speeds, avoid road congestion, etc.
[8] It was around 2.7% on a global basis in 2008 (Fig. 2.7).

can assist the profitability of individual enterprises, particularly capture fishing that use boats with high fuel consumption and intensive farming systems. Besides containing production costs, energy efficiency can also help to make food production less vulnerable to possible interruptions in future energy supplies as well as reduce GHG emissions from the agricultural sector.

It should be noted, however, that any improvements in energy efficiency bear the risk of a 'rebound effect'. This is when reductions in energy demand result in lower costs to the farmer and the money saved is used for purchases in other areas that have an energy component (Barker and Dagoumas, 2009). For example, a fisherman saving fuel by more careful operation of his vessel could use the money saved to purchase a larger outboard motor that then consumes more fuel. While the scale of rebound effect and its duration are the subject of much debate, there is agreement that the phenomenon is real and should therefore be taken into account when estimating overall energy savings.

The following sections cover energy efficiency opportunities for tractors and machinery, irrigation and fertilizers which cut across many agricultural enterprises. These are followed by specific opportunities in dairy farming, sheep and beef, intensive livestock, greenhouse production, fruit production and arable farming (which is also covered in more detail in Chapter 6).

2.5.1   *Tractors and machinery*

Many methods of reducing tractor fuel consumption have been well researched and documented. These include better matching of tractor and machinery size; controlling tractor passes within "tramlines"; selecting tractor and harvester engines with higher fuel efficiencies; early retirement of high fuel consuming machinery before end-of-life; improving engine maintenance regimes; correcting tire pressures; and implementing training programs on tractor and machinery operation, repairs and maintenance. Additional benefits can also result. For example, ensuring correct operation of tractor hydraulics and added ballast to optimize wheel slip during high draught activities such as sub-soiling can result in 10% lower fuel use, 20% savings in time, and reduced soil damage by avoiding excess wheel slip (CAE, 1996)[9]. The development of "precision farming" techniques, including using information technology sensors for monitoring of soil moisture and nutrients and GPS systems on tractors for accurate application of agri-chemicals and fertilizers on crops and pastures adjusted for variations in soil fertility, and can have both direct and indirect energy saving benefits (McBratney *et al.*, 2006).

Many low-GDP countries are already becoming well advanced in farm machinery use, so any fuel efficiency initiatives implemented for farming systems in high-GDP countries could well produce similar results. However, for low-GDP nations, energy efficiency may be a lower priority than food productivity and energy access. For example, farmers in Bangladesh could not afford fuel efficient tractor designs made in India but were able to gain access to cheaper Chinese multi-application engines that then boosted food production as a result (Biggs and Justice, 2011).

2.5.2   *Irrigation*

Irrigation normally involves pumping large volumes of water for use in high-pressure sprinkler irrigation systems, but can also involve less energy-demanding low-pressure trickle or drip-feed systems, and flood irrigation systems. Sprinkler irrigation is usually a high energy-consuming activity for pumping water of around 10–20 GJ/ha/year of electricity or diesel fuel, whilst fruit or outdoor vegetable production systems are usually on a smaller scale and may only consume around half that energy demand. Farmer focus is typically on maximizing irrigation effectiveness in terms of crop productivity and meeting rotation schedules using minimum time and effort rather than on saving energy.

---

[9]Much of the text in this section, and some other sections following, is based on practical experiences and analyses undertaken in New Zealand a decade or more ago. However, the principles involved still apply today and the lessons learned can also be transferred to the agricultural sector of many other countries.

Mechanical irrigation systems should be designed to use water as efficiently as possible, especially in the many regions where water supplies are constrained. For many present irrigation system designs, crop plants often take up less than 50% of the irrigation water applied with losses to evaporation and percolation (FAO, 2011b). Precision irrigation that can provide reliable and flexible water application based on intelligent control systems, along with deficit irrigation and wastewater reuse, could become a major platform for sustainable crop production intensification (SCPI)[10] (FAO, 2011b; Godfray *et al.*, 2010). Hence there is potential to improve water use efficiency, to reduce water run-off, and lower evaporative and infiltration losses that would all result in less electricity and/or diesel fuel inputs used for pumping.

Water supply competition can limit resource consents for taking water for irrigation, so there are potential synergies between energy efficiency and reducing water wastage – "Saving water also saves energy!". A range of initiatives to improve the efficiency of irrigation include measuring of optimum soil moisture content and improved scheduling. Energy savings from existing irrigation systems can result from improving basic operating conditions, mending leaks due to lack of maintenance, and replacing worn or improperly sized pumps. For example, if a pump is working only at 20% efficiency instead of at its design specification of 30%, then the energy demand, and hence GHG emissions, will be increased by 50% above the baseline (Nelson *et al.*, 2009). Both water and energy inputs can be reduced by altering crop sowing dates to avoid anticipated periods of water deficit and mulching operations, as well as by adopting sensor-based, demand-led irrigation systems. The greater deployment of solar PV and wind-powered irrigation systems can be managed carefully in combination with water use efficiency measures to support the moves towards more sustainable agricultural production systems.

2.5.3 *Fertilizers*

As an example, a typical energy requirement for manufacturing, delivery and distribution of super-phosphate, was calculated to be 2.3 GJ/t (Wells, 2001). Therefore, as well as aiming to reduce the direct energy used by machinery for on-farm transport and application of fertilizers, the prime variable of influence on the farm should be to reduce the indirect energy by adjusting the quantity of fertilizer applied per hectare and reducing wasteful application. However, reducing fertilizer application rates simply to reduce energy inputs makes little sense if the productivity of the farm is threatened as a result.

To reduce direct energy during application, ground spreading of high analysis fertilizer should be encouraged where practical to do so since it takes two to five times more fuel to top-dress fertilizer from the air. Ground spreading of fertilizer is usually applied using truck spreaders that consume 1.5–1.8 L/ha whereas aerial application requires 7–8 L/ha for an application rate of 250 kg/ha. The energy required for aerial topdressing can constitute a significant fraction (up to 30%) of the total energy input. However, ground spreading may not be an option on some properties due to the hilly terrain.

More accurate fertilizer use from ground or air can be achieved by improving the precision and timing of applications using engineering and computer-aided technologies such as biosensors for soil fertility monitoring, trace gas detection, and GPS. In high-GDP countries, a combination of these techniques has achieved significant reduction in fertilizer use (Schneider and Smith, 2009). In the USA, for example, this has resulted in around a 30% reduction compared with the total application levels in 1979 (Heinberg and Bomford, 2009).

There has been a trend to replace bulk fertilizers (such as super phosphate and urea) with high-analysis, NPK fertilizers. This reduces the fuel used both in application and transport. However, any energy saving may be offset by the increased indirect energy consumed during the manufacture of these fertilizers. Mineral fertilizer manufacturers have demonstrated various

---

[10]SCPI aims to produce more from the same area of land while reducing negative environmental impacts and increasing contributions to natural capital and the flow of environmental services.

Table 2.4.   Comparison of electrical power usage on dairy farms in four selected countries in 1996 (latest data available) to illustrate the wide variations in the dairy farm sector depending on the farm management system employed (CAE, 1996).

	New Zealand	Australia	UK	USA
Average herd size [number of cows]	150	108	71	47
Average electricity demand [kWh/cow/year]	112	175	313	718
Average annual milk production/cow [L]	3176	3519	5115	3356
Electricity intensity [kWh/1000 L milk]	35.3	43.7	59.5	102.3
Electrical demand [%]				
Vacuum pumping	19.4	19	30.0	28.0
Milk cooling	19.2	30	30.2	21.0
Water heating	32.3	41	37.2	25.0
Water pumping	17.5	0	0.0	9.1
Lighting/effluent pumping/compressors etc.	11.6	10	2.6	16.9

options to reduce energy inputs per unit of fertilizer produced and delivered. In addition, farmers who reduce the quantity of fertilizers applied to crops and pastures as a result of more precise applications also reduce their indirect energy inputs. A shift towards organic fertilizers, including the use of nitrogen-fixing plants, can also reduce indirect energy inputs. This will serve to lower GHG emissions per unit of output and could possibly avoid excess nitrates being discharged into aquifers and surface waters.

2.5.4   *Dairy farms*

The total energy inputs for a typical dairy farm come from fuel for tractors and equipment (15–20%); electricity (25–45% depending if irrigated using electrically driven pumps or not); and indirect energy embedded fertilizers (~35%); and in buildings, roads, machinery, infrastructure etc. (10–20%). However, dairy farms vary widely in terms of size of herd; feeding concentrates in the milking shed or not; rainfed or irrigated; whether owning and operating tractors and machinery or using contractors; effluent and manure disposal systems; and whether the cows are housed all-year-round, seasonally or grazed outside all year, with perhaps some conserved hay, silage or green crops grown for when pasture is in short supply. Therefore, there is a wide range of energy input intensities (Table 2.4).

A typical dairy farm relies on electricity and liquid fuels as direct energy inputs (Fig. 2.9).

Electricity: A good potential for energy efficiency gains stems from management of electricity demand in the milking shed for water heating, refrigeration, and motive power for pumping, air compressors and running other machinery (Fig. 2.10). Power demand at milking times can create peak load spikes on the power distribution lines. Hence contracts with electricity network companies may include increased tariff charges when specified peak kVA demand levels are exceeded (Wells, 2001). This is another driver for use of energy efficiency measures.

Washing milking equipment in place consumes between 150 and 900 L of hot water heated to 85°C every day, the volume depending on farm size and whether carrying out one or two hot washes per day. This can consume 15,000–25,000 kWh of electricity per year. Simple steps to conserve electricity include insulation of hot water cylinders and the use of solar water heaters. A cold-water wash for one milking per day also saves on electricity heating costs but there have been some concerns over the bacterial contamination of the milk resulting from cold-water washes.

Milk cooling tends to be an inefficient operation that can be improved by lagging the milk storage vat to prevent re-warming, especially if the milk vats are placed outside in the sun. Pre-cooling to ensure the milk reaches the vat below 18°C can be achieved either through upgrading of existing plate heat exchangers, incorporation of ice bank technology that uses off-peak power, or by modifying the vat set-up.

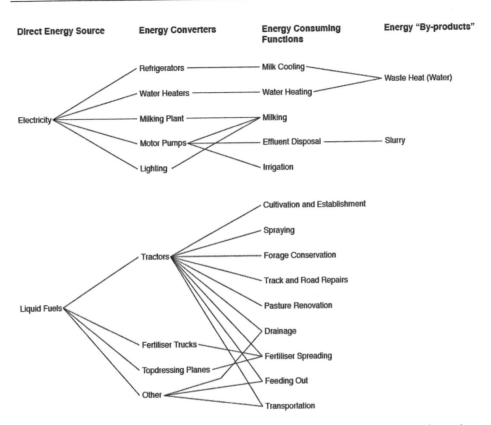

Figure 2.9. Direct energy flows for a typical dairy farm enterprise and useful energy output by-products. (CAE, 1996).

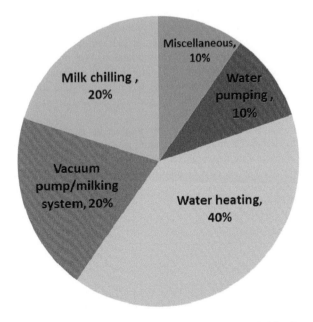

Figure 2.10. Breakdown of typical electricity use on a typical non-irrigated dairy farm (CAE, 1996).

Heat recovery technology presents an opportunity to save up to 30% electricity use by using the heat removed from the milk during the cooling process and from the waste hot water from the cleaning system to preheat the fresh water flowing into the hot water cylinder. Heat exchange systems are marketed as a complete product with good payback periods depending on the size of the herd milked and the number of milkings per day.

Improving energy efficiencies of vacuum pumps can be obtained either by altering the pump placement or by installing a variable speed drive pump. More closely matching the variable load to pump output can provide at least a 50% power saving, with some milking shed systems saving up to 75% of electricity demand. Additional benefits include reduced vacuum pump wear, lower noise, lower maintenance costs, and improved vacuum reserve capacity.

Pumping water for use in the dairy shed and stock drinking water supply can be a significant energy user on dairy farms where bore or stream water is used. Energy used by water pumps can be minimized by regular maintenance and any leaks should be repaired. Wastewater from the sheds and associated yards is normally pumped to an aerobic/anaerobic lagoon for treatment. In most instances no mechanical agitation, recycling or aeration is applied and the treated effluent is eventually pumped on to the land. The systems tend not to always work successfully and energy inputs to improve future aerobic treatment may be required.

Methane recovery from animal wastes through anaerobic digestion produces biogas that can be combusted to generate heat and electricity. This seems an attractive proposition and it has been attempted many times, but corrosion of equipment and the requirements for high labor inputs on a daily basis are major constraints. The New Zealand Standards Association produced guidelines for "*On-farm biogas production and utilization*" in 1987. Much of the technical information still applies today yet there have been relatively few anaerobic digestion plants built on farms since then.

Overall there are good opportunities to reduce the electricity demand through educating farmers and encouraging them to undertake low capital investments to save costs such as installing a timer set to turn on the water heater just in time to heat the water before needed at the end of the morning milking rather than storing it in a cylinder for 24 hours with subsequent losses. Such timing of electricity use can also avoid peak loads and lead to lower tariffs if using time-of-use metering. Other demand-side response examples include avoiding refrigeration loads at peak times and carrying out irrigation at night which also results in less water loss from evaporation.

Liquid fuels: To fuel tractors and machinery, dairy farms require around 7 L of diesel/cow/year, of which 2–3 L is for forage conservation, 1–2 L for cultivation, 1–2 L for feeding out hay and silage, and the rest for pasture renovation and miscellaneous farming operations (CAE, 1996). A typical hay making system consumes around 23–26 L/ha of diesel fuel (around 5.5 L/t), whereas silage requires around 38 L/ha (2.8 L/t). Assuming hay has a moisture content of 15% (wet basis) and silage 45%, then on a per dry ton basis, hay consumes ~6.5 L/dry t and silage ~5 L/dry t. It is likely baled silage will be more fuel-efficient.

Pasture renovation by over-sowing uses approximately 0.7 L of diesel/cow/year. A range of suitable seed drills has become available and pasture renovation by this method has increased in recent years. Where over-sowing has replaced full cultivation followed by re-sowing, 15–30 L/ha of diesel can be saved. In addition, pasture quality has improved since new grass varieties can be easily incorporated into existing pasture by such over-drilling techniques. Consequently, there is less need to cultivate and grow forage crops that were often grown immediately before establishing a new pasture. The fuel that would have been consumed for the establishment of new forage crops has therefore been saved.

2.5.5   *Sheep and beef farms*

Sheep and beef farming systems are generally not energy intensive due to the limited use of buildings and machinery compared with other farm enterprises. However, some farms still have high energy demanding activities such as feedlots, forage conservation, feeding-out, irrigation, pasture management and green fodder cropping (Fig. 2.11).

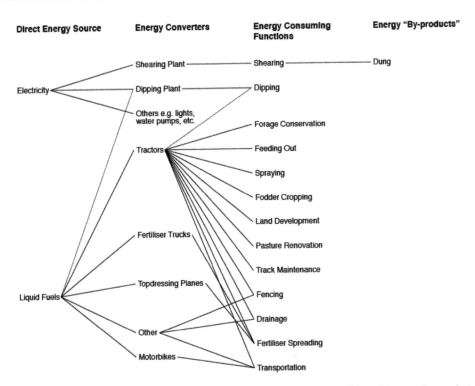

Figure 2.11.   Direct energy flows for a sheep and beef enterprise and potential useful energy by-products (CAE, 1996).

The overall energy intensity of a typical sheep and beef farming system was calculated to be 1.09 GJ/ha/year of which 73% of total energy inputs were indirect, almost half being for the manufacture, transport and application of fertilizer (Table 2.5).

The energy intensity of sheep and beef production systems is strongly correlated to the stocking rate; hill country farms with low stock units (su) (3–4 su/ha) had low energy intensities (0.3–0.5 GJ/ha), whereas low country farms with higher stocking rates (9–10 su/ha) had higher energy intensities (1.8–2.3 GJ/ha) due to increased requirements for supplementary conserved feed needing more tractor fuel for cultivating forage crops and hay and silage making. This relationship is subject to the law of diminishing returns per unit of additional input.

Electricity: Consumption on a typical sheep and beef farm is generally low, ranging from 3000 to 7000 kWh/year or around 1.0–1.5 kWh/su (CAE, 1996). Water pumps for stock water reticulation systems are the main electrical energy consumers with an energy intensity of about 12.5 kWh/ha. Shearing consumes only ~0.02 kWh per sheep shorn (CAE, 1996) but with variations occurring between individual shearers ranging from 35–60 sheep shorn per kWh.

Liquid fuels: Transport is the main diesel and gasoline consuming activity for both off-farm movement of livestock and feed using trucks plus on-farm movement, especially of farm staff using motorbikes and utility vehicles (Fig. 2.12). Tractor use is mainly for fertilizer application (if not done by aircraft), cultivation, forage conservation and pasture renovation. Fuel efficiency of tractors was discussed above.

2.5.6   *Intensive livestock production and fishing*

The major energy-demanding activities for both pig and poultry production are to maintain temperatures of environmentally controlled buildings, provide feed, and remove animal wastes from the buildings (Fig. 2.13). Animal housing for pigs and poultry can be designed and operated

Table 2.5.   Energy inputs and energy intensity of sheep and beef farming systems (CAE, 1996).

Energy input	Energy intensity [GJ/ha/year]
Direct energy:	
Fuel	0.29
Electricity	0.01
Indirect energy:	
Fertilizer	0.36
Machinery	0.10
Water supply	0.07
Contract cartage of livestock	0.07
Fencing materials	0.06
Building materials	0.04
Field contractors	0.03
Agri-chemicals	0.03
Shearing equipment	0.02
Other	0.01
Total	1.09

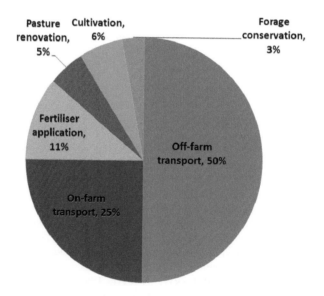

Figure 2.12.   Liquid fuel demands for a typical sheep and beef farm (CAE, 1996).

efficiently to conserve the energy inputs needed to maintain optimum temperatures and humidity levels (GoS, 2011). Computer-controlled feeding of intensively housed livestock can help to reduce waste feed and hence lower overall energy demand. Opportunities also exist to reduce energy inputs needed for water and space heating, drying, feed storage and conveying equipment.

Of the total energy input into intensive pig and poultry enterprises, only around 14–16% comes from direct energy inputs, including electricity for lighting (for layers, to encourage egg production by controlling day length), automatic feed conveying, ventilation, heating, and liquid fuels mainly used for transport (Fig. 2.14). The major indirect energy input is for the processing, production and transport of high protein animal feed, bought-in from other arable crop growers. Maintaining optimal temperature control of the building can ensure a more efficient feed-to-meat

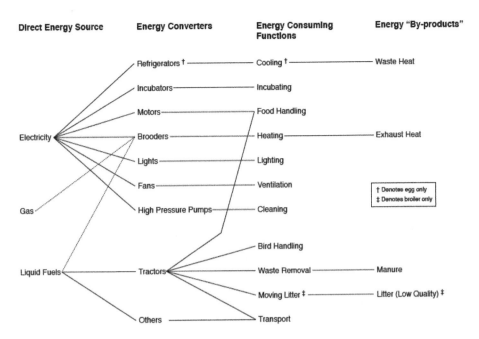

Figure 2.13.   Energy inputs into a pig and poultry farm and useful energy by-products (CAE, 1996).

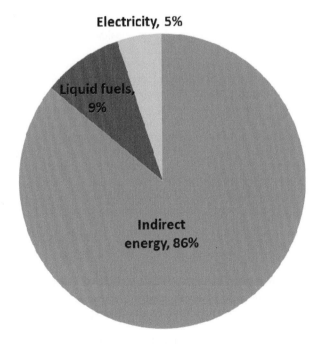

Figure 2.14.   Energy inputs into typical intensive pig and poultry production enterprises (CAE, 1996).

conversion rate as less energy is needed to maintain body heat. Thus electricity consumption and cost can, to a limited degree, be used to offset feed costs and related energy inputs.

Electricity: Consumption on a pig and poultry farm can show seasonal peaks in summer due to the increased use of air conditioning and fans for ventilation, and a peak in winter where heating is required. The climate within a piggery or poultry house can be affected by ambient temperatures, heat released by the livestock, building insulation levels, ventilation rates and the amount of remaining excreta. Maintaining temperatures at optimum levels to avoid stress for the animals results in improved productivity and fewer health problems. Any energy-saving methods considered should bear this objective in mind.

Infrared lamps to maintain the body temperature of younger livestock heat only the animals rather than the whole building. As the livestock grow and mature, the height of the lamps can be raised or the heat output reduced. Quartz linear lamps are favored in large, open, and draughty situations. Good under-floor heating of a brooder or farrowing house designed to deliver 300–400 W/m^2 is an alternative option. Energy costs could be minimized by using off-peak power tariffs, and energy could be saved by use of thermostats and the insulation of the floors and walls of the pens.

Effective building designs can reduce the need for artificial lighting and temperature control by making use of natural sunlight and providing shelter from wind. Correct building orientation, window placement and good insulation may reduce the need for energy consumption. Ventilation, either by natural or forced airflow using electrically powered axial fans, is necessary to remove the body heat. A heat pump could be a more energy efficient option and also be used for cooling in the summer months, though introducing fresh air is preferred for animal health.

Water supplies typically require a 0.5–1 kW electric pump where a good water source is readily available. A high-pressure water supply with a greater capacity motor (around 2–3 kW) would be necessary for animal housing wash down purposes. Animal manure removal is an important activity in maintaining a livestock production system. Options for disposal include recycling by applying to crops as fertilizer and soil conditioners or using as feedstock in an anaerobic digestion process to produce biogas that can help meet the heat and power demand of the enterprise. Although technically feasible, biogas production has rarely proved to be cost effective at the farm scale without government subsidies, except, perhaps, for very large operations.

Fishing: Costly fuel bills for capture fishing can be reduced by using less energy-intensive fishing methods, improving methods of fish stock location, using information and communication technologies to optimize fishing and market decisions, as well as by improving vessel and gear design (World Bank, 2009). Acceptance and understanding of new technologies by boat operators is essential, but so are behavioral changes regarding operation and maintenance. Fuel is wasted by driving the vessels excessively fast and inadequately maintaining them, and hence reducing the working life of the boats. Changes in technology such as propeller designs are of relevance to vessel operators who are either considering the purchase of a new boat or overhauling and re-equipping an existing vessel (Wilson, 1999).

In aquaculture systems, heat recovery may be feasible in intensive hot water systems and optimizing air and water management can also reduce energy demands.

2.5.7   Greenhouse production

Energy use in the greenhouse industry can be influenced by a number of factors including management, location, seasonal production, type and age of greenhouse, the crop/s being grown, and their heating and lighting requirements. Depending on crop type, location, seasonal production and target markets, a crop may or may not require heating during its production cycle.

Energy demand is linked to temperature, humidity, light and water supply, with the aim being to optimize the growth of high quality plant products. Energy costs in heated greenhouses can be partly compensated for by better quality and increased crop productivity but benefits also result from decreased energy intensity per unit of food produced. Energy savings can be achieved by relatively straightforward options such as improving lighting and ventilation system designs, reducing heat losses, improving boiler operation and maintenance, fully using the floor area for plant production, accurately calibrating temperature and humidity sensors, and installing wind

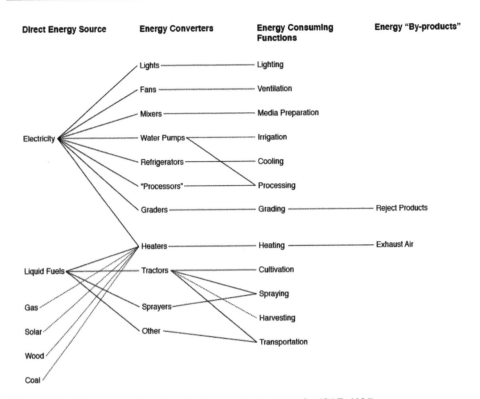

Figure 2.15.   Direct energy flows for a typical greenhouse enterprise (CAE, 1996).

shelters and moveable thermal screens to help reduce heat losses by acting as a barrier against thermal radiation loss at night. Thermal screens, however, do carry a risk of increasing humidity and associated diseases, and even when fully pulled back can reduce light penetration by giving extra shading effects on the crop, with some consequent loss of plant productivity (CAE, 1996).

Greenhouse design determines the amount of energy inputs needed to maintain optimum light and temperature conditions. Design of the structure, orientation, shelter and location can all influence the heating energy consumption of a greenhouse operation as can using double cladding of glazing or plastic. The additional capital costs involved can be offset by energy savings of up to 40%, but payback periods vary with specific conditions. The overall energy intensity of a greenhouse ranges between 2600 and 7000 MJ/m^2, with heating generally being the main influencing variable (Fig. 2.15).

When growing tomatoes or cut flowers, energy inputs can vary from ~20–45 MJ/kg of produce (~1000–2000 MJ/m^2/year) (CAE, 1996), the range reflecting climatic variations and the need for additional heating, mainly required at night during winter periods, though to maintain optimum temperatures for plant growth it may also be necessary at other times. Heat is also used as a management tool by tomato growers to avoid fruit skin cracking as a result of water condensing on the fruit on a hot summer morning.

The heating sources used by greenhouse growers in heated houses around the world are typically coal and natural gas with lower contributions coming from heating oil, LPG, biomass, and electricity. Methods of heat transfer from the boiler include lagging pipes carrying hot water to heat the air, direct heating of circulating air, and root zone heating. Soil bed warming systems encourage plant propagation in a nursery and can be more efficient than heating an entire greenhouse. Direct combustion of LPG, natural gas or biomass can be undertaken in flueless heaters during the day to also enhance growth by increasing in-house carbon dioxide concentrations. Some of the heat produced can be stored in large, insulated water tanks for use during the cooler night periods. Geothermal heat and waste heat from industries or thermal power plants can also be utilized if greenhouses are constructed near to the heat source.

Fuel switching, for example from coal to biomass heating fuels, can reduce GHG emissions but would not reduce the energy demand. In China, fruit and vegetable production in small, heated greenhouses has been integrated with biogas production at the domestic scale using organic wastes from the household and animals as feedstocks (CNSS, 2011).

Electricity: Approximately 30% of greenhouses use some form of ventilation and air circulation systems both to increase yields and provide a more efficient use of energy through a more uniform environment by reducing humidity-related crop diseases. The main method of increasing ventilation and air circulation is to install fans with capacity of around 28 to 65 kW/ha for glasshouses and 5 to 46 kW/ha for plastic houses (CAE, 1996). Electricity consumption by buried cables is relatively low for soil warming at around $1–1.5$ kWh/m^2/day, so it is difficult to implement any energy saving measures. Heat pumps could be an alternative energy efficient option as they can double as dehumidifiers and also be used for cooling in the summer months as well as heating.

Poor lighting conditions can reduce plant growth, increase growth periods and waste heating fuel. Daylight is the major limiting factor to constrain out-of-season crop production, so artificial lighting can be used to compensate. However, capital and running costs are high. Lighting using incandescent bulbs of 60–150 W has been traditionally preferred by growers because this produces a greater source of red light to which plants are sensitive. However, these are being replaced cost effectively by around half the number of fluorescent or high-pressure mercury 18 W lamps to achieve a similar effect. Most greenhouse irrigation systems are low pressure so are relatively low consumers of energy when used for pumping.

Growing crops hydroponically in nutrient solution rather than in soil or other growing media, avoids the energy used for both soil cultivation and steam or chemical sterilization of the soil. However, some electrical input is needed for circulating the nutrient solution and sterilizing it by ultra violet light, pasteurization or ozonation. Warming the nutrient solution coupled with accepting a lower ambient in-house air temperature can reduce overall heating requirements.

Production of mushrooms indoors is also an energy intensive system. Energy demand can be reduced by high insulation levels in buildings and efficient air-to-air heat pumps to maintain optimum air temperatures (CAE, 1996).

2.5.8   Fruit production

Energy inputs into the fruit-producing farming systems can be up to four times higher per hectare than conventional arable cropping operations (CAE, 1996). Most direct energy inputs occur within the orchard as fuel for tractors, hydraulic ladders, and forklifts (Fig. 2.16) with electricity demand varying depending if grading and packing the fruit is undertaken on site or the fruit is transported to a central packing shed.

A recent detailed survey of apple orchardists in New Zealand (Frater, 2010) showed that direct energy contributed around 38% of farm level energy intensity (Fig. 2.17) with the total average energy input at around 55 GJ/ha/year or around 1.45 MJ/kg fruit. Based on life-cycle analysis methodology, post-harvest processes added 0.51 MJ/kg, packaging 1.46 MJ/kg, and refrigerated shipping to export markets across the world 4.24 MJ/kg.

Liquid fuels: Orchards are relatively fuel intensive at around 850 L/ha (CAE, 1996). A quarter of this fuel is consumed during spray applications being the largest consumer of fuel in an orchard. Air blast sprayers are commonly used in spite of low operation efficiency with often only 20–30% of the chemical reaching the target. More efficient sprayer designs tend to require lower energy inputs but are more costly and need increased maintenance and operating skills, so have not become popular.

Using more concentrated sprays may reduce energy demand due to lower application volume rates. Reducing the number of spray applications needed in a season is now being advocated in integrated pest management systems which would also reduce the need for tractor passes

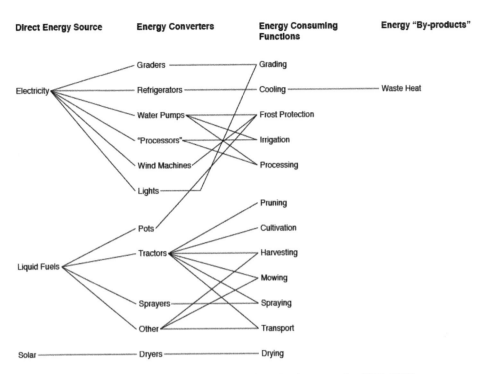

Figure 2.16. Direct energy flows through a typical fruit production enterprise (CAE, 1996).

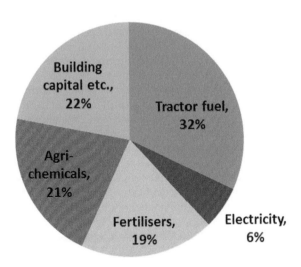

Figure 2.17. Breakdown of direct and indirect energy inputs for a typical New Zealand apple orchard (Frater, 2010). Note: Tractor fuel includes liquid fuels used in other field equipment.

through the orchard. These techniques involve spraying only when required rather than at fixed intervals, which has been the traditional pest control method. Tank mixing of approved chemicals and simultaneous application of more than one spray, where appropriate, would also reduce the number of passes and hence save fuel.

Table 2.6.   Direct liquid fuel energy flows and activities in kiwifruit, grape and pip fruit enterprises (CAE, 1996).

Field activity	Kiwifruit [L/ha]	Grapes [L/ha]	Pip fruit [L/ha]
Vine/tree spraying	120	35	225
Inter-row cultivation	–	16	–
Mowing	80	–	60
Trimming and pruning	20	20	65
Weed spray	15	11	5
Topdressing	10	3	10
Shelter maintenance	35	2	10
Harvest	40	44	100
Materials handling	20	8	35
Product transport	60	18	130
On-farm staff transport	390	28	210
Total	790	220	850

Field operations are the major energy user on an orchard and these vary with type of fruit growing enterprise (Table 2.6). They include vine trimming and pruning, mowing, top dressing, tree shelter belt maintenance, materials handling, and harvesting operation, which for kiwifruit is usually manual, for grapes is mainly by machine, and for pip-fruit is a combination using hydra-ladders (elevators based on hydraulic controls) that are also commonly used for pruning and thinning operations.

Frost protection (not shown in Table 2.6) is not a large energy user as it is not a common occurrence, but may be a vital activity for some orchards. Wind machines powered by diesel generators have been installed in orchards and vineyards to mix the layers of air and avoid frost conditions developing which has met with some success. The use of helicopters to force warmer air down is more energy intensive using on average 160 L/h, but the cost for this could be warranted if the crop can be saved from frost damage. Sprinkler irrigation systems can be used in some instances to form a thin ice layer to protect delicate tissue. Compared with other alternatives this is a low-energy input system but can result in water-logged soils in some regions. Where the water is supplied by an electric pump, a suitably sited thermostat could be used for startup only when the temperature deviates from the desired range. Low water pressure trickle irrigation systems with drip feed nozzles and low water demand are normally installed in orchards so it is unlikely that growers will install overhead sprinklers purely for energy efficiency purposes unless the cost of other methods rises significantly.

Electricity: Crop cooling directly after harvest can improve the quality of fruit for both domestic and export markets, so there has been an increase in on-farm cool stores and packing facilities that has increased the overall electricity demand. Cool stores and refrigeration units can produce high peak load demands of around 800 kW per facility and around 200 kW when the system is idle. Cooling methods include ambient air-cooling, refrigeration, ice banks, hydro cooling and vacuum cooling. Load shifting potential away from times of peak power capacity also has potential using cool stores.

2.5.9   Cropping (see also Chapter 6)

The main direct energy inputs into an arable cropping system are liquid fuels, most of which are consumed during field operations. Electricity is used for crop handling, drying, and especially where irrigation is undertaken (Fig. 2.18). Total energy inputs can range from around 5 GJ/ha for rainfed arable farms growing cereals to over 20 GJ/ha for irrigated farms growing potatoes and vegetables (CAE, 1996). How much post-harvest processing (including washing, grading, drying, packing, etc.) occurs behind the farm gate or in centralized processing plants affects the energy demand of each agricultural enterprise.

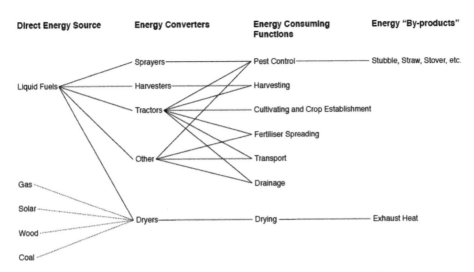

Figure 2.18.   Direct energy flows for a typical arable cropping enterprise and useful energy by-products (CAE, 1996).

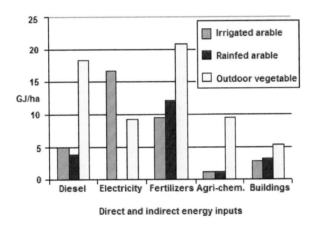

Figure 2.19.   Typical direct and indirect inputs for arable and vegetable crop production (Barber, 2004).

Fuel use and type vary between irrigated arable, rainfed arable and vegetable production systems (Fig. 2.19). The direct energy embedded in fertilizers, agri-chemicals, buildings and infrastructure is also significant. Vegetable production is usually more energy intensive per hectare then arable crops (except where irrigated) due to the larger land areas involved in the latter.

Conservation agriculture is an approach to manage agro-ecosystems and give improved and sustained productivity, increased profits, and food security while preserving and enhancing the resource base and the environment (FAO, 2011c). This broad concept aims to improve farm management by using crop rotations to enhance the soil nutritional status as well as to lower the demand for inorganic nitrogen, reducing pests, minimizing soil disturbance by avoiding tillage and improving soil quality. Reduced energy inputs are usually a co-benefit. Historic carbon losses through conventional cultivation are estimated to be between 40 and 80 GtC and are increasing by a rate of 1.6 ± 0.8 GtC/year, mainly in the tropics (GoS, 2011). Increasing soil carbon content over the long term by the addition of biochar (Section 2.1.3) is under evaluation, as are low-labor input perennial crop-based production systems that can be self-regulating in terms of pest and disease management as well as conserving soil fertility and moisture.

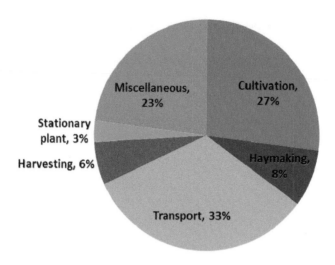

Figure 2.20.   Main liquid-fuel end uses on a typical arable cropping farm (CAE, 1996).

Liquid fuels: Most of the diesel fuel on an arable farm is used for tillage (70–80%), with conventional methods of plowing and cultivation consuming over 50 L/ha. Mouldboard plowing alone can account for around 40% of the total fuel used as it consumes ~15–18 L/ha varying with cultivation depth and soil type. Conservation tillage techniques have been identified as being more fuel-efficient and can reduce fuel consumption for cultivation practices by up to 60–70% as well as improve soil water retention, reduce soil erosion by incorporating crop residues into the surface, and minimize soil carbon losses (Baker *et al.*, 2007). Land clearing should be viewed as a one-off activity that can consume between 5 and 15 GJ/ha of liquid fuels in large crawler tractors, consumption varying with the terrain and type of existing scrub and forest cover. Very little clearing occurs today.

Tractor selection and design are important factors when trying to maximize fuel efficiency on a farm. Tractor selection is generally based on the main power consuming activities, making tractor use inefficient for lighter tasks. Regular maintenance of both tractors and implements is essential for energy efficient operation. The use of tractors for transport, cultivation, tillage, crop establishment, and the transport of fertilizer, seed and grain are the main fuel-consuming activities on an arable farm (Fig. 2.20). Non-specific tractor use and on-farm vehicle transport are also main energy consuming activities at around 0.8 GJ/ha, including travelling between separate farm blocks and undertaking farm track maintenance.

Electricity: Consumption varies with the different arable farming types and farming systems. A relatively small input is at times required for conditioning and refrigeration of the harvested crop before storing at low moisture content or low temperature. Using natural drying is the most cost and energy efficient method, however, in wet seasons adverse weather conditions may result in crop losses. Some harvested cereals are therefore dried by forced ventilation methods, normally using heated air. Grain can be transported to central processing installations (generally fuelled by natural gas, oil or coal) or, prior to storage, dried on the farm in small scale dryers using LPG, oil, electricity, wood or coal. Drying generally requires a heat input of 1–2.5 GJ/t grain when dropping the moisture content by 12 to 14% (CAE, 1996). A wide range of dryer types is available including batch and continuous dryers (25–30 kWh/t), dehumidifiers and solar drying.

2.6   CONCLUSIONS

The world is not yet on course to achieve a sustainable energy future. Global energy supply under business-as-usual will continue to be dominated by fossil fuels for several decades. Reducing the resultant GHG emissions will require a transition to zero- and low-carbon technologies. This could happen over time as business opportunities and co-benefits are identified, but will require policy intervention with respect to the interrelated issues of subsidy removal of fossil fuels; minimizing related environmental impacts; maintaining security of energy supply, and achieving the goals for sustainable development.

Secure energy services are fundamental to achieving sustainable development. In many developing countries to date, provision of adequate, affordable and reliable energy services has been insufficient to reduce poverty levels and improve the overall national standard of living. To provide such energy services for everyone in an environmentally sound way and to help provide food security, will require major investments in the energy-supply chain, conversion technologies and infrastructure (particularly in rural areas).

Future agricultural production could be constrained by fossil fuel prices and availability, particularly in low-GDP countries where imported fossil fuels are a high burden on total GDP. The correlation between energy prices and food prices is therefore of concern. Energy demand for food production can be reduced by either lowering energy intensities or by changing the volume and mix of the commodities produced to those with lower energy inputs (e.g. using soybean protein to displace animal product protein).

Farming costs are becoming more dependent on, and linked more closely to, fossil fuels. Poorer people, are those most vulnerable to fossil fuel price fluctuation and spikes, whether they are small-scale producers or consumers of staple foods. Future high and volatile fossil fuel prices, global or regional energy scarcities, and increasing GHG emissions are the key drivers as to why the global agri-food sector needs to become more "energy-smart" (see Chapter 6).

There are many technologies available for reducing the energy intensity of agricultural production and fisheries. Cheap and available fossil fuels have historically been a constraint on utilizing methods for improving energy efficiency in this sector. More recent higher oil and gas prices in some regions, coupled with the growing understanding of the need for all sectors to reduce their carbon footprints, should see a more rapid uptake of energy efficient technologies throughout the food supply chain, greater deployment of renewable energy systems especially on-farms, and related behavioral changes in the near future.

REFERENCES

ADB: Improving energy security and reducing carbon intensity in Asia and the Pacific. Asian Development Bank, Manila, Philippines, 2009.

Baker, C.J., Saxton, C.E., Ritchie, W.R., Chamen, W.C.T. & Reicosky, D.C.: *No tillage seeding in conservation agriculture*. 2nd edition, CABI Publishers, Wallingford, Oxfordshire, UK, 2007, http://bookshop.cabi.org/?site=191&page=2633&pid=1970 (accessed March 2013).

Barber, A.: Seven case study farms: total energy and carbon indicators for New Zealand arable and vegetable production. Agrilink New Zealand Limited, Auckland, 2004, www.agrilink.co.nz/Portals/Agrilink/Files/Arable_Vegetable_Energy_Use_Main_Report.pdf (accessed March 2013).

Barker, T. & Dagoumas, A.: The global macroeconomic rebound effect of energy efficiency policies: an analysis 2012-2030 using E3MG, May, 2009, http://www.cambridgeenergy.com/archive/2009-05-14/Barker%20&%20Dagoumas_Rebound_14052009V3.pdf (accessed March 2013).

Biggs, S. & Justice, S.: Rural development and energy policy; lessons for agricultural mechanization in South Asia. Occasional paper #19, Observer Research Foundation, New Delhi, India, 2011, www.observerindia.com/cms/export/orfonline/modules/occasionalpaper/attachments/occ_rural_1296292421217.pdf (accessed March 2013).

Bogdanski, A., Dubois, O., Jamieson, C. & Krell, R.: Making integrated food/energy systems work for people and climate – an overview. Environment and Natural Sources Management working paper 45, Food and Agriculture Organization of the United Nations, Rome, Italy, 2010, http://www.fao.org/docrep/013/i2044e/i2044e00.htm (accessed March 2013).

CAE: *Energy efficiency – a guide to current and emerging technologies*, Vol. 2, *Industry and Primary Production*. Centre for Advanced Engineering, University of Canterbury, Christchurch, New Zealand, 1996.

Canning, P., Charles, A., Huang, S., Polenske, K. & Waters, A.: Energy use in the U.S. food system. US Department of Agriculture, Economic Research Service, Washington, DC, 2010, http://www.ers.usda.gov/AmberWaves/September10/Features/EnergyUse.htm (accessed March 2013).

Chen, C., Hill, J.K., Ohlemüller, R., Roy, D.B. & Thomas, C.D.: Rapid range shifts of species associated with high levels of climate warming. *Science* 333:6045 (2011), pp. 1024–1026.

Chum, H., Faaij, A., Moreira, J., Berndes, G., Dhamija, P., Dong, H., Gabrielle, B., Goss, A., Lucht, W., Mapako, M., Masera, O., Cerutti, T., McIntyre, T., Minowa, T. & Pingoud, K.: Bioenergy, Chapter 2 in IPCC Special Report on Renewable Energy Sources and Climate Change Mitigation. Cambridge University Press, Cambridge United Kingdom and New York, USA, 2011, http://srrenipcc-wg3de/report/IPCC_SRREN_Ch02pdf (accessed March 2013).

CNSS: 4-in-1 biogas systems- sanitation and acceptance issues. China Node for Sanitation Systems, May, 2011, http://www.ecosanres.org/pdf_files/4-in-1_Household_Biogas_Project_Evaluation-20110620.pdf (accessed March 2013).

DEFRA: The 2007/2008 agricultural price spikes – causes and policy implications. Global Foods Market Group, a cross-Whitehall group of UK government officials, 2010, http://cap2020.ieep.eu/assets/2010/1/22/HMT_price_spikes.pdf (accessed March 2013).

FAO: How to feed the world in 2050, Food and Agricultural Organization of the United Nations, Rome, Italy, 2009a, www.fao.org/fileadmin/templates/wsfs/docs/expert_paper/How_to_Feed_the_World_in_2050.pdf (accessed March 2013).

FAO: Climate change implications for fisheries and aquaculture – overview of current scientific knowledge. FAO Fisheries and Aquaculture Technical Paper 530, Rome, Italy, 2009b, http://www.uba.ar/cambioclimatico/download/i0944e.pdf (accessed March 2013).

FAO: Energy-smart food for people and climate. Food and Agricultural Organization of the United Nations, Rome, Italy, 2011a, http://www.fao.org/docrep/014/i2454e/i2454e00.pdf (accessed March 2013).

FAO: Save and grow – a policy maker's guide to the sustainable intensification of smallholder crop production. Plant Production and Protection Division, Food and Agricultural Organization of the United Nations, Rome, Italy, 2011b.

FAO: An international consultation on integrated crop-livestock systems for development – the way forward for sustainable production intensification. Integrated Crop Management, 13-2010 ISSN 1020-4555, Rome, Italy, 2011c.

Fischer, G: World food and agriculture to 2030/50: how do climate change and bioenergy alter the long-term outlook for food, agriculture and resource availability? Proceedings of expert meeting on *How to Feed the World in 2050*, Economic and Social Development Department, Food and Agriculture Organization of the United Nations, Rome, Italy, 2009, ftp.fao.org/docrep/fao/012/ak972e/ak972e00.pdf (accessed March 2013).

Frater, T.G.: *Energy in New Zealand apple production*. PhD Thesis, Massey University Library, New Zealand, 2010, http://hdl.handle.net/10179/2295 (accessed March 2013).

GEA: Global energy assessment – towards a sustainable future. IIASA, Vienna, Austria, 2012, http://www.iiasa.ac.at/web/home/research/researchPrograms/Energy/Home-GEA.en.html (accessed March 2013).

GEF: Climate change – a scientific assessment for the GEF. 2012, http://stapgef.org/sites/default/files/Climate%20Change-A%20Scientific%20Assessment%20for%20the%20GEF_2.pdf (accessed March 2013).

Geoghegan, T., Anderson, S. & Dixon, B.: Opportunities to achieve poverty reduction and climate change benefits through low-carbon energy access programmes. Ashden Awards for Sustainable Energy, GVEP, International and IIED. April, 2008, http://www.ashdenawards.org/files/reports/DFID_report.pdf (accessed March 2013).

Godfray, C., Beddington, J.R., Crute, I.R., Haddad, L., Lawrence, D., Muir, J.F., Pretty, J., Robinson, S., Thomas, S.M. & Toulmin, C.: Food security: the challenge of feeding 9 billion people. *Science* 327:5967 (2010), pp. 812–818.

GoS: Foresight project on global food and farming futures. Synthesis Report C12: Meeting the challenges of a low-emissions world, UK Government Office for Science, London, UK, 2011, http://www.bis.gov.uk/assets/bispartners/foresight/docs/food-and-farming/synthesis/11-632-c12-meeting-challenges-of-low-emissions-world.pdf (accessed March 2013).

Hansen, J.: *Storms of my grandchildren*. Bloomsbury Press, London, UK, 2009.

Heinberg, R. & Bomford, M.: The food and farming transition – towards a post-carbon food system. Post Carbon Institute, Sebastopol, CA, 2009, http://www.postcarbon.org/files/PCI-food-and-farming-transition.pdf (accessed March 2013).

Heller, M.C. & Keoleian, G.A.: Life cycle-based sustainability indicators for assessment of the US food system. Center for Sustainable Systems, University of Michigan, Report CSS00-04, 2000, http://css.snre.umich.edu/css_doc/CSS00-04.pdf (accessed March 2013).

Huber, M. & Knutti, R.: Anthropogenic and natural warming inferred from changes in Earth's energy balance. *Nature Geosciences* 5 (2012), pp. 31–36.

IEA: Renewable energy heating and cooling. International Energy Agency IEA/OECD, Paris, France, 2007a, http://www.iea.org/publications/free_new_Desc.asp?PUBS_ID=1975 (accessed March 2013).

IEA: Energy security and climate policy: assessing interactions. International Energy Agency, IEA/OECD, Paris, France, 2007b.

IEA: Cities, towns and renewable energy – YIMFY – Yes. In: My front yard. International Energy Agency IEA/OECD, Paris, France, 2009.

IEA: World Energy Outlook 2010. International Energy Agency, IEA/OECD, Paris, France, 2010.

IEA: World Energy Outlook 2011. International Energy Agency, IEA/OECD, Paris, France, 2011.

IEA: World Energy Outlook, 2012. International Energy Agency, IEA/OECD, Paris, France, 2012.

Index Mundi: Historical commodity prices. 2012; Available for updating at http://www.indexmundi.com/commodities (accessed March 2013).

IPCC: 4th Assessment Report, Synthesis Report. Intergovernmental Panel on Climate Change, 2007, http://www.ipcc.ch/publications_and_data/ar4/syr/en/contents.html (accessed March 2013).

IPCC: Special Report on Renewable Energy and Climate Change Mitigation. Intergovernmental Panel on Climate Change, Working Group III, 2011, http://srren.ipcc-wg3.de/report/IPCC_SRREN_Ch09 (accessed March 2013).

IPCC: Managing the risks of extreme events and disasters to advance climate change adaptation. Special Report, Intergovernmental Panel on Climate Change Working Group II, Cambridge University Press, New York, USA, 2012, www.ipcc-wg2/gov/SREX (accessed March 2013).

Jeffery, S., Verheijen, F.G.A., van der Velde, M., & Bastos, A.C.: A quantitative review of the effects of biochar application to soils on crop productivity using meta-analysis. *Agricul. Ecosyst. Environ.* 144:1 (2011), pp. 175–187.

Kim, G.R.: Analysis of global food market and food-energy price links – based on systems dynamics approach. Hankuk Academy of Foreign Studies, South Korea, 2010, http://www.scribd.com/doc/44712712/Analysis-of-Global-Food-Market-and-Food-Energy-Price-Links (accessed March 2013).

Lehmann, J. & Joseph, S.: *Biochar for environmental management: science and technology.* Earthscan Publications, London, UK, 2009.

McBratney, A., Whelan, B., Ancev, T. & Bouma, J.: Future directions of precision agriculture. *Precision Agricult.* 6:1 (2006), pp. 7–23.

Mitchell, C., Sawin, J.L., Pokharel, G.R., Kammen, D., Wang, Z., Fifita, S., Jaccard, M., Langniss, O., Lucas, H., Nadai, A., Trujillo Blanco, R., Usher, E., Verbruggen, A., Wüstenhagen, R. & Yamaguchi, K.: Policy, Financing and Implementation. In: O. Edenhofer, R. Pichs-Madruga, Y. Sokona, K. Seyboth, P. Matschoss, S. Kadner, T. Zwickel, P. Eickemeier, G. Hansen, S. Schlömer & C. von Stechow (eds): IPCC Special Report on Renewable Energy Sources and Climate Change Mitigation, Cambridge University Press, Cambridge, UK and New York, NY, USA, 2011.

Moomaw, W., Yamba, F., Kamimoto, M., Maurice, L., Nyboer, J., Urama, K. & Weir, T.: Introduction. In: O. Edenhofer, R. Pichs-Madruga, Y. Sokona, K. Seyboth, P. Matschoss, S. Kadner, T. Zwickel, P. Eickemeier, G. Hansen, S. Schlömer & C. von Stechow (eds): IPCC Special Report on Renewable Energy Sources and Climate Change Mitigation, Cambridge University Press, Cambridge, UK and New York, NY, USA, 2011.

Nelson, G.C., Robertson, R., Msang, S., Zhu, T., Liao, X. & Jawajar, P.: Greenhouse gas mitigation – issues for Indian agriculture. Discussion paper 00900, International Food Policy Research Institute, Washington DC, 2009, http://www.ifpri.org/sites/default/files/publications/ifpridp00900.pdf (accessed March 2013).

OECD: Environmental performance of agriculture in OECD countries since 1990. Chapter 1, section 1.4 Energy, 2008, http://www.oecd.org/dataoecd/25/53/40678556.pdf (accessed March 2013).

Peters, G.P., Marland, G., Le Quéré, C., Boden, T., Canadell, J.G. & Raupach, M.R.: Rapid growth in CO_2 emissions after the 2008–2009 global financial crisis. *Nature Climate Change* 2 (2012), pp. 2–4.

REN21: Renewables 2012: Global Status Report. Renewable Energy for the 21st Century, Paris, France, 2012, http://new.ren21.net/Portals/0/documents/Resources/%20GSR_2012%20highres.pdf (accessed March 2013).

Schneider, U.A & Smith, P.: Energy intensities and greenhouse gas emissions in global agriculture. *Energy Efficiency* 2 (2009), pp. 195–206.

Sims, R.E.H., Schock, R.N., Adegbululgbe, A., Fenhann, J., Konstantinaviciute, I., Moomaw, W., Nimir, H.B. & Schlamadinger, B.: Energy supply. In: B. Metz, O.R. Davidson, P.R. Bosch, R. Dave & L.A. Meyer (eds): Contribution of Working Group III to the Fourth Assessment Report of the Intergovernmental Panel on Climate Change, 2007. Cambridge University Press, Cambridge, UK and New York, NY, USA, http://www.ipcc.ch/publications_and_data/ar4/wg3/en/ch4.html (accessed March 2013).

Sims, R., Mercado, P., Krewitt, W., Bhuyan, G., Flynn, D., Holttinen, H., Jannuzzi, G., Khennas, S., Liu, Y., O'Malley, M., Nilsson, L.J., Ogden, J., Ogimoto, K., Outhred, H., Ulleberg, Ø. & van Hulle, F.: Integration of renewable energy into present and future energy systems. In: O. Edenhofer, R. Pichs-Madruga, Y. Sokona, K. Seyboth, P. Matschoss, S. Kadner, T. Zwickel, P. Eickemeier, G. Hansen, S. Schlömer & C. von Stechow (eds): IPCC Special Report on Renewable Energy Sources and Climate Change Mitigation. Cambridge University Press, Cambridge, UK and New York, NY, USA, 2011, http://srren.ipcc-wg3.de/report (accessed March 2013).

Smil V.: *Energy in nature and society – general energetic of complex systems.* MIT Press, Cambridge, MA, 2008.

Smith, S.M., Horrocks, L., Harvey, A. & Hamilton, C.: Rethinking adaptation for a 4°C world. *Phil. Trans. R. Soc.* A 369 (2011), pp. 196–216.

Spielman, D.J. & Pandya-Lorch, R.: Proven successes in agricultural development – a technical compendium to "Millions Fed". International Food Policy Research Institute, Washington DC, 2010, http://www.ifpri.org/publication/proven-successes-agricultural-development (accessed March 2013).

UNDP: The energy access situation in developing countries – a review focusing on the least developed countries and Sub-Sahara Africa. United Nations Development Programme and World Health Organization, 2009, http://content.undp.org/go/newsroom/publications/environment-energy/www-ee-library/sustainable-energy/undp-who-report-on-energy-access-in-developing-countries-review-of-ldcs—ssas.en (accessed March 2013).

UNEP: Towards a green economy: pathways to sustainable development and poverty eradication – a synthesis for policy makers. United Nations Environment Programme, 2011a; www.unep.org/greeneconomy (accessed March 2013).

UNEP: Bridging the emissions gap – a UNEP synthesis report. United Nations Environment Programme, 2011b, http://wwwuneporg/publications/ebooks/bridgingemissionsgap/ (accessed March 2013).

UNFCCC: United Nations Framework Convention on Climate Change, United Nations, FCCC/INFORMAL/84 GE.05-62220 (E) 200705, 1992, http://unfccc.int/resource/docs/convkp/conveng.pdf (accessed March 2013).

WEC: Water for energy. World Energy Council, London, UK, 2010, http://www.worldenergy.org/publications/2010/water-for-energy-2010 (accessed March 2013).

Wells, C.M.: Total energy indicators of agricultural sustainability: dairy farming case study. Ministry of Agriculture and Forestry, Report 2001/3, Wellington, New Zealand, 2001, http://www.maf.govt.nz/mafnet/publications/techpapers/techpaper0103-dairy-farming-case-study.pdf (accessed March 2013).

Wilson, J.D.K.: Fuel and financial savings for operators of small fishing vessels. Fisheries Technical Paper 383, Food and Agricultural Organization of the United Nations, Rome, Italy, 1999.

World Bank: The sunken billions – the economic justification for fisheries reform. The International Bank for Reconstruction and Development and The World Bank, Washington and Food and Agricultural Organization of the United Nations, Rome, Italy, 2009.

CHAPTER 3

Energy in crop production systems

Jeff N. Tullberg

3.1 INTRODUCTION

Energy input levels are arguably the major feature distinguishing between agriculture in the developed and non-developed world, and the single most important factor underpinning greater food production by a smaller number of people. In low resource systems this energy might represent food for human beings or animals, but development usually entails increasing energy levels supplied as fuel or embodied in equipment, fertilizer or agricultural chemicals. In most cases this is derived from low cost fossil fuels, so energy economy has been low on the priority list for farmers and agricultural research organizations in the developed nations.

Cheap fossil energy has freed first-world tillage-based cropping systems from the limitations of animal power, allowing progressively deeper and more vigorous tillage. This chapter presents the argument that increased energy dissipation in the soil is largely a product of the internal contradictions of current mechanization systems. The consequence is increased soil structural degradation and erosion, and reduced water use and fertilizer efficiency. These system effects are particularly noticeable in arid zones with unreliable and sometimes high rainfall intensity, but the consequences in terms of waterway pollution with soil, nutrients and agricultural chemicals are observed almost wherever 'modern', high-technology agriculture is practiced.

This chapter is concerned largely with the opportunities to improve energy efficiency. This is generally expressed as energy per unit output of field crops, because crops are the common basis of more intensification, whether of crop or animal production. The broad distribution of energy inputs to agriculture is noted briefly before considering the efficiency of direct energy use in crop production, particularly land preparation. The embodied energy of agricultural inputs, which is usually much greater than the direct energy, is then considered.

There is a substantial body of literature concerned with both direct and embodied energy inputs, and this often identifies areas of poor efficiency, mechanisms affecting efficiency and strategies for improvement. Some of this literature is reviewed here, but the major focus of this chapter is the interactions and trade-offs amongst direct and embodied energy inputs, and the cropping system implications. Most attention is focused on those aspects where significant improvement appears to be within reach.

Energy aspects of irrigation, protected cropping and animal agriculture are covered in other chapters.

3.2 ENERGY DISTRIBUTION IN FARMING SYSTEMS

The refereed and non-refereed literature includes many surveys of energy use in crop production systems. Extension publications addressed to farmers are usually concerned only with on-farm fuel use, and the cost of the common system options. The same information, together with data on energy embodied in farm inputs is often used now to provide a broader view of farm energy economy. Increasingly, this is presented in the context of the prospects for food security against the background of increasing population, increasing energy costs and climate change.

Table 3.1.   Approximate energy inputs to US farming systems (Pimental, 2009).

Inputs Quantity	Unit	Wheat production			Soybean production		
		Qty [ha^{-1}]	Energy [MJ/ha]	[%]	Qty [ha^{-1}]	Energy [MJ/ha]	[%]
Labor	h	7.8	1005	6	6	1005	9
Machinery	kg	50	3873	23	20	1507	14
Diesel	L	100	4187	24	33.8	1859	18
Nitrogen	kg	68.4	4580	27	3.7	247	2
Phosphorus	kg	33.7	599	3	37.8	653	6
Potassium	kg	2.1	25	0	18.8	201	2
Lime	kg	0	0	0	2000	2352	22
Seeds	kg	60	913	5	56	1884	18
Ag chemicals	kg	4.5	1696	10	1.7	712	7
Electricity	kWh	14	50	0	10	121	1
Transport	Mg.km	198	281	2	150	67	1
TOTAL			17208	100		10608	100
Yield	kg	2900	44660		2600	38740	
MJ output/input*			2.60			3.65	

*Embodied energy: Machinery = 77.5 MJ/kg; Wheat = 15.4 MJ/kg; Soybean = 14.9 MJ/kg.

Pimental (2009) provides a well-known example of this broader view, providing a number of tables illustrating the energy balance of a number of common crops in varying environments. Table 3.1 has been adapted from this, and shows the contrast between the energy inputs to wheat and soybean production in the United States. This is useful background to the present discussion in illustrating:

- the major mechanisms of energy use in modern conventional crop production
- the significance of nitrogen fertilizers, and its supplementation by legume crops
- the importance of 'embodied' energy of machinery, fertilizer, agricultural chemicals.

The total anthropogenic energy input to cropping (i.e. 10–20 GJ/ha/a) is small in comparison to the solar constant (approximately 1 kW/m^2 at the earth's surface on a clear day, or approximately 100 GJ/ha/a). If the crops are used for biofuel production however, this input energy is often of similar magnitude to the energy content of the fuels (ethanol or biodiesel) produced using current technology, accounting for the small output/input ratios of energy crops (Shapouri et al., 2002).

It is interesting to note that that the energy ratios of developed-world cropping are usually better than those of the developing world (Pimental et al., 2009), despite their much greater fossil energy requirements. This also points to an opportunity for improvement, because the very high levels of diesel use are largely associated with tillage. Since the 1960s, scientists have recognized that tillage, despite its universality in most environments at that time, was not essential to crop production. Minimizing or eliminating tillage should provide a substantial reduction in on-farm fuel energy requirements, but the effect on embodied energy is less clear. Weed management, seeding and crop establishment in uneven, compacted surface soil are the major challenges of no-till farming.

Without tillage, the major tasks of crop production are essentially those of materials handling and transport. Seed and other inputs must be moved from supply to placement, and economic components of crops from in-field detachment to sale or use. In 'no-till' farming the seeding operation is the only one which normally involves significant soil engagement, and even in this operation, analysis suggests that overcoming motion resistance of the machinery absorbs much more energy than soil disturbance for seed placement (Section 3.7.2, below). This points to another opportunity for improving energy efficiency, by controlling field traffic.

Table 3.2. Fuel use in common crop operations (various extension publications).

Operation	Fuel use [L/ha]	
	Range	Mean
Deep tillage (subsoiling)	15–25	20
Plowing (moldboard or disc)	15–20	18
Chisel or sweep tillage	7–12	9
Shallow tillage or seeding	3–10	5
Spraying	0.5–2	1.5
Grain harvesting (combine)	8–15	10
Forage harvesting (chopper)	10–20	15
Cotton harvesting (picker)	12–18	15

Other significant energy economies can be achieved by improved agronomic management, particularly the use of legumes to replace a proportion of applied nitrogen, and more intensive cropping to reduce the opportunities for weed growth. Minimizing tillage, controlling traffic and improved agronomic management all provide valuable system strategies for improving the energy efficiency of crop production. Significant further improvement can be expected as developments in guidance, remote sensing and robotics allow more precise matching of inputs with crop needs.

3.3 INPUT ENERGY EFFICIENCY

3.3.1 *Farm machinery operations*

Taken together, fuel and machinery usually form the largest single energy input to high-technology cropping systems. Fuel is used in all aspects of land preparation, seeding, crop protection, harvesting and transport, and often for crop drying. The fuel requirements of common crop operations taken from a range of publications (largely North American and European) are summarized in Table 3.2. These values demonstrate how easily a traditional tillage program (e.g. subsoiling and plowing followed by 3 shallow seedbed preparation operations) can account for 60–80% of mechanical energy input. Australian data, as quoted by Tullberg (2010), are at the lower end of this range, reflecting the minimum tillage systems found to be economically sustainable in an environment of comparatively low, rainfall-limited crop yield. In both environments most of these energy-intensive operations involve tractor-based equipment, the mechanisms of which deserve some consideration.

3.3.2 *Tractive power transmission*

The farm tractor's position as a symbol of modern agriculture is based on its original success as a better, bigger and more comfortable replacement for draft animals as the power unit for a large range of farm operations. Like the draft horse, its defining characteristic is the ability to transmit power (or energy) to farm machines by traction, usually for tillage. Using both the common labels, with the more precise terminology of the most commonly used standards (ASABE 2011) in *italics*, traction is the ability to generate a *draft force*, or drawbar pull. *Motion* (or rolling) resistance is the *draft force* lost in propelling a tractor or machine on a defined surface and *travel reduction* (slip or wheelslip), is the loss of travel distance under load, expressed as a proportion of the no-slip distance travelled by the system (usually in the same number of wheel or track rotations) without drawbar load under reference conditions.

Tractive efficiency is a statement of the energy efficiency of the traction system in converting axle power into drawbar power. The relationship between these parameters is set out in Equation (3.1):

$$t = (1 - s) \cdot D/(D + R) \tag{3.1}$$

where:
t = tractive efficiency (decimal)
s = wheel slip (decimal)
D = draft force [kN]
R = motion resistance [kN].

Slip and rolling (motion) resistance reflect the reality that deformation must occur for soil to generate resistance to horizontal and vertical forces, a complex phenomenon studied using finite element analysis of soil/tire interactions (e.g. Defossez and Richard, 2002). From a simple conceptual perspective however, the first $[(1 - s)]$ term of Equation (3.1) might be thought of as representing energy loss in slip, dissipated largely in horizontal (shear) deformation of soil beneath the tire/soil contact patch. Similarly, the second $[D/(D + R)]$ term might represent energy loss in motion resistance dissipated largely in vertical (compressive) soil deformation.

ASABE (2011) presents a complete family of traction prediction equations, but the only other parameter important to the present discussion is *dynamic ratio* or coefficient of traction, which is defined as draft force (or drawbar pull) expressed as a decimal ratio or of the vertical force on the traction device. Several of the parameters noted above are commonly expressed as percentages, but precise determination of the denominator for each can be difficult except in the laboratory. In the field, it is often necessary to use approximate values which are either readily available from tractor specifications or easily measured in the field.

Common approximations include using pto (power take-off) power as the basis for calculating tractive efficiency, tractor mass as the basis for coefficient of traction, and self-propelled travel on the same surface without drawbar pull as the basis for slip. Power losses between pto and tractor axles, weight addition and transfer effects of implements, and variability of surface conditions respectively contribute errors to these approximations. They are, nevertheless necessary in the absence of practical alternatives, and are likely to be relatively small compared with the variability generally found in field performance.

Much engineering effort has been invested in assessing the relative advantages of different traction systems, such as pneumatic tire construction, steel ("caterpillar") tracks and rubber tracks (or belts). Such studies usually take the optimal value of tractive efficiency for any given surface soil condition as the criterion of performance for that soil condition. Results are often presented as performance characteristics plotted against slip, perhaps because slip monitors are standard equipment in many larger tractors, and most farmers recognize slip as a negative factor affecting tractor performance. Common presentations of this characteristic show tractive efficiency, rather than the total loss characteristic (its converse) which is illustrated in Figure 3.1.

The outcome of many such studies, largely in the USA, has been reviewed by Zoz and Grisso (2003), the essentials of which might be summarized in the ideas that traction on non-rigid field surfaces is more efficient when:

- Soil surfaces are stronger and more resistant to deformation.
- Contact pressures are smaller and more uniform.
- Contact areas are longer, rather than wider.
- Slip is within a limited range in most field conditions: 7–15% for tires, and 4–10% for steel or rubber tracks.

The long contact area provided by tracks (steel or rubber) ensures that less forward motion (and thus energy) is lost in deforming soil backwards to produce draft force. Effects on tractive efficiency in most conditions are nevertheless small because track systems have greater "internal" motion resistance (friction), and less uniform contact pressures. Similarly, while a tractor is usually

Table 3.3. Typical tractive efficiency values [drawbar/pto power] for common tractor types.

Tractor type	Tractive condition			
	Concrete	Firm	Tilled	Soft
2-Wheel drive	0.87	0.72	0.67	0.55
4-WD (unequal wheels)	0.87	0.76	0.72	0.64
4-WD (equal wheels)	0.88	0.77	0.75	0.70
Tracked	0.88	0.76	0.74	0.72

from Agricultural Machinery Management Data ASABE D497.7 MAR2011.

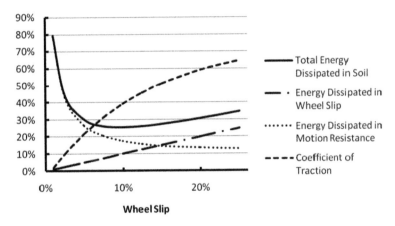

Figure 3.1. Tractive energy dissipation characteristics (Generalized from ASAE D.208).

more efficient when working on stronger soil surfaces, any fuel advantage for soil-engaging operations will be negated by greater draft requirements.

Typical values of tractive efficiency for a range of tractor drive types from the simplest two-wheel drive (2-WD) to the most elaborate (tracked), in a range of traction conditions from the most favorable (concrete) to the most difficult (soft), are set out in Table 3.3. The data illustrate the common observation that tractor drive type has significant power performance effects only in the least favorable soil conditions. As the majority of modern farm tractors are 4-WD, it is not unreasonable to suggest that in most normal operating conditions, tractive (or energy) efficiency of the tractive system (v. pto power) is commonly about 75%.

3.3.3 *Efficiency of tractor-powered tillage*

Tractor handbooks often advise operators that optimal tractive performance occurs within a relatively narrow range of coefficient of traction, for which slip (7–15% for tires, 4–10% for steel or rubber tracks), is usually the best indicator. The basis for this advice is the relationship between tractive performance parameters illustrated in Figure 3.1, which is a graphical representation of traction performance on a typical field surface based on the widely accepted approach of ASAE (2011).

According to Tullberg *et al.* (1983) most tractors operate most of the time at slip levels at or below the optimal range, even when performing heavy tillage. This is old data, but the reason for sub-optimal slip – field variability – is still equally valid. Wheel tractor operation over a large proportion of field area at – for instance – 5% slip might be energy-inefficient. It is nevertheless more efficient (in practical and economic terms) than risking grossly excessive slip, or perhaps bogging, in small difficult areas that exist in many fields. The effect is likely to be more pronounced in minimal or no-tillage systems, where seeding is the major tractive operation, and draft forces are generally smaller, but the consequences of difficult traction conditions more severe.

Energy lost between the axles and drawbar of an agricultural tractor must be dissipated in deformation of the soil or the tire. Tire deformation is largely elastic, so energy loss in belted radial ply tires is generally <1.5% according to a variety of commercial and scientific sources. Soil deformation, on the other hand is largely plastic, and must account for most of the energy losses in traction on soft soil. Conceptually, these losses might be partitioned into those of slip (dissipated largely in horizontal shear deformation of soil), and as those of rolling resistance (dissipated largely in vertical, compressive, soil deformation). Figure 3.1 is presented in terms of energy losses, rather than efficiency, to emphasize this distribution of energy dissipation in the soil.

Using values read from Figure 3.1, for instance, minimum tractive energy loss (i.e. optimum tractive efficiency) occurs at wheel slip levels of 7–12%, but a wheel tractor might typically operate at about 5% slip, when total tractive energy loss is 30% (i.e. tractive efficiency is 70%). If only 5% of tractive energy is absorbed in wheel slip, most of the remaining 25% loss must be absorbed in rolling resistance, which is largely vertical compressive deformation of soil. The approximation might be crude, but it does show how approximately 25% of tractor energy output can be dissipated in soil compaction.

Whether assessed in terms of conventional tractive efficiency, or in broader system terms, the use of tractive power for tillage or seeding is an inefficient process. The inefficiency of operation on soft soil can be reduced to some extent, but not eliminated, by the use of more efficient (and expensive) tractive systems. It can be more substantially reduced by keeping all heavy wheels on permanent, compacted traffic lanes.

3.4   LAND PREPARATION BY TILLAGE

3.4.1   *Tillage equipment*

Tillage equipment is often categorized as primary (heavy) and secondary (intermediate or light). A traditional tillage program starts with the deeper primary tillage, usually with plows (moldboard, disc or chisel), to disrupt compacted soil and bury residue. Concern about subsoil compaction has recently stimulated wider use of deep tillage equipment (subsoilers or rippers) in both tilled and no-till systems. Deep tillage might be the obvious response to deeper compaction, but demonstrably beneficial effects are rare (Kirkegaard *et al.*, 2008), and deep-tilled soil is more susceptible to subsequent re-compaction. It can reasonably be argued that the most common impact of deep tillage is to transfer a compaction problem into deeper soil horizons, where it eventually becomes inaccessible to any economic tillage process

Secondary operations (with cultivators, harrows etc.) are carried out with multiple tines or smaller discs operating at progressively smaller depths to produce a fine, level seedbed. The need for more rapid secondary tillage has stimulated the development of rotary and reciprocating harrows. These pto-powered units are designed for shallow surface finishing, in contrast to the horizontal-axis rotary hoe, which can be used for both primary and secondary work.

Use of the pto – rather than traction – allows more efficient transmission of power directly to the soil. This might be desirable from an energy standpoint, but there is much anecdotal evidence that excessive use of pto-powered tillage units is damaging to soil structure. Farmers sometimes characterize these machines as "beating the soil into submission".

The literature records many attempts to reduce the energy requirements of tillage equipment, and some approaches have been demonstrably successful, at least in terms of reducing energy input. These include:

• adjustments to tine shank and cutting blade rake angles (Godwin and Spoor, 1977).
• shallow leading tines for deep tillage (e.g. Kirby and Palmer, 1992).

- low-friction coatings on blades and moldboards (Loukanov *et al.*, 2005).
- vertical or longitudinal vibration of soil engaging tools (Shahgoli *et al.*, 2009).

Only the first two have been widely adopted, probably on account of the longevity and/or cost considerations of using the others.

Most tillage research has problems with the definition of soil conditions before and after tillage, and quantification of tillage effects. Parameters such as mean disturbed soil cross-section might be a reasonable criterion for subsoiling, and mean weight diameter of aggregates for harrowing, but neither provides a complete description of machine effects. Engineering aspects of these problems can be overcome by use of precisely described and monitored soils in laboratory tillage bins, and these have played a major role in the establishment of a theoretical basis for soil/tool interactions (e.g. Gill *et al.*, 1994).

Field application remains difficult as soil properties such as texture, packing state and moisture content vary unpredictably, spatially both across fields and down the soil profile, and root density is also likely to influence failure characteristics. The fundamental issue is one of relating readily quantifiable soil properties to the broad objective of optimizing crop performance and most field studies still rely on qualitative descriptions supplemented by data such as penetrometer resistance profiles, infiltration measurements and detailed assessments of crop moisture extraction profiles.

3.4.2 *Tillage objectives and functions*

The fundamental objective of soil management for cropping is an effective condition for seed germination, emergence, and optimal crop development. The following tillage objectives are noted by a number of sources (e.g. Murray *et al.*, 2006):

- To develop a good level seedbed (fine particles in seed zone) and rootbed (good structure for rain infiltration, air exchange, and minimum resistance to roots).
- To manage (usually by burial) plant residues that can adversely affect the following crops by harboring insects or disease or by interfering with planting or furrow irrigation.
- To control weeds and unwanted plants.
- To incorporate and mix fertilizers, manure, pesticides, or soil amendments.
- To move soil from one layer to another, removing rocks, profiling soil for planting, irrigating, drainage, or harvesting operations.

The usefulness of these functions has been questioned by observers who point out that soil conditions for crop establishment are often as good as or better before harvesting than after completion of "land preparation" tillage. Crop residues are also demonstrably more useful when left in place to protect the soil surface, weeds can be controlled by herbicides or other non-soil engaging means, and fertilizers incorporated at more appropriate times.

Considered in an energy system context, tillage presents a series of contradictions, particularly as embodied in the modern conventional tillage-based agriculture found in much of the developed world. The problem can be largely attributed to incremental development of cropping practice from relatively low-impact harvesting, shallow plowing and harrowing with human or animal power, to apparently similar processes with heavier equipment, producing greater soil impact requiring deeper tillage and greater energy levels. The consequences are:

- Tires (or tracks) of harvesters and transport equipment leave a series of depressed, compacted strips of soil, a condition unsuitable for easy planting of the next crop. A significant proportion of the energy requirement of harvesting is used to compact the soil (Botta *et al.*, 2007), which subsequently requires greater tillage energy to penetrate the compacted soil.
- Modern conventional farming tillage programs are designed to break up compaction, level the surface, kill weeds and bury crop residues. Tillage requires large energy inputs but much of the energy is dissipated firstly compacting soil beneath tractor tires, and secondly in breaking it up again (Tullberg, 2000).

- Tillage operations normally start deep (primary tillage), and become progressively shallower, with (secondary) operations being used to produce a fine, level surface. Total tractor tire tread width however usually represents 15–40% of tillage implement width, so after completion of tillage and seeding a large proportion of field area (60–120%) has been wheeled by tractors, compacting the profile beneath the depth of each successive tillage operation. The negative effects include a substantial reduction in infiltration rate (Li *et al.*, 2001), which in turn promotes run-off, nutrient loss and denitrification.

The situation might be summarized in the proposition that energy wasted in soil damage, largely (but not exclusively) at harvest, is the major rationale for undertaking energy-inefficient tillage programs. Tillage apparently rectifies visible aspects of that damage, but it rarely extends deep enough to disturb all compacted horizons, and deprives the soil surface of crop residue protection. No-tilled soils are also almost universally better able to support crop production, as demonstrated in the increased yields and reduced inputs found in studies carried out in most of the world's cropping regions.

The improvements in energy efficiency achieved with improved tractive systems or better-design tillage equipment appear minor compared with the effects of avoiding even one tillage operation. Minimizing or dispensing altogether with tillage is the most obvious step in reducing the energy requirements of agriculture, almost regardless of the level of technology applied.

3.5   EMBODIED ENERGY

3.5.1   *Machinery*

Energy embodied in machinery can be determined from the mass of the different materials of construction and their fabrication and transport energy requirements. This is often simplified by using published values of the energy intensity (i.e. embodied energy/mass ratios) for common equipment categories. For farm equipment these are usually between 50 and 80 MJ/kg (Zentner *et al.*, 2004) including repairs over life. Energy intensity can also be expressed per unit purchase price (as MJ/$), or sometimes as a percentage of operating energy when depreciated over expected machine life.

Superficial assessment of farm enterprises often notes the high levels of machinery investment in relation to its utilization (annual, or whole-of-life), and infer a lack of rational asset allocation. Energy embodied in purchased machinery is also often greater than the direct fuel energy input according to several authors (e.g. Doets *et al.*, 2000). Increased utilization is an obvious key to improving both economic and energy efficiency, but difficult to achieve given the seasonality of farm operations. A wheat seeder, for example, might be used for only 10–12 days/year, because this is the expected duration of effective wheat seeding, so utilization is limited to 150–200 h/year, or <2000 hours over a 10-year life).

Crop production depends on processes which occur on a timescale determined by biology and the environment. This dictates a brief optimum time for most crop production operations with a yield (or sometimes cost) penalty for untimely operations occurring either side of this optimum (Whitney, 1988). The recent literature is largely silent on the magnitude of this penalty, which obviously varies with operation, crop and environment. A rational cropping equipment capacity selection process nevertheless depends on some means of incorporating timeliness considerations. This can be explicit (ASABE 2011) or more frequently implicit in the statements such as the one above, that "10–12 days per year . . . . is the expected duration of effective wheat seeding".

An example of this rational approach applied to investment in a typical grain seeding operation is presented in the Box 3.1. This shows how the 'time cost' of wheat seeding implied by the seeding capacity investment decision corresponds with field measurements of the consequences of delay at seeding. It is not suggested that this explicitly reflects a farmers' machine capacity decision-making processes, but that the outcome certainly appears to be rational in economic terms. Similar reasoning might be used to demonstrate the rationality of the embodied energy investment in terms of the energy penalty (yield losses) associated with extended seeding time.

Large economic and embodied energy investment is a direct consequence of the low levels of equipment utilization. Equipment contract and sharing arrangements are often suggested to improve utilization, but local arrangements are problematic when most farmers within a district all need to seed (or harvest) at the same time. Substantial improvements can be achieved only when cooperative arrangements or contractors can extend equipment operating seasons by working across a range of climate zones.

This is relatively common with harvesting equipment, where the task is easily specified and performance (losses, damage etc.) readily assessed. Wide harvester fronts (headers) can also be rapidly removed and towed for road transport. This allows grain harvest contractors to operate across a range of latitudes over a long season in (e.g.) Australia, North America and China, but is still comparatively rare with seeders. This is perhaps a function of the greater crop/soil design and adjustment specificity of seeders – particularly no-till seeders – compared with harvesters, and the greater difficulty of assessing their suitability (for particular situations) and field performance. Seeders are also less costly, and wide no-till seeders can be difficult to fold for road transport.

Box 3.1: The value of seeding time

(This example uses round-figure values typical of Australian broadacre grain production, but similar reasoning can be applied in most cropping environments).

A farmer has invested A\$ 200,000 in a winter crop seeder with a capacity of 6.5 ha/h (approximately 100 ha/day), to allow completion of a 1000 ha winter crop seeding program in 10 days. Winter crop seeding is 60% of the justification for ownership of a tractor, also valued at A\$ 200,000, so it is reasonable to calculate the total investment in the ability to seed winter crops in 10 days as: A\$ 200,000 (seeder) + 60% of A\$ 200,000 (tractor) = A\$ 320,000.

Within any particular machine type group, cost is broadly related to capacity, so this farmer would have been aware of alternatives such as investing (e.g.) A\$ 240,000 or A\$ 400,000 to buy units with a capacity of 75 or 125 ha/day and completing seeding in 13 or 8 days respectively. The important consequence of the capacity investment decision is the proportion of crop seeded at the optimum time. In difficult seasonal conditions this might also represent the proportion of area successfully planted. *Seeding time clearly has a value.*

This ownership decision indicates that 100 ha/day is the farmer's best estimate of the optimal capacity for that operation. In other words, over the long-term, the farmer believes that buying less capacity will cost more (in lost crop income) than it saves (in investment costs), and vice-versa. *Investment in machine capacity implies an assessment of the cost of time.*

In this case, if the annual investment (interest + depreciation) cost of these units is (e.g.) 20% of their purchase costs, the annual cost of owning the capacity to complete seeding in 10 days is 20% of A\$ 320,000 = A\$ 64,000 p.a. The investment decision thus implies a winter crop seeding time cost of A\$ 64,000/10 days = A\$ 6400 per day. If the average yield from that crop is 4 t/ha worth A\$ 150 per ton, total income expected from the 1000 ha is A\$ 600,000. The *implied average cost of time* for this operation is thus A\$ 6400/A\$ 600,000 or approximately *1.1% per day of crop value.*

This daily loss of crop value corresponds with the results of field measurements of yield decline in the region for which this example is typical (Tullberg and Rogers, 1982) suggesting that the farms winter crop seeding capacity is close to the optimum. This simple approach is conceptually similar to that of ASABE (2011) which provides a quantitative approach to machine capacity selection based on crop, region and machine specific "timeliness coefficients".

Table 3.4.   Energy embodied in fertilizers (from Zentner, 2004).

Fertilizers	[MJ/kg]
Nitrogen	75.63
Phosphate (P_2O_5)	9.53
Potassium (K_2O)	9.85
Sulfur	1.12

3.5.2   *Fertilizer*

Fertilizer – particularly nitrogen – is often identified as a major energy cost of modern agriculture. Commonly used values for energy embodied in fertilizers are quoted in Table 3.4, and these values are generally comparable to those quoted by Pimental (2007) and other sources. Fertilizer manufacture has been estimated to consume >1% of world energy production, which is largely a consequence of the energy intensity of ammonia production. Some of the subsequent processes of nitrogen fertilizer manufacture are, however exothermic. This energy is captured in some more recent plant designs, reducing energy and emissions/unit product by a factor of 4 compared with older plants, according to Wood and Cowie (2004). These authors cite studies of emissions (and energy) embodied in phosphate and potassium fertilizers, illustrating major variability.

Poor efficiency of nitrogen and phosphate fertilizer use is also a matter of concern, and global estimates by Smil (1999) suggested that only 50% of nitrogen applied as fertilizer is found in harvested crops. The balance is lost via two major pathways: in denitrification to atmospheric nitrogen and its oxides (Dalal *et al.*, 2003) or as nitrates in run-off or leachate to waterways or groundwaters (Masters *et al.*, 2008). Almost regardless of the loss mechanism, pollution by nitrogen and other fertilizer components represents serious environmental and/or human health threats in many regions of the world, specifically:

- Fertilizer nitrogen and phosphorus have both been identified as the major cause of algal blooms and eutrification of waterways and coastal waters. Nitrate levels at or exceeding world health organization recommendations are also a threat to water supplies in parts of the developed and the developing world. Improved fertilizer use efficiency thus has many benefits in addition to those of greater energy efficiency in food production.
- Denitrification is an additional environmental concern because nitrous oxide (a powerful greenhouse gas) is produced under near- waterlogged soil conditions when nitrate and carbon is present in soil organic matter or as crop residue. Methane, another significant greenhouse gas, is absorbed in small quantities by healthy aerated soils, but produced in rather larger quantities when soil is totally waterlogged or flooded (Dalal *et al.*, 2003). Warmer environments exacerbate this problem.

In addition to reducing productivity, waterlogging is also the precursor to runoff in which nutrients are lost to waterways both as solutions and in association with eroded soil particles, so both these loss mechanisms are important when high-intensity rainfall (or irrigation) occurs after fertilizing (Masters *et al.*, 2007).

Artificial fertilizers are now responsible for a large proportion of world food supply, but their use has become common only since the 1950's. Nitrogen is also supplied to soil in small quantities by rainfall, but legumes can supply much larger quantities. In undisturbed soils free living N-fixing bacteria have also been observed adding nitrogen to the soil in significant amounts. A recent summary of leguminous crop and pasture studies in Australia for instance, quoted annual N fixation levels of over 100 kg/ha (Peoples *et al.*, 2001).

Legume nitrogen is not all available to the next crop, some being lost by similar pathways as fertilizer N, but the broad value of including legumes in rotations has been demonstrated in

Table 3.5. Energy embodied in herbicides (from Zentner *et al.*, 2004).

Product	Embodied energy	
	Active ingredient [MJ/kg]	At usual rates [MJ/ha]
Glyphosate	511	225
Paraquat	538	292
Diquat	75	141
2,4-D Dicamba	336	37
Trifluralin	167	233
Metolachlor + atrazine	313	1045

a number of environments (Hargrove 1985). They are particularly attractive as cover crops to provide post-harvest soil protection that can be killed before seeding the next cash crop. This technique, often called "green" or "brown" manuring is being used increasingly by growers in a number of countries. Cover crops can also be used to absorb excess nitrate.

Animal manures improve soil organic matter levels and have a significant advantage in providing slow release of nutrients over an extended period. They can substantially reduce the requirement for energy-intensive fertilizer, but transport and distribution energy requirements can be large for these materials of relatively large mass and volume per/unit nutrient. Soil compaction issues from manure distribution can also be significant.

Soil 'health', or soil 'quality' are ill-defined ideas, but both are generally acknowledged to depend on improvements in soil structure, organic matter and biota levels. More healthy, or better-quality soils have better structure, greater levels of organic matter and larger, more diverse populations of soil biota. They are better able to hold and release crop nutrients in a controlled way (Szabolcs, 1994). There is also evidence of their positive contribution to – for instance – nitrogen availability from soil fauna, such as ants and termites (Evans *et al.*, 2011).

A number of authors have noted that long-term sustainability demands the recycling of all nutrients, including those consumed by human beings, but this faces major challenges as urbanization increases transport distances and redistribution energy requirements. It is already occurring on a relatively small scale, but heavy metal and biotic pollution and hygienic considerations of sewage sludge is a significant issue.

3.5.3 *Agricultural chemicals*

Reduced on-farm fuel use in no-till systems can be offset to some extent by the energy embodied in herbicide inputs, and values for a number of common herbicides are quoted in Table 3.5. The original source is a 1999 Canadian publication but recent, well supported data on this topic are rare, perhaps because manufacturing technology is proprietary. Values quoted by Helsel (2006) are similar but smaller, perhaps reflecting improvements in herbicide manufacturing technology.

It is interesting to note that the embodied energy of a number of herbicides, expressed per hectare, is large and of similar magnitude to the direct energy requirement of light tillage. No-till advocacy rarely acknowledges the embodied energy of herbicides in tillage system comparisons, but it should be noted that the values in Table 3.5 are for overall herbicide application. They take no account of the scope to reduce active application rates by improved technology and better targeted application, or the opportunities to avoid herbicide use by improved crop system management (Friedrich and Kassam, 2012).

Long term evidence shows that herbicide use in well-designed no-till systems is not greater than in conventional tillage systems, casting doubt on the claim that tillage is a complete solution to weed management (Lindwall and Sonntag, 2010). Despite this, tillage is still the most common on-farm response to the issues of weeds with multiple herbicide tolerance, and is identified as such in many prescriptions for integrated weed (or crop) management as a means of slowing the development of difficult weeds.

In the absence of new herbicide modes of action, the development of herbicide tolerance in weeds represent a major threat to no-till systems as currently practiced in most high-resource regions, and thus a threat to the longer-term sustainability of agricultural production and food security. Arguments about blame attribution are common, but few informed people disagree with the proposition that, as a society with sustainability and food security challenges, we have a collective responsibility for the gross overuse, and hence rapid devaluation of excellent herbicides like glyphosate. The important issue is weed management in an environment of increasing populations of herbicide-resistant weeds.

Physical weed control systems such as flame, steam, microwave energy or high-pressure water jets might be economic for highly-targeted weed control, but these all appear too energy-intensive to replace overall herbicide spraying. In this context, very shallow, non-inverting thin-blade physical weed control could be seen as the least detrimental option for addressing the immediate problem, particularly when practiced interrow, in-crop. In terms of energy or sustainability it is certainly preferable to deeper or more vigorous tillage.

The efficacy of current herbicides can be extended by preventing the reproduction of any herbicide "escapes". These are the weeds that have developed some tolerance to a herbicide, and so survive normal application. This can now be achieved in a second operation using sprayer-mounted optical weed detection technology ('weedseekers') to deliver spot applications of different herbicides to those escapes. In the longer term, the destruction of herbicide escapes might be seen as the first major opportunity for large-scale use of robotics in cropping, using swarms of small autonomous vehicles to search for, identify and deliver herbicide or physical weed treatments.

3.6   MORE ENERGY-EFFICIENT CROPPING SYSTEMS

3.6.1   *General considerations*

Energy output/input ratios might be important in energy crop production, but of limited relevance when the objective is human nutrition. Energy efficiency on the other hand is important on economic and environmental grounds. While the fuel energy characteristics of tillage-based agriculture can be improved in a number of ways, the effects are usually small compared with those of reducing or completely avoiding tillage. Tillage-based cropping was almost universal until the 1990s, so topics such as 'tractor-implement matching', 'improving tractive efficiency' and 'tillage implement fuel comparisons' were common in the rural press, extension service literature and farmer discussions. Over these years, it is reasonable to assume that these systems were optimized at least to farmer's satisfaction. More recent improvements, discussed in Section 3.3.1, might reasonably be regarded as marginal in comparison to those of system change.

The fundamental requirement of effective cropping systems is to provide good soil conditions for seed germination, emergence, and crop development, combined with adequate nutrition and effective weed control. Conditions for effective crop establishment have been summarized by Murray *et al.* (2006), in the context of soil preparation and seeder performance.

Relevant issues include the requirement for seed placement:

- at the correct depth
- in close contact with moist soil

Smoothing of soil surfaces by tillage might facilitate better and more uniform seed placement when seeder frame height is used to control depth, but is less relevant to seeders with individual row depth control. Fine seedbeds might also improve seed–soil contact, but seed-firming and in-row press-wheels can achieve the same objective. It is also useful to note that soils are often in a better condition for seed placement before they are disturbed by harvest and post-harvest operations.

These are the observations that encouraged scientists and farmers to consider no-till production systems. Early adopters saw the potential benefits of reducing soil degradation and fuel and machinery costs. They also had to confront issues such as those noted above, together with questions of residue distribution and seeding into residue-covered soil with varying compaction levels. Effective use of herbicide weed control, and its integration with agronomic measures also requires a great deal of new learning.

No-till reduces the number of machinery operations, but soil compaction is still a serious issue for no-till systems, particularly where agriculture uses heavy machines, an issue that becomes obvious more rapidly in heavier soils and more humid environments. Surface compaction from equipment wheels presents a challenge for no-till seeding machines, and probably accounts for much of their weight and power requirement. These surface effects can be avoided by delaying machine operations until the soil has dried, but this entails a significant timeliness penalty and does not address the ongoing yield effects of compaction deeper in the profile.

Compaction deeper in the profile (e.g. 20–60 cm) is produced by heavy axle loads. Mechanical solutions are extremely expensive and difficult, and natural amelioration occurs only slowly whether by physical or biological mechanisms. These processes occur quite rapidly in the surface layers, but more slowly at greater depth, taking almost 2 years and 3 crops to reach 20 cm in an undisturbed 'self-mulching' soil (McHugh *et al.*, 2009). It can reasonably be argued that growers only see the effects of surface compaction, because deeper compaction is universal in cropped fields of highly mechanized systems. Surface soil compaction can be reduced, but not eliminated by additional investment in very-low-pressure tires or tracks; the impact at depth is also reduced, but still substantial (Ansorge and Godwin, 2006).

3.6.2   *No-till and conservation agriculture*

Conservation agriculture (CA) has been given a more precise definition by FAO (2001), but it is used here as the omnibus term covering all systems which attempt to improve the sustainability of productive cropping by reducing or eliminating tillage. Scientists have long known that tillage is not an essential element of cropping, and have documented the close association between intensive tillage, burial of surface residue, soil erosion and degradation. Alternatives to tillage emerged in the 1950s with the commercial production of the first herbicides and no-till seeders. Agricultural research and extension services started to demonstrate the economic and soil conservation benefits of minimizing or eliminating tillage in the 1970s. Despite early adoption by a few enthusiasts however, large-scale adoption was slow with deep-seated opposition from ingrained beliefs that "working the soil" – tillage – was an essential aspect of good farming (see Box 3.2).

The energy, time, and cost benefits of reducing tillage are nevertheless large and easily demonstrated, particularly in grain production systems. The soil conservation benefits of main-taining surface cover are also clear, and particularly convincing to farmers in erosion-susceptible environments (e.g. Huf and Agnew, 1992).

The outcome has been an initially gradual adoption process in regions such as Australasia and the Americas, where increasing numbers of farmers usually:

- started by replacing conventional tillage with non-inverting "stubble-mulch tillage"
- progressively replaced weed control tillage with herbicide operations
- replaced or modified seeders to operate effectively through greater crop residue
- gained 'system' understanding of no-till cropping: e.g. herbicide and agronomic weed management, harvester residue distribution and no-till seeder setting and adjustment
- formed and joined 'no-till' groups, often aided by government agencies
- began adopting controlled traffic farming and precision agriculture.

In the case of Australia for example, the majority of Australian grain farmers appeared to become convinced that minimum or no-tillage was the best option during the 1990s. This view was reinforced by increasing fuel prices and reducing herbicide costs so by 2010 > 80% of Australian grain was produced in no-till (Kearns and Umbers, 2010). Broadly similar processes appear to be occurring at different rates almost throughout the developed world, but most rapidly in areas where soil conservation issues are obvious to farmers and the community. Adoption of no-till has been slower in more humid areas, particularly Europe.

Box 3.2: Dictionary definitions of "cultivation" mention two apparently contrasting sets of ideas:

- raising and care of crops, including land preparation by tillage *and*
- greater education and refinement of the human condition.

The intrinsic association between tillage and civilization in the English language is a reminder that soil preparation by tillage – human or animal-powered –was an essential precursor to reliable large-scale food production; and this in turn was a prerequisite of the development of many of the attributes of civilization: permanent settlement, urbanization and education.

This might be part of the explanation for a semi-mystical attachment to tillage, sometimes reinforced by the same fear or suspicion of agricultural chemicals which has motivated the organic food movement.

Information resources on no-till systems are now plentiful, but the evidence suggests that most farmers adopting no-till still experience some short-term yield reduction. This can be ascribed to learning curve effects covering a range of problems including nutrient deficiencies as soil adjusts to surface (rather than buried or burned) residue, soil compaction and delays attributed to all the small issues of a new system. Despite major improvements in seeder design, no-till crop establishment uniformity and early vigor is often still comparatively poor. Mentoring and expert guidance from experienced no-till farmers can reduce or avoid these problems, but it is still relatively uncommon. Despite these issues, few no-till adopters willingly revert to full tillage systems.

As farmers have become accustomed to no till they increasingly look to minimize soil disturbance, which probably accounts for the trends towards use of disk, rather than tine-type openers. Observation suggests however that these units have become progressively heavier over the years and offer a further opportunity for reducing energy requirements (see Box 3.3).

Box 3.3: Researchers and farmers remark on the greater 'softness' of soil after a few years of no-till, and discussion of disc seeding systems usually notes their minimal soil disturbance and low draft (e.g. Ashworth *et al.*, 2010). Despite this evidence, disc-type no-till seeders have continued to get heavier in highly mechanized systems. Mass per row is now commonly 300–500 kg (before adding seed, fertilizer, hopper and distribution system), because downforce (weight) has been the practical answer to disc penetration problems. Although Radford *et al.* (2000) showed that soil strength at 50 mm was increased by a factor of 3–5 by one wheeling by a 10 t axle, it is rarely noted that these penetration problems occur largely in tractor/harvester wheel tracks,

Field traffic is usually uncontrolled, so every planter unit must be capable of penetrating and firming seed in wheeltracks, so manufacturers normally provide the same downforce on all row units. It follows that downforce not required for disc penetration in soft soil must be carried on the row-unit depth and firming wheels. Inefficiency occurs because these small-diameter, rigid wheels have a rolling resistance of 20–30% of downforce, which is similar to the force ratio (draft/downforce) of seeder discs themselves (Tice and Hendrick, 1986).

This ensures that unit draft and power requirement for *all* rows is similar to that of those rows operating in wheeltracks, even though wheeltracks may occur only over a small proportion of operating width. The outcome is unnecessary weight, frame strength and tractor power. If, for example, 400 kg and 130 kg/row is required to penetrate wheeltrack and non-wheeled soil, respectively, the draft penalty for each row in non-wheeled soil is approximately 0.8 kN.

At typical operating speeds (9 km/h), the power penalty is 2.0 kW/row in non-wheeled soil (drawbar), or almost 4 kW (engine). When wheel tracks are 30% of machine width, this unnecessary weight and draft results in an 80% increase in tractor weight and power requirement.

Personal observation indicates that larger grain farmers have generally been the first adopters of conservation agriculture in all regions of mechanized farming, but intensive root crop and vegetable producers still show little interest. In the developed world, this lack of interest is largely a consequence of a price-competitive market system demanding product uniformity and continuous supply. The resulting system combines use of very precise planting or trans-planting equipment demanding fine, level, residue-free conditions, with once-over harvesting using very heavy equipment. With harvesting times dictated by markets, rather than soil conditions, the outcome is often a post-harvest soil condition in which no-till planting could not be contemplated.

The first step in improving outcomes in vegetable cropping might be standardization of harvest and transport vehicle wheel track gauge width to ensure the wheels or tracks of these heavy units do not travel on cropping soil. This is moderately difficult for most individual crops, but much more challenging when considering the range of different crop/machine combinations necessary for a sensible rotation. In these circumstances change is unlikely to occur soon unless forced by external pressures such as environmental campaigning or government regulation.

Conservation agriculture-related improvements in rainfall use efficiency, crop yield and soil organic matter are generally greater and more reliable in the sub-humid and arid environments common in the developing world. An important consequence is that CA is now seen by most international development agencies as essential to improving food security and reducing poverty. This development approach has been led by the U.N. Food and Agriculture Organization, which defines essentials of conservation agriculture as no-till, permanent soil surface protection and crop rotation (FAO, 2001).

Major issues of conservation agriculture (CA) in grain crops, worldwide include:

- reliance on natural amelioration of soil compaction (see Box 3.4) in the absence of tillage. There is evidence of increased resilience of CA soil, but this is challenged by the use of increasingly heavy machinery with its deleterious effects on productivity in many environments of the developed world. (Soane and Ouwerkerk, 1994). Wheel ruts from wet harvests can leave farmers little choice but repair by tillage, which is the rationale for advertising subsoilers as the 'essential no-till implement'.
- the development of increasing herbicide tolerance in weed species. Effective responses include more intense and strategic rotation of crops and cover crops, non-soil engaging mechanical control and a diversity of herbicide modes of action. The problem can be ameliorated to some extent by the early identification and treatment of weeds when more herbicide-susceptible (usually when young). High-technology options for weed management in this environment are discussed in Section 3.5.3, but when faced with an unexpected problem, farmers can often see tillage as the immediate solution.
- Very high levels of crop residue can present a challenge to seeding equipment, particularly in humid environments. The opposite problem has more serious consequences in a number of arid environments where subsistence farmers on limited land are forced to use all residues and animal manures for fuel. This removes soil surface protection and eliminates any nutrient or organic matter return, accelerating a cycle of soil impoverishment.

Box 3.4: Traffic-induced soil compaction

Vertical compression and shear are most evident at the soil surface as depressed wheel tracks and "clodiness" in soil subsequently disturbed by shallow tillage or seeding, but the effects of vertical load can be detected at substantial depth. Ansorge and Godwin (2006) for instance showed that traffic by wheels and tracks carrying weights typical of modern combine harvesters produced vertical soil movement >5 mm 0.6 m beneath the surface (and only 0.1 m above the rigid floor of the soil tank). Horizontal deformation was more evident near the surface.

Compaction increases soil bulk density by reducing soil pore volume. It also usually interrupts connectivity between soil pores. This restricts and can almost totally inhibit the movement of water and air within the soil, with severe consequences for infiltration, aeration and internal drainage (Li *et al.*, 2001). It results in increased runoff during rainfall and greater duration of waterlogging afterwards, factors which are commonly observed in tractor and harvester wheel tracks. Runoff is one of the major pathways of nutrient loss and extended waterlogging accelerates denitrification loss of fertilizer N. It also increases nitrous oxide emission from wheeled soil by a factor of 5–7 (Ruser *et al.*, 1998).

Compaction also affects soil aggregates and increases soil strength, as measured by penetrometer, reducing the ability of plant roots to explore the soil profile. Both factors inhibit soil biological activity ("soil health") and damage crop productivity (Pangnakorn *et al.*, 2003). Farmers tend to identify "compaction" only when severe problems are evident, but tractor tires must always compact any non-rigid soil under their tires, and tire width is usually a substantial proportion of the width of any tillage (or seeding) implement. Kuipers (1994) tabulated total annual traffic intensity (wheeled area/field area) values ranging from 2.5 (winter wheat), to 5.4 (potatoes), but these are for tillage-based systems. They might still be accurate for (e.g) intensive root or vegetable production, but summing the wheel width/working widths percentages of planter, sprayer and harvester in no-till farming usually suggests a value of about 0.5 (i.e. 50% of field area wheeled per crop) is common now.

Energy is dissipated in compacting soil, which in turn requires significantly greater energy to disturb, so the energy lost in inefficient traction also increases the energy requirement of tillage and seeding equipment. Tullberg (2000) monitored the draft of chisel and sweep tines operating in soil behind the wheels of a 5 t tractor and noted that it was often greater (by a factor of approximately 2) than the draft of identical tines on the same toolbar working in soil not affected by wheels.

These data, combined with measurements of typical tire/implement width ratios and tractive efficiency, was the basis for the conclusion that in tractor tillage operations, approximately 50% of tractor output energy is used in the process of first compacting, and then disrupting soil in its own wheel tracks. This value is probably smaller in no-till systems, where compacted soil is disturbed only by the seeder, but undisturbed compaction beneath seeding depth will still reduce soil porosity, aeration and rainfall infiltration rates.

3.6.3   *Controlled traffic farming*

Controlled traffic farming (CTF) has its origins in ancient systems of zone or bed farming in which traffic areas (originally human) remained separated from production areas. It is based on the commonsense observation that "wheels work best on roads, but crops grow best in soft soil" (Vermeulen *et al.*, 2010). Efficient traffic and traction by wheels or tracks depends on soil being deformed or 'compacted' to the extent that it can resist wheel loads. This state is generally too compact for efficient crop production, an issue particularly obvious in highly mechanized systems using tractors and harvesters with masses often in excess of 8 Mg and 20 Mg, respectively.

Permanent traffic lanes are the obvious solution, but such systems require all machines to conform to uniform (or modular) operating and wheel track gauge widths (transverse distance

Figure 3.2.   Seeding immediately behind the harvester in Queensland controlled traffic farming (photograph
courtesy of Mr J. Reddy).

between wheel centerlines). Many different arrangements are possible, but the expense of harvesters and inflexibility of their track gauges is often the decisive factor. This is the case in
Australian grain farming, where controlled traffic farmers commonly use a track gauge of 3 m,
with operating widths of 9 m for seeders and harvesters, and 27 m for the sprayer. With all heavy
load-bearing tires of 0.5 m width, the permanent traffic lanes occupy <12% of field area in this
arrangement (Tullberg *et al.*, 2007).

Accurate guidance is also necessary to ensure all equipment remains on the traffic lanes, and
this was a difficult issue for early adopters of CTF in the 1980s and 90s. The problem has now been
largely resolved by the development and widespread of '2 cm autosteer' (precise field navigation
systems based on real-time kinetic correction of satellite location signals from terrestrial base
stations or networks). These systems are now almost universally used by CTF farmers in tractors
and seeders, harvesters and sprayers.

Controlling traffic eliminates the no-till farmer's dilemma of trading off the certainty of severe
soil damage from wet harvesting against the risk of total crop loss following delay. CTF is almost
universally combined with no-till because it also improves timeliness in herbicide application,
facilitates precision row/interrow weed management, and often enables the residue-tolerant, disease suppressive practice of interrow planting. With residue left standing as high as possible,
and residue effectively spread behind the harvester, this enables immediate re-seeding (Fig. 3.2),
which increases opportunities for double-cropping with economic or cover crops.

CTF deals with a major issue of no-till systems by eliminating compaction in the cropping area,
allowing CTF farmers to reliably capture the energy benefits of conservation agriculture. Driving
on permanent, non-planted, compacted traffic lanes significantly reduces power requirements
(Tullberg, 2000), but in system efficiency terms it is also important that compacted traffic lanes
allow field operations to take place more rapidly after rain. Because traffic lanes are usually
slightly depressed relative to the field surface, system design for traffic lane drainage can optimize
field access and also provide surface drainage.

The reduced energy requirement and improved timeliness of CTF reduces the machinery capacity required to complete cropping operations within the optimum period. If seeding, for example,
can be started 2 days earlier when the operation has previously been completed in 10 days, seeding equipment (= seeder + share of tractor) capacity, cost, weight and embodied energy can be
reduced by almost 20% within the same seeding timeliness constraints.

The other important impact of controlled traffic is the improvement in soil porosity, connectivity
and internal drainage. This has been shown to enable a substantial (~50%) improvement in the
soils rainfall infiltration rate (Li *et al.*, 2001), biological activity (Pangnakorn *et al.*, 2003) and
capacity to store plant available water (McHugh *et al.*, 2009). These are the factors likely to be
responsible for greater yields from controlled traffic systems (e.g. Tullberg *et al.*, 2007), which
occur despite the nominal reduction in cropped area.

Controlling traffic also has a significant indirect impact on energy requirements, because a
major improvement in infiltration rates will also reduce the duration of waterlogging and volume

of run-off, in turn reducing the opportunity for nutrient loss in run-off or by denitrification. There is good quantitative evidence of both effects (Masters *et al.*, 2008; Tullberg *et al.*, 2011), but their magnitude in any one season will vary greatly with conditions. Some CTF/no-till farmers claim to have reduced nitrogen inputs by 30% while simultaneously increasing yields.

To date, CTF based on existing equipment is the optimal embodiment of conservation agriculture principles, but its wider adoption is constrained by the limited availability of compatible equipment systems. Track gauges for harvesters, for instance, are rarely similar to those of tractors, tend to be inflexible and vary considerably between crops. The situation is tolerable in grain production, because most larger harvesters have a 3 m track gauge, and tractors can be modified to match this. Gauges of 3 m and more might be acceptable in rural Australia and the American/Canadian midwest, but narrow roads and road traffic regulations make such systems unacceptable in more densely populated countries.

Track gauge and operating width compatibilities are an almost insuperable problem for vegetable production, faced with a great variety of different crops and harvesting systems within their rotations. Ownership of these expensive and specialized harvesting machines by different contractors and processors further complicates the issue. Controlled traffic systems are likely to be improved at some point in the future by the introduction of wide-span or "gantry" farming equipment, which should be equally applicable to intensive or extensive cropping systems. It has been the objective of substantial research and (to date unsuccessful) commercial development since the 1960s (Chamen *et al.*, 1994).

3.6.4   Precision and high-technology

3.6.4.1   Precision agriculture

'Precision agriculture' (PA) is a term coined in the 1980s when the potential to link yield monitoring equipment and satellite-based spatial location systems was first realized, and it became possible to monitor the variation in yield between different areas of a field. Wide variations have clear implications for nutrient requirements and seeding density, and point to the opportunity for spatial optimization. Since that time most grain harvesters are now marketed with yield mapping capability, and similar developments are at an earlier stage in cane, cotton, root crop and forage harvesters.

'Variable rate technology' was also developed in response to yield mapping in the 1990s, and is now widely available as an option on many spraying, seeding and fertilizing units. Spatial resolution of yield maps is limited by harvester operating width, and early satellite data were also poor in this respect, so precision agriculture has tended to focus on management optimization within large (several hectares) zones. Improvements in satellite imagery, the development of low-cost 'unmanned aerial vehicles' for agricultural surveillance, and ground vehicle-mounted remote sensing units might change this approach.

Over the same period, the increased sophistication of remote sensing technologies, particularly aerial or satellite imagery, has improved the capacity for spatial detection of differences in crop (or pest) activity at different growth stages. When combined with human observation ('ground truthing') this presents a valuable opportunity to target inputs and avoid wasteful 'whole-field' treatments.

Precision agriculture is now a major research topic, with a substantial literature, but detection of variation has often proved easier than the timely identification of an appropriate response, and farmer adoption is still limited. Considerable progress has nevertheless been made on topics such as nutrient distribution, and there are several reports of commercial, on-farm economy in fertilizer application in the range of 15–30% (e.g. Bongiovani and Lowenberg-Deboer, 2004).

3.6.4.2   Precision guidance

Controlled traffic farming requires a greater degree of precision than PA to maintain all wheels on permanent traffic lanes. Many farmers now use guidance systems ('autosteer') of varying precision but CTF farmers usually have the more precise '2 cm autosteer'. The '2 cm' labeling usually indicates positioning within 2 cm ± 1 standard deviation, and precision at the implement can be significantly less. Practically achievable precision and reliability of these systems nevertheless continue to improve with developments such as networked correction signals and implement steering.

This level of repeatable spatial precision provides an opportunity for 'high-precision' farming allowing extensive agriculture to manage crop row and inter-row areas to an extent previously only achievable with manual labor. Applications include:

- more rapid post-harvest establishment of double or cover crops with "interrow seeding" between the rows of standing residue from the previous crop
- more precise 'band' (or shielded band) application of crop chemicals to the crop row or interrow area, as an alternative to overall spraying
- more precise and economical in-crop fertilizer application providing the opportunity to improve synchronicity of nitrogen supply with crop demand
- more timely weed- and pest management applications with the chance to reduce application frequency and dose rates of chemicals.
- relay cropping, in which the successor crop is seeded between the rows of the standing but senescent, unharvested, previous crop.

The combination of precision and timeliness could be valuable in reducing the opportunity for fertilizer and crop chemical loss and pollution when heavy rain occurs before crop absorption of those materials. This is a common situation in systems where all fertilizer is applied at seeding and is largely available for the next 4–6 weeks. Better alignment of fertilizer supply with crop demand can minimize these losses, and this can sometimes be achieved with precise (sometimes foliar) applications of liquid nutrient formulations delivered at high speed by self-propelled sprayers.

Interrow seeding can reduce carryover of root diseases and also exploit row-specific zones of better soil moisture conditions. Avoiding existing rows of residue also simplifies the seeder task and can maintain (e.g.) standing cereal stubble to provide physical support for a pulse crop. Interrow seeding and relay cropping are both valuable in facilitating opportunistic planting, allowing farmers to take advantage of the unreliable rainfall events of semi-arid environments.

Each of these measures provides significant energy and environmental benefit. All are possible within any cropping system, but considerations of timeliness, machine operating stability and accuracy make their large-scale application much easier and more practicable in systems using controlled traffic, preferably combined with no-till.

3.6.4.3 *Robotics*

Search engines provide ample evidence of the number of groups in the USA, Europe and Australia currently developing autonomous equipment for cropping. Many envisage a large number of small, relatively inexpensive robots replacing the current small numbers of large and expensive field units (Blackmore, 2005). Weed control applications are cited in most cases, but several also mention seeding or harvesting. Some cite the avoidance of soil compaction, rather than labor or economics, as their major justification. This probably assumes that robot mass will be less than some critical level responsible for compaction.

Small autonomous machines with spatial capability can already follow rows, distinguish (with variable reliability) between crop plants and weeds, and precisely target control measures (Slaughter *et al.*, 2008). Coordination between multiple small robots is an issue, but for weed control herbicide quantities per unit area are small, so logistics would not be a problem. Logistics becomes more difficult for seeding and fertilizer application, and very challenging for harvesting. This indicates the need for larger vehicles as described by Sherear *et al.* (2010) who notes that these involve serious liability issues, which are the major impediment to commercialization.

If the weight/capacity relationship of autonomous units is similar to current equipment, it suggests similar cost/capacity, energy and soil impact parameters. Power from on-board solar panels sounds attractive, but the greater rolling resistance of small-wheeled units might be propulsion challenge. Beneficial effects on specific energy requirements will depend on the ability of autonomous equipment to achieve better results, perhaps by using different mechanisms for (e.g.) placing seed or removing grain.

3.6.5 Cropping system energy comparisons

This chapter would be incomplete without a comparison of the overall energy effects of using some of the cropping system options discussed above. This considers annual grain cropping systems typical of eastern Australia, and attempts to combine brevity, transparency and objectivity. It is presented in Table 3.6, where each column sets out comparative data for representative cropping systems, follows a similar format to Table 3.1.

Table 3.6 section a) sets out the total fuel used by each machine, based on the frequency of use noted in Table 3.6 d) and fuel requirements of conventional operation (out in Table 3.2), modified on a conservative basis for changes resulting from no-till and/or controlled traffic. These have been modified used values. These and other inputs are totalled and expressed in energy terms in section b), together with the conversion factors. 'Other components' in section b) represents the energy value of seeds, electricity and transport (from Table 3.1). Overall totals are presented in section c), together with crop yields, output energy and the energy ratios. In addition, the energy ratios are re-calculated to illustrate the effect of using legumes to supply 60% of nitrogen fertilizer.

The cropping system systems options considered are defined below:

- *Mulch tillage:* shallow, non-inverting conventional tillage. Often the first, erosion-motivated step away from conventional tillage, and interestingly requiring approximately roughly 50% of the fuel of noted in Table 3.1 for US wheat production with conventional tillage.
- *Early No-till:* replacement of routine weed control tillage operations with herbicide, sometimes retaining occasional chisel tillage, which is progressively eliminated. Yield unchanged.
- *Mature No-till:* no regular tillage, but occasional tillage operations used after wet harvests. Better herbicide strategy and agronomic measures reduce herbicide use. Yield improvement ~10%.
- *CTF, No-till:* reduces fuel requirements of all operations, avoids wet harvests issue farmers, further 10% yield improvement and improved nitrogen fertilizer efficiency.
- *Precision CTF No-till:* optimizing the timeliness and precision of CTF to reduce herbicide applications. Further small improvements in yield (~5%) and in nitrogen efficiency, using split application with additional applications via sprayer.

Cropping energy ratios are of dubious relevance to food production, but of great importance when considering energy cropping. The idea of "growing your own fuel" might appear attractive to farmers, but the input/output ratios of established biofuel systems are poor, particularly those based on conventional production of grain ethanol. Energy ratios for better systems are included in Table 3.6, simply to illustrate the potential for better systems to improve the raw (on-farm) energy increment from 2.5–5 up to values in excess of 10.

Data presented in Table 3.6 demonstrates the combined effects of reducing energy input and increasing productivity that comes from reducing firstly 'loosening' soil disturbance (no-till), and then progressing to reduce 'compacting' soil disturbance. The process was described by one Australian advocate of no till farming, saying "we spent many years trying to reduce tillage, before we started trying to control traffic. In retrospect, the reverse process would be much better, because controlling traffic makes no-till so much easier".

A factor not considered in Table 3.6 is the increased capacity to intensify production (using either 'economic' or 'cover' crops) with system improvements that reduce the time that between harvesting one crop and seeding the next. It is this increased production that has been the major factor improving the economics of controlled traffic, no-till farming in Australia, but the increased biomass production also provides a sustainability benefit.

Soil carbon sequestration is often cited as one of the major opportunities to reduce carbon dioxide content of the atmosphere, and mitigate global warming. In warmer, semi-arid environments sequestration opportunities of cropping agriculture are limited, but no-till usually has a

Table 3.6. Tillage and traffic effects in grain production.

System			Mulch tillage	Early no-till	Mature no-till	CTF no-till	Precision CTF no-till
a) Fuel use		Units					
Chisel/sweep plow		L/ha	18	4.5	3	0	0
Cultivator shallow tillage		L/ha	18	0	0	0	0
Sprayer		L/ha	1.5	7.5	6	3	2.5
Seeder		L/ha	5	6	5	3	3
Grain harvester		L/ha	10	10	10	8	8
b) Totals							
Total fuel	Input	L/ha	52.5	28	24	14	13.5
Labor time	Input	h/ha	7.8	4.2	3.3	2.6	2.4
Total equipment mass	Input	kg/ha	50	40	40	30	25
Fertilizer nitrogen	Input	kg/ha	60	60	55	45	38
Wheat yield	Output	kg/ha	4000	4000	4500	5000	5000
c) Energy values							
Labor time	129	MJ/h	1006	542	426	335	310
Machinery	77.5	MJ/kg	3875	3100	3100	2325	1938
Fuel energy	40	MJ/L	2100	1120	960	560	540
Nitrogen	67	MJ/kg	4020	4020	3685	3015	2546
Herbicide	200	MJ/ha	200	1000	800	600	400
Other components	1236	MJ/ha	1236	1236	1236	1236	1236
Total input energy		MJ/ha	12437	11018	10207	8071	6969
Output energy	15	MJ/kg	61600	61600	69300	77000	77000
Output/input energy ratio			5.0	5.6	6.8	9.5	11.0
Energy ratio if 60% N produced by legumes			5.9	6.8	8.3	11.7	13.5
d) Operations			Frequency of use				
Chisel/sweep plow			2.0	0.5	0.3	0.0	0.0
Cultivator shallow tillage			3.0	0.0	0.0	0.0	0.0
Sprayer			1.0	5.0	3.0	3.0	2.5

Fuel use based on values in Table 3.2, modified for no-till & CTF operation

positive effect (although this might just be to reduce the rate of soil carbon loss). Regardless of the situation, increased biomass production under the combined effects of controlled traffic and no-till must increase the rate of organic carbon input (as crop residues) and reduce the rate of oxidation loss.

3.7 CONCLUSION

Energy efficiency in crop production will be of increasing importance as world population and the costs of convenient fossil energy increase. Simultaneously, some current high-energy food production systems are degrading the productivity of existing farmland while urban development is reducing the availability of high-quality land for food production. In the absence of catastrophic impacts on population, increased crop production must be a food security imperative, and unless some new energy source is discovered, this must be accomplished with reduced energy inputs.

Improved efficiencies will obviously be required in all the major groups of agricultural energy requirement: fuel for field operations and transport, the embodied energy of machinery, and the embodied energy of fertilizer and agricultural chemicals. Information set out in this chapter shows that the first requirement of improved energy efficiency is to minimize unnecessary soil disturbance and nitrogenous fertilizer use. Tillage is often seen as the major soil disturbance problem, but that ignores the system issue of disturbance in the opposite sense – disturbance – soil compaction – that creates the conditions where farmers see tillage as necessary.

No-till systems have been shown to improve productivity in most seasons, often generating some resilience to compaction effects. No-till performs much better, however, in a controlled traffic farming environment where seeders and crops do not have to cope with a significant proportion of recently-compacted soil. In addition to dispensing with most reasons for tillage, CTF increases the efficiency of all field operations and promotes soil structural and biological amelioration to improve crop productivity. CTF no-till is the basis of a significant reduction in the power requirement, robustness and weight of equipment for no-till seeding, allowing reduction of the energy embodied in equipment.

Embodied energy of fertilizers and agricultural chemicals is influenced both by improved targeting of these inputs, and by machinery effects on the structural and biological environment of the soil. Controlled traffic and precise guidance can improve the temporal and spatial precision of fertilizer placement in these systems and reduce the opportunity for nutrient losses. It also facilitates improved timeliness and row-interrow precision which can increase the efficacy of spraying and reduce active ingredient requirements. Similarly, improved aeration and greater biological activity in controlled traffic/no till improves the soil's nutrient buffering capacity, while improved internal drainage reduces the frequency and duration of any waterlogging events, and hence nutrient loss in denitrification or run-off.

Most farmers want to be good custodians of their land, but they are also small business people, responsive to the economic environment. They are also predominantly 'conservative' in their general reluctance to change (rather than adapt) trusted and reliable production systems even when these are steadily degrading the soil resource. This reluctance is greatest when pervasive degradation effects are universal, and hence largely invisible, and economic incentives to change are limited. To this extent, government fuel subsidies for farmers might reasonably be regarded as a long-term disservice to the extent it reduces the incentive to examine more energy-efficient systems.

REFERENCES

Ansorge, D. & Godwin, R.J.: High axle load – track – tire comparison. In: R. Horn, H. Fleige, S. Peth & X. Peng (eds): Soil management for sustainability, *Advances in Geoecology* 38. Catena Verlag GmbH, Germany, 2006.

ASABE: Agricultural machinery management data ASAE D497.7: 2011. American Society of Agricultural and Biological Engineers, St Joseph, MI, 2013.

Ashworth, M., Desboilles, J. & Tola, E.K.: Disc seeding in zero till farming systems. Western Australian No-Tillage Farmers Association, Perth, WA, Australia, 2010.

Blackmore, B.S., Stout, W., Wang, M. & Runov, B.: Robotic agriculture – the future of agricultural mechanization? In: J. Stafford, V (ed): *5th European Conference on Precision Agriculture*. Wageningen Academic Publishers, The Netherlands, 2005, pp. 621–628.

Bongiovani, R. & Lowenberg-Deboer, J.: Precision agriculture and sustainability. *Precis. Agr.* 5 (2004), pp. 359–387.

Botta, G.F., Pozzolo, O., Bomben, M., Rosatto, H., Rivero, D., Ressia, M., Tourn, M., Soza, E. & Vazquez, J.: Traffic alternatives for harvesting soybean (*Glycine max* L.): Effect on yields and soil under a direct sowing system. *Soil Till. Res.* 96 (2007), pp. 145–154.

Chamen, W.C.T, Audsley, E. & Holt, J.B.: Economics of gantry- and tractor-based zero-traffic systems. In: B.D. Soane & C. van Ouwerkerk (eds): *Soil compaction in crop production*. Elsevier, Amsterdam, The Netherlands, 1994.

Dalal, R.C., Wang, W., Robertson, G.P. & Parton, W.J.: Nitrous oxide emissions from Australian agricultural lands and mitigation options: a review. *Aust. J. Soil Res.* 41 (2003), pp. 165–195.

Defossez, P. & Richard, G.: Models of soil compaction due to traffic and their evaluation. *Soil Till. Res.* 67:1 (2002), pp. 41–64.

Doets, C.E., Best, G. & Friedrich, T.: Energy in conservation agriculture. Sustainable Development and Natural Resources Division, Food and Agriculture Organization of the United Nations (FAO), Rome, Italy, 2000.

Evans, T.A, Dawes, T.Z, Ward, P.R. & Lo, N.: Ants and termites increase crop yield in a dry climate. *Nat. Commun.* 2 (2011), article number 262.

FAO: The economics of conservation agriculture. Food and Agriculture Organization of the United Nations, Rome, Italy, 2001, http://www.fao.org/DOCREP/004/Y2781E/Y2781E00.HTM (accessed September 2013).

Friedrich, T. & Kassam, A.: No-till farming and the environment: do no-till systems require more chemicals? *Outlooks Pest Manag.* 23:4 (2012), pp. 153–157.

Gill, W.J., Schafer, R.L. & Wismer, R.D.: Soil dymanics and soil bins. In: W.J. Chancellor (ed): Advances in soil dynamics. *ASAE Monograph* 12, American Society of Agricultural Engineers, St Joseph, MI, 1994.

Godwin, R.J. & Spoor, G.: Soil failure with narrow tines. *J. Agric. Eng. Res.* 22:4 (1977), pp. 213–222.

Hargrove, W.L.: Winter legumes as a nitrogen source for no-till grain sorghum. *Agron. J.* 78:1 (1985), pp. 70–74.

Helsel, Z.R.: Energy in pesticide production and use. *Enc. Pest Manag.* 1:1 (2006), pp. 1–4.

Huf, S. & Agnew, J. (eds): Rain to grain. Queensland Department of Primary Industries Information Series Q194039, Brisbane, QLD, Australia, 1992.

Kearns, S. & Umbers, A.: Farm practices baseline report. Grains Research and Development Corporation, Canberra, ACT, Australia, 2010.

Kirby, M. & Palmer, A.L.: Investigating shallow leading tines for deep ripping. *The Australian Cotton Grower* 13, 1992, pp. 10–13.

Kirkegaard, J., Angus, J., Swan, A., Peoples, M. & Moroni, S.: Ripping yarns: 25 years of variable responses to ripping clay soils in south-eastern Australia. 2008, http://www.regional.org.au/au/asa/2008/concurrent/managing-subsoils/5934_kirkegaardja.htm /accessed September 2013).

Kuipers, H. & van de Zande, J.C.: Quantification of traffic systems in crop production. In: B.D. Soane & C. van Ouwerkerk (eds): *Soil compaction in crop production*. Elsevier, Amsterdam, The Netherlands, 1994.

Li, Y., Tullberg, J.N. & Freebairn, D.M.: Traffic and residue cover effects on infiltration. *Aust. J. Soil Res.* 39 (2001), pp. 239–247.

Lindwall, C.W. & Sonntag, B. (eds): Landscape transformed: the history of conservation tillage and direct seeding. Knowledge Impact in Society c/o Johnson-Shoyama Graduate School of Public Policy, University of Saskatchewan, Canada. 2010.

Loukanov, I.A., Uziak, J.J. & Michalek, J.: Draught requirements of enamel coated animal drawn mouldboard plough. *Res. Agr. Eng.* 51, 2005:2 (2005), pp. 56–62, http://www.agriculturejournals.cz/publicFiles/57239.pdf (accessed September 2013).

Masters, B., Rohde, K., Gurner, N., Higham, W. & Drewry, J.: Sediment, nutrient and herbicide runoff from canefarming practices in the Mackay Whitsunday region: a field-based rainfall simulation study of management practices. Queensland Department of Natural Resources and Water for the Mackay Whitsunday Natural Resource Management Group, Australia, 2008.

McHugh, A.D., Tullberg, J.N. & Freebairn D.M.: Controlled traffic farming restores soil structure. *Soil Till. Res.* 104 (2009), pp. 164–172.

Murray, J.R., Tullberg, J.N. & Basnet, B.B.: Planters and their components: types, the tributes, functional requirements, classification and description. *ACIAR Monograph* 121, Australian Centre for International Agricultural, Canberra, ACT, Australia, 2006.

Pangnakorn, U., George, D.L., Tullberg, J.N. & Gupta, M.L.: Effect of tillage and traffic on earthworm populations in a vertosol in south-east Queensland. *International Soil Tillage Research Organisation Proceedings*, University of Queensland, Brisbane, QLD, 2003.

Peoples, M.B., Bowman, A.M., Gault, R.R., Herridge, D.F., McCallum, M.H., McCormick, K.H., Norton, R.M., Rochester, I.J., Scammell, G.J. & Schwenke, G.D.: Factors regulating the contributions of fixed nitrogen by pasture and crop legumes to different farming systems of eastern Australia. *Plant Soil* 228:1 (2001), pp. 29–41.

Pimental, D.: Energy inputs in food crop production in developing and developed nations. *Energies* 2 (2009), pp. 1–24.

Radford, B.J., Bridge, B.J., Davis, R.J., McGarry, D., Pillai, U.P.,Rickman, J.F., Walsh, P.A. & Yule, D.F.: Changes in properties of a vertisol and responses of wheat after compaction with harvester traffic. *Soil Till. Res.* 54 (2000), pp. 155–170.

Radford, B.J., Yule, D.F., McGarry, D. & Playford, C.: Amelioration of soil compaction can take 5 years on a vertisol under no till in the semi-arid subtropics. *Soil Till. Res.* 97 (2007), pp. 249–255.

Ruser, R., Flessa, H., Schilling, R., Steindtl, H. & Beese, F.: Soil compaction and fertilisation effects on nitrous oxide and methane fluxes in potato fields. *Soil Sci. Soc. Am. J.* 62 (1998), pp. 1587–1595.

Shahgoli, G., Saunders, C., Desbiolles, J. & Fielke, J.: The effect of oscillation angle on the performance of oscillatory tillage. *Soil Till. Res.* 104:1 (2009), pp. 97–105.

Shapouri, H., Duffield, J.A. & Wang, M.Q.: The energy balance of corn ethanol: an update. Agricultural Economics Report 34075, United States Department of Agriculture, Economic Research Service, 2002.

Sherear, S.S., Pilta, S.K. & Luck, J.D.: Trends in the automation of agricultural field machinery. *Club of Bologna Proc.*, 2010, www.clubofbologna.org/ew/documents/KNR_Sherear.pdf (accessed September, 2013)

Slaughter, D.C., Giles, D.Y. & Downey, D.: Autonomous robotic weed control systems: a review. *Comput. Electron Agr.* 61:1 (2008), pp. 63–78

Smil, V.: Nitrogen in crop production: an account of global flows. *Global Biogeochem. Cy.* 13 (1999), pp. 647–662.

Soane, B.D. & Ouwerkerk, C. van: Soil compaction problems in world agriculture. In: B.D. Soane & C. van Ouwerkerk (eds): *Soil compaction in crop production*. Elsevier, Amsterdam, The Netherlands, 1994.

Szabolcs, I.: The concept of soil resilience. In: D.J. Greenland & I. Szabolcs (eds): *Soil resilience and sustainable land use*. CAB International, Wallingford, UK, 1994, pp. 33–39.

Tice, E.M. & Hendrick, J.G.: Disc coulter operating characteristics. *ASAE Paper* 86-1534. American Society of Agricultural Biological Engineers, St Joseph, MI, 1986.

Tullberg, J.N.: Traffic effects on tillage energy. *J. Agr. Eng. Res.* 75:4 (2000), pp. 375–382.

Tullberg, J.N.: Tillage, traffic and sustainability. *Soil Till. Res.* 11 (2010), pp. 26–32.

Tullberg, J.N. & Rogers, I.L.: Time costs in extensive agriculture. *Proceedings Agricultural Engineering Conference*, Armidale. Institution of Engineers, NSW, Australia, 1982, pp. 57–61.

Tullberg, J.N., Rickman, J.F. & Doyle, G.J.: Reliability and the operation of large tractors. *Transactions of the Institution of Engineers Australia* 8:2 (1983), pp. 58–62.

Tullberg, J.N., Yule, D.F. & McGarry, D.: Controlled traffic farming—From research to adoption in Australia. *Soil Till. Res.* 97 (2007), pp. 272–281.

Tullberg, J.N., McHugh, A.D., Khabbaz, G., Scheer, C. & Grace, P.: Controlled traffic/permanent bed farming reduces GHG emissions. *Proceedings WCCA* 2011, Brisbane. 2011.

Vermeulen, G.D., Tullberg, J.N. & Chamen, W.T.C.: Controlled traffic farming. In: A.P. Dedousis & T. Bartzanas: Soil engineering. *Springer Soil Biology Series*, 2010.

Witney, B.: *Choosing and using farm machines*. Longman Scientific and Technical, Harlow, UK, 1988, p. 175.

Yule, D.F. & Chapman, W.: Controlled traffic farming – more productivity, sustainability and resilience *WCCA*, Brisbane, QLS, 2011, p. 179, http://aciar.gov.au/theme1, (acesssed September 2013).

Zentner, R.P., Lafond, G.P., Derksen, D.A., Nagy, C.N., Wall, D.D. & May, W.E.: Effects of tillage method and crop rotation on non-renewable energy use efficiency for a thin Black Chernozem in the Canadian Prairies. Soil Till. Res. 77 (2004), pp. 125–136.

Zoz, F.M. & Grisso R.D.: Traction and tractor performance. ASAE Distinguished Lecture Series No 27, ASAE Publication Number 913C0403, American Society of Agricultural and Biological Engineers, St Joseph, MI, 2003.

CHAPTER 4

The fossil energy use and CO_2 emissions budget for Canadian agriculture

James Arthur Dyer, Raymond Louis Desjardins & Brian Glenn McConkey

4.1 INTRODUCTION

This chapter presents an estimate of the farm energy budget in Canada and discusses the role of farm energy use in greenhouse gas (GHG) emissions from agriculture. Although the discussion and examples cited are for Canada, many of the principles should apply to mechanized agriculture in most other countries. Energy use in agriculture is at a nexus point in a number of global challenges. The research and policy challenges identified in this chapter are addressed under five energy issues: GHG emissions, energy supply, global food security, biofuel feedstock crops and adaptation to climate change (CC).

4.1.1 *Energy use issues*

The policy linkages to these issues are summarized as follows:

	Energy use issues	*Research and Policy challenges*
1.	GHG emissions	quantify and mitigate the carbon footprints of farm systems
2.	Energy supply	ensure affordable energy supply to farms
3.	Food security	enable mechanization of food crop production
4.	Biofuel crops	compare energy use by food and feedstock
5.	CC adaptation	quantify carbon-footprints of emerging farm systems

4.1.1.1 *GHG emissions*
Since agriculture accounts for about 11% of the total Canadian GHG emissions (Janzen *et al.*, 2006), it is essential to understand all GHG sources within the sector. While quantifying the carbon footprint of farm systems is the first policy challenge, it is the direct result of first quantifying farm energy use. The contribution of farm energy to the GHG emissions budget of agriculture has been overshadowed by other GHGs, namely, manure and enteric methane, nitrous oxide and soil carbon loss. This neglect has happened mainly because GHG emissions from energy use are not considered part of the agricultural emissions for inventory reporting (Jaques *et al.*, 1997). Although farm use of fossil energy is a relatively small GHG emissions term, it is highly variable and in many cases manageable (Desjardins *et al.*, 2005; Dyer and Desjardins, 2009a), and deserves attention as a mitigation opportunity. Future energy technologies could make our energy supply networks less fossil dependent or reduce the emission costs of farm inputs. How farm fossil CO_2 emissions might respond to these changes is another reason for monitoring farm energy use.

4.1.1.2 *Energy supply*
Although new fossil fuel sources are being tapped, this expanding supply is not growing as fast as global demand for fossil energy (Panzica, 2013). This means that food production will increasingly have to compete with other human activities for fossil energy. An affordable supply of energy to

farms cannot be ensured without first understanding farm energy use. Farm energy has proven to be difficult to measure because of two problems. First, the energy budget is composed of a broad range of terms. This is most true for farm field operations, even though they are all dependent on diesel fuel. The second reason is that it is difficult to extrapolate spatially across the diversity of farming systems from specific farm site energy estimates. In spite of the apparent dominance of a few major cash crops (wheat, canola, and feed grains for livestock), there is a diversity of farm types in Canada and each type has a different energy budget, both qualitatively and quantitatively (Statistics Canada, 2009). This diversity is further complicated by the dependence of farm energy consumption on farm management decisions and preferred practices on individual farms.

4.1.1.3   *Food security*

During 2011, the global human population surpassed seven billion and is projected to reach nine billion by 2050 (TEE, 2011). There is little doubt that this increase will put more pressure on the agriculture sector to produce more food and to intensify production practices. To increase food production, the energy demand for important food crops will also increase (Pimentel, 2009). Greater mechanization is one of the means to achieve this intensification. But greater mechanization cannot be achieved without ensuring an affordable supply of fossil fuels to farmers. Increased food production will require additional nitrogen fertilizer (Pimentel and Pimentel, 2008). The manufacture of nitrogen fertilizer requires natural gas, which releases fossil CO_2 (OEE, 2008). While global warming could bring new areas into production for higher food value crops, climate change will most likely also mean the loss of much arable land, as well.

4.1.1.4   *Biofuel crops*

Proponents have touted biofuels as a renewable energy source and as a means of reducing fossil CO_2 emissions (IEA, 2004; Klein and LeRoy, 2007). Detractors of this industry have challenged whether biofuel production results in a net energy increase or lower GHG emissions than conventional fossil fuels (Patzek, 2004; Searchinger et al., 2008). Other critics question whether biofuel feedstock production is the best use of arable land and whether such land use diversion threatens global food supply (Karman et al., 2008). In some situations however, diverting land from livestock to feedstock production may lead to a net reduction in agricultural GHG emissions (Dyer et al., 2011a). An objective, quantitative comparison of the energy use for the production of food and feedstock crops would contribute greatly towards resolving this debate.

4.1.1.5   *CC adaptation*

With CC now widely acknowledged to be underway (Borenstein, 2012; IPCC, 2007), food production systems can be expected to shift and many land uses will change. Strategic decisions that reduce the negative impacts and enhance the opportunities will be needed to help agriculture adapt to CC (Dolan et al., 2001). While the search for ways to mitigate GHG emissions must continue, at least some adaptation to CC will be inevitable. In the future, farming systems that are now minor could emerge as important international food providers. For example, horticultural crops could emerge as main stream food industries, particularly at higher latitudes, as global warming progresses. Crops that are currently isolated to small pockets such as the Annapolis Valley of Nova Scotia (Bailey et al., 2008a) may become more widespread. Conversely, increasing climate variability may necessitate increased use of irrigation or greenhouses capable of year-round operation. To identify emerging food production systems and ensure that they are sustainable, the carbon footprint must be part of the overall environmental assessment of each farm system.

4.1.2   *Defining the farm energy budget*

The first objective in this chapter was to define the farm energy budget. Dyer and Desjardins (2009a) identified six farm energy terms that depend on decisions made inside the farm gate and the type of energy consumed. Farm field operations were grouped as a single term. Farm inputs, including fertilizers and other field chemicals, and farm machinery, defined two of these terms.

#	Category name & #	Operation	Code	#	Category name & #	Operation	Code
					Apply soil amendments (Cat 4)		
	Seed annual crops (Cat 1)			11		Spread manure	Org-N
1		Plowing	PlowG	12		Apply fertilizer	CommN
2		Disking	DiskG		Harvest perennial forage (Cat 5)		
3		Seeding	SeedG	13		Mowing	Mow-F
	Harvest annual crops (Cat 2)			14		Raking swath	RakeF
4		Combine	Combg	15		Load hay wagon	LoadF
5		Move grain	CartG	16		Bale hay	BaleF
6		Swath grain	Swath	17		Chop halage/silage	ChopF
	Weed control (Cat 3)			18		Blow into wagon	BlowF
7	Row crop	Cultivate	CultR	19		move out of field	CartF
8	Anual crops	Spray crops	ChemG	20	Re-seed perennial forage (Cat 6)		SeedF
9	Summerfallow	Cultivate	CultS	21		(5-year interval)	PlowF
10		Spray	ChemS	22		{see Cat 1}	DiskF

Code = the symbol used to identify each operation in the F4E2 output files.
Cat = operational category number; # = field operation number (1 to 22)

Figure 4.1.   A scheme for organizing the 22 field operations simulated in the F4E2 model into six operational categories of field work.

The remaining three terms are gasoline, heating fuels and farm electricity. Gasoline is mainly used for farm-owned transport, particularly the so-called pickup truck. Based on the analytical methodologies required for spatial disaggregation, Dyer *et al.* (2013) separated these six terms into three groups.

4.1.2.1   *Group 1*

Because of the complexity of field operations and their dependence on so many farm machinery management decisions (Dyer and Desjardins, 2003a; 2005a), diesel fuel used in farm field work was designated as one group. Historically, a small percentage of farm tractors were powered by gasoline (Dyer and Desjardins, 2003b), but in modern agriculture, the diesel engine has become the overwhelmingly dominant choice for powering farm machinery. Although mechanized production of field crops can include up to 22 field operations, not all farms carry out all of these operations (Dyer *et al.*, 2010a). These field operations have been incorporated into the Farm Fieldwork and Fossil Fuel Energy and Emissions (F4E2) model (Dyer and Desjardins, 2003a) which will be discussed in more detail below. For modeling purposes, these operations can be divided into several categories. The operational categories for this chapter, which were modified from the previous scheme suggested by Dyer and Desjardins (2009b), are shown in Figure 4.1.

   Category 1 includes seeding and tillage of annual crops, with the specific operations being plow, disk and seed. Category 2 includes the harvesting of annual crops, which involves swathing (cut and windrow), combining and carting the grain off the field. Swathing is optional depending on whether field drying time is needed or possible, but is more popular in western Canada. Category 3 includes all of the weed and pest control operations in both annual crops and in summer-fallow. The two main operations are cultivating and spraying, but cultivating is only possible in row crops. Category 4 involves the application of soil amendments, including fertilizers, manure and lime (lime is assumed to be applied with other fertilizers). Category 5 includes the range of harvesting options and operations associated with harvesting perennial forages. Category 6 involves the reseeding of perennial forages at multi-year intervals (usually 5 years). Categories 5 and 6 are discussed in detail with the F4E2 output presented below. The relationships among these operations and the six field work categories are outlined in Figure 4.1.

4.1.2.2   *Group 2*

The supply of chemical fertilizers and pesticide sprays was integrated into a single energy term. The fossil energy to supply these chemicals was determined from a direct conversion of the weight of consumption of these chemicals (Dyer and Desjardins, 2007). Nitrogen fertilizers are the most energy-intensive chemical inputs to manufacture (Dyer and Desjardins, 2009a), and have available sales records in Canada (Korol, 2002). So, although this conversion was based on the natural gas to manufacture just nitrogen fertilizer, it was indexed to include the weight of additional farm chemicals as well as the nitrogen fertilizer (Dyer *et al.*, 2013). Although the coal required to manufacture farm machinery depends as much on farm level decisions as field operations (Dyer and Desjardins, 2006a), farm machinery supply was grouped with farm inputs. The energy for manufacturing was similarly indexed upward to include other supply functions such as the transport of that machinery to the farm. Both farm inputs were considered by Dyer and Desjardins (2009b) to be indirect energy because the energy defined by these two terms was not actually consumed on the farm.

4.1.2.3   *Group 3*
Gasoline, heating fuels and electrical power constitute the third group. All three energy terms had to be determined empirically. One of the distinctive features of this group was the high dependence on farm type. In contrast, the terms in Groups 1 and 2 are linked to field crop production, regardless of whether those crops are fed to livestock, or used as biofuel feedstock or food production (Dyer *et al.*, 2013). Electrical power was a partial exception to the need for empirical determination because of a semi-empirical index of the CO_2 emissions from this term based on farm types (Dyer and Desjardins, 2006b). This index demonstrated the correlation, at least for this energy term, between energy consumption and farm types, particularly among livestock farms. The heating fuels term, as defined by Dyer and Desjardins (2009a), was a combination of three fuels, including furnace-oil, liquid propane (LPG) and natural gas.

4.1.2.4   *Excluded energy terms*
Several energy terms were excluded from the farm energy budget described by Dyer and Desjardins (2009a). The most critical shortcoming of this farm energy budget was the exclusion of long range transport of products to market. The difficulty with fuel use for long range transport was in defining the distances to market, as the network of collection points would be different for each farm commodity. Most farmers do not pay their own shipping costs (except for required inputs) and would not track those distances or the fuel costs. To quantify this term, information must be gathered on a commodity specific basis, rather than from farm level surveys. Otherwise, the energy-per-km shipping costs could be estimated using established foodmile methodology (Dyer *et al.*, 2011b; LifeCycles, 2010; Xureb, 2005).

   For the most part, the exclusions are only important for farm systems that are minor in Canada. For example, most of Canada's arable land is under rainfed agriculture. Hence, irrigation is not a major energy user in Canada. Similarly, on-farm storage, particularly where cooling is required for perishable products, is not used on most Canadian farms because Canadian farm products consist mostly of grains, oilseeds or livestock carcasses. The dairy industry is the one exception. However, the energy used for this activity is likely included empirically in the electrical energy term (CAEEDAC, 2001; Tremblay, 2000). Whereas Canada has a viable vegetable greenhouse industry, it is small in comparison to Canada's other farming systems and most of these greenhouse businesses do not operate year-round (Dyer *et al.*, 2011b). Since horticulture was identified above as potentially a newly emerging food industry under CC-adaptation, irrigation, cooling systems, and heated and lighted greenhouses could become much more common across Canada.

4.2   METHODOLOGY

4.2.1   *Modeling farm energy consumption*

The most heavily relied upon tool in Canada for learning more about farm energy use has been statistical surveys. However, such surveys are expensive to carry out and thus infrequently repeated.

Extraction of essential information from farm surveys is often restricted by the need to protect confidentiality of individual farmers. The alternative to surveys is to develop models of farm operations and decision making. Models are the only practical way of representing the dynamics of commodity-specific production systems (Janzen *et al.*, 2006). Although such models are not completely independent of empirical data, they allow that data to be extrapolated to periods and regions other than where and when these data were collected. Models can also help to design and streamline future surveys by identifying questions that are helpful in describing the farm energy budget and eliminating those that are not.

The most comprehensive source of farm energy use information in Canada is the 1996 Farm Energy Use Survey (FEUS) of Canada (CAEEDAC, 2001; Tremblay, 2000). The FEUS provided commodity-specific estimates for the three energy terms in Group 3, the terms for which detailed modeling algorithms were not available (Dyer *et al.*, 2013). Due to confidentiality constraints, the FEUS only allowed data about energy type to be grouped by farm type for Canada as a whole. The FEUS allowed these energy data to be adjusted for the shares of each fuel type that were used in farm households instead of farm use, although this distinction could not be applied to farm types (Tremblay, 2000). The FEUS was critical to the F4E2 model because it provided data on the consumption of diesel fuel in Canadian agriculture against which the model was verified (Dyer and Desjardins, 2003a).

4.2.2   Computations for field operations

For farm field operations, the controlling factors are the field mechanics and farm machinery management parameters that have traditionally fallen under the mandate of agricultural engineering. As a result, a wide ranging set of mechanical coefficients have been measured for most of the listed field operations (ASAE, 1971; 1982/84; 2000). For the F4E2 simulation model (Dyer and Desjardins, 2003a; 2005a) these coefficients were incorporated as predictive equations for the dynamics of field operations.

The basic F4E2 model operates as an interactive farm level system and generates estimates for assumed farm field areas and the applicable operations selected from those shown in Figure 4.1. We have designated the version of the F4E2 model based on conversational (keyboard) inputs as the console mode. In this mode of operation, F4E2 simulations have been used to demonstrate fossil CO$_2$ emissions and energy consumption on assumed model farms (Dyer and Desjardins, 2007). Another mode of operation was to use F4E2 output as a meta-model for farm energy use (Dyer and Desjardins, 2005b). The estimates presented in this chapter also represent the meta-model mode of application. This mode involved reducing simulations for model farms and typical assumed field sizes, crop choices and operational decisions to a per-ha basis. One set of farm simulations was run for each province in Canada (with the four Atlantic region provinces combined as one region or province).

On most typical Canadian farms, field operations involve a fleet of tools and implements. The need for so many field operations in modern crop production makes this the most complex of the six terms in the farm energy budget. The computations involved in farm machinery management which were used in the F4E2 model are well established methodology and have been described in detail in numerous agricultural engineering sources (Finner, 1973; Jacobs and Harrell, 1983; Stout, 1984). Because their incorporation into the F4E2 model was described by Dyer and Desjardins (2003a), only a generalized review of these relationships is given in simplified form here. This description is not meant to be a guide to actual computations.

A typical computation for any farm operation involves field size, implement width, the resistance of the implement, work time available, operating speed and the power available to pull or operate the implement. In this context, an implement can include any tractor-towed or self-propelled piece of farm equipment. These terms are not all independent variables. For example, if the implement width is not known, it can be determined from the size of the field, the operating speed and work time (Wu *et al.*, 1986) as follows.

$$\text{Width} = \text{Fieldarea}/(\text{Worktime} \times \text{speed}) \tag{4.1}$$

The work time available to carry out any farm operation is a fundamental concept in timeliness (Edwards and Boehlje, 1980; Hunt, 1972). Timeliness refers to the completion of critical operations, such as spring seeding or fall harvesting within the available weather windows (Dyer and Baier, 1980). These windows allow the seeding of crops early enough to maximize the growing season, or the harvesting of ripened crops before they are weather-damaged. Both windows are limited by the frequency of days with trafficable soil, or workdays (Dyer and Murray, 1985). Soil that is trafficable is dry enough to support heavy farm machinery without damage to soil structure (Dyer and Baier, 1979). If the implement widths are known, then the work time can be computed from Equation (4.1).

The resistance of the implement is usually measured as a force per unit of width, but is occasionally measured as force per unit of crop product (typically for harvesters). Implements involve three types of resistances (Jacobs and Harrell, 1983). For tillage implements, there is a draft resistance when they are pulled through the soil. For any implement that requires an internal mechanism, such as combines, hay balers or seeders, there is internal mechanical resistance, whether it is self-propelled or powered by the tractor's power take-off (pto). Any implement with wheels, from wagons to combines, as well as the tractors themselves, also has rolling resistance, or the resistance of a wheel rolling over a rough surface. For tractors, or the power sources (engines) in any self-propelled implement, the required power is the product of the implement width, the rated resistance of the implement on a per-unit of width basis and the operating speed.

$$\text{Power} = \text{Width} \times \text{resistance} \times \text{speed} \tag{4.2}$$

Fuel consumption, fossil energy or fossil CO_2 estimates, are all determined by re-integrating power estimates by work time. While fuel energy can also be determined directly from width, resistance and field area (force times distance), without computing power, there are advantages to this intermediate power computation. Based on tractor power simulations from the F4E2 model, Dyer and Desjardins (2006a) derived an index for the energy required to supply farm machinery. A reasonable approximation of this index for Canada was that the farm machinery supply term is 70% of the farm field operations term (Dyer and Desjardins, 2009a). Computed tractor power could also be used to verify F4E2 simulations against reported tractor sizes in terms of their hours of use and their fuel consumption (Dyer and Desjardins, 2006a).

4.2.3   *Response to tillage systems*

One of the main drivers of energy consumption in Group 1 (field operations) has been the transition that has been going on in Canada in tillage practices. With fewer passes prior to seeding to prepare the soil and lighter draft implements, the fuel to complete spring field work has dropped appreciably, particularly in western Canada (Dyer and Desjardins, 2005a). The F4E2 model is directly sensitive to this change and tillage systems are an important input to F4E2 calculations.

Figure 4.2 shows the composition of tillage system choices in each province for the census years 1996, 2001 and 2006. This chart illustrates the dramatic change in tillage that has taken place in western Canada and helps to explain sharp decreases in the field operations energy term over the three census years in that region.

4.2.4   *Converting energy use to fossil CO$_2$ emissions*

To satisfy three of the five energy use issues and policy challenges discussed above, farm energy terms had to be converted to fossil CO_2 emissions. A different conversion was required for each of the six farm energy terms (Dyer *et al.*, 2013). For diesel, coal and gasoline, the conversion factors were 70.7, 86.2 and 68.0 Gg{CO_2}/PJ (Neitzert *et al.*, 1999), respectively. To convert the

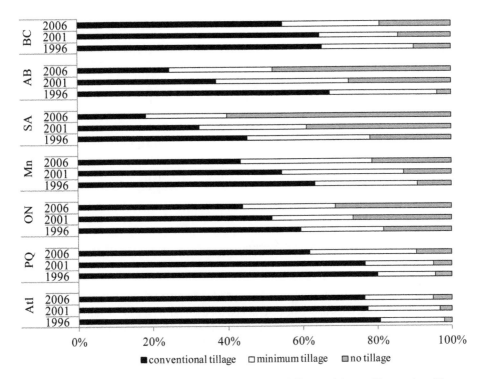

Figure 4.2.   Percent of farmland in Canada under conventional tillage, minimum tillage and no tillage systems in each province over three census years. The abbreviations for provinces in are as follows: Atl = Atlantic Provinces, PQ = Quebec, ON = Ontario, MN = Manitoba, SA = Saskatchewan, AB = Alberta and BC = British Columbia.

F4E2 estimates of fuel consumption for tractors and self-propelled implements to work energy and fossil CO$_2$ emissions, the conversions from Table 4.1 for diesel fuel are used, since almost all modern farm equipment is fueled by diesel oil.

Based on a fertilizer manufacturing analysis by Nagy (2001), Dyer and Desjardins (2007) used 57.9 Gg{CO$_2$}/PJ as the conversion factor for fossil CO$_2$ emissions from fertilizer supply. Since this input term was not assessed in the FEUS, an additional direct calculation was needed for this term. The energy consumption rate (based on natural gas) for fertilizer manufacture and supply of nitrogen fertilizer was 71.3 MJ/t{N}. This coefficient was derived from a set of coefficients compiled by Nagy (2001) for a range of fertilizer types. Although this conversion was for just nitrogen supply, it was indexed to include other farm chemicals, mainly phosphate and potash fertilizers, and allowed for a very small additional share of the input energy for the supply of pesticides. Consequently, Dyer and Desjardins (2009a) defined this CO$_2$ emissions term as chemical inputs, rather than fertilizer supply.

The resulting estimate of CO$_2$ emissions associated with the supply of farm chemical inputs was 4.05 t{CO$_2$}/t{N} (Dyer and Desjardins, 2009a; 2009b). This estimate was 9% below the same period average fossil CO$_2$ emissions for this term by Janzen *et al.* (1999). Snyder *et al.* (2007) reported CO$_2$ to nitrogen conversion rates that were the same as the 4.05 t{CO$_2$}/t{N} conversion used by Dyer and Desjardins (2009a; 2009b) for Nebraska and 10% higher for Michigan. The Canadian Ammonia Producers have proposed that increased use of ammonia-based nitrogen fertilizers could lower the carbon footprint of this chemical input (CAP, 2008). While this process may consume less natural gas to manufacture than other forms of nitrogen fertilizer, more scientific monitoring is needed to verify this claim before incorporating it into national GHG inventory reports.

Table 4.1. Energy conversion units for the six fuel types used in the Canadian farm energy budget.

Fuel type	Heat/vol	Density	CO_2/vol	CO_2/energy
	[TJ/ML]	[g/L]	[t/kL]	[t/TJ]
Gasoline	34.7	642.6	2.4	67.7
Kerosene	37.7	695.2	2.6	67.7
Propane	25.5	416.7	1.5	59.8
diesel	38.7	745.8	2.7	70.7
	[TJ/GL]	[g/m^3]	[t/ML]	[t/TJ]
Natural gas	37.8	512.5	1.9	49.7
	[TJ/kt]		[t/t]	[t/TJ]
Coal	27.7		2.387	86.2
Heating fuel conversion from the 1996 FEUS* for grains and oilseeds				61.0

*, FEUS = 1996 Farm Energy Use Survey of Canada.

Since heating fuel includes three separate fossil fuels (CAEEDAC, 2001; Tremblay, 2000), the associated CO_2 emission rates had to be determined for each farm type in the same way as energy consumption rates for heating fuel were determined. This was done by converting the set of fuel and farm type estimates for this energy term and converting them to CO_2 emissions, using 59.8, 61.0 and 67.7 Gg{CO_2}/PJ, for LPG, natural-gas and furnace-oil (Neitzert *et al.*, 1999), respectively. The conversion factor for each fuel and farm type was the product of these CO_2 emissions and energy consumption amounts found for each type in the 1996 FEUS (Tremblay, 2000), which were converted to a single rate for the combined heating fuel term of 64.1 Gg{CO_2}/PJ in the production of grains and oilseeds.

Dyer *et al.* (2013) recognized that there are great differences among provinces in the dependence of coal-based electrical power generation (NRCan, 2005). This was a departure from the single average conversion factor for CO_2 emissions for the consumption of electric power used by Dyer and Desjardins (2009a). The single factor conversion was based on 22% of Canadian electricity generation being from coal-fired plants. In this chapter, a conversion factor for each province was computed separately using the provincial percent of coal generation from each province. These percent thermal generating factors were 15, 96, 76, 1, 16, 0 and 59%, respectively, for British Columbia, Alberta, Saskatchewan, Manitoba, Ontario, Quebec and the Atlantic Provinces, respectively (NRCan, 2005). Weighting for the share of electrical energy used in the Canadian agriculture sector (CAEEDAC, 2001; Tremblay, 2000) in each province resulted in a revised national estimate for thermal generation of the electrical power used in agriculture of 36% thermal energy.

4.2.5 *Interfacing farm energy use with other GHG emission models*

The earliest application of F4E2 was in a sensitivity analysis of fossil CO_2 emissions from farm field operations related to several important farm level decisions in Canada (Dyer and Desjardins, 2003b). An earlier version of the farm energy budget described in this chapter was integrated with nitrous oxide emission estimates from required nitrogen fertilizers to determine the GHG emissions budgets for 21 major field crops in Canada (Dyer *et al.*, 2010b). In this same study, soybeans and canola were evaluated as potential feedstock crops for increased biodiesel production. These tests indicated a net positive impact on fossil CO_2 emissions for using both of these crops as biodiesel feedstock.

Output from F4E2 has been incorporated as a sub-model in two more broadly reaching models in the Canadian agriculture sector. The first application (Dyer *et al.*, 2010a) was in the database

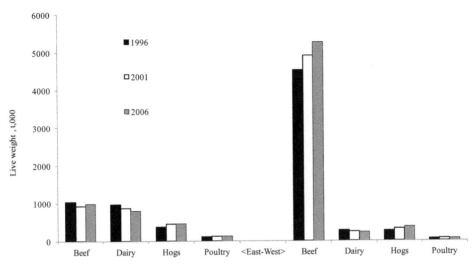

Figure 4.3.   Total live weight of livestock in eastern and western Canada divided at the Manitoba-Ontario border over three census years.

of the Canadian Economic and Emissions Model for Agriculture (CEEMA) which defines field crop GHG emissions at the Census Agricultural Regions (CAR) scale (Kulshreshtha *et al.*, 2000). This application was recently widened to include all six terms in the farm energy budget (Dyer *et al.*, 2013). The second application was in the Unified Livestock Industries and Crops Energy and Emissions System (ULICEES) model (Vergé *et al.*, 2012) which operates at a provincial scale. The energy sub-model in ULICEES also incorporates the complete farm energy budget. With this energy sub-model, ULICEES has been used to quantify the avoided GHG emissions by taking land out of livestock production and seeding it to biofuel feedstock crops (Dyer *et al.*, 2010a) and the GHG emissions from cropland in Canada that was not used to support livestock production (Dyer *et al.*, 2011c).

With increasing and inelastic food demand, energy use and GHG emission intensities (where intensity is the ratio of energy or GHG emissions to food production) are important indicators in defining viable farm energy policies. These interfaces between the farm energy budget and the other farm GHG emission models just described have relied on intensity based indicators. Because higher yields lower GHG emission intensity, yields have the same importance as GHG emissions in defining the sustainability of farming systems. For example, based on emission intensity, the high yielding nature and high value of horticultural crops may result in a smaller carbon footprint, than many grain and oilseed crops.

4.3   FARM ENERGY USE CALCULATIONS

The rest of this chapter presents the complete farm energy budget quantitatively (at least for the energy terms described here) for the seven provinces and three census years: 1996, 2001 and 2006. The first demonstration only deals with the energy used to produce crops, regardless of the use of those crops. The second demonstration is of the energy budget within the ULICEES model and its response to each of the four major Canadian livestock types. Figure 4.3 shows the distribution of livestock in eastern and western Canada (divided at the Manitoba-Ontario border). Livestock types in this figure are compared on the basis of total live weight, rather than by population differences because of the different body weights of livestock and the complexity of their respective lifecycles.

Table 4.2.　Areas in each of six major agricultural land uses, or crop categories, over seven provinces for computing the farm energy and greenhouse gas budgets in Canada.

Provinces	Annuals	Row crops	Legumes	Hay	S-fallow[1]	Pasture
	[ha × 1000]					
1996						
Atlantic	224	34	32	201	1	148
Quebec	830	425	185	882	9	519
Ontario	2496	1514	569	1018	20	1012
Manitoba	3959	191	128	749	324	2010
Saskatchewan	13437	857	857	1089	4432	6327
Alberta	7593	235	230	1924	1437	8530
B.C.	179	31	31	348	39	1412
2001						
Atlantic	220	38	36	232	2	55
Quebec	1120	477	210	730	5	183
Ontario	2699	1713	661	958	14	313
Manitoba	3923	252	183	792	256	383
Saskatchewan	14095	1604	1604	1282	3132	1406
Alberta	7473	363	357	2255	1236	2231
B.C.	205	36	36	413	37	233
2006						
Atlantic	217	40	38	232	2	54
Quebec	1129	508	222	804	4	147
Ontario	2700	1829	702	961	12	303
Manitoba	3891	276	201	810	127	498
Saskatchewan	13349	1836	1836	1612	2429	1962
Alberta	7578	405	399	2043	906	2484
B.C.	198	38	38	388	26	246

[1] summer-fallow, no crops planted to conserve moisture.

4.3.1　Land use areas

The first set of data to examine is the inputs that drive the farm energy budget (Table 4.2). Except for harvesting where fuel consumption can depend on crop yields, the F4E2 model and the indices for the other energy terms do not respond to changes in individual crops. These estimates are, however, sensitive to the more general land use differences. These land use categories required for F4E2 simulations include all land in annual crops, hay (perennial forage), and summer-fallow. The annual crops must be broken down by whether or not they are row crops, as well as whether or not they are also pulse crops. The row crop distinction determines whether a cultivator can be used for weed control, rather than just spraying herbicides. The pulse crop distinction determines whether fertilizer, particularly nitrogen, needs to be applied because pulses are legumes which can fix atmospheric nitrogen. Hence, nitrogen applications are lowered by higher portions of the annual crop areas being in pulses.

The spreading of manure was restricted to just cropland that was in the land base that supports livestock production, since it was assumed that manure would not be available on farms that do not produce livestock. Hence, a feedback from the ULICEES model was needed to define the percentage of annual cropland that supports livestock in each province. A specific function for manure application was one of the later upgrades to F4E2 and was described in more detail by Dyer and Desjardins (2005b).

Table 4.3. Fossil CO_2 emissions from Canadian field crop production over three census years where emission sources (column headings) correspond to the field operation categories shown in Figure 4.1.

Provinces	Till & seed	Harvest annuals	Control weeds	Apply soil amendments	Harvest forage	All operations
	[Gg CO_2]					
1996						
Atlantic	32	11	1	14	7	65
Quebec	114	42	11	60	34	260
Ontario	247	138	42	90	58	575
Manitoba	370	153	11	164	36	733
Saskatchewan	1015	464	35	507	54	2075
Alberta	641	315	24	223	80	1282
B.C.	29	8	2	12	12	63
2001						
Atlantic	32	11	1	15	8	67
Quebec	136	56	12	70	29	302
Ontario	245	149	51	92	58	595
Manitoba	348	152	13	159	38	709
Saskatchewan	911	485	61	527	64	2048
Alberta	593	309	41	273	93	1310
B.C.	33	9	2	13	14	72
2006						
Atlantic	32	11	1	16	8	67
Quebec	129	57	14	61	32	292
Ontario	231	150	56	98	60	595
Manitoba	316	151	14	156	39	675
Saskatchewan	700	458	72	516	80	1826
Alberta	497	313	48	278	85	1221
B.C.	31	9	2	9	13	64

4.3.1.1 *Land use*

Since the F4E2 model simulates the work and fuel required to control weeds on the land that is in summer-fallow, this was a required area input to the model and is included in Table 4.2. F4E2 simulations also account for the differences in fuel use between harvesting hay and combining annual crops. Hence, the areas in harvested perennial forages are included in Table 4.2. Improved pasture is also shown in Table 4.2. Even though this area accounts for no farm fuel use in F4E2 computations, it gives an appreciation of the overall balance in the use of arable land in Canada. Although not shown in this table, yields for the generalized crop categories are required inputs to the calculations, specifically for the fuel use estimates for harvesting.

4.3.1.2 *Farm field operations*

In Table 4.3 the energy terms of the field operations are presented as fossil CO_2 emissions, rather than as energy consumption because the most important application of F4E2 to date has been in defining fossil CO_2 emissions (Dyer *et al.*, 2010a; 2010b; 2012; Vergé *et al.*, 2012). These emission estimates can be converted back to fossil energy use with the diesel coefficient in Table 4.1. The 22 operations simulated in F4E2 could not all be practically shown in this table. Hence, the columns represent the categories of operations described above and shown in Figure 4.1.

Harvesting forage (Category 5) includes the harvesting of silage corn, even though this is an annual crop. Harvesting forages involves a mix of (or selection from) operations from five

Table 4.4. The complete energy budget (six terms) for Canadian farms in seven provinces expressed as fossil energy consumption units.

Provinces	Field work	Machinery supply	Ghemical supply	Gasoline	Heating fuel	Electrical power	Total energy
	[TJ] {fossil energy}						
1996							
Atlantic	920	528	1870	72	43	23	3456
Quebec	3681	2113	6289	268	158	86	12595
Ontario	8133	4669	12398	805	475	260	26739
Manitoba	10376	5956	22271	1277	753	412	41045
Saskatchewan	29352	16849	36921	4336	2556	1398	91413
Alberta	18142	10415	30715	2450	1445	790	63956
B.C.	890	511	1920	58	34	19	3431
2001							
Atlantic	942	541	1976	69	40	22	3589
Quebec	4274	2454	6985	349	206	113	14380
Ontario	8422	4835	12131	841	496	271	26995
Manitoba	10034	5760	21531	1223	721	394	39663
Saskatchewan	28969	16630	37865	4392	2590	1417	91862
Alberta	18530	10637	31581	2329	1373	751	65201
B.C.	1013	582	1804	64	38	21	3521
2006							
Atlantic	955	548	2256	69	41	22	3891
Quebec	4137	2375	5401	360	212	116	12601
Ontario	8421	4834	12968	861	508	278	27869
Manitoba	9556	5485	20884	1241	731	400	38297
Saskatchewan	25833	14830	36554	4256	2509	1373	85356
Alberta	17266	9912	30728	2416	1425	779	62526
B.C.	909	522	1006	63	37	20	2559

different harvesting systems based on different moisture contents of the hay when taken off the field (Dyer and Desjardins, 2005a). These systems have different rates of use in different climates, or provinces, because they involve different field drying times. Reseeding perennial forages uses the tillage and seeding calculations from Category 1 (tillage and seeding), but only applies these three coefficients at 5-year intervals (a default period for the console mode of F4E2). Hence, these terms are grouped with Category 1, rather than being shown individually.

4.3.1.3 *Farm energy use budget*
In Table 4.4 all six of the energy terms are shown, including farm field work as a single term. These are presented as energy units, rather than as GHG emissions. This was because each term is powered by a different type of energy. But they all represent fossil fuel consumption (even electricity). Column 1, which is a summary of the CO_2 emissions in Table 4.3 converted to energy use, corresponds to Group 1 of the three energy groups discussed above. Columns 2 and 3 of Table 4.4 show the two input energy terms in Group 2. Columns 4, 5 and 6 are the three empirical energy terms that were derived from the 1996 FEUS (Tremblay, 2000). They correspond to Group 3.

4.3.1.4 *Fossil energy use for livestock production*
To date, one of the most important applications of the farm energy budget has been to compare livestock systems in Canada. The land base on which the required livestock feed is grown was

Table 4.5. Energy budget of the four major livestock industries in eastern and western Canada over three census years.

Livestock type	Field work	Machinery supply	Chemical supply	Gasoline	Heating fuel	Electrical power
	[TJ] {fossil energy}					
Eastern Canada, 1996						
Beef	2247	1290	3432	1449	829	559
Dairy	3331	1912	4018	2272	1258	2959
Hogs	2305	1323	5155	788	1231	1126
Poultry	1423	817	2304	307	1796	641
Western Canada, 1996						
Beef	8802	5053	16482	6329	3618	2442
Dairy	1052	604	1192	637	352	829
Hogs	2298	1319	4459	550	859	786
Poultry	828	475	908	151	883	315
Eastern Canada, 2001						
Beef	1997	1147	3432	1301	744	502
Dairy	2964	1701	4018	2043	1131	2661
Hogs	2711	1556	5155	943	1474	1348
Poultry	1549	889	2304	344	2010	717
Western Canada, 2001						
Beef	9755	5600	16482	6869	3927	2650
Dairy	934	536	1192	570	316	743
Hogs	2750	1579	4459	683	1067	976
Poultry	988	567	908	183	1067	381
Eastern Canada, 2006						
Beef	2082	1195	3432	1377	787	531
Dairy	2686	1542	4018	1883	1042	2452
Hogs	2709	1555	5155	967	1510	1382
Poultry	1531	879	2304	346	2024	722
Western Canada, 2006						
Beef	10203	5857	16482	7363	4210	2841
Dairy	837	480	1192	532	294	692
Hogs	2879	1653	4459	768	1201	1098
Poultry	904	519	908	176	1028	367

defined as the Livestock Crop Complex (LCC) (Vergé *et al.*, 2012). It is the land in the LCC, rather than the livestock themselves, that accounts for most of the energy consumption that can be attributed to livestock production. The small amounts of energy that are attributed directly to the animals are from the Group 3 terms. Because the LCC is restricted to, and defined by, each livestock population, the energy budget calculations had to be integrated with the calculations within the ULICEES model. This integration was specific to each livestock type.

Table 4.5 shows the energy budgets for the four major livestock industries in Canada. Because a separate budget was needed for each of these industries, this application of the energy budget is presented on an east-west scale only, rather than for provinces. The six terms of the energy budget are shown in the same order in which they appeared in Table 4.4. The LCC areas in Table 4.5 were calculated in ULICEES for 2006 based on livestock diets surveyed in 2001 (Elward *et al.*, 2003). The energy estimates in ULICEES for 1996 and 2001 were calculated using the 2006 LCC, but these LCC areas were indexed by the live weight ratios of 1996 or 2001 to 2006 (see Fig. 4.3). This indexing was done separately for each of the four livestock types. This allowed changes in the livestock farm energy budgets to be demonstrated without the impact of changing livestock diets. Hence, the 1996 and 2001 ULICEES estimates should be treated as just approximations for these years.

4.4   RESULTS

This chapter presents large sets of data, but does not attempt to draw original scientific conclusions from those data since most these data have already been published in different formats in various locations (as referenced throughout this chapter). The range in total energy consumption values among the seven provinces demonstrates that there is quite a bit of spatial difference within Canada. The western provinces account for most of the agricultural energy consumption, since they also have most of the farmland. By presenting results from three census years, it is fairly clear that the farm energy budgets are relatively constant, even though minor trends were evident in some provinces.

Table 4.2 gives an appreciation for the dimensions of the field operation categories shown in Table 4.3. Land in annual crops is the largest form of agricultural land use throughout Canada, exceeding both hay and pasture land (not counting unimproved pasture, or rangeland). Except for Quebec and Ontario between 1996 and 2001, the land used for annual crops has not changed appreciably over the 10 years. Land in perennial forage (harvested hay) increased slightly in the west and decreased slightly in the east. Legumes have shown the sharpest rise throughout Canada. In the east this is due to increased soybeans (Hannam, 2006). In the west this is due to the pulse industry doubling in size since 1990 (Pulse Canada, 2002). Row crop areas have also increased: in the east due to corn and in the west due to pulses. Summer-fallow is only a significant form of land use in the Prairie Provinces, due to the more arid climate of western Canada. The area under summer-fallow has been decreasing.

Table 4.3 shows that spring seeding (including tillage) and harvesting of annual crops accounts for most of the fossil CO_2 emissions from farm field operations. This is consistent with annual crops being the dominant form of land use. Weed control (spraying and cultivation) and harvesting hay were the smallest categories of field operations with respect to fossil CO_2 emissions. Applying soil amendments (manure and fertilizers) was surprisingly high, accounting for almost as much fossil CO_2 emissions as harvesting annual crops.

The most noteworthy trend has been the decrease of CO_2 emissions from spring field work in the Prairie Provinces, which is consistent with the increasing use of minimum and no tillage systems in Manitoba (MN), Saskatchewan (SA) and Alberta (AB) shown in Figure 4.2. There is also a decrease in conventional tillage in Ontario (ON) in Figure 4.2, which is reflected in a slight drop in CO_2 emissions from that province in Table 4.3. With less spring tillage, weed control has become a slightly higher energy user. There is little change in the CO_2 emissions from harvesting forage.

Table 4.4 shows that the supply of farm chemicals, predominantly nitrogen fertilizer, is the highest user of fossil fuel energy in Canadian agriculture. This was consistent across all provinces and over the three census years. Field operations were the second highest user of fossil fuel of the six farm energy terms. The supply of farm machinery was the third largest term (being only a constant percentage of field operations). The three terms in the third group (those indexed to the FEUS data) emitted less than 10% of the fossil CO_2 from field crops in Canada (Dyer *et al.*, 2013). The greatest variation among provinces was from the electric power term, due to the provincial differences in the use of coal for generating power. Heating fuels showed the lowest emissions for Canada of the six energy terms.

As would be expected from Figure 4.3, Table 4.5 shows that beef production in western Canada is the dominant energy user of the livestock industries. The supply of farm chemicals was the highest energy user among all six energy terms in the livestock energy budget. Although the Group 3 terms were the lowest energy terms in most cases, gasoline use exceeded machinery supply in the beef and dairy industries. Electrical energy was generally the lowest farm energy term in the beef industry over the three years.

The group 3 terms in the livestock energy budgets (Table 4.5), expressed as a percent of the total energy budget, were almost six times as high as they were in the field crops budgets (Table 4.4). This was because for Table 4.4 only the grains and oilseeds energy use quantities were used in order to keep the livestock operational decisions out of the field crop energy budget. The Group 3 energy terms shown in Table 4.5 include only the energy that the FEUS attributed to the four livestock farm types for these three terms. For these three terms, the FEUS reported that the energy used in the livestock farms was twice as much as was used in the grains and oilseeds farms (Tremblay, 2000).

4.5   DISCUSSION AND CONCLUSIONS

The description of the farm energy budget provided in this chapter covered the different energy terms and the types of energy involved in each term. By also describing (at least generally) the input data required to calculate each term, it was intended to provide guidance on how this energy budget could be interfaced with geographic databases. Whereas these energy use estimates were presented at a provincial scale, this energy budget could be adapted to the spatial scale of other databases, even though the scale might be sub-provincial. This would only require that the inputs outlined in Figure 4.2, Table 4.2 and the related discussions can be extracted from each geographic unit (region) in any such database. This interface has already been demonstrated for the CEEMA database (Dyer et al., 2013).

Farm energy is not a single issue. It is a complex mix of fuel types and farm activities. While all of the terms in the farm energy budget produce fossil CO_2 emissions, there is no single option available for mitigating all of these emissions. For example, of the six energy terms, only two use the same type of fuel, since both thermal electricity and machinery supply rely on coal. The quest for mitigation options for farm energy use as a source of fossil CO_2 also has a human limitation. Farmers have historically had a strong economic incentive to minimize energy consumption. For example, the cost of the diesel fuel that powers farm field work is an incentive to avoid unnecessary field operations, as seen through decreased tillage in western Canada.

Farmers have also had incentives to minimize tractor sizes by matching implement widths to tractors because excessively large tractors represent an unnecessary investment. Under-powered tractors are also unattractive to most farmers because of the so-called timeliness penalties, the loss of crop yield due to delayed or incomplete field operations (as was discussed above). Without this rational decision making, the F4E2 model could not have achieved the level of accuracy (based on the 1996 FEUS) that it did (Dyer and Desjardins, 2003a). However, this also means that, while farm energy use is very responsive to farm management, there is little that policy makers can expect farmers to do that they are not already doing with respect to most farm energy terms.

Some minor saving in energy costs related to field work decisions may be realized in eastern Canada. In western Canada reduced tillage has taken place because it benefits soil quality. Hence, the argument that reduced tillage saves appreciable fuel costs and fossil CO_2 emissions is likely only incidental to western farmers. In the east, however, there has been less uptake of reduced tillage (Fig. 4.2) since the soil management benefits are less certain than in the west. Therefore, the argument that reduced tillage will lower energy costs, particularly with rising diesel prices, may help convince more eastern farmers to use less tillage.

While it represents a high source of fossil CO_2 emissions, the supply and manufacture of nitrogen fertilizer may not offer as much opportunity for reducing the farm related fossil CO_2 emissions as the magnitude of this term would suggest. Applying less than optimal nitrogen fertilizer for crop growth is felt by farmers through lower yields. Thus, there is little incentive for farmers to reduce application rates to save energy. As well, reduced crop yields from inadequate fertilizers may actually raise the GHG emission intensity of these crops. However, since reducing nitrogen fertilizer also reduces nitrous oxide emissions, there is still great potential to reduce the overall carbon footprint of agriculture. As suggested above, there may be an opportunity to reduce the consumption of the natural gas required to synthesize nitrogen fertilizer, but this option would be the responsibility of the fertilizer manufacturing industry (and was beyond the scope of this chapter).

The only other way to achieve a reduction in fossil CO_2 emissions through reducing nitrogen fertilizer is through greater reliance on legume crops, both in animal feed and in food for human consumption (more pulses). These nitrogen fixing crops eliminate the need for chemical nitrogen fertilizer. Animal manure is another non-chemical source of nitrogen, but the animals that provide the manure have their own carbon footprint that must be taken into account, particularly if animal production is encouraged mainly to supply manure.

The other supply term in the farm energy budget, the manufacture of farm machinery, may offer another CO_2 emission reducing strategy. Dyer and Desjardins (2006b) found that the depreciation, or lifespan, of tractors on most farms could only be roughly approximated. In spite of this difficulty, this same study also showed that the fossil CO_2 emissions associated with tractor manufacturing were highly sensitive to the period of depreciation. This suggests that the longer farmers can keep their older tractors running, the lower the source of fossil CO_2 emissions from manufacturing replacement machinery will be.

One of the five energy issues raised in the introduction of this chapter was food security. Global demand for most food products are very likely to increase over the next three or four decades (based on current population projections). Therefore, unlike many discretionary consumer products such as the size or number of cars, reducing food production is not a viable GHG emission mitigation strategy. However, some farm products have lower fossil CO_2 and GHG emission intensities than others. This suggests two things: first that food production is only effectively evaluated on the basis of GHG emission intensity, rather than total GHG emissions; second, different food products must be compared on the basis of GHG emission intensity.

Unfortunately, GHG emission intensities cannot be calculated without first calculating the actual GHG emission quantities. Although the fossil CO_2 and energy use estimates presented in Tables 4.3 and 4.4 are integral values, the land use areas presented in Table 4.2 allow the area based intensities for field crops to at least be approximated. No quick estimation of CO_2 emissions or energy use intensities from livestock (for Table 4.5) is offered in this chapter, for two reasons. First, such estimation is too complex to be achieved without actually running ULICEES to determine each LCC for 1996 and 2001. Second, such estimation would be of little policy value given the small role played by farm energy compared to methane and nitrous oxide emissions from livestock in defining the carbon footprint of livestock production in Canada (Vergé *et al.*, 2012). Two of the five energy issues discussed in the Introduction, GHG emission budgets and biofuels, also represent situations where energy use alone does not define the optimum policy decisions regarding GHG emissions.

The greatest limitation of the estimates presented in this chapter is for the question of finding CC adaptation measures. The scope of the estimates in Tables 4.3, 4.4 and 4.5 are to agronomic crops and the four most dominant types of livestock farms. Hence, only the most common and widespread agricultural systems in Canada fit the farm energy budget described in this chapter. For example, the 20 different farm types in Nova Scotia (Bailey *et al.*, 2008a) may help to explain why Bailey *et al.* (2008a) reported twice as much farm energy use in the Atlantic Provinces as shown in Table 4.4 (Bailey *et al.*, (2008b). As stated above, CC adaptation requires production systems that are better suited to future climates and more capable of supplying food to an expanded global population. Examples include field vegetables (including berries), vegetable greenhouses, orchards and vineyards, small meat animals (eg. rabbits) and possibly even aquaculture.

The energy implications for the increased use of irrigation need to be investigated since future climate variability may dictate more need of this practice throughout Canada. The increase in farm energy use in Nova Scotia between 1990 and 2004 reported by Bailey *et al.* (2008) was not apparent in Table 4.4, even though Nova Scotia has more agriculture than the other Atlantic Provinces. This may reflect the increasing dependence on irrigation in Nova Scotia, particularly for horticultural crops. Irrigation of these crops is also heavily relied upon in Ontario, Alberta and British Columbia (Statistics Canada, 2010). Hence, a term for operating irrigation systems should be included in the Canadian farm energy budget.

Since there is no assurance that this energy budget would apply to minor food production systems, CC adaptation will require that new data be collected from farms that fall under these minor farm type categories. New models for these farm types would be needed, as well. Part of the required policy ground work would be to investigate the array of possible CC adaptation candidates and then to assess their energy and GHG emission budgets. As with the current range of conventional field crops and livestock systems, it is important to determine the farm energy budgets, not just for their contribution to GHG emissions, but also to understand the energy needs and ensure that those needs would be met.

The biggest shortcoming of the energy budget presented in this chapter was the heavy dependence on 1996 survey data (CAEEDAC, 2001; Tremblay, 2000). A repeat of the FEUS on a broader scale would greatly enhance the ability of policy makers to predict energy consumption in Canada. This survey should extend beyond just the actual quantities of energy consumption and include investigation into the fossil energy related decisions made by farmers. This is particularly essential with respect to the selection, acquisition and use of farm machinery (which account for two of the six farm energy terms). As a matter of practical operation, a future console version of the Canadian farm energy budget will integrate the five non-field work energy terms with the F4E2 model.

Decisions regarding the management of farm buildings and farm owned vehicles for local transport (mainly, the distances involved) should also be part of the survey questions about energy use. Whereas an enhanced database of farm energy use would not replace the use of the sort of modeling methods described in this chapter, uncertainties about the accuracy of those models would definitely be reduced through an updated energy database.

The study of Canadian farm energy use described in this chapter, particularly the F4E2 model, offers some insight about how predictive energy use models can be applied. The console mode of operation of F4E2 is more precise than the meta-model mode. The console mode assumed that the operator of the model is a farmer (or is similarly familiar with farm operations) and can always provide the requested input. The meta-model mode cannot be run without data that are broadly available.

Conversely, the meta-model mode of operation can fulfill the need for extensive assessment (extrapolation over time and space) because of this computational simplicity. But another limitation of the meta-model mode is that most of the feedback or iterative computations that could be programmed into the original F4E2 program (Dyer and Desjardins, 2003a) cannot be used in the tabular structure of the meta-model, even when it is applied in a farm-specific situation (Dyer and Desjardins, 2005b). In spite of the limits on applying the console version of F4E2 (i.e. operating farm by farm), it can help to update the meta-model version. It can also be used to solicit information on operational experience from any farmers who can be recruited to run it and the appropriate safeguards are taken to protect their confidentiality. But the meta-model mode of the farm energy budget, as demonstrated by the data shown in Tables 4.3 to 4.5, which can also be presented as emission intensities will continue to be the most policy-relevant application of this information.

REFERENCES

ASAE: Agricultural machinery management data. Agricultural Engineering Yearbook, 1971. pp. 287–294. ASAE D230.2 American Society of Agricultural Engineers (ASAE). St. Joseph, MI, 1971.

ASAE: Agricultural machinery management data. Agricultural Engineering Yearbook, 1982 and 1984. pp. 213–220. ASAE: D230.3&4 American Society of Agricultural Engineers (ASAE). St. Joseph, MI, 1982/84.

ASAE: Agricultural machinery management data. ASAE Standards 2000: EP496.2, pp. 344–349 and D297.4 pp. 213–220. American Society of Agricultural Engineers (ASAE), St. Joseph, MI, 2000.

Bailey, J.A., Gordon, R., Burton, D. & Yiridoe, E.K.: Factors which influence Nova Scotia farmers in implementing energy efficiency and renewable energy measures. *Energy* 33 (2008a), pp. 1369–1377.

Bailey, J.A., Gordon, R., Burton, D. & Yiridoe, E.K.: Energy conservation on Nova Scotia farms: baseline energy data. *Energy* 33 (2008b), pp. 1144–1154.

Borenstein, S.: New figures: More of US at risk to sea level rise. The Associated Press, 14 March 2012, http://www.bostonglobe.com/news/nation/2012/03/14/new-figures-more-risk-sea-level-rise/bSFyWORB5GYrJkc7Gja64I/story.html (accessed May 2013).

CAEEDAC: A review of the 1996 Farm Energy Use Survey (FEUS). A Report to Natural Resources Canada (NRCan) by The Canadian Agricultural Energy End-Use Data Analysis Centre (CAEEDAC), 2001, http://www.usask.ca/agriculture/caedac/pubs/pindex.html (accessed April 2013).

CAP: Benchmark energy efficiency and carbon dioxide emissions. Canadian Ammonia Producers (CAP). Office of Energy Efficiency, Natural Resources Canada, Cat. No. M144-155/2007E-PDF Ottawa, Canada, 2008.

Desjardins, R.L., Vergé, X., Hutchinson, J., Smith, W., Grant, B., McConkey, B. & Worth, D.: Greenhouse gases. In: *Environmental sustainability of Canadian agriculture*. Agri-environmental indicator report series, Report # 2, Agriculture and Agri-food Canada, 2005, pp. 142–148.

Dolan, A.H., Smit, B., Skinner, M.W., Bradshaw, B. & Bryant, C.R.: Adaptation to climate change in agriculture: Evaluation of options 2001. Occasional Paper No. 26. Department of Geography, University of Guelph, Ontario, Canada, 2001, www.c-ciarn.uoguelph.ca/documents/reports (accessed July 2013).

Dyer, J.A. & Baier, W.: Weather based estimation of field workdays in fall. *Can. Agr. Eng.* 21(1979), pp. 119–122.

Dyer, J.A. & Baier, W.: Weather and farm fieldwork. *Can. Agr.* 25:1 (1980), pp. 26–28.

Dyer, J.A. & Desjardins, R.L.: Simulated farm fieldwork, energy consumption and related greenhouse gas emissions in Canada. *Biosyst. Eng.* 85:4 (2003a), pp. 503–513.

Dyer, J.A. & Desjardins, R.L.: The impact of farm machinery management on the greenhouse gas emissions from Canadian agriculture. *J. Sustain. Agr.* 20:3 (2003b), pp. 59–74.

Dyer, J.A. & Desjardins, R.L.: Analysis of trends in CO_2 emissions from fossil fuel use for farm fieldwork related to harvesting annual crops and hay, changing tillage practices and reduced summer-fallow in Canada. *J Sustain. Agr.* 25:3 (2005a), pp. 141–156.

Dyer, J.A. & Desjardins, R.L.: A simple meta-model for assessing the contribution of liquid fossil fuel for on-farm fieldwork to agricultural greenhouse gases in Canada. *J. Sustain. Agr.* 27:1 (2005b), pp. 71–90.

Dyer, J.A. & Desjardins, R.L.: Carbon dioxide emissions associated with the manufacturing of tractors and farm machinery in Canada. *Biosyst. Eng.* 93:1 (2006a), pp. 107–118.

Dyer, J.A. & Desjardins, R.L.: An integrated index for electrical energy use in Canadian agriculture with implications for greenhouse gas emissions. *Biosyst. Eng.* 95:3 (2006b), pp. 449–460.

Dyer, J. A. & Desjardins, R.L.: Energy-based GHG emissions from Canadian Agriculture. *J. Energy Institute* 80:2 (2007), pp. 93–95.

Dyer, J.A. & Desjardins, R.L.: A review and evaluation of fossil energy and carbon dioxide emissions in Canadian agriculture. *J. Sustain. Agr.* 33:2 (2009a), pp. 210–228.

Dyer, J.A. & Desjardins, R.L.: The contribution of farm energy use to past, present and future greenhouse gas emissions from Canadian agriculture. In: C.P. Vasser (ed): *The Kyoto Protocol: Economic assessments, implementation mechanisms, and policy implications*. Nova Science Publishers, Inc., Hauppauge, NY, 2009b.

Dyer, J.A. & Murray, D.R.: Planting and harvesting dates in Ontario. OMAF Factsheet, AGDEX 075, Order No. 85-079, Ontario Ministry of Agriculture and Food, Queens Park, Canada, 1985.

Dyer, J.A., Kulshreshtha, S.N. McConkey, B.G. & Desjardins, R.L.: An assessment of fossil fuel energy use and CO_2 emissions from farm field operations using a regional level crop and land use database for Canada. *Energy* 35:5 (2010a), pp. 2261–2269.

Dyer, J.A., Vergé, X.P.C., Desjardins, R.L., Worth, D.E. & McConkey, B.G.: The impact of increased biodiesel production on the greenhouse gas emissions from field crops in Canada. *Energy Sustain. Dev.* 14:2 (2010b), pp. 73–82.

Dyer, J.A., Vergé, X.P.C., Desjardins, R.L. & McConkey, B.G.: Implications of biofuel feedstock crops for the livestock feed industry in Canada. In: M.E Dos Santos Bernardes (ed): *Environmental impact of biofuels*. InTech Open Access Publisher, Rijeka, Croatia, 2011a, pp. 161–178.

Dyer, J.A., Desjardins, R.L., Karimi-Zindashty, Y. & McConkey, B.G.: Comparing fossil CO_2 emissions from vegetable greenhouses in Canada with CO_2 emissions from importing vegetables from the southern USA. *Energy Sustain. Dev.* 15:4 (2011b), pp. 451–459.

Dyer, J.A., Vergé, X.P.C., Kulshreshtha, S.N., Desjardins, R.L. & McConkey, B.G.: Residual crop areas and greenhouse gas emissions from feed and fodder crops that were not used in Canadian livestock production in 2001. *J. Sustain. Agr.* 35:7 (2011c), pp. 780–803.

Dyer, J.A., Desjardins, R.L., McConkey, B.G., Kulshreshtha, K. & Vergé, X.P.C.: Integration of farm fossil fuel use with local scale assessments of biofuel feedstock production in Canada. In: Z. Fang (ed): *Biofuel economy, environment and sustainability*. InTech Open Access Publisher, Rijeka, Croatia, ISBN 980-953-307-471-4. 270, 2013, pp. 97–122.

Edwards, W. & Boehlje, M.: Machinery selection considering timeliness losses. 1980 *T. ASAE* 23:4 (1980), pp. 810–815 & p. 821.

Elward, M., McLaughlin, B. &Alain, B.: Livestock feed requirements study 1999–2001. Catalogue No. 23–501-XIE, Statistics Canada, 2003.

Finner, M.F.: *Farm field machinery*. American Printing and Publishing Inc., Madison, WI and Agricultural Engineering, University of Wisconsin – Madison, WI, 1973.

Hannam, P.: Canada's soybean value chain. Soy20/20 Project (Peter Hannam, Chair), Guelph, Ontario, Canada, http://www.soy2020.ca/pdfs/canadas-soybean-value-chain.pdf. 2006 (accessed July 2013).

Hunt, D.R.: Selecting an economic power level for the big tractor. *T. ASAE* 15:3 (1972), pp. 414–415 & p. 419.

IEA: Biofuels for transport – An international perspective. International Energy Agency (IEA), Paris, France, 2004, http://www.iea.org/textbase/publications/free_new_Desc.asp?PUBS_ID=1262 (accessed September 2008).

IPCC: The Fourth Assessment Report of the Intergovernmental Panel on Climate Change, Summary for Policymakers. 2007, http://ipcc.ch/publications_and_data/ar4/wg1/en/contents.html (accessed May 2013).

Jacobs, C.O. & Harrell, W.R.: Basic principles of power. In: *Agricultural power and machinery*. McGraw Hill Book Co., New York, 1983, pp. 101–109.

Janzen, H.H., Desjardins, R.L., Asselin, J.M.R. & Grace, B. (eds): The health of our air – Toward sustainable agriculture in Canada. *Agriculture and Agri-Food Canada* Publ. No. 1981 E, 1999.

Janzen, H.H., Angers, D.A., Boehm, M., Bolinder, M., Desjardins, R.L., Dyer, J.A., Ellert, B.H., Gibb, D.J., Gregorich, E.G., Helgason, B.L., Lemke, R., Massé, D., McGinn, S.M., McAllister, T.A., Newlands, N., Pattey, E., Rochette, P., Smith, W., VandenBygaart, A.J., & Wang, H.: A proposed approach to estimate and reduce net greenhouse gas emissions from whole farms. *Can. J. Soil. Sci.* 86 (2006), pp. 410–418.

Jaques, A.P., Neitzert, F. & Boileau, P.: Trends in Canada's greenhouse gas emissions 1990–1995. Environment Canada Report Catalogue no. En49-5/5-8E, Air Pollution prevention Directorate, Pollution Data Branch, Environment Canada, Ottawa, Canada 1997.

Karman, D., Rowlands, D., Patterson, N. & Smith, M.: Technical and policy implications of transportation biofuel regulatory approaches. Final Report to: Natural Resources Canada. By: Carleton University, Ottawa, Canada, 2008.

Klein, K.K. & LeRoy, D.G.: The biofuels frenzy: What's in it for Canadian agriculture? Green Paper prepared for the Alberta Institute of Agrologists. Presented at the Annual Conference of Alberta Institute of Agrologists. Banf, Alberta, March 28, 2007. Department of Economics, University of Lethbridge, Lethbridge, Alberta, Canada, 2007.

Korol, M.: Canadian fertilizer, shipments and trade 2001/2002. Farm Income and Adaptations Policy Directorate, Agriculture and Agri-Food Canada. 2002, http://www.agr.ca/policy/cdnfert/text.html (accessed July 2013).

Kulshreshtha, S.N., Jenkins, B., Desjardins, R.L. & Giraldez, J.C.: A systems approach to estimation of greenhouse gas emissions from the Canadian agriculture and agri-food sector. *World Res. Rev.* 12 (2000), pp. 321–337.

LifeCycles: Calculating food miles. LifeCycles — based in Victoria, British Columbia, 2010, http://lifecyclesproject.ca/ (accessed May 2013).

Nagy, C.N.: Energy and greenhouse gas coefficients inputs used in agriculture. Report to the Prairie Adaptation Research Collaborative (PARC). The Canadian Agricultural Energy End-use Data and Analysis Centre (CAEEDAC) and the Centre for Studies in Agriculture, Law and the Environment (CSALE), 2000.

Neitzert, F., Olsen, K. & Collas, P.: Canada's greenhouse gas inventory – 1997-Emissions and removals with trends. Air Pollution Prevention Directorate, Environment Canada. Ottawa, Canada, 1999.

NRCan: Energy efficiency trends in Canada, 1990 and 1996–2003. Office of Energy Efficiency, Natural Resources Canada. Cat. No. M141-/2003, 2005.

OEE: Canadian ammonia producers benchmarking energy efficiency and carbon dioxide emissions. Office of Energy Efficiency (OEE), Natural Resources Canada, Ottawa ON, in co-operation with the Canadian Fertilizer Institute and the Canadian Industry Program for Energy Conservation, Canadian Industry Program for Energy Conservation, 2008.

Panzica, B.: Oil shifts ahead. Energy and Capital Newsletter, 6 January 2013, www.angelnexus.com (accessed January 2013).

Patzek, T.W.: Thermodynamics of the corn-ethanol biofuel cycle. *Crit. Rev. Plant Sci.* 23 (2004), pp. 519–567.

Pimentel, D.: Energy inputs in food crop production in developing and developed nations. *Energies* 2 (2009), pp. 1–24.

Pimentel, D. & Pimentel, M.H.: Ecological systems, natural resources, and food supplies. In: D. Pimentel & M.H. Pimentel (eds): *Food, energy, and society*. 3rd edn., CRC Press of the Taylor & Francis Group, Boca Raton, FL, 2008, pp. 21–34.

Pulse Canada: Canadian pulse research strategy: building the pulse industry through science – 2001/2002. Canadian Agri-food Research Council, Agriculture and Agri-food Canada, Pulse Research Canada. Saskatoon, SK, Canada, 2002, http://www.specialcrops.mb.ca/CanadianPulseResearchStrategy.pdf (accessed January 2013).

Searchinger, T., Heimlich, R., Houghton R.A., Dong, F., Elobeid, A., Fabiosa, J., Tokgoz, S., Hayes, D. & Yu, T.-H.: Use of U.S. croplands for biofuels increases greenhouse gases through emissions from land use change. *Science* 319 (2008), pp. 1238–1240.

Snyder, C.S., Bruulsema, T.W. & Jensen, T.L.: Green-house gas emissions from cropping systems and the influence of fertilizer management—a literature review. International Plant Nutrition Institute, Norcross, GA, 2007, http://www.ipni.net/ghgreview (accessed September 2013).

Statistics Canada: The financial picture of farms in Canada. 2006 Census of Agriculture, 2009, http://www.statcan.gc.ca/ca-ra2006/articles/finpicture-portrait-eng.htm#A1 (accessed May 2013).

Statistics Canada: Agricultural water use in Canada. Catalogue no. 16-402-X, 2010, http://www.gov.mb. ca/waterstewardship/licensing/wlb/pdf/water_statistics/sc_agricult ural_water_use_in_canada%20-%20 2010.pdf (accessed May 2013).

Stout, B.A.: *Energy use and management in agriculture*. Bretonne Publishers North Scituate, MA, 1984, pp. 80–85.

TEE: The 9 billion-people question – A special report on feeding the world. February 26, 2012. The Economist Editorial (TEE). 15 pp. Available from The Rights and Syndication Dept., http://www.economist.com.rights, 2011.

Tremblay, V.: The 1997 farm energy use survey – Statistical results. Contract Report prepared for The Office of Energy Use Efficiency, Natural Resources Canada, Ottawa, Ontario, Canada, and The Farm Financial Programs Branch, Agriculture and Agri-Food Canada, Ottawa, Ontario, Canada 2000.

Wu, Z., Kjelgaard, W.L. & Persson, P.E.: Machine widths for time and fuel efficiency. *T. ASAE* 27:6 (1986), pp. 1508–1513.

Vergé, X.P.C., Dyer, J.A., Worth, D.E., Smith, W.N., Desjardins, R.L. & McConkey, B.G.: A greenhouse gas and soil carbon model for estimating the carbon footprint of livestock production in Canada. *Animals* 2 (2012), pp. 437–454.

Xureb, M.: Foodmiles: environmental implication of food imports to Waterloo region. Region of Waterloo Public Health, 2005, http://chd.region.waterloo.on.ca/en/researchResourcesPublications/resources/ FoodMiles_Report.pdf. (accessed May 2013).

CHAPTER 5

Energy efficiency technologies for sustainable agriculture and food processing

Lijun Wang

5.1 INTRODUCTION

Because of the increasing energy prices and efforts for the reduction of greenhouse gas emissions, it has become significant to improve the energy efficiency, replace the existing energy intensive operations with new energy efficient ones, and increase the use of renewable energy in the agricultural and food industry (Wang, 2008). The improvement of energy efficiency in the agricultural and food industry should not only be considered to provide economic benefit but also provide benefits for environmental protection, social sustainability, energy supply security and industrial competitiveness. This chapter started with the examination of the energy use in the agricultural production and food processing facilities. The energy-saving opportunities were then identified and energy conservation measures were discussed.

Various emerging technologies including novel thermodynamic cycles (Kuzgunkaya and Hepbasli, 2007; Ozyurt et al., 2004), non-thermal food processes (Brown et al., 2007; Toepfl et al., 2006) and novel heating methods (Nguyen et al., 2013; Yang et al., 2010) have been developed to replace conventional energy intensive systems used in the agricultural and food industry. These emerging technologies provide another potential to reduce energy consumption, reduce production costs and improve the sustainability of agricultural production and food processing. This chapter provided an over review of the working principles, applications, and energy efficiency of these emerging technologies.

5.2 ENERGY CONSUMPTION IN THE AGRICULTURAL PRODUCTION AND FOOD PROCESSING

5.2.1 *Energy consumption in the agricultural production*

The agricultural production requires large amounts of energy inputs for fertilization, mechanization, irrigation, crop propagation and herbicides. Alluvione et al. (2011) analyzed the energy inputs for wheat, maize and soybean using different cropping systems, which are given in Table 5.1. As shown in Table 5.1, the average energy inputs for the production of wheat, maize and soybean using the traditional cropping methods are 15.4, 29.7 and 14.3 GJ/ha, respectively. The fertilization and mechanization are two dominant energy users in the traditional cropping system. The amounts of energy for fertilizers are 58.4, 59.6 and 25.2% of the total energy inputs for the production of wheat, maize and soybean, respectively. The amounts of energy for mechanization are 34.4, 22.2 and 42.0% of the total energy inputs for wheat, maize and soybean, respectively (Alluvione et al., 2011). Diesel fuel is about 93% of the mechanization energy. About 55% of the total mechanization energy is used for seedbed preparation and seeding (Alluvione et al., 2011). Among the three crops of wheat, maize and soybean, the energy input for maize was double those of wheat and soybean (Alluvione et al., 2011).

The total energy outputs from crops include the energy stored in grain, straw and root. The energy outputs from the wheat, maize and soybean are given in Table 5.2. For the traditional

Table 5.1.   Energy inputs of different crops in different cropping systems (adapted from Alluvione *et al.*, 2011).

Crop	Inputs	Energy inputs for different cropping system [GJ/ha]			
		Conventional	Low inputs	Integrated	Average
Wheat	N	6.4	4.8	4.8	5.4
	P	1.1	0.0	0.0	0.4
	K	1.5	0.0	1.4	1.0
	Total fertilization	9.0	4.8	6.2	6.7
	Mechanization	5.3	4.6	5.3	5.1
	Irrigation	0.0	0.0	0.0	0.0
	Crop propagation	0.9	0.9	0.9	0.9
	Herbicides	0.2	0.2	0.2	0.2
	Total	**15.4**	**10.5**	**12.7**	**12.9**
Maize	N	14.6	8.8	9.1	10.8
	P	1.0	0.4	0.0	0.5
	K	2.2	1.6	1.4	1.7
	Total fertilization	17.7	10.8	10.5	13.0
	Mechanization	6.6	4.9	6.5	6.0
	Irrigation	4.1	4.1	4.1	4.2
	Crop propagation	0.3	0.3	0.3	0.3
	Herbicides	0.7	0.1	0.4	0.4
	Total	**29.7**	**20.3**	**21.9**	**23.9**
Soybean	N	1.3	0.0	0.0	0.4
	P	0.9	0.0	0.0	0.3
	K	1.4	1.2	1.4	1.3
	Total fertilization	3.6	1.2	1.4	2.1
	Mechanization	6.0	4.0	5.7	5.2
	Irrigation	3.1	3.1	3.1	3.1
	Crop propagation	0.6	0.6	0.6	0.6
	Herbicides	1.0	1.0	0.8	0.9
	Total	**14.3**	**9.9**	**11.6**	**12.0**

Table 5.2.   Energy outputs of different crops in different cropping systems (adapted from Alluvione *et al.*, 2011).

Crop	Outputs	Energy outputs for different cropping system [GJ/ha]			
		Conventional	Low inputs	Integrated	Average
Wheat	Grain	96.3	96.8	102.1	98.4
	Straw	93.5	102.1	94.6	96.7
	Root	78.7	63.7	73.0	71.8
	Total	**268.5**	**262.5**	**269.7**	**266.9**
Maize	Grain	218.5	201.6	211.3	210.5
	Straw	189.5	171.6	184.6	181.9
	Root	99.3	69.6	49.2	72.7
	Total	**507.2**	**442.8**	**445.2**	**465.1**
Soybean	Grain	65.6	70.7	75.9	70.7
	Straw	76.3	65.4	75.8	72.5
	Root	86.5	81.3	61.9	76.5
	Total	**228.4**	**217.4**	**213.6**	**219.8**

cropping method, the total energy outputs of wheat, maize and soybean are 268.5, 507.2 and 228.4 GJ/ha, respectively. Among the three crops, the total energy output for maize was also double those of wheat and soybean (Alluvione *et al.*, 2011). The energy outputs of grains are 96.3 GJ/ha for wheat, 218.5 GJ/ha for maize and 65.6 GJ/ha for soybean (Alluvione *et al.*, 2011). The ratios between the energy outputs in grains and the energy inputs are 6.25 for wheat, 7.36 for maize and 4.59 for soybean.

In developing countries, the agricultural production systems have changed dramatically and the use of mechanization, chemical fertilizers and pesticides increased substantially with the modernization of agricultural production. The energy consumption of Chinese agriculture had increased by 57.59% from 1991 to 2008 (Lu *et al.*, 2011). The energy input in the Turkish agriculture increased by 133% from 19.6 GJ/ha in 1975 to 45.7 GJ/ha in 2000. The shares of energy inputs for fertilizers and mechanization in the total energy use in 2000 were 50.9 and 38.8%, respectively. The energy inputs for the supply of fertilizers increased by 2.5 times from 1975 to 2000 in the Turkish agriculture. The share of mechanical power used for mechanization in Turkish agriculture increased from 14% in 1975 to 57% in 2000. The increase in mechanical energy increases the consumption of diesel and electricity. On the other hand, the increasing use of fertilizers increased the yields of crops. The total energy output increased from 27.1 GJ/ha in 1975 to 39.1 GJ/ha in the Turkish agriculture in 2000 (Hatirli *et al.*, 2005).

5.2.2   Energy consumption in the food industry

5.2.2.1   Overview of energy consumption in the food industry

Food processes utilize significant amounts of labor, machinery and energy to convert edible raw materials into higher-value food products. In the United States, the food processing industry is one of the largest manufacturers. The shipment value from the food industry was increased from US\$ 538 billion in 2006 to U\$ 646 billion in 2010. The shipment value from the food industry was about 10.72 and 13.15% of the total shipment value from all manufacturing industries in the United States in 2006 and 2010, respectively. The food industry is also a large energy consumer. The total cost for purchasing fuels and electricity in the food industry was US\$ 9.92 billion in 2006, which was 9.57% of the total energy costs for all manufacturing industries (US Census Bureau, 2010). In the European Union, the food and tobacco sector, on average, accounted for about 8% of the total energy demand in the whole manufacturing sector in 2001. In The Netherlands, the food and tobacco sector accounted for about 9% of the total industrial energy demand and 23% of the industrial value added (Ramirez *et al.*, 2006a).

There is a large variation in the shares of commercial energy going into the food production system in developing countries. In some developing countries, the energy consumed in the food processing industry could be very high. In Thailand, the food and beverage manufacturing sector accounted for about 30% of the total energy consumption and generated 16% of the total added value in all manufacturing sectors in 2000 (Bhattacharyya and Ussanarassamee, 2004).

The current energy cost is only about 2% of the total production costs in the food industry in the developed countries (Ramirez *et al.*, 2006a). However, the energy indicator based on economic values, which was defined as the total costs for the supply of energy divided by the total shipment value of food products, was increased from 0.78 cents energy cost/dollar shipment value in 2002 to 1.84 cents energy cost/dollar shipment value in the United States in 2006. The energy indicator for the food industry was increased 2.36 times from 2002 to 2006 while the energy indicator for all manufacturing sectors was increased 2.1 times during the same period (US Census Bureau 2010). It consumed 2.74 MJ of energy for each dollar shipment value in the food industry of the United States in 2002 (US EPA, 2007). Ramirez *et al.* (2006b) found that the energy indicator, which was based on energy consumption in MJ per ton of finished product in the meat industry of four European countries including France, Germany, Netherlands and United Kingdom was increased from 14 to 48% in the 1990s. The possible reason for this increase was the stronger hygiene regulations and increased processed products such as frozen and cut meat products to meet the changes of consumer preferences. Generally, the increase of the energy indicator based

Table 5.3.   Energy use and indicator in different food manufacturing sectors in the United States in 2006 (US Census Bureau, 2006).

Manufacturing sector	Shipment value [Million US$]	Total energy cost [Million US$]	Energy indicator [cents energy cost/ $ shipment value]	Percent of electricity cost among the total energy cost [%]	Percent of total energy cost in the whole industry [%]
3111 Animal food manufacturing	33988	522	1.54	47.1	5.3
3112 Grain and oilseed milling	57667	2198	3.81	37.0	22.3
3113 Sugar and confectionery product manufacturing	28225	577	2.04	36.6	5.8
3114 Fruit and vegetable preserving and specialty food manufacturing	56279	1302	2.31	44.3	13.1
3115 Dairy product manufacturing	75428	1184	1.57	52.4	11.9
3116 Animal slaughtering and processing	149577	2055	1.37	53.4	20.7
3117 Seafood product preparation and packaging	10849	246	2.27	45.5	2.5
3118 Bakeries and tortilla manufacturing	54173	911	1.68	54.7	9.2
3119 Other food manufacturing	71602	926	1.29	52.9	9.3

on the economic values in the food industry could be partially caused by the increase in the prices of major energy sources and tougher government regulations. The changes in food products and processes might also contribute to the increase of the energy indicator.

5.2.2.2   Energy use in different food manufacturing sectors

There are six main food processing sectors in terms of food products, which include (i) grains and oilseeds milling, (ii) sugar and confectionary processing, (iii) fruits and vegetables processing, (iv) dairy processing, (v) meat processing and (vi) bakery processing. In the United States, the meat manufacturing sector is the largest individual sector in the food industry in terms of shipment value followed by grain and oilseed milling, fruit and vegetable, and bakeries and tortilla. The grain and oilseed milling, and the meat manufacturing are two large energy consumers in the food industry as shown in Table 5.3, which consume 22.3 and 20.7% of the total energy input into the whole food industry. Different food manufacturing sectors have different energy indicators. In the United States, the grain and oilseed milling has the highest energy indicator while the meat manufacturing sector has the lowest energy indicator as shown in Table 5.3. It required 3.81 cents to generate 1 dollar of shipment value in the grain and oilseed milling sector, compared to 1.37 cents of energy cost per dollar shipment value in the meat manufacturing sector in the United States in 2006 (Wang, 2008).

However, in other developed countries, the distribution of the energy use in different food sectors may be different. Each of the manufacturing sectors of (i) fruits vegetable, (ii) meat products, (iii) vegetable oil and animal fat, (iv) dairy products, (v) prepared animal feeds, and (vi) grain mill, starches and starch products consumed about 10–15% of the total energy input into the whole food industry in The Netherlands in 2001 (Ramirez et al., 2006a).

In the United States, although the purchased electricity is only about 20% of the total consumed energy, the food industry spent nearly 47% of its energy expense on average to purchase electricity in 2006 because the electricity price per unit of energy content is higher than that of other energy sources. The bakeries and tortilla manufacturing, meat manufacturing and dairy product manufacturing sectors spent a little more than 50% of their energy expenses on purchased electricity.

The costs of purchased electricity in the grain and oilseed milling, and sugar and confectionery sectors were only 37% of their total energy expenses in 2006. The grain and oilseed milling, and sugar and confectionery sectors require high boiler usage. Within the food industry, the dairy product manufacturing, and the meat manufacturing sectors particularly have a high electricity demand while the sugar sector particularly has a high fuel demand (Fritzson and Berntsson, 2006).

5.2.2.3   *Energy use for production of different food products*

The amount of energy used for the production of a given amount of products highly depends on the type of the products. Table 5.4 gives the energy use for producing different food products in The Netherlands in 2001 (Ramirez *et al.*, 2006a). The ranking of food commodities in energy consumption does not follow the same pattern as the volumes processed. It consumed 9385 and 9870 MJ heat to produce each ton of milk powder and whey powder, respectively while production of one ton pasta consumes only 2 MJ of heat. Production of wheat starch consumes 2960 MJ of electricity per ton of product while production of beet pulp consumes only 5 MJ electricity per ton of product. The food processing involves different unit operations such as drying and cooling. The energy consumption for the same operations may vary significantly as the consequence of the type of equipment used, operating practice, ambient temperature, local infra-structure, and the skills of staffs. Drying and freezing consume large amounts of energy. On average, a drying process consumes 6 MJ energy to remove 1 kg of water from products and a freezing process consumes 1 MJ (or 0.3 kWh of electricity) to process 1 kg of food products at $-20°C$.

5.2.3   *Energy sources in the agricultural and food industry*

5.2.3.1   *Energy sources for agricultural production*

The energy inputs for agricultural production include the energy for mechanization and agronomic supplies. On average, the energy inputs for fertilizers, mechanization and irrigation for the production of wheat, maize and soybean using different cropping methods are 47.1, 31.8 and 15.3%, respectively (Alluvione *et al.*, 2011). Farming machinery such as tractor, herbicide sprayer, fertilizer spreader, stalk shredder and baler are widely used in the modern agricultural industry. The field operations of the machinery require diesel fuel and lubricant oil. Direct diesel fuel consumption represents the majority of energy inputs for mechanization and irrigation, which are 92.7 and 96.1% of their total energy used in mechanization and irrigation, respectively (Alluvione *et al.*, 2011). The amounts of energy required for the production of N, P and K fertilizers depend on the technology and the composition of raw materials. The average energy inputs for the production of urea, ammonium nitrate, tripe superphosphate and muriate of potash are 48.5 MJ/kg N, 39.4 MJ/kg N, 16.4 MJ/kg P and 6.3 MJ/kg K, respectively in Europe (Alluvione *et al.*, 2011).

Figure 5.1 shows the energy consumption and energy sources used by the Chinese agriculture from 1991 and 2008 (Lu *et al.*, 201). Petroleum, coal and electricity were three major energy sources used in the Chinese agriculture. As shown in Figure 5.1, the petroleum was the dominant energy source, which was about half of the total energy used in the Chinese agriculture. Electricity only accounted for a small portion of the total energy used in the Chinese agriculture.

5.2.3.2   *Energy sources for food processing*

The main energy sources used in the food manufacturing industry include petroleum, natural gas, coal, renewable energy and electricity. Figure 5.2 shows the delivered energy consumption in terms of the types of fuels in the food processing industry of the United States in 2002 (US EPA, 2007). The natural gas and electricity were two main energy sources used in the food industry of the United States, which were 52 and 21% of the total energy consumed, respectively in 2002. While natural gas remains the largest energy source, consumption of renewable energy sources is projected to grow most rapidly as the food industry becomes more proficient in recovering and utilizing process and agricultural wastes. It should be noted that the food manufacturing industry produces about 9% of the electricity with its onsite power systems, 95% of which are the heat and power cogeneration systems (US EPA, 2007).

Table 5.4. Energy use for production of different food products in The Netherlands in 2001 (adapted from Ramirez *et al.*, 2006a).

Product	Specific electricity consumption	Specific fuels and heat consumption	Unit
Meat sector:			
Beef and sheep	341	537	MJ/ton dress carcass weight
Pig	465	932	MJ/ton dress carcass weight
Poultry	1008	576	MJ/ton dress carcass weight
Processed meat	750	3950	MJ/ton product
Rendering	234	1042	MJ/ton raw material
Fish sector:			
Fresh fillets	129	6	MJ/ton product
Frozen fish	608	6	MJ/ton product
Prepared and preserved fish	482	1062	MJ/ton product
Smoked and dried fish	12	2077	MJ/ton product
Fish meal	684	6200	MJ/ton product
Fruits and vegetables:			
Potatoes product	5722		MJ/ton product
Un-concentrated juice	250	900	MJ/ton product
Tomato juice	125	4789	MJ/ton product
Frozen vegetables and fruits	738	1800	MJ/ton product
Preserved mushrooms	2898		MJ/ton product
Vegetables preserved by vinegar	2178		MJ/ton product
Tomato ketchup	380	1700	MJ/ton product
Jams and marmalade	490	1500	MJ/ton product
Dried vegetables and fruits	1500	4500	MJ/ton product
Crude and refined oil	672		MJ/ton product
Dairy products:			
Milk and fermented products	241	524	MJ/ton product
Butter	457	1285	MJ/ton product
Milk powder	1051	9385	MJ/ton product
Condensed milk	295	1936	MJ/ton product
Cheese	1206	2113	MJ/ton product
Casein and lactose	918	4120	MJ/ton product
Whey powder	1138	9870	MJ/ton product
Starch and starch products:			
Wheat starch	2960	8800	MJ/ton product
Maize starch	1000	2331	MJ/ton product
Potato starch	1425	3564	MJ/ton product
Prepared animal feeds:			
For farm animals	475		MJ/ton product
For pets	2306		MJ/ton product
Sugar:			
Refined sugar	555	5320	MJ/ton product
Beet pulp	5	1820	MJ/ton product
Other products:			
Sweet biscuits	4581		MJ/ton product
Waffles and wafers	3195		MJ/ton product
Soup and broths	7659		MJ/ton product
Pasta	648	2	MJ/ton product
Flour	420	30	MJ/ton product
Cacao beans	6384		MJ/ton product
Non roasted coffee	141	1597	MJ/ton product
Roasted coffee	518	1997	MJ/ton product
Extracts of coffee solid form	15675		MJ/ton product
Beer	19.5	153	MJ/hL product
Mineral water and soft drinks	133	199	MJ/1000 L product
Unsweetened water and soft drinks	120	360	MJ/1000 L product

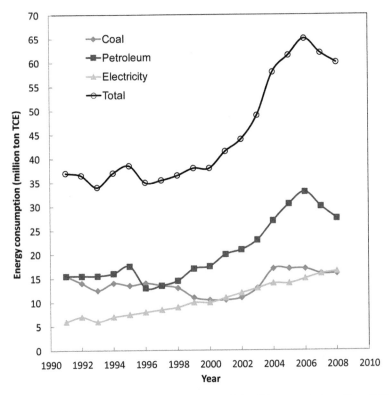

Figure 5.1.   The energy consumption and energy sources of the Chinese agriculture from 1991 to 2008. (Adapted from Lu *et al.*, 2011).

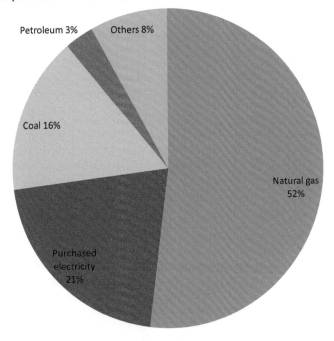

Figure 5.2.   Delivered energy consumption by the type of fuels in the food manufacturing industry in the United State in 2002 (US EPA, 2007).

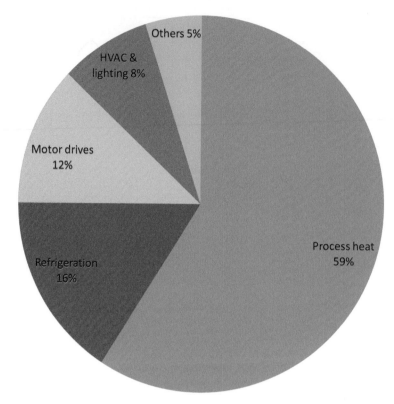

Figure 5.3.　Energy consumption by the end users (adapted from Wang, 2008).

　　The end users of energy in the food industry are process heating, process cooling and refrigeration, machine drive, and miscellaneous users. About half of all energy input is used to process raw materials into products. Fuels are mainly used for process heat and space heating while electricity is used for refrigeration, motor drives and automation. Figure 5.3 gives the energy consumption by the end users. Process heat for thermal processing and dehydration consumes approximately 59% of total energy in the whole food industry. Steam is one of the most important processing medium in food processing facilities. Boiler fuel is nearly one-third of the total energy consumption. Ovens and furnaces are also widely used in thermal processes. Refrigeration uses approximately 16% of total energy in the whole food industry. Motor drives represent 12% of the total energy use. The non-process uses including space heating, venting, air conditioning, lighting and onsite transport only consumes about 8% of the total energy (Okos *et al.*, 1998).

5.2.4　*Energy efficiency in agricultural production and food processing*

There are thermodynamic losses of energy during agricultural production and food processing. The first law of thermodynamics is usually used to trace the flow of energy through a relevant industrial system to determine the primary energy inputs needed to produce a given amount of a product. The energy efficiency is defined as the efficiency based on the first law. However, the analysis based on the first law of thermodynamics does not take into account of the type of an energy source in terms of its thermodynamic quality. The second law of thermodynamic is used to assess the maximum amount of work achievable in a given system with different energy sources, which is termed as the exergy. The exergy efficiency is defined as the efficiency based on the second law (Wang, 2008). Figure 5.4 gives the comparison of the energy and exergy efficiencies

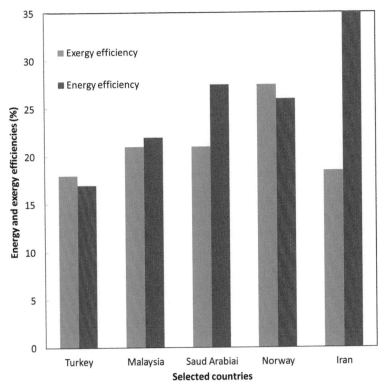

Figure 5.4.   Comparison of energy and exergy efficiencies of the agricultural sectors in different countries (adapted from Ahamed *et al.*, 2011).

of the machineries used in the agricultural sectors of different countries. The energy and exergy efficiencies of the machineries used in the agricultural sectors of different countries were from 17 to 34%, and from 18 to 28%, respectively, depending the types of machineries used in the agricultural sectors. For example, both electrical and diesel engine pumps are used for irrigation. The energy efficiency of an electrical pump is 90% compared to 22% for a diesel engine. However, the exergy efficiency of an electrical pump is only 4.53% compared to 22% for a diesel engine (Ahamed *et al.*, 2011).

Fischer *et al.* (2007) reported that around 57% of the primary energy inputs into the whole industry are lost before reaching intended process activities. Estimates from several studies indicate that on average, savings of 20 to 30% energy can be achieved without capital investment, using only procedural and behavioral changes. Industrial energy consumption can be further cost-effectively reduced by 10 to 20% through well-structured energy management programs that combine energy conservation technologies, operation practices and management practices.

5.3   ENERGY CONSERVATION AND MANAGEMENT IN AGRICULTURAL PRODUCTION AND FOOD PROCESSING

5.3.1   *Energy conservation in agricultural production*

The greenhouse gas (GHG) emissions from the agricultural industry are in the forms of nitrous oxide (N_2O), methane (CH_4) and carbon dioxide (CO_2). The global warming potentials (GWPs) of N_2O and CH_4 are 310 CO_2-equivalent and 25 $CO_{2\text{-eq}}$, respectively. It was estimated that the total agricultural emissions in Danish agriculture accounted for about 24% of the total national GHG emissions. About 63% of the annual GHG emissions in the Danish agriculture in 2010 were from N_2O and CH_4 while the CO_2 emissions from direct and indirect fossil energy use only accounted for 31% of the total emissions. The remaining 6% GHG emission came from the CO_2 emission due to the mining of the soil carbon pools (Dalgaard *et al.*, 2011). Therefore, agricultural mitigation efforts should not only be directed towards substituting fossil fuel with bioenergy but also at substantially reducing methane and nitrous oxide emissions and maintaining soil carbon stocks (Dalgaard *et al.*, 2011).

The increase of agricultural sustainability requires to reduce the agronomic inputs through the conservation tillage, use of disease resistant cultivars, rational use of pesticides and target application of nutrients and to diversify crops. Better manure and fertilizer handling, fodder optimization and optimized cropping and land use can mitigate GHG emissions in the form of N_2O, CH_4 and CO_2 from soil carbon pools (Dalgaard *et al.*, 2011). Balancing N fertilization with actual crop requirements and adopting minimum tillage could reduce the energy consumption of cropping systems by 64.7 and 11.2%, respectively. These two methods have been proved the most efficient techniques to reduce energy consumption in the agricultural production (Alluvione *et al.*, 2011).

The low-input cropping and integrated farming have been introduced to reduce the energy use in the agricultural production. The low input cropping method is to reduce the energy inputs and environmental impacts through minimum tillage, calibration of fertilizer rates on crop off-take, post-emergence herbicides use at low rates and nutrient leaching prevention using winter catch crops. The use of conservation tillage, which can reduce energy use, soil erosion, leaching and runoff of agricultural chemicals and carbon emissions, has been increasing in response to the growing concerns about the impact of agricultural production on the environment (Liu *et al.*, 2013; Plaza *et al.*, 2013). The integrated farming method is to reduce energy inputs and environmental impacts by modulating mineral fertilization and selecting pesticide active ingredients and doses (Alluvione *et al.*, 2011).

The low-input cropping and integrated farming systems can significantly reduce the energy inputs. Compared to the traditional cropping method, the low-input cropping method can reduce the energy inputs by around 31% on average for producing wheat, maize and soybean. The integrated farming method can reduce the energy inputs by 17.5% for wheat, 26.3% for maize and 18.9% for soybean (Alluvione *et al.*, 2011). As shown in Table 5.2, there was no significant difference in total energy outputs for the three crops using the energy efficient cropping methods, compared with those using the traditional cropping method. The energy outputs of wheat and soybean grains using the energy efficient cropping methods were even higher than those using the traditional method. The energy output of maize grain using the energy efficient cropping methods was a slight lower than that using the traditional method.

5.3.2   *Energy conservation in the utilities in food processing facilities*

The prospective energy efficiency improvement measures to be used in the food industry should be technically feasible and economically practical. The improvement of energy efficiency in food processing facilities requires the evaluation of numerous prospective energy conservation measures to increase the energy usage efficiencies of the utilities such as steam, compressed air and electricity.

5.3.2.1   *Energy savings in steam supply*
Most of food processing facilities have large steam and/or hot water demands. Boilers are the largest fuel user to provide steam as process heat for different unit operations such as sterilization, pasteurization, evaporation and dehydration in the food industry. Boilers consume about one third of the total energy use or more than half (50%) of the fuel use (Einstein *et al.*, 2001). Considerable amounts of energy are lost from stack flue gas, blowdown water, steam leaks and poor surface insulation during steam generation and distribution. The savings can easily represent

approximately 20% of total energy use in the steam systems. These savings can also substantially reduce the carbon dioxide emissions (Wang, 2008). The energy savings for a boiler system can be divided into two categories: design and operation optimization, and waste heat recovery. The optimization may include:

- The proper size of a boiler;
- The proper pressure and temperature of steam;
- Optimal amount of excess air; and
- Optimal amount of blowdowns.

Stack heat recovery systems can improve boiler efficiency by as much as 15%. If right equipment is used, up to 78% of the heat stored in blowdown water can be recovered. The heat recovery from blowdown water was considered to save 1.3% of boiler fuel use for small boilers with a payback of 2.7 years (Einstein *et al.*, 2001). Energy can also be saved from the steam distribution system through the steam trap maintenance, condensate recovery, repairing steam leaks and insulation (Wang, 2008).

5.3.2.2   *Energy savings in compressed air supply*

Compressed air is another important processing medium for conveying foods and process control in food processing facilities. Example users of compressed air in a food processing facility are cooking retorts, wrapping machines, color sorters, lid machines, open blowing for water removal and drying of cans prior to labeling, and pneumatic controls. The production of compressed air can be one of the most expensive processes in manufacturing facilities. Proper improvements to compressors and compressed air systems can save 20–50% of the energy consumed by the systems (Mull, 2001). The energy conservation technologies include the use of high-efficiency and variable speed motors, reduction of inlet air temperature, use of cooling or waste heat recovery unit for compressors, reduction of air leaks along the air distribution line, reduction of air pressure and use of localized air delivery system (Wang, 2008).

5.3.2.3   *Energy savings in power supply*

Motor drives and refrigerators are two large electricity users in the food industry, which consumes about 48% and 25% of the total electricity use, respectively (Okos *et al.*, 1998). However, in the meat sector, refrigeration constitutes between 40% and 90% of total electricity use during production time and almost 100% during non-production time (Ramiez *et al.*, 2006b). The mover of a mechanical compression refrigerator is a motor or compressor. Pumps and pumping represent one of the most important aspects of nearly all food processing operations. Pumps are a mechanical system of moving liquid and semi-liquid products in or around a food processing facility. Motors are needed to drive a pump. Cost savings can be achieved by controlling the peak demand, improving the power factor, and reducing the electricity consumption as well. Since the energy loss in a motor is in the range of 5–30% of the input power, it is important to consider the energy conservation technologies for motors. Most motors operate in a fashion that requires both real power due to the presence of resistance and reactive power due to the presence of inductance in the motors. Increase of power factors should be considered for improving electrical efficiency and reducing the energy costs of motors. Motors are designed to operate most efficiently under their rated loads. Therefore, it is an effective way to conserve energy by matching the required loads with the rated loads of motors. The selection of a high power factor and variable speed motor may result in significant savings when purchasing a new motor (Wang, 2008).

5.3.2.4   *Energy savings in heat exchanger*

In food processing facilities, many unit operations such as refrigeration, freezing, thermal sterilization, drying and evaporation involve the transfer of heat between food products and heating or cooling media. Heating and cooling of foods is achieved in equipment of heat exchangers. Heat exchangers also play a key role in waste heat recovery. Several energy conservation technologies

Table 5.5.   Summary of some energy savings identified in a Nestle factory (adapted from Muller *et al.*, 2007).

Measure	Energy type	Energy saving [MWh/year]	Estimated payback [year]
Replacing compressed air usage by dedicated blower	Electricity	166	2
Regulation of HVAC	Electricity	80	negligible
Removing stand-by of air compressors with a VSD unit	Electricity	69	23
Fixed compressed air leakage	Electricity	50	negligible
Insulating pipes of high temperature condensate return	Fuels	338	1.5
Vacuum production in dryer	Fuels	150	1
Regulation of steam user	Fuels	50	negligible

including heat transfer enhancement, fouling removal, optimization of heat exchanger design, and optimization of heat exchanger network have been used to improve the energy efficiency of heat exchangers (Wang, 2008).

5.3.2.5 *Energy savings by recovering waste heat*

Any processing air, vapor and water effluent streams above the ambient temperature may be an energy source. Boiler flue gas, boiler blowdown water, steam condensate, exhaust gas from dryers and ovens, cooling air and water from air compressor and large motors, and vapor from cookers are the examples of waste heat sources. It has been estimated that up to 50% of the energy consumed in the United States was discharged as waste heat to the environment (Mull, 2001). By recirculation and recovery of waste heat, the energy consumption of food processing facilities could be cut by 40%. Recovery of waste heat can not only save money due to the reduced energy consumption and capacity requirement for energy conversion equipment, but also prevent thermal pollution to environment. The economic benefits of waste heat recovery include the reduction of energy costs for purchased fuels and capital costs for energy conversion equipment with less capacity. However, the economics of a waste heat recovery system depends on utilization, quantity and quality of the recovered waste heat, and the heat transfer equipment for waste heat recovery (Wang, 2008). As heat with end-use temperatures below 200°C is usually the dominant thermal energy demand in the food industry, part of the thermal energy demand can be easily supplied by the waste heat recovered from a higher temperature source nearby if available.

Table 5.5 is an example to show the potential energy savings identified for a Nestle factory in Switzerland (Muller *et al.*, 2007). In the Nestle facility, insulating of high-temperature condensate return pipes and using a low-pressure air blower to replace the use of high-pressure compressed air for the sealing operation of process units can save 1127 GJ/year (or 338 MWh/year) fuels and 166 MWh/year electricity, respectively.

5.3.3 *Energy conservation in energy-intensive unit operations of food processes*

5.3.3.1 *Energy savings in thermal food processing*

Thermal processes such as pasteurization and sterilization, chilling and freezing, and evaporation and drying are energy intensive unit operations used in the food industry for food preservation and safety. It is a real need for food manufacturers to combine food safety and quality with energy conservation. The energy saving opportunities for each unit operation include three aspects: (i) improvement of energy efficiency for existing units, (ii) replacement of energy-intensive units with novel units, and (iii) use of renewable energy sources, particularly food processing wastes.

Thermal pasteurization of liquid foods such as milk and fruit juices is a well-established and effective means of terminal decontamination and disinfection of these products. Thermal processing is also an important method of food preservation in the manufacturing of canned foods and retortable pouches. The basic function of a thermal process is to inactivate pathogenic and food spoilage causing microorganisms in foods. Simpson *et al.* (2006) developed a mathematical model to estimate total and transient energy consumption during heating of retortable shelf-stable foods. According to the results from Simpson *et al.* (2006), retort insulation can reduce 15–25% of current energy consumption depending on selected conditions. Furthermore, in batch retort operations, maximum energy demand occurs at the venting step that only lasts for the first few minutes of the process cycle while very little energy is needed thereafter to maintain the process temperature. The increase in the initial temperature of food products can reduce the peak energy demand in the order of 25–35%. Also it is customary to operate the retorts in a staggered schedule so that no more than one retort is vented at any one time. Thus operating practice can also reduce the peak energy demand during retorting.

5.3.3.2   *Energy savings in concentration, dehydration and drying*
Concentration, dehydration and drying are a common unit operation in food processing facility to lower the moisture content of foods in order to reduce water activity and prevent spoilage or reduce the weight and the volume of food products for transport and storage. Dehydration and drying are an energy-intensive unit operation because of the high latent heat of water evaporation and relatively low energy efficiency of industrial dryers. The typical temperature during air drying is between 65 and 85°C. Loss of moisture from food products and high temperature processing during air drying may cause undesirable effects on the textural properties and nutritional values of the products. Other common drying methods include microwave drying, freezing drying and vacuum drying.

A drying process is a simultaneous heat and mass transfer operation. The energy required for evaporation of water, which is dependent of temperature and pressure, is in the range of 2.5–2.7 MJ/kg. However, total energy input into conventional dryer is in the range of 4–6 MJ/kg of removed water depending on the thermal efficiency of the drying systems. Several studies have been conducted on exergy analyses of food drying (Akpinar, 2004; Akpinar *et al.*, 2005; 2006; Colak and Hepbasli, 2007; Corzo *et al.*, 2008; Dincer and Sahin, 2004; Midilli and Kucuk, 2003; Ozgener and Ozgener, 2006). The energy and exergy efficiencies of a pasta drying process were found to be 75.5–77.09%, and 72.98–82.15%, respectively (Ozgeber and Ozgener, 2006). In order to increase the energy efficiency, the air leaving the dryer can be re-circled back to the dryer. The exergy efficiency decreases with the increase in air temperature and velocity. Corzo *et al.* (2008) conducted exergy analyses of thin layer drying of coroba slices. At drying temperatures from 71 to 93°C and drying air velocities from 0.82 to 1.18 m/s, the exergy efficiency of the thin layer drying of coroba slices was in the range from 97% to 80%. The exergetic efficiency for drying red pepper slices in a convective type dryer varied from 97.92% to 67.28% at the inlet temperature from 55 to 70°C and a drying air velocity of 1.5 m/s (Akpinar, 2004).

Several methods have been used to improve the energy and exergy efficiencies of the evaporation, dehydrating and drying process. Mechanical processes such as filtration and centrifugation can be used to remove as much water as possible before evaporation and drying. Evaporation is an energy intensive unit operation. Mechanical recompression evaporation is the most commonly used for concentration of dilute solutions in food processing. Membrane technology has a potential to reduce overall energy consumption in combination with the evaporation technology (Cassano *et al.*, 2011; Kumar *et al.*, 1999; Onsekizoglu *et al.*, 2010). A multiple-effect evaporator system is a simple series arrangement of several evaporators, which use steam to remove product moisture by evaporation. The evaporated water vapor from food products is collected and used as the steam for the next evaporator in the series. This collection and reuse of vapor result in smaller energy requirements to remove product moisture. The greater the number of evaporators in the series, the smaller is the energy consumption. It was found changing from four effect evaporators to five evaporators could save 20% energy. However, the number of evaporators in series should also be determined by the economics of the process (Wang, 2008).

5.3.3.3   *Energy savings in refrigeration and freezing*

Food processing facilities make heavy use of refrigeration. It is estimated that the refrigeration system uses as much as 15% of the total energy consumed worldwide. In the whole food industry of the USA, about 25% of the electricity is used for process cooling and refrigeration (Okos *et al.*, 1998). The dairy sector and the meat sector are likely to be the highest and second users of refrigeration, respectively. Generally, energy conservation for refrigeration unit operations can be achieved by improved insulation, best practice, and use of novel refrigeration cycles powered by waste heat.

Air blast chillers or freezers are widely used in the food industry. The fans of air blast chillers or freezers add heat load to chillers or freezers during operation. Since the heat generated by fans increases with required air load, it is critical to optimize the air velocity to minimize the heat generation and maximize the refrigeration effect during air blast chilling or freezing of food products. During air blast chilling/freezing, the heat transfer from the cold medium of air to the inside of foods must pass two layers of thermal resistance: the external resistance to heat convection between the cold air and the food surface, the internal resistance to heat conduction in the solid foods. Biot number, *Bio*, is the ratio of the internal resistance to the external resistance, which is expressed as (Singh and Heldman, 2007):

$$Bio = \frac{hl}{k} \tag{5.1}$$

where *l* is a characteristic dimension of the food body, *m*, which is the radius for round shaped body and half of the thickness for a flat shaped body; *h* is the surface convective heat transfer coefficient, $W/m^2/°C$; and *k* is the thermal conductivity of foods, $W/m/°C$.

There are:

- *Bio* ≤ 0.1, negligible international resistance to heat conduction,
- 0.1 < *Bio* < 40, finite internal resistance and external resistance, and
- *Bio* ≥ 40, negligible external resistance to heat convection (Singh and Heldman, 2001).

According to Equation (5.1), the increase of air velocity will increase the value of *h* and thus the *Bio* number. For cooling or freezing of big food items such as beef with a big characteristic dimension, *l*, a small air velocity should be used. Higher velocities can only lead to small reduction in the cooling time but require a large increase in fan energy and generate more extra heat load. For cooling or freezing of small food items such as peas, a high air velocity should be used. Air impingement, water spray or water immersion chiller can also be used to achieve a high surface heat transfer coefficient for small food items (Wang, 2008).

5.4   UTILIZATIONS OF ENERGY EFFICIENCY TECHNOLOGIES IN AGRICULTURAL PRODUCTION AND FOOD PROCESSING

While the improvement of the energy efficiency of the existing facilities is still the main consideration for the decrease of energy costs, novel energy conservation technologies such as heat pump, supercritical-fluid processing, non-thermal sterilization and pasteurization processes, and thermal energy-powered refrigeration cycles have been introduced into food processing facilities. Capital investments should incorporate with cutting-edge and energy efficient technologies and processes as the energy prices continue to increase.

5.4.1   *Application of novel thermodynamic cycles*

5.4.1.1   *Heat pump*

Heat can only be transferred from a high temperature region to a lower temperature region in nature. If work is done on a thermodynamic cycle, the cycle can transfer heat in the opposite direction. A system, which transfers heat from a low temperature region to a higher temperature region, is called a heat pump. A heat pump operates like a refrigerator or an air conditioner. When the working substance changes from a liquid to a gas by evaporation or boiling, it absorbs heat. When it changes back from a gas to a liquid by condensation, it releases heat. The condition for the phase changes of the substance is determined by its temperature-pressure relationship. A liquid can be boiled at a low temperature to absorb heat if its pressure is low while a vapor can be condensed at a high temperature to release heat if its pressure is high. A heat pump uses these properties of a working substance such as a refrigerant to transfer heat from a low-temperature source to a higher temperature sink. Coefficient of performance (*COP*), is the ratio of the heat delivered to the sink to the power used by the compressor motor, is used to determine the efficiency of an ideal heat pump, which is given by:

$$COP = \frac{Q_L}{W} = \frac{Q_L}{Q_H - Q_L} = \frac{T_H}{T_H - T_L} \tag{5.2}$$

where Q is the heat, W is the work done on the heat pump cycle, T is the temperature in Kelvin and the subscripts of H and L are denoted to high temperature and low temperature, respectively. If a heat pump is used to increase the temperature of a heat source from 343 to 423 K, the *COP* of the ideal heat pump is 5.3 [$COP = 423/(423 - 343) = 5.3$] according to Equation (5.2). The *COP* of a heat pump decreases as the temperature of the heat source decreases and the temperature of the heat sink increases. Also, due to the restriction of existing compressor design, the temperature of waste heat supplied to the heat pump should be lower than 110°C. Since most of compressor used in a heat pump cycle has only 65–85% efficiency, there is frictional energy loss through the cycle, and heat exchangers including the condenser and evaporator require a temperature difference for heat transfer, the *COP* of an actual heat pump is much lower than that of an ideal one. However, since the efficiency for electricity generation from fossil fuel is lower than 35%, the *COP* of a heat pump that consumes electricity to recover heat should be greater than 3 for economically attractive applications (Wang, 2008).

Pasteurization and sterilization usually has a heating step and a cooling step. A heat pump can be used to couple the energy flow between the heating unit and cooling unit. Ozyurt *et al.* (2004) designed a liquid to liquid heat pump for the pasteurization of milk as shown in Figure 5.5. The hot pasteurized milk is cooled by the evaporator of the heat pump while the cold raw milk is heated by the condenser of the heat pump. For the pasteurization temperature at 72°C and coagulation temperature at 32°C, the measured *COP* of the heat pump ranged from 2.3 to 3.1. The heat pump system can save 66% of the primary energy compared to traditional plate and double jacket milk pasteurization systems (Ozyurt *et al.*, 2004).

Heat pumps have been used to increase the drying efficiency of convectional air dryer as shown in Figure 5.6 (Wang, 2008). A heat pump dehumidifier dryer functions in a manner similar to a refrigerator. It consists of a condenser for high-temperature heat exchange, a compressor, an evaporator for low-temperature heat exchange, and an expansion value to decrease the pressure of working liquid. A fan is usually used to provide air movement in the drying chamber. In a heat pump dehumidifier, the evaporator is used to remove the moisture in the moist air exiting the drying chamber, and the condenser is used to increase the temperature of the dry air from the evaporator. The hot dry air is then sent back to the drying chamber (Adapa and Schoenau, 2005; Kiatsiriroat and Tachajapong, 2002). A heat pump can also be used to extract heat from a low-temperature energy source such as geothermal energy through its evaporator and upgrade the extracted heat to a high-temperature heat source at its condenser for drying as shown in Figure 5.7 (Kuzgunkaya and Hepbasli, 2007). A review of heat pump systems for drying application was given by Goh *et al.* (2011).

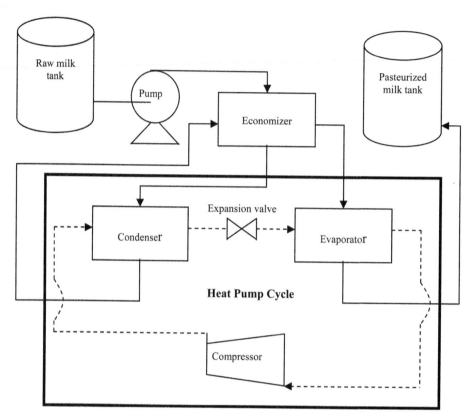

Figure 5.5.   Schematic diagram of a liquid-liquid heat pump system for pasteurization of milk (Ozyurt *et al.*, 2004).

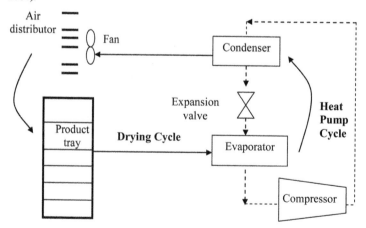

Figure 5.6.   Schematic diagram of a typical heat pump dehumidifier dryer (adapted from Wang, 2008).

5.4.1.2   *Novel refrigeration cycles*

Novel refrigeration cycles based on liquid-liquid absorption, liquid-solid adsorption and fluid ejection offer potential energy saving opportunities for food refrigeration. Novel refrigeration cycles such as absorption-refrigeration and adsorption-refrigeration cycles can be powered by low-grade waste heat or other renewable energy sources such as geothermal energy and solar energy at a low temperature (e.g., 70°C) (Sun and Wang, 2001).

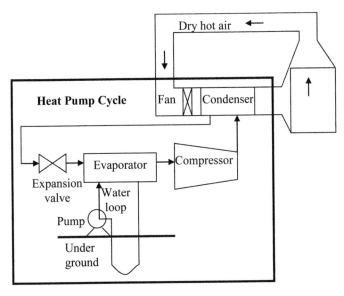

Figure 5.7.   Schematic diagram of an air-water heat pump system for drying (Kuzgunkaya and Hepbasli, 2007).

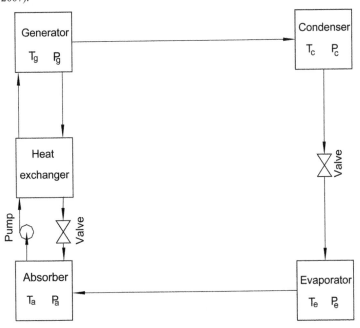

Figure 5.8.   The absorption refrigeration cycle (adapted from Sun and Wang, 2001).

An absorption-desorption cycle is based on that the partial pressure of refrigerant vapor is a function of the temperature and concentration of a refrigerant solution. LiBr-H_2O and H_2O/NH_3 are two main working solutions used in absorption refrigeration cycles (Sun and Wang, 2001). In an absorption cycle as shown in Figure 5.8 (Sun and Wang, 2001), low-pressure refrigerant vapor from the evaporator is absorbed by a strong absorbent solution in the absorber. The pump increases the pressure of the weak solution absorber and delivers the high-pressure weak solution to the generator, where heat from a high-temperature source vaporizes the refrigerant vapor in

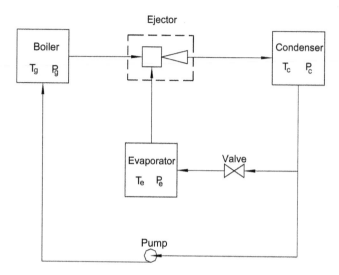

Figure 5.9. The ejector refrigeration cycle (adapted from Sun and Wang, 2001).

the weak solution. The strong solution then returns to the absorber through a throttling valve to reduce the pressure of the strong solution. The high-pressure refrigerant vapor generated from the weak solution flows to the condenser where the vapor is condensed into liquid. The liquid refrigerant then flows into the evaporator through a throttling valve to reduce its pressure. In order to improve cycle performance, a solution heat exchanger is normally added to the cycle. The heat exchanger is an energy saving component but not essential to the successful operation of the cycle. High-temperature heat source is added in the generator to change the weak absorbent-refrigerant solution to a strong absorbent-refrigerant solution and low-temperature heat from the product to be chilled is added to the refrigerant in the evaporator. The heat of the cycle received is rejected to the outside of the cycle in the condenser and absorber. The coefficient of performance (COP) for an absorption refrigeration machine, which is the ratio of heat absorbed by the evaporator to the heat added in the generator, is in the region of 0.6–0.9. These values are much smaller than that of a mechanical vapor compression machine with a typical value of 3. However, it is difficult to compare two systems only in terms of COP values since electricity used in the vapor compression cycle is much more expensive than the low-grade thermal energy used in the absorption cycle.

An ejector-refrigeration system is shown in Figure 5.9 (Sun and Wang, 2001). This cycle is sometimes referred to as an ejector-compression refrigeration system. The ejector cycles may be powered by thermal energy. In an ejector-refrigeration cycle, low-grade thermal energy is supplied to the boiler, where the liquid refrigerant is vaporized. The vapor is used as the driving fluid of the ejector, where the refrigerant vapor is suctioned and mixed with the driving fluid to form a single stream. The stream is further compressed to the condenser pressure through another part of the ejector with a special configuration. Emerging from the ejector, the fluid undergoes a temperature reduction and then condenses in the condenser with the condensation heat being rejected to the environment. Finally, the condensate is partly pumped to the boiler and partly expands via a throttling valve and evaporates in the evaporator to produce the necessary cooling effect. The evaporated vapor is entrained by the ejector to complete the cycle.

A typical adsorption refrigeration system is shown in Figure 5.10. Like a mechanical vapor compression cycle, adsorption refrigeration cycle is also based the evaporation of a liquid refrigerant to generate cooling effect. An adsorption refrigeration cycle consists of an evaporator, a condenser and an adsorber. An adsorber filled with adsorbent replaces the compressor of a mechanical vapor-compression refrigeration cycle and the absorber of an absorption refrigeration cycle to accelerate the evaporation of a refrigerant. The commonly used adsorbents include activated carbon, silica gel, zeolite and metal hydride. The key part of an adsorption refrigeration cycle is the working

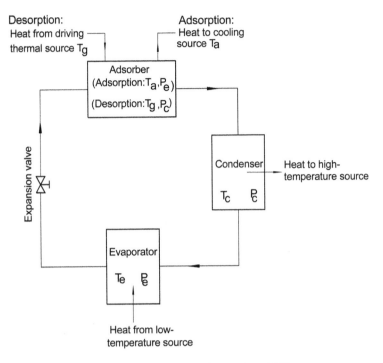

Figure 5.10.   The adsorption refrigeration cycle (adapted from Wang, 2008).

pairs such as zeolite-water or activated carbon-methanol. The amount of refrigerant, which can be charged by the adsorbent is usually determined by the properties of a refrigerant-adsorbent working pair, the temperature of the adsorbent and the partial pressure of refrigerant vapor. A whole adsorption refrigeration cycle can be divided into two stages. During the charging or adsorption period, the cold and fresh adsorbent bed is able to adsorb the refrigerant vapor from the evaporator to generate cooling effect. During the discharging or desorption period, the saturated adsorber is heated by thermal energy to discharge the refrigerant vapor from the adsorber. The refrigerant vapor is delivered to the condenser and is condensed back to a liquid phase. The liquid refrigerant flows to the evaporator for next cycle. The refrigeration occurs only during the charging period. The main technical issue related to an adsorption refrigeration cycle is the low heat transfer rate through a solid adsorbent bed because of its low thermal conductivity. The low heat transfer rate lead to a long cycling time, low *COP* and a large size, compared to a mechanical vapor compression unit to generate the same amount of cooling effect. Enhancement of heat transfer through an adsorbent bed is critical to make the adsorption refrigerators economically viable (Wang, 2008).

5.4.1.3 *Heat pipes*
A heat pipe is a pipe through which heat passes if there is a small temperature difference between its two ends. A heat pipe is illustrated in Figure 5.11. A typical heat pipe consists of an evacuated metal tube. A certain amount of working fluid is placed in wick of the inside of the tube. One end of the tube serves as the evaporator and the other end is the condenser. Heating the evaporator end of the tube causes the liquid to evaporate and the vapor travels along the tube to the other end. At the other end, which is the condenser, vapor condenses and releases its latent heat. The liquid travels back to the original end along the wick by capillary action to complete a cycle. Since a heat pipe uses latent heat rather than sensible heat to transfer heat from one location to another location, it is a highly efficient approach to transfer heat. A heat pipe can be used to quickly remove heat from a hot medium or transfer heat into a cold medium. The type of heat pipes is

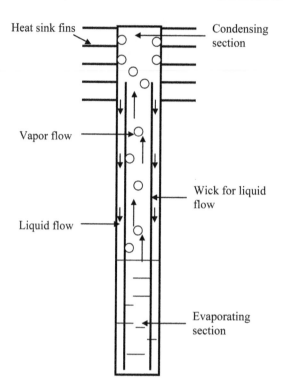

Figure 5.11.   Schematic drawing of a heat pipe (Adapted from Wang, 2008).

determined by the method by which the condensate returns back to its hot end. The condensate return can be achieved by gravity or capillary action in a wick (Ketteringham and James, 2000).

A heat pipe can transfer heat several hundred times fast than a solid copper rod of the same diameter. Heat pipes can dissipate substantial quantities of heat with a small temperature difference. This is achieved as a result of good heat transfer through mass transfer with phase change of a working fluid inside the pipe. Heat pipes have been considered as a potential to improve cooking and cooling operations by providing a fast path for heat transfer between the surface and the thermal center of foods. The use of heat pipes in cooking and cooling of foods can significantly save processing time and thus energy (James *et al.*, 2005; Ketteringham and James, 2000). Heat pipes have also been used to recover waste heat as heat pipes operate almost isothermally with a small temperature difference between the lower and the upper ends. Srimuang and Amatachaya (2012) and Chaudhry *et al.* (2012) provided reviews of the applications of heat pipe heat exchangers for heat recovery.

5.4.2   *Application of non-thermal food processes*

Thermal processes are usually considered to be energy intensive. Also the slow heat transfer through food products is usually a limiting factor for thermal treatment of food products. Non-thermal pasteurization techniques including food irradiation, pulsed electric field treatment and high pressure processing have also been developed to replace the conventional thermal sterilization and pasteurization processes for saving energy, and improving product quality and safety. These non-thermal processing times are usually very short. For example, a food during high-pressure processing is exposed to a pressure up to 1000 MPa for a few minutes. Pulsed electric field treatment is based on the delivery of pulses at high electric field intensity of 5–55 kV/cm for few milliseconds. Food irradiation lasts for several seconds to several minutes. Most

Table 5.6.   Inactivation of *E. coli* in milk by heat and alternative preservation technologies (adapted from Lado and Yousef, 2002).

Preservation process	Product	Treatment conditions	Targeted bacteria	Log count decreased
Heat	Milk	63°C, 16.2 s	*E. coli*	5.9
Gamma irradiation	Milk	10 kGy	*E. coli*	7.0
High-pressure processing	Milk	500 MPa, 25°C, 5 min	*E. coli*	5.9
Pulsed electric field	Milk	22.4 kV/cm, 330 μs	*E. coli*	4.7

alternative preservation processes can achieve the equivalent effect of thermal pasteurization but not sterilization as shown in Table 5.6 (Lado and Yousef, 2002).

5.4.2.1   *Food irradiation*

Irradiation pasteurizes foods by exposing them to very high-energy electrons ($\lambda = 10^{-11}-10^{-13}$ m) such as X-rays, electron beams and gamma rays, which are similar to light waves ($\lambda = 4 \times 10^{-7}-7 \times 10^{-7}$ m), ultraviolet waves ($\lambda = 10^{-7}$ m) or microwaves ($\lambda = 10^{-3}-1$ m) but have shorter wave length (Farkas, 2006). Food irradiation is a cold process, which can damage the DNA of living cells effectively so that the living cells become inactivated. Compared with thermal pasteurization, food irradiation is a more efficient pasteurization method for solid foods without causing significant changes in taste and quality of the products. The energy used for food irradiation is very small. The dose of food irradiation is usually less than 10 kGy (energy equivalent to 10 kJ/kg of foods). The increase of food temperature due to irradiation is less than 3°C (Loaharanu, 1996).

5.4.2.2   *Pulsed electric fields*

Application of an external electrical field to a biological cell induces an electrical potential across the cell membrane. If the electrical potential exceeds a critical level, local structural changes of the cell membranes will occur. As a consequence, a drastic increase in membrane permeability occurs, which impairs on the irreversible loss of physiological control systems and therefore causes the cell death. Pulsed electric fields treatment has been used to pasteurize liquid foods (Heinz *et al.*, 2003). Application of pulsed electric fields treatment at low temperatures has a potential to preserve food products with a fresh-like character and high nutritional value. The application of pulsed electric fields for treatment of liquid foods at 30°C requires a specific energy input of 100 kJ/kg or more (Heinz *et al.*, 2003). Pulsed electric fields treatment is usually considered to have a higher energy input than a thermal process with heat recovery capacity. When operating at elevated treatment temperature and making use of synergetic heat effects, the pulsed electric field energy input might be reduced close to the amount of 20 kJ/kg required for a conventional thermal pasteurization process with 95% of heat recovery (Toepfl *et al.*, 2006). Due to high production costs, commercial applications of pulsed electric fields treatment as an alternative to traditional thermal process have not yet been accomplished (Heinz *et al.*, 2003). The investment costs of a pulsed electric fields unit with a production capacity of 5 ton/h are estimated to be in the range of 2 to 3 million US$ (Toepfl *et al.*, 2006).

5.4.2.3   *High-pressure processing*

A high-pressure process inactivates microbial by targeting on the membranes of biological cells. During pressurization, water and acid molecules show increased ionization. This ionization change in living cells causes the major killing effect on living cells during pressurization. Although atomic bonds are barely affected by a high pressure, alternation of proteins or lipids can be observed when exposed to a high pressure. Lethal damage occurs when alternation of proteins or lipids occurs in the membranes of biological cells (Manas and Pagan, 2005; Toepfl *et al.*, 2006). Theoretically, the compression work and energy required for the resulted temperature increase due to pressurization (3°C/100 MPa) is about 52 kJ/kg and 70 kJ/kg upon compression of pure water up to 600 MPa, respectively.

5.4.2.4   *Membrane processing*

Separation processes play an important role in food processing facilities for concentration of liquid foods, recovery of by-products from processing waste water, removal of contaminants from liquid foods, and treatment of waste water. Evaporation is widely to concentrate dilute streams such as milk, fruit juices and sugar juices in the food industry. Evaporation is an energy intensive unit operation. Mechanical vapor recompression is the most common technology used to improve the energy efficiency of an evaporation process. Membrane filtration, which is driven by pressure difference, is an alternative method to energy intensive separation processes with a phase change such as distillation and evaporation for concentrating dilute streams with less energy consumption. The membrane filtration has a potential to save 30–50% of energy used by distillation and evaporation (Kumar *et al.*, 1999). Membrane filters can also be used to remove suspended solids, dissolved solids and microbes from a liquid solution. In the food processing industry, the membrane technologies can be used to:

- Concentrate dilute liquid foods such as milk and juice;
- Remove microbes in liquid foods for cold pasteurization;
- Recover dissolved solids such as sugars from processing waste water; and
- Treat processing waste water.

Microfiltration and ultrafiltration have been used to remove bacteria and spores from milk, liquid egg white and juice at a low temperature (Carneiro *et al.*, 2002; Mukhopadhyay *et al.*, 2010; Walkling-Ribeiro *et al.*, 2011). Membrane treatment is an efficient technique to inactivate microbial in apple juice and pineapple juice (Carneiro *et al.*, 2002). The average removal efficiency of *Salmonella enteritidis* in liquid egg white was found to be $6.8 \log_{10}$ CFU/mL (Mukhopadhyay *et al.*, 2010). The reduction of the native microbial load in milk by microfiltration at a transmembrane flux of 660 L/h/m^2 was found to be $3.7 \log_{10}$ CFU/mL, which is close to $4.6 \log_{10}$ CFU/mL for thermal pasteurization at $75°$C for 24 s (Walkling-Ribeiro *et al.*, 2011).

5.4.2.5   *Supercritical fluid processing*

Supercritical fluids such as supercritical carbon dioxide can be used to remove moisture from foods (Brown *et al.*, 2008). Supercritical CO_2 has a low critical temperature and pressure ($31.1°$C and 7.3 MPa). Therefore, drying with supercritical CO_2 can be operated at a much lower temperature than conventional air drying. However, since supercritical CO_2 is a non-polar solvent, the water solubility in supercritical CO_2 is 4 mg/g at $50°$C and 20 MPa and 2.5 mg/g at $40°$C and 20 MPa (Sabirzyanov *et al.*, 2002). Therefore, small quantities of polar co-solvents such as ethanol are usually added into the supercritical CO_2 to increase the solubility of polar water in supercritical CO_2. Supercritical CO_2 drying has been found to generate more favorable re-hydrated textural properties than the air dried equivalents (Brown *et al.*, 2008).

5.4.3   *Application of novel heating methods*

5.4.3.1   *Microwave and radio frequency heating*

Microwave and radio frequency (*RF*) are widely used in the food industry. Microwave energy is electromagnetic waves with frequencies between 300 MHz and 300 GHz. Radio wave is a rate of oscillations in the range of about 3 kHz to 300 GHz, which are alternating currents to carry radio signals. Microwaves are non-ionizing electromagnetic waves and commercial microwaves heating applications use frequencies of 2450 MHz, sometimes 915 MHz in USA and 896 MHz in Europe. During conventional heating, the energy is transferred to the material through convection, conduction and radiation of heat from the surfaces of the materials. However, during microwave and radio frequency heating, the energy is delivered directly to the product through molecular interaction with an electromagnetic and electric field.

Microwave heating has been used for high temperature-short time (HTST) sterilization/ pasteurization of foods, particularly thick food items. Because of the low thermal conductivity, it is impossible to achieve HTST treatment for food items with a several-centimeter thickness by conventional heating methods. Microwave energy can penetrate into the food items and cause a rapid temperature rise to pasteurization temperatures. Huang and Sites (2007) used a microwave heating system for in-package pasteurization of ready-to-eat meats. They observed that the inactivation rate of *Listeria monocytogens* during microwave pasteurization was 0.41, 0.65 and 0.94 log(CFU/pk)/min at the surface temperature of 65, 75 and 85°C, respectively. The overall rate of bacterial inactivation for the water immersion pasteurization at the same surface temperatures was only 30–75% higher than that of microwave in-package pasteurization. However, microwave pasteurization is much faster than the water immersion pasteurization.

Lakshmi *et al.* (2007) found although the absorption of microwave energy by water is 86–89%, the conversion efficiency of electrical energy to microwave energy is only about 50%. The total thermal efficiency was only around 44%. During microwave heating, moisture loss may occur due to the evaporation of moisture at a high temperature. In this case, part of the thermal energy delivered to the foods releases from moisture evaporation rather than increases the temperature of the foods.

Although the main frequencies of RF used for industrial heating lie in the range 10–50 MHz, only selected frequencies at 13.56, 27.12 and 40.68 MHz are permitted for industrial, scientific and medical applications. As frequency and wavelength are inversely proportional, the wavelengths of RF are much longer than those of microwave. The wavelength of 27.12 MHz RF is 11 m, compared to 0.12 m for microwave at 2450 MHz. The penetration depth of wave energy is proportional to wavelength. During RF heating, electromagnetic power can penetrate much deeper into samples than microwave heating. The RF heating can minimize the surface overheating and offer more uniform heating over the sample geometry due to both deeper level of power penetration and also more uniform field patterns than microwave heating (Marra *et al.*, 2009a).

5.4.3.2 *Ohmic heating*
Ohmic heating is a rapid and relatively uniform heating method. The basic principle of ohmic heating is that electrical energy is converted to thermal energy within an electrical conductor. Like microwave heating, the increase of food temperature during ohmic heating is caused by the heat generated inside an electrically conductive food material when a current is applied across the material (Jun and Sastry, 2005; Marra *et al.*, 2009b). The main advantage of ohmic heating is the rapid and uniform heating process, which will minimize the losses of the structure, nutritional and sensorial properties of a food product. It can be used as a continuous in-line heater for cooking and sterilization of viscous and liquid food products. Ohmic heating has high energetic efficiency. The electric energy of ohmic heating was completely dissipated as heat with an efficiency of 94.2%. A small amount of heat is lost from the heating chamber to surrounding air, which may slightly reduce the overall efficiency of the system (Nguyen *et al.*, 2013).

The electrical conductivity is a function of temperature, ionic concentration and solid content. For most aqueous materials, the electrical conductivity increases linearly with temperature. It is important to consider the changes in electrical conductivity and system performance during ohmic heating (Icier and Ilicali, 2005). The efficacy of ohmic processing can be influenced by the conductivities of individual components within the food and their behavior and interactions during the heating process. Conductivity measurements on pork cuts indicated that lean is highly conductive compared to fat and addition of fat to lean reduced the overall conductivity (McKenna *et al.*, 2006).

5.4.3.3 *Infrared radiation heating*
Infrared (IR) radiation is energy in the form of electromagnetic wave and is more rapid in heat transfer than heat convection and conduction. The IR energy can be generated by gas-fired and electrically heated emitters (Yang *et al.*, 2010). Infrared radiation can achieve contactless heating. Advantages of infrared over convective heating with hot air or water are higher heat transfer coefficients (Jaturonglumlert and Kiatsiriroat, 2010). Infrared radiation has a very low penetration depth. However, it can heat up the surface rapidly and heat is transferred inside the materials by thermal conduction. Therefore, IR can be used to dry food products and deactivate bacteria and mold on the surface of food products (Trivittayasil *et al.*, 2011; Yang *et al.*, 2010).

5.5　SUMMARY

Energy conservation is vital for the sustainable development of agricultural and food industry. Reduced energy consumption through conservation can benefit not only energy consumers by reducing their energy costs, but also the society in general by reducing the use of energy resources and the emission of many air pollutants such CO_2. To develop a sustainable society, much effort must be devoted not only to discovering renewable energy resources, but also to increasing the energy efficiency of devices and processes utilizing these resources. The agricultural production consumes large amounts of energy for fertilization, mechanization, irrigation, crop propagation and herbicides. Agricultural mitigation efforts should not only be directed towards substituting fossil fuel with bioenergy but also at substantially reducing methane and nitrous oxide emissions and maintaining soil carbon stocks. The increase of agricultural sustainability requires to reduce the agronomic inputs through the conservation tillage, use of disease resistant cultivars, rational use of pesticides and target application of nutrients and to diversify crops. Food processing industry is the fifth biggest consumer of energy in the United States. Because of the increasing energy prices and efforts for reduction of CO_2 emission, it has become significant to improve the energy efficiency and replace the existing energy intensive unit operations with new energy efficient processes. Energy efficiency improvement and waste heat recovery in the food industry has been a focus in the past decades. Replacement of conventional energy intensive food processes with novel technologies such as novel thermodynamic cycles, non-thermal and novel heating processes may provide another potential to reduce energy consumption, reduce production costs and improve the sustainability of food production. Some novel food processing technologies have been developed to replace traditional energy intensive unit operations for pasteurization and sterilization, evaporation and dehydration, and chilling and freezing in the food industry. Most of the energy conservation technologies can readily be transferred from other manufacturing sectors to the food processing sector.

REFERENCES

Adapa, P.K. & Schoenau, G.J.: Re-circulating heat pump assisted continuous bed drying and energy analysis. *Int. J. Energy Res.* 29 (2005), pp. 961–972.

Ahamed, J.U., Saidur, R., Masjuki, H.H., Mekhilef, S., Ali, M.B. & Furqon, M.H.: An application of energy and exergy analysis in agricultural sector of Malaysia. *Energy Policy* 39 (2011), pp. 7922–7929.

Akpinar, E.K.: Energy and exergy analyses of drying of red pepper slices in a convective type dryer. *Int. Comm. Heat Mass Transf.* 31 (2004), pp. 1165–1176.

Akpinar, E.K., Midilli, A. & Bicer, Y.: Energy and exergy of potato drying process via cyclone type dryer. *Energy Conver. Manag.* 46 (2005), pp. 2530–2552.

Akpinar, E.K., Midilli, A. & Bicer, Y.: The first and second law analyses of thermodynamic of pumpkin drying process. *J. Food Eng.* 72 (2006), pp. 320–331.

Alluvione, F., Moretti, B., Sacco, D. & Grignani, C.: EUE (energy use efficiency) of cropping systems for a sustainable agriculture. *Energy* 36 (2011), pp. 4468–4481.

Bhattacharyya, S.C. & Ussanarassamee, A.: Decomposition of energy and CO_2 intensities of Thai industry between 1981 and 2000. *Energy Econ.* 26 (2004), pp. 765–781.

Brown, Z.K., Fryer, P.J., Norton, I.T., Bakalis, S. & Bridson, R.H.: Drying of foods using supercritical carbon dioxide-investigations with carrot. *Innov. Food Emerg. Technol.* 9 (2008), pp. 280–289.

Carneiro, L., dos Santos Sa, I., dos Santos Gomes, F., Matta, V.M. & Cabral, L.C.M.: Cold sterilization and clarification of pineapple juice by tangential microfiltration. *Desalination* 148 (2002), pp. 93–98.

Cassano, A., Conidi, C. & Drioli, E.: Clarification and concentration of pomegranate juice (*Punica granatum* L.) using membrane process. *J. Food Eng.* 107 (2011), pp. 366–373.

Chaudhry, H.N., Hughes, B.R. & Ghani, S.A.: A review of heat pipe systems for heat recovery and renewable energy applications. *Renew. Sustain. Energy Rev.* 16 (2012), pp. 2249–2259.

Colak, N. & Hepbasli, A.: Performance analysis of drying of green olive in a tray dryer. *J. Food Eng.* 80 (2007), pp. 1188–1193.

Corzo, O., Bracho, N., Vasquez, A. & Pereira, A.: Energy and exergy analyses of thin layer drying of coroba slices. *J. Food Eng.* 86 (2008), pp. 151–161.

Dalgaard, T., Olesen, J.E., Petersen, S.O., Petersen, B.M., Jørgensen, U., Kristensen, T., Hutchings, N.J., Gyldenkærne, S. & Hermansen, J.E.: Developments in greenhouse gas emissions and net energy use in Danish agriculture: How to achieve substantial CO_2 reductions? *Environ. Pollut.* 159 (2011), pp. 3193–3203.

Dincer, I. & Sahin, A.Z.: A new model for thermodynamic analysis of a drying process. *Int. J. Heat Mass Transf.* 47 (2004), pp. 645–652.

Einstein, D., Worrell, E. & Khrushch, M.: *Steam systems in industry: Energy use and energy efficiency improvement potentials.* Lawrence Berkeley National Laboratory, Paper LBNL-49081, 2001, http://repositories.cdlib.org/lbnl/LBNL-49081 (accessed August 2013).

Farkas, J.: Irradiation for better foods. *Trends Food Sci. Technol.* 17 (2006), pp. 148–152.

Fischer, J.R., Blackman, J.E. & Finnell, J.A.: Industry and energy: challenges and opportunities. *Resource: Eng. Technol. Sustain. World* 4 (2007), pp. 8–9.

Fritzson, A. & Berntsson, T.: Efficient energy use in a slaughter and meat processing plant-opportunities for process integration. *J. Food Eng.* 76 (2006), pp. 594–604.

Goh, L.J., Othman, M.Y., Mat, S., Ruslan, H. & Sopian, K.: Review of heat pump systems for drying application. *Renew. Sustain. Energy Rev.* 15 (2011), pp. 4788–4796.

Hatirli, S.A., Ozhan, B. & Fert, C.: An econometric analysis of energy input-output in Turkish agriculture. *Ren. Sustain. Energy Rev.* 9 (2005), pp. 608–623.

Heinz, V., Toepfl, S. & Knorr, D.: Impact of temperature on lethality and energy efficiency of apple juice pasteurization by pulsed electric fields treatment. *Innov. Food Sci. Emerg. Technol.* 4 (2003), pp. 167–175.

Huang, L. & Sites, J.: Automatic control of a microwave heating process for in-package pasteurization of beef frankfurters. *J. Food Eng.* 80 (2007), pp. 226–233.

Icier, F. & Ilicali, C.: Temperature dependent electrical conductivities of fruit purees during ohmic heating. *Food Res. Int.* 38 (2005), pp. 1135–1142.

James, C., Araujo, M., Carvalho, A. & James, J.: The heat pipe and its potential for enhancing the cooking and cooling of meat joints. *Int. J. Food Sci. Technol.* 40 (2005), pp. 419–423.

Jaturonglumlert, S. & T. Kiatsiriroat, T.: Heat and mass transfer in combined convective and far-infrared drying of fruit leather. *J. Food Eng.* 100 (2010), pp. 254–260.

Jun, S. & Sastry, S.: Modeling and optimization of ohmic heating of foods inside a flexible package. *J. Food Proc. Eng.* 28 (2005), pp. 417–436.

Ketteringham, L. & James, S.: The use of high thermal conductivity inserts to improve the cooling of cooked foods. *J. Food Eng.* 45 (2000), pp. 49–53.

Kiatsiriroat, T. & Tachajapong, W.: Analysis of a heat pump with solid desiccant tube bank. *Int. J. Energy Res.* 26 (2002), pp. 527–542.

Kumar, A., Croteau, S. & Kutowy, O.: Use of membranes for energy efficient concentration of dilute steams. *Appl. Energy* 64 (1999), pp. 107–115.

Kuzgunkaya, E.H. & Hepbasli, A.: Exergetic performance assessment of a ground-source heat pump drying system. *Int. J. Energy Res.* 31 (2007), pp. 760–777.

Lado, B.H. & Yousef, A.E.: Alternative food-preservation technologies: efficacy and mechanisms. *Microbes Infect.* 4 (2002), pp. 433–440.

Lakshmi, S., Chakkaravarthi, A., Subramanian, R. & Singh, V.: Energy consumption in microwave cooking of rice and its comparison with other domestic appliances. *J. Food Eng.* 78 (2007), pp. 715–722.

Liu, Y., Gao, M., Wu, W., Tanveer, S.K., Wen, X. & Liao, Y.: The effects of conservation tillage practices on the soil water-holding capacity of a non-irrigated apple orchard in the Loess Plateau, China. *Soil Tillage Res.* 130 (2013), pp. 7–12.

Loaharanu, P.: Irradiation as a cold pasteurization process of food. *Vet. Parasitol.* 64 (1996), pp. 71–82.

Lu, Y., Mu, H. & Li, H.: An analysis of present situation and future trend about the energy consumption of Chinese agriculture sector. *Procedia Environ. Sci.* 11 (2011), pp. 1400–1406.

Manas, P. & Pagan, R.: Microbial inactivation by new technologies of food preservation. *J. Appl. Microbiol.* 98 (2005), pp. 1387–1399.

Marra, F., Zell, M., Lyng, J.G., Morgan, D.J. & Cronin, D.A.: Analysis of heat transfer during ohmic processing of a solid food. *J. Food Eng.* 91 (2009a), 56–63.

Marra, F., Zhang, L. & Lyng, J.G.: Radio frequency treatment of foods: Review of recent advances. *J. Food Eng.* 91 (2009b), pp. 497–508.

McKenna, B.M., Lyng, J., Brunton, N. & Shirsat, N.: Advances in radio frequency and ohmic heating of meats. *J. Food Eng.* 77 (2006), pp. 215–229

Midilli, A. & Kucuk, H.: Energy and exergy analyses of solar drying process of pistachio. *Energy* 28 (2003), pp. 539–556.

Mukhopadhyay, S., Tomasula, P.M., Luchansky, J.B., Porto-Fett, A. & Call, J.E.: Removal of *Salmonella enteritidis* from commercial unpasteurized liquid egg white using pilot scale cross flow tangential microfiltration. *Int. J. Food Microbiol.* 142 (2010), pp. 309–317.

Mull, T.E.: *Practical guide to energy management for facilities engineers and plant managers.* ASME Press, New York, 2001.

Muller, D.C.A., Marechal, F.M.A., Wolewinski, T. & Roux, P.J.: An energy management method for the food industry. *Appl. Therm. Eng.* 27 (2007), pp. 2677–2686.

Nguyen, L.T., Choi, W., Lee, S.H. & June, S.: Exploring the heating patterns of multiphase foods in a continuous flow, simultaneous microwave and ohmic combination heater. *J. Food Eng.* 116 (2013), pp. 65–71.

Okos, M., Rao, N., Drecher, S., Rode, M. & Kozak, J.: *Energy usage in the food industry.* American Council for an Energy-Efficient Economy, 1998, http://www.aceee.org/pubs/ie981.htm (accessed August 2013)

Onsekizoglu, P., Bahceci, K.S. & Acar, M.J.: Clarification and the concentration of apple juice using membrane processes: a comparative quality assessment. *J. Membrane Sci.* 352 (2010), pp. 160–165.

Ozgener, L. & Ozgener, O.: Exergy analysis of industrial pasta drying process. *Int. J. Energy Res.* 30 (2006), pp. 1323–1335.

Ozyurt, O., Comakli, O., Yilmaz, M. & Karsli, S.: Heat pump use in milk pasteurization: an energy analysis. *Int. J. Energy Res.* 28 (2004), pp. 833–846.

Plaza, C., Courtier-Murias, D., Fernandez, J.M., Polo, A. & Simpson, A.J.: Physical, chemical, and biochemical mechanisms of soil organic matter stabilization under conservation tillage systems: a central role for microbes and microbial by-products in C sequestration. *Soil Biol. Biochem.* 57 (2013), pp. 124–134.

Ramirez, C.A., Blok, K., Neelis, M. & Patel, M.: Adding apples and oranges: the monitoring of energy efficiency in the Dutch food industry. *Energy Policy* 34 (2006a), pp. 1720–1735.

Ramirez, C.A., Patel, M. & Blok, K.: How much energy to process one pound of meat? A comparison of energy use and specify energy consumption in the meat industry of four European countries. *Energy* 31 (2006b), pp. 2047–2063.

Sabirzyanov, A.N., Il'in, A.P., Akhunov, A.R. & Gumerov, F.M.: Solubility of water in supercritical carbon dioxide. *High Temp.* 40 (2002), pp. 203–206.

Simpson, R., Cortes, C. & Teixeira, A.: Energy consumption in batch thermal processing: model development and validation. *J. Food Eng.* 73 (2006), pp. 217–224.

Singh, R.P. & Heldman, D.R.: *Introduction to food engineering.* 3rd edition, Academic Press, San Diego, CA, 2001.

Srimuang, W. & Amatachaya, P.: A review of the applications of heat pipe heat exchangers for heat recovery. *Renew. Sustain. Energy Rev.* 16 (2012), pp. 4303–4315.

Sun, D.W. & Wang, L.J.: Novel refrigeration cycles. Chapter 1 in: D.W. Sun (ed): *Advances in food refrigeration.* Leatherhead Publishing, Leatherhead, UK, 2001, pp. 1–69.

Toepfl, S., Mathys, A., Heinz, V. & Knorr, D.: Review: potential of high hydrostatic pressure and pulsed electric fields for energy efficiency and environmentally friendly food processing. *Food Rev. Int.* 22 (2006), pp. 405–423.

Trivittayasil, V., Tanaka, F. & Uchino, T.: Investigation of deactivation of mold conidia by infrared heating in a model-based approach. *J. Food Eng.* 104 (2011), pp. 565–570.

Unruh, B.: Delivered energy consumption projections by industry in the annual energy outlook. 2002, http://www.eia.gov/oiaf/analysispaper/industry/consumption.html (accessed August 2013).

US Census Bureau: *2010 Annual survey of manufactures.* 2010, http://factfinder2.census.gov (accessed August 2013).

US Environmental Protection Agency (US EPA): Energy trends in selected manufacturing sectors: Opportunities and challenges for environmentally preferable energy outcomes. 2007, http://www.epa.gov/sectors/pdf/energy/report.pdf (accessed August 2013).

Walkling-Ribeiro, M., Rodriguez-Gonzalez, O., Jayaram, S. & Griffiths, M.W.: Microbial inactivation and shelf life comparison of 'cold' hurdle processing with pulsed electric fields and microfiltration, and conventional thermal pasteurization in skim milk. *Int. J. Food Microbiol.* 144 (2011), pp. 379–386.

Wang, L.J.: *Energy efficiency and management in food processing facilities.* Taylor and Francis, Boca Raton, FL, 2008.

Yang, J., Bingol, G., Pan, Z., Brandl, M.T., McHugh, T.H. & Wang, H.: Infrared heating for dry-roasting and pasteurization of almonds. *J. Food Eng.* 101 (2010), pp. 273–280.

CHAPTER 6

Energy-smart food – technologies, practices and policies

Ralph E.H. Sims & Alessandro Flammini[1]

6.1 INTRODUCTION

The world food supply "from paddock-to-plate" consumes energy at all stages along the chain. It is heavily dependent on fossil fuels for agricultural production, transport, food processing, storage, cooking etc. The world will have to produce 60% more food by 2050 than today to meet the growing demands from continuing population growth and higher protein diets according to projections by the Food and Agricultural Organisation of the United Nations (FAO, 2009a). Therefore, what will eventuate under business-as-usual is an ever greater reliance on:

- *direct* fossil fuel inputs used at the operational level on the farm or fishery for harvesting, land preparation, irrigation etc. (e.g. diesel for tractor, trucks and boat fuels, electricity for water pumping, and heat for crop drying) as well as for processing, preparing and cooking food, and
- *indirect* fossil fuel inputs not directly consumed at the operational level (e.g. natural gas for manufacturing fertilizers, pesticides; coal to produce heat for steel manufacturing for agricultural machinery, cement for buildings and oil products for construction equipment for buildings and roading).

However, the long-term security of future supplies of fossil fuels is uncertain in terms of the possible dwindling of reserves, the fluctuating prices, and the greenhouse gas emissions resulting from their combustion (see Chapter 2, this volume).

The "Green Revolution" that helped overcome food shortages in OECD countries after World War II was largely successful due to the high inputs available of cheap energy, as well as benefitting from plant breeding and improved crop and animal management. Fossil fuels were utilized to increase mechanization, produce fertilizers and agri-chemicals, develop specialist intensive buildings for animals, power irrigation systems, transport products long distances to markets etc. Gaining similar productivity and efficiency gains today in agricultural production for low-GDP[2] countries may not be so easy given the future insecure supply of fossil fuels and their relatively high and fluctuating prices. The main goals, therefore, should be to reduce the current dependence of the industrial food supply chain on these direct and indirect fossil fuel inputs and to avoid

---

[1] Although this chapter is based on a recent UN FAO report that the authors were closely involved in researching and writing, the views expressed here do not necessarily reflect those of the UN FAO and the responsibility for any errors or inaccuracies in this chapter lies with the authors of it.

[2] Modern agri-food systems are evolving rapidly in some developing countries as for example, in China where supermarkets are starting to dominate the food supply chain (Vorley, 2011). Therefore, with regard to energy and food linkages it is no longer practical to simply classify countries using standard comparisons such as OECD or non-OECD, developed or developing, traditional or conventional, and subsistence or industrialised. For this analysis, it was helpful to be able to group countries to represent the major differences in the agri-food supply chains that exist. Therefore, it was decided to use the terms "high-GDP" and "low-GDP". "High-GDP" loosely groups together the top 50 or so countries measured in terms of their gross domestic product on a purchasing power parity basis divided by their population. "Low-GDP" comprises the remaining 176 or so nations. (http://www.indexmundi.com/g/r.aspx?v=67).

creating a similar dependence for subsistence farmers as they move towards modern farming practices.

Energy efficiency and renewable energy linked with the food supply chain have the potential to improve current practices. Subsistence farmers, family fisherfolk supplying local markets, or corporate farming companies supplying global supermarket chains can all benefit, as can small and large food processing enterprises, distribution companies and food retailers.

This chapter is largely based on an overview project *Energy-Smart Food for People and Climate* as undertaken by the authors (and others) for the FAO and launched at the 17th Conference of Parties of the UN Framework Convention on Climate Change (UNFCCC) held in Durban in December 2011 (FAO, 2011a). Becoming "energy-smart" implies using low-carbon energy systems in an efficient manner and improving energy access whilst enabling the achievement of food security and development goals. Energy-smart food is inexorably linked with the efficient use of water for agricultural production and processing purposes such as livestock water supplies, crop irrigation, vegetable washing, meat processing etc.

Total energy-related costs as a share of the total production costs vary widely with the food product and agricultural management practices employed, but the proportion is usually relatively high, particularly for agri-food systems in high-GDP countries. For example, in the USA, energy-related costs ranged from 10% of total crop production costs for soybean to 31% for maize (DEFRA, 2010a). In low-GDP countries, agricultural development can be constrained by fossil fuel prices particularly in countries where imported fossil fuels are a high burden on total GDP. The correlation between energy prices and food prices is therefore of concern as fossil fuel prices rise or fluctuate. Farming costs are becoming more dependent on and linked to the fossil fuel market and the poorest small-scale subsistence producers and consumers of staple foods are the most vulnerable to price fluctuations and spikes. Future high and volatile fossil fuel prices, global energy scarcities, and the need to limit GHG emissions are the key drivers for the global agri-food sector to become more "energy-smart".

6.1.1   *The key challenges*

The global agricultural and fishing industries are dependent on natural energy flows from the sun and the various forms of chemical energy stored biologically in the soils and oceans which are essential for plant growth. Additional energy inputs (also termed energy subsidies) assist these natural processes in order to make a given area of land or unit of water produce more food than it would do naturally. Here is where humans have learned to intervene. Basically, industrialized agriculture has evolved into adding auxiliary energy ("subsidized energy") to the system. In essence, as well as using natural sunshine during the process of photosynthesis, agriculture converts inedible (exo-somatic) energy from fossil fuels to produce edible (endo-somatic) energy – that is, the food we consume. The various means of achieving this vary widely between countries and cultures.

Auxiliary energy can be in the form of human labor, draught animal power, renewable energy, and liquid, gaseous or solid fuels that can be combusted in engines to produce heat, mechanical energy or electricity. Meeting the global food demand of a growing world population over the past century has, at least in part, been achieved by significantly increasing the fossil fuel inputs along the entire agri-food chain, but this high dependency of the food system on fossil fuels is now becoming cause for concern. By 2030 it is expected that as a result of continued population and economic growth the global demand for energy will rise by 40% (IEA, 2011), water use by 40% (WEF, 2011) and food demand by 50% (Bruisma, 2009). These increasing demands will have to be met in the context of an already stressed natural resource asset, limited availability of productive land area and climate change impacts from extreme weather events. Agriculture and the entire agri-food supply chain contribute approximately 22% of total annual anthropogenic GHG emissions, plus an additional 15% contributed from land use changes, including deforestation to gain more agricultural land area (FAO, 2011a; IPCC, 2007a). Additional risks from potential climate change impacts on food supply security means that the resilience of the agri-food sector requires careful evaluation. Analysis of probable climate change impacts on agricultural productivity up to 2050 has shown that negative effects could lead to a reduction in food availability and human well-being, particularly in developing regions (IPCC, 2007b; Spielman and Pandya-Lorch, 2010).

The global economy will have to make a major transition from business-as-usual to address these challenges. However, there have been few signs over recent decades of a major transition towards different patterns of food production that are more focused on rural development, eco-logical compatibility and quality (Arizpe, *et al.*, 2011). The magnitude, complexity, and need for urgent action, explains the current importance being given to the food/energy/water nexus. The unsustainable supplies of fossil fuels and water will inevitably result in impacts on land use, land acquisitions and the environment at local, national and global levels. "Doing more with less" is the related concept captured by the FAO in its proposed "*Save and Grow*" paradigm shift (FAO, 2011b). To achieve this, the global agri-food sector will require action to be taken by all stakeholders regarding innovative agricultural and fisheries practices, technology development, new policies and institutional arrangements at all levels.

Making the transition to a low-carbon, "green economy" will improve human well-being and social equity while significantly reducing environmental risks and ecological scarcities (UNEP, 2011). Pertinent to sustainable development issues covered in this chapter are the use of more ecologically-friendly farming methods that improve yields significantly for subsistence farmers; improving freshwater access and efficiency of use; promoting improved resource and energy efficiency; substituting fossil fuels with low-carbon resources and clean energy technologies; and reducing food losses along the supply chain. However, it must be noted that the primary production sector[3] is complex relying on healthy soils, a high nutrient status, adequate supplies of water, careful management of resources, and a reliable flow of energy inputs.

Any attempts to reduce inputs that would impact negatively on the productivity, processing activities, or quality of food, should not be promoted. The energy inputs wasted through losses of related resources (such as water leakage, soil erosion, food losses and/or wastes) need to be addressed. Diminishing the volumes of food losses along the whole supply chain will lead to reductions in demand for land, water, energy inputs and greenhouse gas (GHG) emissions[4].

Addressing the energy status of the agri-food sector highlights options for reducing energy intensity[5] and increasing local use of renewable energy resources. Such energy sources arising "from" the sector, being produced on-farms or in processing plants, include wind, solar, small-hydro and bioenergy (for example produced from crop residues) that can be used either on-site to substitute for purchased direct energy inputs, or be sold for use off-site to earn additional revenue for the farm or processing plant owner.

Water supplies are also an essential component of food production. Only a broad systemic approach to the "water-energy-food security nexus resources platform"[6] will lead to the cross-sectoral systems solutions needed to achieve efficiency improvements in agriculture, energy and water use. Careful management of land and water resources can maximize returns and therefore provide green economy benefits such as poverty reduction, economic growth, and job creation.

Solutions for a future "Green Economy" also highlight increasing international and national concerns about reliable access to clean water. This is primarily a consequence of poor management and inequalities in distribution. Currently around 0.9 billion people lack access to safe drinking

---

[3] Primary production here includes cropping, pastoral-fed and intensive livestock, aquaculture and fishing.
[4] This chapter focuses on the mitigation of energy-related CO_2 emissions whereas more detailed analyses of mitigating CH_4, N_2O and HFCs produced from the agri-food sector appear elsewhere; see for example, the UK's Foresight Project on Global Food and Farming Futures (GoS, 2011) and the USEPA (2006) report on non-CO_2 gases.
[5] Energy intensity concerns the amount of energy used per unit of food produced. It better highlights the links between energy and production than does the the amount of energy per unit area of land.
[6] http://www.water-energy-food.org/en/home.html

water and 2.6 billion people lack access to adequate sanitation. In addition, 1.3 billion people lack access to electricity, 2.7 billion have no access to modern and healthy forms of cooking, close to 1 billion people are undernourished, and many ecosystems are stressed, as evidenced by 25% of agricultural land that has become degraded. The projected increase in population, mainly urban and relatively wealthy, will put significant additional pressure on ecosystems leading to further degradation, especially in light of the fact that urban and wealthy societies consume natural resources at a very high and usually unsustainable way.

Energy, water and food demands are interconnected, with growing trade-offs occurring as demands are set to increase. Water is a key component for the extraction, transport and processing of fossil fuels and the cooling operation of thermal power stations. There is also growing competition for water supplies between hydro-power generation and irrigation demands (at both the large- and mini-scales) which is exacerbated by climate change impacts on seasonal supplies, snowpack melts, glacier recession, droughts, floods etc. Water is also essential for biomass production.

Forest production and the wood product processing industry are not included in this chapter except where woody biomass by-products can be used to provide useful energy for the agri-food sector or where agro-forestry features in the rural landscape. International trade and "food miles" will not be discussed in detail, nor issues relating to the impacts of energy use and farm management on water quality, soil nutrient status, groundwater supplies, biodiversity, or the sustainability of a farm enterprise, unless there is a direct relationship with energy supply technologies.

Implementing energy-smart food systems could improve efficiencies of water use (through wise active water management) as well as reduce overall energy consumption, encourage deployment of alternative energy sources, and enable agricultural production systems to produce food and energy simultaneously in a water-efficient way. This will help provide greater security for food, water and energy supplies, meet energy access targets, identify climate-smart solutions, and implement systems that are sustainable and resilient to future climate change.

The four key elements relating to energy-smart food production, sustainable agriculture and rural development are:

- having the right energy mix to address the ambition to produce 60% more food by 2050;
- consolidating rural livelihoods in a sustainable way by provision of basic energy services;
- providing affordable and secure energy systems (through renewable energy systems when possible) now and in the future; and
- addressing how the whole agri-food industry could become more sustainable and energy-smart.

This implies using low-carbon energy systems in an efficient manner, providing greater energy access, and enabling the achievement of national food security and sustainable development goals. Specific objectives of this chapter are therefore to:

- outline the current status of agriculture and the agri-food supply chain across the entire spectrum;
- illustrate methods by which small and large-scale farmers and fishers can make their management practices less vulnerable to future energy supply interruptions and price fluctuations whilst also reducing GHG emissions;
- discuss how access to modern energy services could be provided to subsistence farmers with resulting increases in productivity, reduced losses and improved livelihoods;
- consider how renewable energy systems, including bioenergy from local biomass resources, can be deployed to benefit the sector by:
  o supplying useful direct energy inputs for farms and fisheries;
  o selling excess energy carriers off-farm to gain additional farm revenue;
  o supplying energy to the agri-food processing industry;
  o providing transport biofuels for use by the sector; and
  o identifying gaps in the knowledge where further research would be beneficial;

- highlight the links between energy and food losses, hence reinforcing the case to minimize food losses along the supply chain and at consumption; and
- briefly discuss policies that can be employed at various levels to improve knowledge and capacity as well as to develop the resilience of the sector so it can adapt to possible future energy supply constraints and climate change impacts.

The key challenges identified and discussed throughout this chapter are as follows:

- The international goal of increasing the volume and quality of global food supplies to meet the ever-growing demand could be constrained by the projected higher costs, insecurity and limited reserves of global oil and natural gas supplies (IEA, 2011), coupled with the global consensus of the need to reduce GHG emissions.
- Agri-food systems, from small local to large scales, will be required to produce more food, essentially through increased productivity per land area. Over time, improving access to energy systems for subsistence farmers and rural communities will be needed. Competing land use and energy access are discussed in more detail on Chapter 2 (this volume).
- The global agri-food system will need to provide sufficient, secure and "climate-smart" food supplies[7] during the next few decades. The losses along the food supply chain, currently around one third of all food produced, will need to be reduced by appropriate investments and policies.
- Trying to become more "energy-smart" along the agri-food chain by reducing the high dependence on fossil fuel inputs, will require enabling policies, increased public awareness, education, behavioral change and significant investments in clean energy technologies and efficient water use.

6.1.2 *Scales of agricultural production*

Agricultural systems are complex and diverse. They range between basic subsistence small hold-ings growing food for their family's own consumption to large commercial, corporate farms supplying huge supermarket chains across the world. All levels of this spectrum are dependent on energy inputs:

- Industrialized agriculture has developed in recent decades where cheap fossil fuel options and electricity have been widely available, and could rely on the high energy intensity that can be provided by fossil fuels. This has led to family farms merging into larger, corporate-owned properties, often growing monocultures, and with the field layout changed to accommodate large machinery used to enhance performance and productivity. Demand for low-input, agri-chemical free, organic food continues to grow, mainly in OECD countries, but this form of farm management does not necessarily have lower energy inputs per kilogram of food produced compared with high-input systems.
- Human labor and draught animal power continue to provide energy at the subsistence farm scale as they have done for centuries in most low-GDP countries. As well as feeding their families, many small-scale farmers provide fresh food to local community markets, but this traditional situation is evolving in many places so that these small farmers also supply raw food products to local and regional processing plants. To achieve this successfully, various educational and aid campaigns have been undertaken to increase productivity by supplying higher yielding seed varieties (including genetically modified options), improving farm management skills by capacity building, increasing the availability and use of artificial fertilizers and agri-chemicals, and encouraging the uptake of mechanization to reduce drudgery. This, in turn, has reduced demand for human labor and encouraged the drift to urban environments such that for the first time, now over half the world's population lives in cities. This is estimated to increase to 75% of global population being urban dwellers by 2050 (Smith, 2012).

---

[7] See FAO Climate-smart Agriculture initiative at http://www.fao.org/climatechange/climatesmartpub/

Table 6.1.   Simple typology of typical "small" and "large" scale farms and fisheries based on qualitative assessments of unit scale, levels of production intensity, labor demand, direct and indirect fossil fuel dependence, investment capital availability, food markets supplied, and energy intensity (FAO, 2011a).

Scale of farm/fishery	Overall input intensity	Human labor units	Animal power inputs	Fossil fuel dependence	Capital availability	Major food markets	Energy intensity
Subsistence level	Low	1–2	Common	Zero	Micro-finance	Own consumption	Low
Small family unit	Low/ medium	2–3	Possible	Low/ medium	Limited	Local fresh/ process/own use	Low/ high
	Medium/ high	2–3	Rarely	Medium/ high	Limited	Local fresh/ regional process/ own use	Low/ high
Small business	Low/ medium	3–10	Rarely	Medium/ high	Medium	Local/regional/ export	Low/ high
	Medium/ high	3–10	Never	High	Medium	Local/regional/ export	
Large corporate business	High	10–50	Never	High	Good	Regional process/ export	Low/ high

Differentiation between "small" farm and "large" farm enterprises is difficult since there are no clearly defined boundaries. Supplying food processors or large supermarket retailing companies with fresh products is feasible at all farm scales, other than subsistence levels. To do so, small or large producers usually have to invest in modern storage facilities that often require electricity and/or fossil fuel inputs. Good roads and/or rail links are also important components for getting the food products to the markets. A simple classification is provided to illustrate the concepts of "large" and "small" as discussed throughout this chapter (Table 6.1). The table does serve to illustrate that no strong correlation exists between input intensity and fossil fuel dependency. However, it is realized that many exceptions to the typology exist, such as small enterprise tea plantations employing many pickers or small family fishing boats having relatively high fossil fuel dependence and hence a large share of the costs. Energy efficiency improvement opportunities exist on most farms regardless of scale, whether it be changing the type of light bulbs, maintaining water pumps, or insulating livestock buildings. Full details of energy efficiency technologies on farms are provided in Chapter 2 (this volume).

6.1.2.1   *Subsistence*
The most basic forms of small-scale, agricultural and fishing activities produce food solely for family and local consumption. They use very few energy inputs, usually only human labor and draught animal power that are not included in world energy statistics, partly because they are diffuse but also because energy balance data is unavailable on the total additional food and feedstuffs needed to offset the energy input for the work outputs obtained. Energy access and livelihoods are the main priorities for subsistence farmers and fishers but they are often constrained by finance availability.

6.1.2.2   *Small family farms*
Family units that do not employ additional labor typically include small plot market gardens; organic vegetables, fruit and rice growers; pig and poultry enterprises; privately-owned fishing boats; and small dairy herds from a few cows up to 100 or more, depending on the degree of

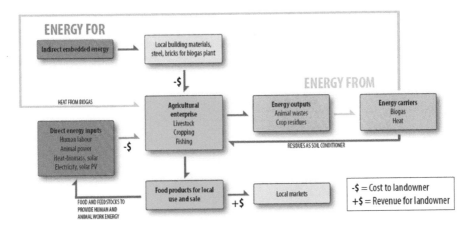

Figure 6.1.   Energy and money flows through a typical, small-scale, family-managed, farming enterprise with relatively low inputs and the opportunity to produce some renewable energy for use on-site. Outputs are mainly fresh food for local market consumption, though local food processing companies may also be supplied.

modernization. Energy efficiency options usually exist for managers of such small enterprises to consider, except perhaps those dependent only on human and animal power. In addition, small mixed farms commonly utilize non-fossil fuel forms of direct energy such as solar heat for crop drying, local fuelwood resources for heating of dwellings in colder climates, on-farm produced biogas for cooking, and electricity for lighting generated from solar photovoltaic (PV) systems (Fig. 6.1).

6.1.2.3   *Small businesses*
These tend to be family operated but at a slightly larger scale than family units and often employing several staff. They are usually privately-owned, and the owner/managers have opportunities to reduce fossil fuel dependence as a result of energy efficiency improvements as well as from renewable energy installations on-farm for local supply (similar to those outlined in Fig. 6.1). Some of the renewable energy generated could also possibly be sold off-farm (see below). Local community benefits may also result.

6.1.2.4   *Large farms*
At the other extreme end of the spectrum, corporate[8] agri-food production systems are highly dependent on the purchase of high direct energy inputs throughout the supply chain (Fig. 6.2). Examples include owners of fishing trawler fleets, beef feedlots, sugar production and processing companies, and palm oil plantations. Large farm estates can be owned and managed by the processing mill company that they supply, or by a grower-owned co-operative when co-benefits are more likely to flow to the local community. Large corporate organizations usually have access to finance for capital investment, such as for advanced energy efficient and renewable energy technologies, with the energy carriers produced (electricity, gas, heat, liquid fuels) being either used on-farm or sold off-farm to earn additional revenue.

A transition of the millions of subsistence fisheries and farms to more advanced, primary production enterprises will usually require considerable investments in machinery, fertilizers, irrigation, transport and infrastructure, all of which are dependent on energy inputs. Should more low-GDP country governments and aid agencies choose to encourage this transition, both global

[8]"Industrialized", "market-based" "commercial" and "multi-national" are terms used synonymously with "corporate" to describe modern, large-scale, agri-food systems that produce food, fish, animal feed or fibre.

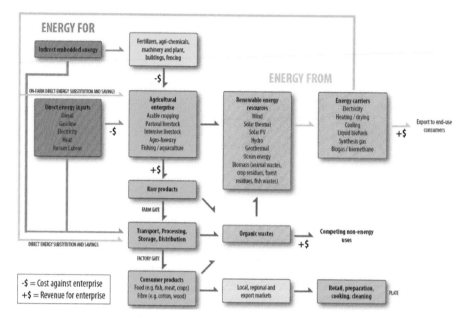

Figure 6.2.   Energy flows through a large-scale, high input, corporate farming enterprise with raw food and fiber products mainly supplying local and regional processing plants, supermarket chains or exported. Renewable energy resources, when available from farms and processing plants, can be utilized on-site to substitute for purchased direct energy inputs to meet production and processing demands, or exported off-site so that revenue is then earned from both food and energy products.

energy demands and GHG emissions will be increased. Therefore, aiming to industrialize existing low-input agricultural systems simply by increasing fossil fuel subsidies into business-as-usual operations may no longer be a feasible and justifiable option. Leapfrogging to more efficient and sustainable production systems, including the use of renewable energy, (resulting in a greater portion of oil, coal and gas reserves then being available for key applications where no other economic options exist such as feedstocks for the chemical industry), could be the most viable solution for the agri-food sector during the next few decades.

6.2   ENERGY INPUTS AND GHG EMISSIONS

The end-use energy demand of the global agri-food chain is around 32% of the present total global final energy demand (Fig. 6.3). High-GDP countries consume over 50% of the global energy demand for the agri-food sector, and with a higher average energy/capita rate of approximately 35 GJ/capita/year compared with low-GDP countries at around 8 GJ/capita/year. High-GDP countries also have a higher share of energy inputs in the processing and distribution stages of the supply chain (FAO, 2011a).

Changes in total GHG emissions from the agri-food sector over time can be useful indicators of system efficiency, particularly when linked with the costs of production and supply. Anthropogenic GHG emissions totaled over 44 Gt $CO_{2\text{-}eq}$[9] in 2006 of which emissions from primary production (including a share of energy for electricity demand) were around 14% of the total (Fig. 6.4). Land

---

[9]The unit ton of carbon dioxide equivalent (t $CO_{2\text{-}eq}$) accounts for the varying warming effects and half lives in the atmosphere of different gases so is used to compare the global warming potential of greenhouse gases and to assess their cumulative impacts.

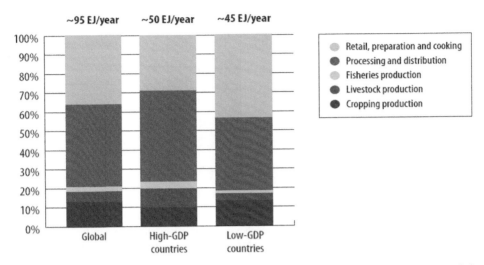

Figure 6.3.   Indication of global, shares of end-use energy demands throughout the agri-food supply chain showing total final energy for the sector and a breakdown between high-GDP and low-GDP countries. (Based on Giampietro, 2002; Smil, 2008; IEA, 2010; Woods *et al.*, 2010; GoS, 2011 and others). [Note: These analyses were based on the range of data available as presented in the literature but these were at times unreliable, incomplete and out of date since agri-food energy use patterns are rapidly changing. In addition, there is no standard collection methodology for food-related energy and GHG data.]

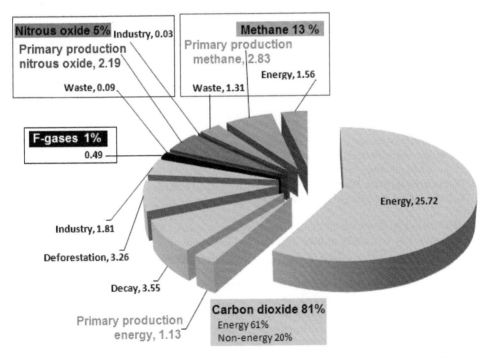

Figure 6.4.   Global anthropogenic greenhouse gas emissions by gas in 2006 totaled 44.17 Gt $CO_{2\text{-eq}}$ of which emissions from primary production (methane, nitrous oxide and carbon dioxide) were around 14% of the total; land use and land use change accounted for around 15%; and energy approximately 65% (mostly 25.72 Gt CO_2). (Based on IPCC, 2007a).

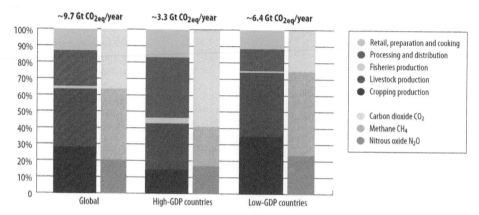

Figure 6.5.   Global shares of anthropogenic greenhouse gas emissions along the agri-food supply chain with breakdown by gas and also for high- and low-GDP countries. (Based on Giampietro, 2002; USEPA, 2006; Smil, 2008; IEA, 2010; Woods *et al.*, 2010; GoS, 2011 and others). [Note: Some of the data presented are uncertain, particularly for low-GDP countries where possibly over two thirds of total agri-food GHG emissions originate but are not always accurately recorded.]

use and land use change accounted for around an additional 15% and energy approximately 65% (mostly CO_2). In primary production, emissions result from:

- *carbon dioxide* from fossil fuel combustion used on farms and in fishing vessels, plus electricity inputs, contributes around ~3–4% of total GHG emissions (plus those arising from the oxidation of soil carbon, but noting that carbon released from biomass above or below ground, or from respiration from aquatic organisms, is usually balanced by photosynthetic carbon uptake);
- *methane* from agriculture accounts for nearly half of total anthropogenic methane and is over 6% of total GHG emissions mainly from ruminant digestion (goats, sheep, cattle, deer), paddy rice cultivation, anaerobic soils and sediments, plus small quantities from some aquaculture sites;
- *nitrous oxide* from agriculture contributes ~5% of total GHG emissions, most coming from the action of soil bacteria on ammonium and nitrates originating from inorganic fertilizer use, manure and other organic wastes, crop residues and nitrogen fixing plants; and
- *fluorocarbons* (or the family of F-gases) that give a small contribution, mainly from refrigerant leakages along the cool chain (GoS, 2011).

GHG emissions from the entire agri-food chain, including landfill gas produced from food wastes, presently contribute around 22% of total emissions (Fig. 6.5). The share of methane emissions in low-GDP countries is around double those from high-GDP countries mainly due to paddy rice emissions (IPCC, 2007a). Conversely, a greater share of CO_2 emissions arises in high-GDP countries, mainly as a result of fossil fuels combusted to power farm machinery and transport vehicles and to generate heat and electricity for food storage and processing. Overall on a per capita basis, the average emissions in high-GDP countries are approximately 2.2 Gt $CO_{2\text{-eq}}$ per capita being around double those per capita in low-GDP countries.

GHG emissions arise at all stages of the food system. They are associated mainly with the consumption of fossil fuel-based energy used directly in food production, the synthesis of nitrogen fertilizers, soil preparation, and other emissions associated with husbandry or decomposition of organic matter (including food wastes). Emissions on-farm are a mix of carbon dioxide, nitrous oxide and methane, while in the post-farm operations emissions are mainly constituted by carbon dioxide associated to energy use. Nitrous oxide and methane are powerful greenhouse gases with

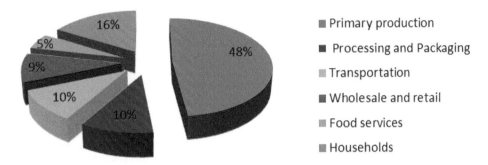

Figure 6.6.   GHG emissions arising along the U.K. agri-food supply chain (FAO elaboration of 2009 data in DEFRA, 2011).

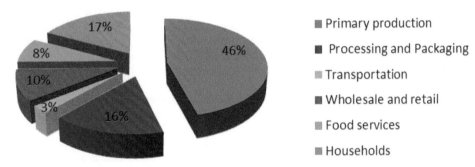

Figure 6.7.   GHG emissions along the agri-food chain of the USA (FAO elaboration based on USDA 2009 data).

global warming potentials much higher than carbon dioxide over a 100 year timeframe. However, when measuring these emissions using carbon dioxide equivalents the energy component associated with carbon dioxide emissions is most relevant, especially in post-harvest operations of the food chain. Beyond the farm gate, energy is expended in transport, storage, processing, retailing and preparation. GHG emissions associated with fisheries and aquaculture production are relatively modest and mainly associated with fuel used in fishing vessels and feeds used in aquaculture. Identifying the sources of emissions is important for targeting interventions.

At the household level in the USA, Weber and Matthews (2008) calculated the total of 8.1 t $CO_{2\text{-eq}}$ food-related GHG emissions per average household per year stemmed from CO_2 (44%), nitrous oxide (32%) and methane (23%) with a minor contribution from CFCs and other industrial F-gases from refrigerants (1%). In the UK, Garnett (2008) showed 18% of UK total GHG emissions came from the agri-food sector of which around half arose from agricultural production and the remainder from beyond the farm gate including food processing, packaging, transport, retail, commercial food services, preparation and food waste disposal (Fig. 6.6).

A similar result can be derived from data provided by the United States Department of Agriculture (Fig. 6.7). These results depend on a number of factors, including the definition of the boundaries of the food system. For example, the inclusion of dishwashing, throw-away plates and cutlery, or international food trade could significantly change the overall picture. For example, food net trade in the U.K. food system would be responsible for around 24% of total emissions of the total food supply chain, bringing down the relative proportion of emissions attributable to farming to just 32% (FAO elaboration of 2007 data in DEFRA, 2010b).

Estimates of global fishery sector GHG contributions based on current total energy requirements include 40–90 Mt $CO_{2\text{-eq}}$/year from fuelling fishing vessels, 35–40 Mt $CO_{2\text{-eq}}$/year from aquaculture activities, and 10–15 Mt CO_{2-eq}/year for post-harvest and processing. Other estimates include 3–4 Mt CO_{2-eq}/year from the air freighting of 435,000 t of fresh fish, and up to 340 Mt CO_{2-eq}/year for non-air transport, though this is difficult to quantify and estimates vary widely (GoS, 2011; FAO, 2009b).

Measures for changing agricultural processes and practices to reduce GHGs (Garnett, 2008) include stimulating carbon sequestration by increasing the use of conservation agriculture systems (such as no-till cropping); incorporating more organic matter into soils; optimizing nutrient use by more precise dosage and timing of applications; integrating aquatic systems with agricultural systems; improving productivity outputs per unit of GHG generated; employing anaerobic digestion of manure, slurry, fish and animal wastes; and reducing the carbon intensity of fuel and raw material inputs through improvements in energy efficiency, selection of materials and use of renewable energy.

The use of water-source, ground-source or air-source heat pumps also have a number of applications in agriculture. Based on a simple refrigeration process, a heat pump can acquire low temperature heat from the air, water, soil etc. and upgrade it to a higher, more useful, temperature. For example, it can take the heat from milk, thereby chilling it, and then use this heat to provide hot water for cleaning the milking plant. Or it can take heat from the air or ground and concentrate it for heating pig or poultry sheds or greenhouses in winter, as well as operate in reverse to provide cooling in summer. When used to replace oil, natural gas, LPG or direct electrical resistance for heating applications within the farming industry, due to the high efficiency (coefficient of performance), a reduced carbon footprint usually results (DARDNI, 2013).

In relation to improving energy efficiency, cost-effective reductions in energy use could be achievable at all stages of the food supply chain in most countries (IPCC, 2007c; 2007d). Individual businesses need to decide how best to reduce their specific energy demands and related GHG emissions as these are determined by local circumstances. For example, selection of new equipment should not be made only on the initial capital cost, but should include on-going energy inputs assessed through life cycle analysis evaluations. Many successful case studies exist of energy demands being reduced at all stages along the agri-food chain with just a few examples given in the sections below to illustrate some potential options. Further details are given in Chapter 2.

Good knowledge of an entire system can be more effective in identifying the most efficient alternative rather than taking a demand-side or a supply-side approach. For example, to improve energy efficiency along the agri-food supply cool chain, it may be better to optimize heat loads during the food processing phase than to improve the energy efficiency of refrigeration cooling systems (Cleland, 2010).

6.2.1   *Energy inputs for primary production*

Ambitions to increase global food supplies through increased productivity of crops, animals and fish resources may be partly constrained by the limited future availability of cheap fossil fuel supplies. Small-scale agricultural and fishery production systems in low-GDP countries may not be able to emulate past efforts of high-GDP countries in achieving desirable productivity increases if to do so will depend on increased reliance on fossil fuels.

The major energy consuming technologies for cropping, livestock production and fishing are briefly outlined below. Major variations exist depending on the type and scale of enterprise such as cropping (see Chapter 3) or dairying where, for example, the direct energy inputs of an extensive, unsubsidized, grazing enterprise in Australia at 2–3 GJ/ha can be compared with intensive, subsidized, dairy farming systems in The Netherlands at 70–80 GJ/ha (Smil, 2008).

6.2.1.1   *Tractors and machinery*

In 2005, the 27 million four-wheel tractors then operating in the world (around two thirds of them in high-GDP countries) consumed ~5 EJ (125 billion liters) of diesel fuel for land development, on-farm transport and field operations (Smil, 2008). A further 1.5 EJ/year was consumed for the manufacture and maintenance of tractors and implements. For the numerous small, pedestrian-operated two-wheel designs used mainly by small farmers in low-GDP countries, additional fuel demand is not known but illustrates the benefits that the availability of cheap fossil fuels, often through government subsidies, has been able to deliver at the small farm scale over recent decades. A life-cycle analysis (LCA) to compare GHG emissions from using animal power for cultivation (including ruminant methane emissions) versus using these small mechanical cultivators running on gasoline would be an interesting exercise.

A successful model is from Bangladesh where all farmers have access to the 350,000 two-wheel tractors operating there today, but only around 3–4% of farmers own one. Most farmers employ service providers to hire a machine through a highly developed supply chain (Sims and Baudron, 2012). The deployment of small, mobile, demountable, multi-purpose diesel engines, used for powering irrigation pumps, small boats, tractors, trucks, electricity generators and simple food processing equipment, revolutionized food production and increased productivity significantly after the devastating 1987 floods (Steele, 2011). Public policy changes to import regulations enabled innovative, Chinese-made, farm equipment to be imported that could be easily repaired by local mechanics and sold more cheaply than the more sophisticated and expensive machinery manufactured in India. Therefore Chinese technology developments enabled the agromechanization of Bangladesh to occur. Based upon this experience, both Nepalese and Indian machinery manufacturers have recognized business opportunities by copying the concept that is now being deployed mainly into low-cost, farm machinery markets in rural communities. Farm services have expanded as a result of the versatility and transportability of this equipment (Biggs and Justice, 2011).

In sub-Saharan Africa, approximately 65% of cultivation is still carried out using hand-tools, 25% by draught-animal powered technologies (Sims and Baudron, 2012) and the remaining 10% using tractors (but expected to reach 25% by 2030). To increase the level of agricultural mechanization will require access to affordable and reliable fuel supplies together with suitable financing arrangements, ownership agreements, hiring opportunities for tractors when used off-farm, availability of spare parts, maintenance and repair services, skill upgrading and education (Ashburner and Kienzle, 2011).

6.2.1.2 *Irrigation*
Around two thirds of global water supplies used for irrigation are drawn from underground aquifers where extraction rates exceed the recharging rates. This, plus the high cost of water desalination, is why some countries have considered reducing irrigated crop production and relying more on imported grain.

Mechanical pumping of water on to approximately 10% of the world's arable land area, of around 300 Mha, consumes around 225 PJ/year for powering the pumps plus 50 PJ/year of indirect energy for manufacture and delivery of irrigation equipment (Smil, 2008). Irrigated land provides around 40% of global cereal supply due to achieving higher yields than rainfed systems and giving the option for double, or even triple, cropping practices (FAO, 2011b) with better production efficiency. In Africa only 4% of cropland is irrigated, mainly due to a lack of available financial investment. In India, irrigation practices have increased yields but resulted in around 3.7% (58.7 Mt CO_{2-eq}) of the country's total GHG emissions in 2000 (Nelson *et al.*, 2009). Energy intensive electricity pumping in deep wells accounted for two thirds of the total demand, with projections for it to rise to 87 % in 2050 as the shallower water reserves deplete.

6.2.1.3 *Fertilizers*
Applications of nitrogen (N), phosphorus (P) and potassium (K) macro-nutrients have contributed significantly to crop yield increases in recent decades and this demand will probably continue to expand, mainly in low-GDP countries. In 2000, energy embedded in inorganic fertilizer manufacture was around 7 EJ globally (GoS, 2011; Giampietro, 2002; Smil, 2008). Nitrogen fertilizer production alone accounts for about half of the fossil fuel used in the primary production sector and significant amounts of nitrous oxide can be emitted during the production of nitrates (GoS,

Table 6.2.   Total on-farm energy inputs (including direct energy for tractor fuels, heating, lighting, motors and indirect energy for feed, buildings and equipment) per unit of animal food product (Smil, 2008).

Food product	Animal feed conversion	Total direct and indirect energy inputs
Chicken meat	4.2 kg/kg edible meat	25–35 MJ/kg meat
Pork	10.7 kg/kg edible meat	25–70 MJ/kg meat
Beef (feedlots)	31.7 kg/kg edible meat	80–100 MJ/kg meat
Laying hens	4.2 kg/kg eggs	450–500 MJ/year
Dairy milk	0.7 kg/liter milk	5–7 MJ/liter of fresh milk
Fish (trawler capture)		5–50 MJ/kg (mainly liquid fuel inputs)
Shrimps		107–121 MJ/kg

2011). Average annual N, P, K applications range from virtually zero in sub-Saharan Africa to 500, 50, 100 kg N, P, K/ha respectively in double-cropped Chinese rice fields (Smil, 2008). The uptake by crops of N applied by fertilizers tends to be inefficient, being for instance as low as around 26–28% of the total nitrogen applied in fertilizers for cereals and 20% for vegetables in some regions of China (Miao *et al.*, 2011). More efficient methods of transport and application to reduce wastage, and hence to also lower both direct and indirect energy inputs, are covered in Chapter 6.

6.2.1.4   *Livestock*
Intensive livestock enterprises usually rely on bought-in animal stockfeed delivered to the farm which can be a significant component of total energy inputs (Table 6.2). Regional differences are evident, with low-GDP countries consuming ~1 MJ fossil fuel per MJ of energy in the animal feed product, whereas high-GDP countries consume ~4.3 MJ/MJ (Giampietro, 2002). Extensive pastoral systems for sheep, goats, deer and cattle tend to have lower energy inputs than intensive livestock systems, but any hay and silage conserved on-farm, forage crops grown for grazing or carting (sometimes termed zero-grazing), or stockfeed purchased, also rely on some indirect energy embedded in the feed.

6.2.1.5   *Protected cropping in greenhouses*
Fruit, vegetable and flower production in peri-urban areas using intensive greenhouse designs with closed cycle system, hydroponic or aeroponic cultures (delivering water and nutrients without soil) rely on relatively high direct energy inputs. Artificial lighting and seasonal heating can consume up to approximately 40 MJ/kg of fresh product such as tomatoes or peppers (FAO, 2011b). The total area of simple shade-houses is increasing in countries such as China and South Korea since the energy inputs are relatively low when compared with energy-intensive heated greenhouses. In general, crops grown in glass or plastic cladded greenhouses can have energy intensity demands around 10 to 20 times that of the same crops when grown in open fields (Saunders and Hayes, 2009).

6.2.1.6   *Fishing and aquaculture*
In 2008, 44.9 million people were directly engaged full time or part time in capture fisheries and aquaculture, more than double the number in 1980. This represents 3.5% of the 1.3 billion people economically active in the broad primary production sector worldwide (SOFIA, 2010). Capture fishing is one of the most energy-intensive methods of food production. Boats are relatively high fuel consumers and most owners are acting to reduce energy demand since fuel costs, typically 15–20% of total costs, can equate to 50% of catch revenue for some enterprises (FAO, 2009b). Small boats mostly have inefficient engines with high fuel consumption that cannot be easily improved,

but there is little data on their use. Indirect energy inputs for boat building and maintenance is around 10% of fuel energy consumption (Smil, 2008).

The global fishing fleet captures around 80–90 Mt of fish and invertebrates each year consuming around 620 liters of fuel per ton of catch (Tyedmers *et al.*, 2011). Small-scale enterprises produce around half of the total fish catch using a fleet made up of about 4.3 million small vessels of which two thirds are powered by internal combustion engines that rely on fossil fuels and the rest, mainly located in Asia and Africa, use sails and oars (SOFIA, 2010).

Aquacultural enterprises (fish farming and mariculture) produce around 55 Mt of product per year and are expanding. Some enterprises, such as shrimp farming, rely on direct energy for pumping and aerating water as well as on indirect energy for producing and delivering the fish feed. Total energy demand is ~50 PJ/year of direct energy plus ~300 PJ/year for indirect energy, which can equate to ~100–200 MJ/kg of fish protein (Smil, 2008). However, energy input estimates vary widely in the literature.

6.2.1.7 *Forestry*

Energy inputs per hectare over the life of a plantation forest are relatively low, mainly involving machinery use at harvest, replanting and possibly some fertilizer applications. Forest management, timber production, transport distances and processing methods into pulp and paper or timber products vary widely with location, terrain and region. Hence a large range of energy inputs per ton of logs produced and processed results.

6.2.2 *Energy inputs for secondary production*

6.2.2.1 *Drying, cooling and storage*

Crop drying and curing can be one of the more energy-intensive on-farm operations. For example in Zimbabwe, tobacco curing used to account for over half the total national on-farm energy demand (FAO, 1995). Cereals are normally dried artificially after harvest and prior to storage and transport in order to maintain quality. Electricity, natural gas or liquefied petroleum gas (LPG) can be used to provide heat at around 0.5–0.75 GJ/t in order to dry wet grain down to an acceptable moisture content for storage at around 14% moisture content (wet basis). Solar heat can be used for drying grain or fruit, either naturally in the open air or in solar-heated facilities. Innovative drying technologies, such as di-electric drying, can lead to major energy savings.

Storing food can involve between 1 and 3 MJ/kg of retail food product (Smil, 2008). Refrigerated storage, including during transport, can account for up to 10% of the total carbon footprint for some products when electricity inputs, manufacture of cooling equipment, and GHG emissions from lost refrigerants are all included (Cleland, 2010). The total refrigeration component of the carbon footprint for the UK food supply chain can be broken down into transport (24%), retail (31%), and domestic (40%) applications, with the remaining 5% coming from embedded energy in equipment manufacturing. Drying and cooling are not always practiced in low-GDP countries where post-harvest losses, including from pests, can be high.

6.2.2.2 *Transport and distribution*

Recent analyses of energy demand by the transport, distribution and retail components of the food supply chain are beginning to appear through life cycle analysis (LCA) literature[10]. LCA comparisons can provide useful indicators of the impacts from current food production options and the potential impacts from changing demand, but care is needed in interpreting the results[11]. Several LCAs have been undertaken to calculate total GHG emissions associated with individual food products. For example, Williams *et al.* (2007) examined seven products for the UK market

---

[10]For example, a set of LCA publications is available from http://www.fcrn.org.uk/research-library/lca
[11]LCA methodology varies widely in terms of attribution of GHGs to co-products and the boundaries assumed for accounting inputs. Some analyses include GHG emissions from land use change, others do not.

Figure 6.8. Road transport of agricultural products in low-GDP countries can be constrained by the cost or availability of suitable vehicles.

and concluded that food imports from countries where productivity is greater, and/or where refrigerated storage requirements are lower, could lead to a smaller carbon footprint than having a preference for local production.

In 2000, over 800 Mt of global food shipments were made (Smil, 2008), equating to over 130 kg per person. Transport and distribution at various stages along the agri-food chain are therefore vulnerable sectors to fluctuating fossil fuel prices. Transport, under specific circumstances, can account for up to 50 to 70% of the total carbon footprint of some food products (for example when transporting fresh fruit or fish by road to markets several hundred kilometers away). In low-GDP countries however, typical transport distances to markets can be a lot less where poor quality roads or lack of suitable vehicles constrain travel (Fig. 6.8).

Globalization in the past two decades appears to have increased the average travel movement of food products by 25%. Locating production and handling of food physically closer to areas of high population density can help to reduce energy inputs for road transport to markets, often termed "food miles" (Heller and Keoleian, 2000). Taking an extreme case, the average US household consumes around 5 kg/day of food with the average transport distance per kg totaling 8 240 km[12] (Weber and Matthews, 2008). Trips by householders to purchase food can account for an additional 1–4 MJ of vehicle energy inputs per kg of food purchased.

Producing specific crop and animal products in locations where productivity is naturally the highest before transporting them over long distances can at times outweigh any energy savings from reduced food miles since transport by ship or rail have relatively low MJ/t km (Table 6.3). For this reason, for the average household in the USA, "buying local" would reduce GHG emissions by only around 4–5% at the most (Weber and Matthews, 2008) because American diets are particularly diversified and a change giving preference to local seasonal fruits and vegetables would have a major impact on emissions. The trend in some high-GDP countries towards buying fresh or minimally processed food at "farmers markets" that sell only local produce may save relatively

---

[12]The total freight transport distance to meet food demand from farm to retail (including transport of seeds, fertilizers and feed for livestock), was approximately 1200 billion t-km, or around 15,000 t-km/household/year. 1997 data updated to 2004, including imported food that is likely to have increased the average distance to markets since 1997.

Table 6.3. Shares of global freight transport by mode and relative energy intensity (Bernatz, 2010; Smil, 2008; Heinberg and Bomford, 2009).

Travel mode	Energy intensity [MJ/t km]	Share of global freight transport [% of total t km]	Share of local distribution transport [% of total t km]
Rail	8–10	29%	16%
Marine shipping	10–20	29%	Not applicable
Inland waterway	20–30	13%	19%
Road trucks	70–80	28%	62%
Trolley, cycle, tractor	Variable	Data not available	3%
Aviation	100–200	1%	0% (domestic only)

little energy on transport, but could save energy on processing and packaging of supermarket goods (Bomford, 2011). Overall, total global GHG emissions from transport of food remain far smaller than those emissions from primary production (Weber and Matthews, 2008).

When fresh fruit is exported, international shipping is an important component of the total energy demand, being up to 45% when apples are transported from New Zealand to Europe for example (Cleland, 2010). Frater (2011) undertook detailed surveys of various stakeholders and calculated that 7.67 MJ of energy were consumed per kilogram of apples exported from New Zealand (1.45 MJ in the orchard; 0.51 MJ during post harvest; 1.46 from packaging; and 4.24 MJ for shipping).

Air transport is relatively costly in terms of both energy intensity and economic cost, so it is rarely used for food products. For example, only 0.5% of fresh fruit imported by the USA from many countries is shipped by air (Bernatz, 2010). Long-distance transport of food by ship is more energy efficient and hence gives lower GHGs (between 10 and 70 g CO_2/t km) than rail (20–120 g CO_2/t km) or road (80–180 g CO_2/t km) (IMO, 2008).

6.2.3 Food processing

The total amount of energy needed for processing and packaging has been calculated to lie between 50 and 100 MJ/kg of retail food product (Smil, 2008). The food processing industry requires energy for heating, cooling, and electricity with the total demand being around three times the direct energy consumed behind the farm gate (White, 2007). In addition, energy is embedded in the packaging which can be relatively energy-intensive due to the use of plastics and aluminum. It accounted for around 5% of total weight of supermarket food purchases in the UK with ~60–70% of that being recyclable (LGA, 2009). For processing fish, the direct energy demand for ice making, canning, freezing, drying/curing and producing fish meal and fish oil by-products is ~0.5 PJ/year.

The low energy efficiency of smaller-scale food processing plants in many low-GDP countries enables the application of improved technologies and measures to yield considerable environmental and economic benefits, even though energy bills are typically only 5–15% of total processing costs. Simple, general maintenance on older, less-efficient processing plants can often yield energy savings of 10–20% for little or no capital investment. Medium-cost investment measures (such as optimizing combustion efficiency, recovering the heat from exhaust gases, and selecting variable or the optimum size of high efficiency, electric motors) can give typical energy savings of 20–30%. Higher savings are possible but usually require greater capital investment in new equipment (IPCC, 2007c).

6.2.3.1 Preparation and cooking

Food storage and preparation in households in high-GDP countries consume electricity for the operation of refrigerators, freezers, microwave ovens and appliances (~40% of total household food-related energy); heat for cooking on stoves and ranges (~20%); and for heating water and operation of dishwashers (~20%) (Heller and Keoleian, 2000). Washing dishes in hot water consumes around 2–4 MJ/kg of food consumed (Smil, 2008). Cooking typically requires 5–7 MJ/kg of food.

In low-GDP countries cooking using fuelwood on open fires that are only 10–15% efficient can require 10–40 MJ/kg when boiling rice and up to 200 MJ/kg when cooking beans that take around eight times longer (Balmer, 2007). Using improved biomass stove designs of around 20–40% efficiency would reduce the fuelwood energy demand accordingly. In rural China, Nepal and elsewhere, domestic-scale anaerobic digesters can produce about 24 MJ of biogas per day, sufficient to cook three meals and provide lighting for a household of five people (Bogdanski et al., 2010a).

Overall, the total energy demand to provide food on the table can be a significant share of a nation's total consumer energy supply. For example, it is ~15.7% in the USA (Canning et al., 2010), ~20% in UK (GoS, 2011) and as high as ~30% in New Zealand which is a food exporting country. Low-GDP countries can have relatively higher shares in spite of the lower energy demands for transport and food processing. For example in some African countries, the share of national energy used for the agri-food chain can be as high as 55%, of which demand for primary production is typically around 10%, for transport and processing 15%, and for cooking and preparation up to 75%.

6.3   THE HUMAN DIMENSION

Energy efficiency opportunities exist at all stages along the agri-food chain for both large and small-scale farming systems in high- and low-GDP countries (FAO, 2011a; also see Chapter 2). Although more energy efficient technologies are being deployed, behavioral change is the main determinant of successful reductions in energy demands and GHG emissions. Most farmers appreciate that on-farm energy efficiency improvements can be deemed successful only if productivity does not decline as a result. But there are many win-win solutions where energy can be saved at the same time as productivity, and hence revenue, increases. For subsistence farmers, improved energy access can lead to better livelihoods but even under conditions of limited energy supply, using the available energy smartly is imperative to enable affordability of supply.

To help reduce overall demands by the agriculture sector for energy, as well as for water and land, avoiding food losses along the agri-food chain could be a significant response. Arguments for dietary change away from meat products, and to reduce obesity, have also been promulgated to reduce GHG emissions from the sector. These issues are discussed below.

6.3.1   Food losses and wastage

Approximately one third of the food produced in the world is not consumed. Food losses occur at all stages along the supply chain and equate to around 1200 Mt per year (Gustavsson et al., 2011). If losses and wastage could be reduced at all stages along the agri-food supply chain, including on-farm, when fishing, and during storage, cooking and consumption, this would lower total energy demands and resulting GHG emissions of the sector, whilst reducing competition for land, water and other resources used.

In low-GDP countries, post-harvest activity, particularly improved food storage, is argued by many practitioners as being the main factor in helping farmers to increase their income. This is because of the large amount of food products that can be lost soon after harvest in many of those countries. In sub-Saharan Africa, South Asia and South-East Asia where food is relatively scarce, losses are only around 6–11 kg/capita/year (Gustavsson et al., 2011). In these low-GDP regions, prepared food is rarely wasted but considerable food losses occur earlier in the supply chain due to poor harvesting techniques, inadequate storage and drying facilities, and the inferior quality of the transport infrastructure, packaging and market systems. Access to reliable and affordable energy

Figure 6.9.   Natural exposed drying of harvested millet and other cereal crops can lead to heavy losses from birds, vermin and when handling.

is often the key to ensuring good quality post-harvest operations are put in place. For example the co-benefits from using enclosed solar crop dryers as opposed to traditional, exposed, open-air drying with high losses (Fig. 6.9) could help subsistence farmers improve their livelihoods or feed extra family members.

Food waste in European and North American countries is around 95–115 kg/capita/year (Gustavsson *et al.*, 2011) which can best be addressed at the consumption level. Most of this waste occurs due to the deterioration of fresh produce as a result of mismatching supply with demand, poor purchase planning, careless preparation, and consumer behavior, such as food rejection due to stringent quality standards not meeting expectations or the "use-before" date on a supermarket product having expired, or leaving prepared food unconsumed by those who can afford to by over-catering or taking too large a portion.

Raising awareness to avoid food losses and waste could lower food costs and benefit international goals to reduce related energy inputs, GHG emissions, competition for land use, poverty and hunger (UNEP, 2011). The quantified global data on food losses and wastage throughout the agri-food chain was used to assess the energy embedded in the food losses that are therefore, in effect, wasted (Fig. 6.10). Overall, the direct and indirect energy embedded in global food losses is thought to equate to around 38% of the total amount of final energy consumed annually by the whole agri-food chain. However, some data are uncertain, so the analysis can only be treated as a strong indication that the energy embedded in wasted food is significant. In the USA for instance, losses from the farm gate to the plate account for about 2% of the total national annual energy consumption (Cuellar and Weber, 2010).

Public campaigns aiming at policy makers and the general public[13] have begun to draw attention to the problem of food losses and waste. Investments by both public and private sectors to help reduce losses in crop, fish and livestock production systems could reduce supply chain risks,

---

[13] See for example www.tastethewaste.com and "We feed the world" (2006) at http://video.google.com/videoplay?docid=-7738550412129841717

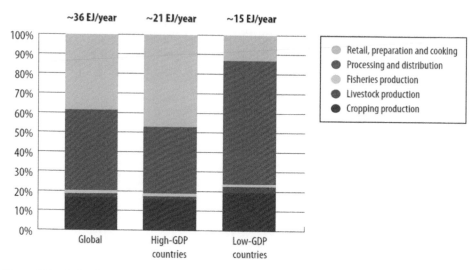

Figure 6.10.   Indicative shares of energy inputs embedded in food products that are lost along the agri-food supply chain with around 60% occurring in high-GDP countries mostly at the retail, preparation and cooking stage compared with low-GDP countries where most losses occur during processing and transport including post-harvest storage. [Notes: This FAO (2011a) assessment used food loss data from Gustavsson *et al.*, 2011 and global energy data from Giampietro, 2002; GoS, 2011; IEA, 2010; Smil, 2008; and others. Accumulative energy losses were taken into account[14]].

improve food product quality, and reduce GHG emissions per unit of consumption. Avoiding a share of post-harvest losses would reduce the total costs of food production and the related GHG emissions per unit of consumption arising from the treatment and disposal of food wastes.

In many low-GDP countries, financial and technical constraints exist when attempting to optimize harvesting techniques and improve storage facilities and the packaging, infrastructure and marketing components of the agri-food chain. Reducing food harvest and storage losses through education of small farmers could have a positive impact on their livelihoods in a relatively cost-effective manner.

Food losses in high-GDP countries can be exemplified by the situation in the United Kingdom where around 13 M dry tons of food and drink wastes are thrown away each year (NSCA, 2006). Approximately 80% of this comes from the retail and domestic sectors. The value of food wasted by the UK domestic sector equates to over £10 billion annually, with an estimated cost for the average household of £250–400 per year. Most of the wasted food ends up in landfill sites, though more anaerobic digesters are being developed to convert this resource to biogas (Environment Agency, 2012). Potatoes, bread and apples are the most wasted by quantity with salads the highest wasted as a proportion of the total produced.

Around two thirds of wasted food is edible with half of it untouched, and therefore classified as "avoidable" waste. The remainder is either "unavoidable" such as inedible coffee grounds or apple cores; "unavoidable due to preference" such as removing bread crusts from sandwiches, fat

---

[14]The accumulative energy concept can be illustrated as follows. If 1 kg of wheat grain is lost during the harvesting operation then that equates to say 10 MJ of energy wasted during the agricultural production process. However, if 1 kg of bread is baked then not consumed, the energy inputs for producing the wheat, then drying, storing, transporting, processing into flour, packaging, distributing and baking it all have to be included in the total embedded energy of the bread being thrown away. This could exceed say 100 MJ including the 10 MJ from growing and harvesting the wheat.

from red meat, or deliberately over-catering for a social event; or "unavoidable due to cooking method" such as peeling potatoes before cooking as mashed or roasted as opposed to serving as baked potatoes with skins intact.

The food choice expectations of people living in high-GDP countries have become dependent on affordable refrigeration systems across the whole food supply chain. Developing similar systems for low-GDP countries may prove to be challenging. Avoiding dependence on refrigeration is difficult where economic development depends on exporting perishable food to high-GDP countries. This requires reliable cold chains to be in place. Possible solutions focus on bulk preservation and transport only to local markets, or the use of passive evaporative-cooling technologies rather than active cooling that depends on an electricity supply being available. Stand-alone solar chillers have good potential once they become economically viable. Overall, to reduce food supply chain losses in low-GDP countries, post-harvest storage and technologies would need to be appropriate, simple, prevent pest infestation, and, where possible, use local renewable sources of energy (see Section 6.4).

6.3.2 *Changing diets*

The average daily energy intake from food consumption required for a human is around 9 MJ, varying with age and activity and whether having a sedentary or laboring lifestyle. Average food production and availability in sub-Saharan Africa is below 8.5 MJ/day/capita compared with high-GDP countries where it is nearly double this at around 15.7 MJ/day/capita (Smil, 2008). So in high-GDP countries around 50% more food is produced than the average national daily consumption requirements. Obesity results from consuming some of this excess food, but obviously considerable food wastage is also occurring.

Significant energy demand reductions for food supply could in theory be achieved by changing human diets away from animal products which could also help lower GHG emissions from the agriculture sector. However, this would need to be socially acceptable. Purchasing food that is based on energy efficient management systems and is locally produced, supplied only when in season, needing to be only lightly cooked, and with a low content of livestock products would result in reductions in overall energy demands for the product (Schneider and Smith, 2009). GHG intensities vary widely with different food groups with red meat on average being around 150% higher (in terms of $CO_{2\text{-}eq}$/kg) than chicken or fish. Changing to a low red meat and milk product diet can therefore be an effective means of lowering the carbon and energy footprint of a household (Weber and Matthews, 2008). A household changing its consumption of red meat and dairy products to vegetable-based protein sources for just one day a week, could produce similar GHG mitigation benefits as obtained by buying all their weekly food from local providers, thus avoiding the energy used for transport. In practice the reverse is happening as upper and middle income classes in low-GDP countries are tending to move towards a more western diet with higher animal produce content.

However, the argument that all meat consumption is bad is rather simplistic (Godfray *et al.*, 2010):

- There are significant differences in the production efficiency and hence energy used in the production process of the major classes of meat consumed by people (for example, 8 kg of cereals for animal feed are needed to produce 1 kg of beef meat, whereas only 4 kg are needed for pork and 1 kg for chicken). Moreover, through better rearing or genetically improving animal breeds, it may be possible to increase the conversion efficiency of animal feed intakes with which meat is produced.
- A significant proportion of livestock is still grass-fed using pasture land which is often not suitable for arable cropping. In addition, pigs and poultry are often fed on human food wastes in subsistence farming systems.
- In low-GDP countries, meat and milk represent the most concentrated source of some vitamins and minerals, which is important for individuals such as young children. Livestock are also widely used for animal power for cultivating, for transporting goods and carrying people. They

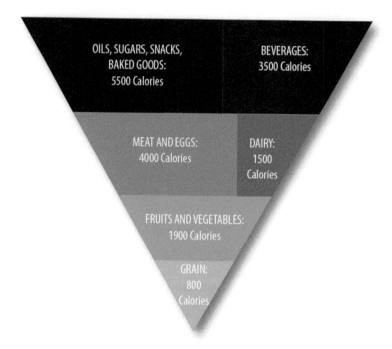

Figure 6.11.   Representation of daily per capita energy input from food intake for a typical food diet in the USA (Bomford, 2011).

also provide a local supply of manure, can be a vital source of income, and are of huge cultural importance for many poorer communities. So food co-products can provide added benefits.

- Meat produced from intensive livestock farms situated close to cities can be an efficient way of providing the high amount of food energy needed by urban societies. Since the energy and protein contents of meat are higher than in a similar mass of cereals, smaller amounts need to be transported to the urban centers and less energy is needed to provide the same amount of protein. Urban societies could be less sustainable if relying solely on a vegetarian diet.

On the other hand, reduction in energy use through dietary change can also be achieved by eating minimally-processed or whole foods which require less energy for processing and packaging. Most of the agri-food system is used to provide highly-processed, high-calorific value foods, with only a small fraction of calorie intake in typical high-GDP diets coming from unprocessed grains, fruits, and vegetables (Fig. 6.11). The analysis excludes pet food which can add a significant carbon footprint for some families (Vale R. and Vale B., 2010).

6.3.3 *Modern energy services*

The provision of energy services can do much for food production and food security. At present, almost 3 billion people have limited access to modern energy services for heating and cooking and 1.4 billion have zero or limited access to electricity (see Chapter 2). The increase in global food prices in 2008, partly due to increased world energy prices, hit low-GDP countries the hardest. In the poorest households, food can account for 50 to 80% of total expenditure compared with 7–15% in the average household in high-GDP countries. In most low-GDP countries, subsistence farming/fishing is the most practiced type of food production. Where energy can be provided in an economic manner, resulting increases in productivity can result together with reduced food losses from better storage and improved livelihoods.

6.4   RENEWABLE ENERGY SUPPLIES FROM AGRICULTURE

New and renewable sources of energy stand at the centre of global efforts to induce a paradigm shift towards green economies, poverty eradication and ultimately sustainable development.
Ban-Ki Moon, UN Secretary General, 22 August, 2011

Renewable energy can enhance access to reliable, affordable and clean modern energy services, is particularly well-suited for remote rural populations, and in many instances can provide the lowest cost option for energy access (IPCC, 2011a). Using local renewable energy resources along the entire agri-food chain can help to improve energy access, allay energy security concerns, diversify farm and food processing revenues, avoid disposal of waste products, reduce dependence on fossil fuels and GHG emissions, and help achieve sustainable development goals.

It is well documented that climate change will most likely impact on future food production, that the agri-food sector will need to adapt, and that investments aimed at improving agricultural adaptation will inevitably favor some crops and regions above others. South Asia and southern Africa are the regions that, without sufficient adaptation measures being put in place, will likely suffer greater adverse impacts on several of the staple crops that their populations depend upon for food security (Lobell *et al.*, 2008).

In addition, climate change will likely have impacts on the technical potential of renewable energy resources and their geographic distribution (IPCC, 2011d). For example,

- increased cloud cover could reduce solar radiation levels but probably not significantly overall;
- the global technical potential of hydropower could increase slightly due to changes in precipitation but with substantial variations across regions and countries;
- changes in the regional distribution of wind energy resources are expected; and
- energy crop productivity could be affected by changes in precipitation, soil conditions, and atmospheric CO_2 concentration levels, probably with a small overall global impact but with considerable regional differences that will be difficult to assess.

Expanding the deployment of renewable energy technologies in rural areas can result in economic, social and environmental co-benefits at the local level while also providing energy directly to the agri-food system. Co-benefits include local development, employment opportunities, improved livelihoods, social cohesion, up-skilling of local trades people, improved health due to reduced air pollution, reduced drudgery, a better gender balance for work responsibilities, and an enhanced sense of community spirit (IPCC, 2011b). Renewable energy can also provide basic services for non-agricultural activities in rural areas including domestic lighting, cooking, entertainment, information and communication, and motive power. This provides the opportunity to trigger the development of new businesses, especially in rural areas.

The use of more efficient biomass cooking stoves can reduce the demand for traditional fuel-wood by half compared with open fires (IPCC, 2011a) but not all programs to introduce them have succeeded due to the informal character of the fuelwood supply chain. In addition, new stove designs have to be culturally acceptable, so cooking in the heat of the day using a solar oven, for example, may be rejected compared with cooking in the cooler evenings using fuelwood. Traditional biomass cooking stoves may be less efficient, less healthy and more laborious than solar or biogas designs, but they are often more affordable, which is critical for the poorer sections of rural populations (Geoghegan *et al.*, 2008; UNDP, 2009). The dissemination of improved designs of domestic stoves therefore mainly succeeds when micro-finance is available for the capital investment needed. Around one quarter of the 2.7 billion people who rely on traditional biomass for cooking (plus another 0.3 billion who rely on coal) now use improved stove designs in 166 million households, two-thirds being in China (UNDP, 2009). Other biomass fuel options under development to give more efficient and healthy domestic cooking applications include ethanol gels and DME (dimethyl ether) (IPCC, 2011a).

Producing biofuels locally at the small scale, for example using raw, filtered vegetable oils directly as fuels for simple diesel engines (ideally after processing the triglyceride oils to esters), can at least enable isolated communities to utilize agricultural machinery, generate electricity, and transport food products to local markets.

Energy access can play an important role for food production at the small farm holder level by increasing human labor output. Energy poverty implies that a substantial amount of time is spent collecting fuelwood from long distances to meet basic household energy needs of cooking and, in colder climates, heating. Access to alternative fuels can result in time and labor savings from collecting less fuel wood, usually by women. The extra time available can then be utilized for greater food production, other productive activities such as weaving, or social community activities. The cost effective relationship(s) between energy access expansion and sustainable agri-food production systems in low-GDP countries could be better ascertained by quantifying the time and labor that renewable energy provision could save.

Providing access to modern energy services for those without, by applying the most energy efficient and appropriate technologies, must be balanced with affordability. Newly introduced systems ideally should enhance current crop yields as a result, improve storage and processing activities, minimize future investment, operating costs and make human labor most effective. The trade-off between realizing these co-benefits and the possible higher cost of the renewable energy depends on local resource availability (see below). Biomass, wind, solar, hydro, geothermal and ocean energy resources are geographically widely distributed throughout rural areas and islands. These can be converted into the full range of energy carriers (electricity, heat, cold, liquid biofuels and gaseous biofuels including biogas) where it is economically viable and socially acceptable to do so. Wind, solar and wave resources are variable, both daily and seasonally, which needs to be taken into account when designing an energy supply system. The energy used when manufacturing renewable energy technologies and transporting them to the site, (and hence the carbon footprint where fossil fuels are used) is usually relatively small giving an energy payback for most wind systems of months of operation rather than years[15].

The transition away from fossil fuels to renewable energy systems has begun, but it will take time for the full effects to take place so the dependence on some fossil fuels will continue for many years. The affordability of new technologies based on low annual incomes also needs to be carefully considered. Relatively high capital investment costs for installing small wind turbines, mini-hydro schemes, solar PV systems, anaerobic digesters and small bioenergy CHP plants may require micro-financing arrangements to be made available for small farmers by national and local governments, aid agencies or the private sector. Renewable energy deployment in rural areas provides considerable opportunities, but the need for flexibility and trade-offs between the continued use of fossil fuels, renewable energy substitution and improved energy efficiency measures should be recognized.

6.4.1   *Renewable energy resources*

Where good solar, wind, hydro, geothermal, or biomass resources exist, these can be used to substitute for direct fossil fuel energy inputs to generate heat or electricity for use on-farm or by aquaculture enterprises. If energy is produced on the farm that exceeds demand, then it can be exported off the property to gain additional revenue for the owner of the enterprise. Such activities can result in rural development benefits for landowners, small industries and local communities[16]. There are also many examples of municipalities in rural regions enhancing rural development by successfully attracting businesses due to the availability of local renewable energy resources.

The land area required for renewable energy projects is usually relatively small, with the exception of biomass energy crops. Large wind farms typically use ~5% of total land area for

---

[15]The IPCC Special Report on Renewable Energy has energy payback details in chapters 2 to 7 and many other reference sources exist such as http://www.ncbi.nlm.nih.gov/books/NBK44130/
[16]A recent OECD project (OECD, 2012) assessed the links between renewable energy and rural development and evaluated such opportunities using 15 case studies in 10 OECD countries. The experiences could well equally apply to non-OECD countries.

foundations and roading; large solar PV arrays could extend over several hectares so are more often located on building rooftops; and small hydro run-of-river projects usually need only a small area of land alongside the stream for the turbine house. Bardi (2004) calculated that the fraction of land needed to displace global fossil fuel use with solar and wind energy technologies would use around 1.5% of the ~50 Mkm^2 land area currently used for agriculture and this would have "minimal impact on food and textile agricultural production".

The potential for bioenergy systems to reduce GHG emissions is the subject of debate due to possible impacts on land use change, consequential indirect emissions, and competition with food for land and water. There is concern that producing biomass will become so attractive in response to increasing carbon prices that people will be evicted from their lands, that rainforest and other sensitive ecosystems will be destroyed to allow for biomass plantations, and that food prices will increase significantly (Azar, 2011). Conversely, it is argued that diversification of markets (such as corn or wheat being sold for milling, animal feed or biofuel feedstock), could provide economic stimulus for increased investment in capital and skills. Agro-forestry practices that produce biomass linked with food production can have benefits (such as the mitigation of saline soils in Australia). Energy crop management can also help ensure soil fertility is maintained and in some cases be enhanced for future food production.

The interaction between biomass production and food prices is also a controversial issue as potentially volatile energy markets can impact on food prices which would then have implications for sustainable development (IPCC, 2011a). There are also concerns that carbon linked with bioenergy diverts it from being returned to the soil and removal of biomass from the land also results in soil nutrient depletion. This can restrict the volumes of biomass available from a given site, particularly for conservation agriculture and organic farming systems that avoid the use of inorganic fertilizers. Integration of energy and food production from biomass crops is technically feasible under many situations but needs to be managed carefully in a sustainable manner. Detailed analysis on the sustainability of biomass use is being undertaken by such organizations as FAO[17], IEA Bioenergy[18], the Roundtable on Sustainable Biofuels[19] and the Global Bioenergy Partnership[20].

Small biomass heat systems used by subsistence farmers, are not discussed in detail in this chapter. Nor are large-scale, centralized renewable energy systems such as large hydro dams (which currently generate around 20% of global electricity (IEA, 2011), concentrating solar power installations, and large geothermal plants since they are usually not integrated into agricultural systems. However, businesses involved in the manufacture of fertilizers, agricultural machinery etc. have the opportunity to purchase "green renewable energy" sourced from low carbon sources (such as hydro, wind or synthesis gas from biomass) in order to reduce their dependence on fossil fuels. They can also improve the energy efficiency of the manufacturing process and hence reduce the embedded energy in their products.

6.4.2 *Renewable energy systems*

Renewable energy can be used throughout the agri-food sector either directly to provide energy supplies on-site or indirectly as a result of being integrated into the existing conventional energy supply system (Fig. 6.12). The availability of a reliable and affordable energy supply can become an essential component for sustainable development. Reducing the dependence of the agri-food system on fossil fuels by substituting renewable energy, at least in part, is feasible for on-farm and aquaculture production as well as for transporting raw food feedstocks, processing food, distributing finished products, and cooking. In low-GDP countries, renewable energy also presents an opportunity to provide much needed, basic energy services.

---

[17]For more information see www.fao.org/bioenergy
[18]www.ieabioenergy.com/LibItem.aspx?id=6770
[19]http://rsb.epfl.ch/ and Ismail *et al.* (2011).
[20]www.globalbioenergy.org

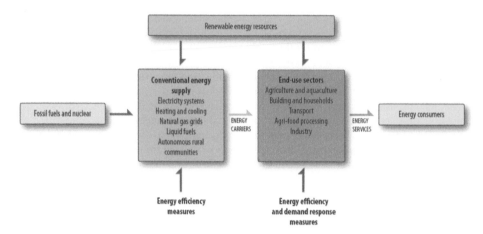

Figure 6.12.   Renewable energy resources can be used directly on-site by the end-use sectors of the agri-food chain or indirectly through integration with conventional energy supply systems that are mainly based on fossil fuels and nuclear power. Energy services delivered to energy consumers contain varying shares of renewable energy. (Based on IPCC, 2011c).

In locations where good renewable energy resources exist, farmers, fishers and agri-food processing businesses have the opportunity to install technologies to generate wind power, solar power, micro-hydro-power or, possibly in the future, electricity from ocean energy resources. At the decentralized scale, solar thermal, biomass and geothermal resources can be utilized to supply both heating and cooling. Detailed assessments of each technology, together with issues concerning integration into existing and future energy supply systems, sustainable development, costs and potentials, and supporting policies, are discussed in detail in the IPCC report "Renewable Energy and Climate Change Mitigation" (IPCC, 2011d).

Given the importance of providing adequate energy to reduce food losses at post-harvest stages in low-GDP countries, significant attention has been given to the possibility of using renewable energy beyond the farm gate (see for example, GIZ, 2011). Solar energy has been successfully used to provide both dry and cold storage of food, as has biomass combustion to provide heat. One example of the latter concerns the use of wood to dry spices in Sri Lanka that has diversified income streams and increased revenue to a range of local stakeholders operating within the spice production market chain. As well as selling the fuelwood by-product from pepper plants to the spice dryers, small scale growers are now also able to sell mature spices after they have been dried and preserved (FAO, 2009c).

At present, renewable energy meets over 13% of global primary energy demand (Chapter 2), although almost half of this comes from traditional biomass used inefficiently for cooking and heating. Many scenarios show there is good potential for the share of modern renewable energy to rise to over 70% by 2050 (IPCC, 2011d) but it should be noted that for electricity generation, there are large differences in capacity factors of the various generation technologies. No electricity generation plant operates 100% of the time due to the needs for maintenance or the unavailability of the resource. Capacity factors (actual generation capacity compared with nominal generation capacity) vary from around 95% for the best nuclear power plants to around 10% for solar PV systems in less sunny regions. For example, a 5 MW solar PV plant, or a 2–3 MW wind plant, would annually produce an equivalent amount of electricity (MWh/year) as a 1 MW coal plant. These differences should be taken into account when comparing renewable energy with other energy sources (Smil, 2012).

The levelized costs of renewable energy from many technologies (calculated over their lifetime) are typically higher than present average prices for electricity, heat and transport fuels (Fig. 6.13). However, costs for several renewable energy technologies are likely to continue to decline due

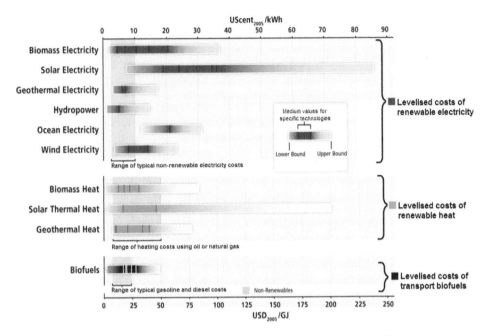

Figure 6.13. The costs of electricity, heat and liquid biofuels produced from renewable energy sources can be higher than when produced from conventional fossil fuels, but under specific circumstances, some renewable technologies are already competitive (shown where the horizontal bars overlap with the vertical range bars representing conventional wholesale electricity, heating and gasoline/diesel transport fuel costs). (Based on IPCC, 2011d).

to mass production and learning from experience. In many specific situations, they are already economically competitive. For example, in remote rural regions with no electricity grid access, autonomous renewable energy systems are already competitive where they can avoid expensive grid connection costs.

6.4.2.1 *Biomass and bioenergy*

The costs of delivering biomass supplies to a bioenergy conversion plant (in terms of $/GJ delivered) vary widely depending on scale, average transport distance, and type of biomass. They are therefore site specific (IPCC, 2011e). For example, a project "Cogen for Africa", funded by the Global Environment Facility (GEF) and aiming to assist African countries implement efficient combined heat and power (CHP[21]) bioenergy systems, showed the potential is high. Mauritius already obtains close to 40% of its total electricity supply from CHP systems (Karekezi and Kithyoma, 2006).

In food processing plants where biomass is already collected on-site as a co-product component of the main food production process (such as kernels and bunches from palm oil production), costs when used for energy purposes can be relatively low (~US$ 0–2/GJ) or even negative where waste disposal costs are avoided. Food processing plants such as sugar or rice mills often utilize their biomass by-products (bagasse or rice husk ligno-cellulosic materials left after processing) for CHP generation usually used on-site, but in some instances, excess electricity is produced which can be sold to the electricity grid to generate an additional revenue-earner for the company. Wet processing wastes, such as tomato processing rejects, skins and pulp wastes from juicing,

---

[21]Combined heat and power (CHP) or cogeneration is the use of a heat engine or a thermal power station to simultaneously generate both electricity and useful heat, thus significantly improving the overall efficiency.

can be used as feedstocks in anaerobic digestion plants for biogas production (or in hydrothermal processing plants to produce bio-oil and co-products, though not yet fully commercial). The biogas obtained can be used to generate heat and/or power for use on-site or, after cleaning the mix of gases to biomethane by removing carbon dioxide and hydrogen sulfide, be injected into the gas grid or compressed as a vehicle fuel (NSCA, 2006). Constraints to exporting biomethane by injection or electricity include whether the existing grid passes near the farm or processing plant which if so, could avoid high connection costs. The seasonal nature of some food processing plants operations has an impact on energy provision and will require specific contractual arrangements under a power purchase agreement with the electricity utility.

With careful management, biomass can be produced sustainably, not compete with food for land use or water, and avoid increasing both direct and indirect GHG emissions. Biomass originates from several steps along the agri-food supply chain including animal wastes, crop and forest residues, food processing by-products, and food wastes from retailers, households and restaurants. For use on farms, collection and storage of animal wastes and crop residues such as baled cereal straw add to the delivered biomass costs (~US$ 2–4/GJ) (Sims, 2003). Biomass resources are flexible in that they can be used on-site if and when needed to provide direct energy inputs; processed on-site into energy carriers for sale; sold off-site for collection and use at community-scale district heating or anaerobic digestion CHP plants; or sold off-site and accumulated into larger volumes from within a region to supply commercial scale, liquid biofuel production plants. Competition also exists for other uses of the biomass such as feedstocks for bio-materials and bio-chemicals.

At the smaller-scale, market analysis of 15 case studies in 12 countries in Latin America, Africa and Asia (FAO, 2009c) confirmed that biomass arising from on-farm residues can be used to produce useful heat, power and biofuels for local use[22], contribute to rural livelihoods, reduce imported fossil fuel dependence, and offer new opportunities for rural communities without impacting on local food supply security. Energy crops are being purpose-grown in some countries to provide biomass for conversion to liquid transport biofuels (such as corn, sugarcane and oilseed rape) but also for co-generation of heat and power. Competition for the limited biomass resource, as well as for land and water resources between food and biofuels, are on-going concerns. New crop varieties are being developed specifically for commercial use to improve yields and reduce energy and water consumption.

Purpose grown energy crops have higher delivered costs since production, harvesting, transport and storage costs need to be included (~US$ 5–10/GJ or higher). Modern combustion, gasification and pyrolysis thermo-chemical conversion technologies are largely mature, although improvements in performance and conversion efficiencies are continually being sought. This is also the case for bio-chemical conversion processes such as anaerobic digestion and ligno-cellulosic enzymatic hydrolysis. The results of analysis from demonstration and commercial plants show costs are wide ranging and site-specific.

Interest in using aquatic plants, macro-algae and micro-algae as feedstocks for liquid biofuels production has developed recently, given their carbon sequestration potential. Harvesting of aquatic plants can help reduce excessive nitrogen and phosphorus levels in lakes and coastal waters resulting from eutrophication by urban or agricultural nutrients. Oil yields per hectare can be several times higher than those for vegetable oil crops, but numerous demonstration projects have confirmed that separation of cell mass and usable substrate is still costly and the system requires relatively high energy inputs. Algae-based bio-refinery systems and seaweed production to assimilate dissolved nutrients combined with intensive fish or shrimp culture in integrated multi-trophic aquaculture systems can offer a viable option in the future (Soto, 2009; Van Iersel *et al.*, 2010; Thomas, 2011).

---

[22]Examples include electricity generation from jatropha oil-fuelled engines; charcoal briquette production; afforestation; ethanol production and stoves; wood-fired dryers; biogas from sisal fibre production residues.

6.4.2.2   *Non-biomass renewable energy*

Solar, wind, hydro and geothermal energy resources, where available, can be used locally to contribute to energizing rural communities, particularly for those currently with limited energy access. These resources can also provide sustainable energy services for the agri-food processing and transport sectors. Thus many small and medium enterprises involved in the primary and secondary production sectors, can diversify their business activities and earn additional revenue. As well as GHG mitigation potential, various other co-benefits resulting from renewable energy project deployment can be considered to be drivers for supporting policies being implemented by local, regional, state and national governments (see Section 6.5). These co-benefits include realizing improvements in air pollution, health, energy access, energy security, water use, capacity building and employment opportunities.

Indication of direct energy inputs and relative intensities of energy demand needed "for" a range of primary production enterprises and the possible renewable energy resources available "from" each enterprise type are given in Table 6.4. Solar, wind and hydro resources vary with location (such as mean annual wind speeds tending to be higher on hills than flat arable land) so there are many exceptions to the typical examples given in the table. Woody biomass residues from forests and small woodlots are not included nor are human labor, draught animal power, or traditional biomass.

6.4.3   *The potential for energy-smart agriculture*

6.4.3.1   *A landscape approach to farming systems*

Large-scale, high input farming systems often involve crop monocultures and isolated specialized enterprises such as dairy farming or intensive pig production. In contrast, integrated food-energy systems can demonstrate improved energy use efficiencies in primary production, at times without costly capital investments (Bogdanski *et al.*, 2010a). Linking food production with natural resource management and poverty reduction in a value chain can result in rural development opportunities to meet the needs of local communities. To date this "landscape approach" has tended to lack an energy component. Hence, energy-smart solutions can include simultaneously producing food and energy as a means of achieving sustainable production systems in rural communities (Fig. 6.14). When renewable energy is produced on farms in parallel with food production, integrated food-energy systems (IFES) result.

Several successful examples of IFES already exist globally (Bogdanski, 2010b). They can function at the small-farm scale to give self-sufficiency at the household or village level but can also be relevant to large-scale farming operations. Combining food and energy production on the same block of land using multiple cropping or agro-forestry systems, possibly linked with livestock and fish production, is one possible IFES approach. Another approach could be to maximize the on-farm synergies between food production and renewable energy through using biomass by-products (such as crop residues and animal wastes) for bioenergy applications, as well as utilizing wind, solar and hydro resources where locally available (Bogdanski *et al.*, 2010a). The agricultural manufacturer, New Holland, advocated a future concept for an "*Energy Independent Farm*" which generates renewable electricity on-site, some of which is used for electrolysis to produce hydrogen fuel for tractors and trucks (Rodriguez, 2011). At the smaller scale traditional family farming in China, Ho (2011) outlined an integrated food and energy "*Dream Farm*" based around organic farming and anaerobic digestion that minimizes waste and optimizes the sustainable use of resources. The potential to reduce the national energy demand was calculated to be 14%, or even higher if options for energy savings are included. An "*Integrated Energy Farm*" concept has also been proposed where a central business and living area is surrounded by land producing food, biomass and other renewable energy sources (El Bassam, 2010). This concept can go beyond cropping and livestock and also include aquaculture for food, feed and energy purposes (Van Iersel *et al.*, 2010).

Table 6.4. Energy inputs, demand intensities, renewable energy resources and exporting energy potentials for a range of typical primary production enterprises (Table based on IPCC, 2011c).

Type of enterprise	Direct energy inputs	Energy demand intensity	Potential renewable energy sources	Energy export potential
Arable (e.g. wheat, maize, cassava, rice, palm oil, sugarcane).	Tractor fuel (diesel). Electricity for irrigation, storage facilities, conveying. Heat for drying (LPG, gas).	High diesel for machinery. High energy demand if irrigated; Low to medium for conservation agriculture. Low heat demand and seasonal.	Crop residues for heat and power generation and possibly biofuels. Biofuels and biogas from energy crops. Solar if good sites.	Biofuels (or feedstocks). Solar power. (Wind and hydro-power are less likely on flat to undulating arable land)
Vegetables – large-scale for processing (e.g. potatoes, onions, carrots).	Tractor fuel (diesel). Electricity for irrigation, grading, conveying, cooling, ventilation, storing.	High diesel for machinery. High power demand if irrigated and for post-harvest chillers.	Dry residues for combustion. Wet residues for anaerobic digestion or hydrothermal treatment. Solar and wind possible.	Heat and biogas mainly used on-site. Solar power possible.
Market garden vegetables (small-farms).	Gasoline or diesel for 2 or 4 wheel tractors. Electricity for washing, grading.	Medium tractor fuel demand. Low power for post-harvest; medium for cool stores.	Residues and rejects for small biogas plant or bio-oil production for use on-site – but small scale and seasonal.	Low.
Protected cropping – greenhouses.	As for market garden, plus heat and power for lighting, irrigation.	Low for machinery. High heat demand in winter.	Some residues for combustion. As for market gardens.	Low.
Fruit orchard (e.g. pip fruit, bananas, pineapples, olives).	Tractor fuel (diesel or gasoline). Electricity for drip irrigation, grading, cool stores.	Medium fuel. Medium electricity if irrigated and on-site post-harvest storage.	Pruning residues for combustion. Reject fruit for biogas.	Low.
Dairying (large-scale of more than 50 cows).	Diesel for tractors. Electricity for milking, pumping, cooling, irrigation, lighting. Heat for water, pasteurizing.	High electricity, especially if irrigated. Medium fuel for machinery. Low heat.	Manure for biogas. Waste heat from milk cooling. Solar thermal. Solar and wind if good sites.	Heat and power from biogas. Solar or wind power.
Pastoral livestock (e.g. sheep, beef, deer, goat, llama).	Diesel or gasoline for machinery. Electricity for shearing, refrigeration.	Medium if some pasture conserved. Very low power demand.	Wind and small-hydro if hill country. Forest residues. Solar if good sites.	Wind and hydro power on hill country.
Intensive livestock (e.g. pigs, poultry, calves).	Electricity for lighting, ventilation, water pumping. Tractor fuel (diesel).	High if mainly housed indoors. Medium to low if partly outdoors. High if feed grown, low if bought-in.	Manure for biogas. Poultry litter for combustion. Solar and wind for water pumping.	Heat and power from biogas and poultry litter at community scale.
Capture fishing –trawlers	Marine fuel (diesel or fuel oil). Electricity for refrigeration, ice.	High fuel. Low electricity.	Reject catch. Fish process wastes for biogas, oils.	Low.
Capture fishing – small boats	Diesel, gasoline, 2-stroke fuel. Electricity for refrigeration, ice.	Low/medium fuel. Low power demand.	Fish process wastes for biogas.	None.
Aquaculture – fish farms on-shore or off-shore.	Diesel, gasoline, 2-stroke fuels for service boats. Heat. Electricity for refrigeration, ice, water pumping, aeration.	Low fuel if on-shore; medium if off-shore. Low heat for warm water. Low to medium electricity depending on type of enterprise.	Process wastes for biogas. Geothermal or solar thermal heat. Ocean energy – e.g. wave, tidal, and ocean current systems.	Low. Electricity from ocean energy possible in future.

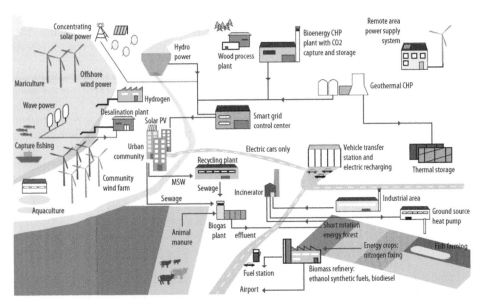

Figure 6.14.   A conceptual integrated food energy system (IFES), shown in a landscape/seascape perspective aiming towards a future sustainable and secure agri-food supply system in both high-GDP and low-GDP countries (based on IEA, 2009).

IFES could provide a compromise between mono-cultural production (that aims to maximize profit in the short term) and the older, long established "mixed farming" systems that integrate pasture, livestock and crop production on the same property. To gain the benefits from both approaches, IFES could possibly evolve into a large-scale regional system combining food and energy production techniques across several neighboring farms, thereby benefitting from both division of labor for specialized tasks and the integration of systems with nutrient cycling. Such an approach could support rural development ambitions in both developed and developing countries seeking food and energy security at the same time.

Agricultural and aquacultural management approaches can involve crop rotations, organic farming techniques, crop/livestock integration (FAO, 2011c) and combined agriculture/aquaculture production systems. These stem from better use of natural resources including water, substitution of inorganic fertilizer inputs by animal wastes and crop residues, and greater reliance on human and animal labor inputs. Organic farming techniques typically rely on lower energy inputs per hectare, particularly for poultry and horticultural crops (Williams *et al.*, 2010). However, the more useful indicator of energy per unit of food output (MJ/kg) depends on the productivity achieved which is very variable. Some analyses of organic farms have shown lower energy demands but this may be partly offset by increased human labor inputs (Ziesemer, 2007).

Many examples of low external energy-dependent farming systems involve inter-cropping rather than conventional mono-culture, but care is needed when interpreting results since a full LCA approach has not always been employed by the researchers. Results from comparative studies similar to those outlined below therefore tend to vary and long-term trends remain unclear:

- In the UK an integrated cropping system monitored over 5 years used 8% less energy than conventional cropping mainly due to reduced cultivations. However, since crop yields declined, energy inputs per unit of output (MJ/kg crop yield) were similar (Bailey *et al.*, 2003).
- A 12 year assessment of a low input, crop rotation system in Italy demonstrated approximately 30% lower energy inputs compared with conventional cropping due to the need for less fertilizers (Nassi o di Nasso *et al.*, 2011). Crop and energy yields were not significantly affected by management intensities.

- A 15 year study in Spain (Moreno *et al.*, 2011) showed that organic crop farming used a third of the energy compared with conservation and conventional farming systems, mainly due to low or zero use of agri-chemicals, though direct energy inputs were higher due to increased cultivation for weed control.
- In a Canadian pasture study (Khakbazan *et al.*, 2009) inorganic fertilizer accounted for 93% of the total energy inputs of a grass-only pasture system. An unfertilized alfalfa-grass pasture system was the most energy efficient since it relied on leguminous nitrogen. Fertilized grass-only or alfalfa-grass pastures improved pasture productivity, but did not improve profitability compared to unfertilized pastures. Adding alfalfa seed to the grass mix at the time of sowing, with no added fertilizer, was economically the best pasture improvement strategy.
- Cereal crops grown in rotation with white clover crops gave lower yields than conventional crops in New Zealand (Nguyen *et al.*, 2005), though energy intensities (MJ/kg) were lower on two of the three sites.
- Crop rotations in China (Jianbo, 2006) showed improved energy ratios from a) introducing a *Paulownia* tree crop into a wheat/peanut rotation (the energy values of the non-food agro-forest products being included in the evaluation) and b) using a tea/bean/green crop rotation by replacing inorganic nitrogenous fertilizers needed for tea mono-cultures with leguminous bean and green crops.
- Integrated agriculture-aquaculture systems in the north east Himalayan region of India using animal manures as fish feed over 4 years (Bhatt and Bujarbaruah, 2011) gave best energy ratios for poultry/fish and cattle/fish combinations, but since the crops grown for animal feed were not included in the analysis, the results are of limited value.
- In Vietnam, an on-farm comparison of 11 agriculture-aquaculture farm systems showed rice/fish systems with higher energy inputs gave lower MJ/kg food production, mainly due to achieving higher yields of fish than in a lower input orchard/fish system (Phong *et al.*, 2011).
- A study of several Chinese and Vietnamese systems (Bogdanski *et al.*, 2010b) concluded that the long tradition of integrated agriculture is the reason for successful scale-up of IFES, particularly those including biogas schemes. Constraints include lack of investment finance and continued technical support.
- Energy intensity analysis of primary production systems and methods of reducing energy dependency should be considered in parallel with other long term sustainability factors such as biodiversity, soil quality, improved livelihoods and health, and in particular, with increased productivity since any yield reduction will tend to increase the energy intensity (MJ/kg of food produced).

6.4.3.2   *Institutional arrangements and innovative business models*

The scale and complexity of the challenge related to the food/energy/water nexus implies that it will have to be addressed through pluralistic, multi-stakeholder and inclusive institutional arrangements including financial schemes, technical support services, and novel business models. Financial arrangements between farmers and energy operators and the resulting division of labor can be effective in ensuring quality and facilitating the scaling up of "smart" farming systems. One example links on-farm irrigation with electricity demand response solutions to reduce regional peak power levels over short term periods. Demand response systems can reduce the need to build more power generation plants to meet the increasing peak demands as well as avoid the costly upgrading of transmission lines so they can carry higher capacity. When particularly high periods of peak power occur, farmers can be paid by the electricity utility for specific appliances to be automatically turned off by "shedding load"[23]. Suitable "interruptible loads" include cool stores and irrigation pumps that can be automatically turned off for up to a few hours before being turned back on again without detrimental effects. This is particularly the case for high-power demanding

---

[23]One leading company in this area, with a good description of demand response, can be found at www.enernoc.com

pumps (up to 100 kW) on advanced overhead sprinkler irrigators. Using variable application rates through automatic control of independent solenoid valves on each sprinkler to independently vary the water application required for each sprinkler can precisely control water application variations depending upon varying crops and soil types[24].

Other examples of division of labor whereby farmers can concentrate on what they do best and other people deal with the energy inputs, include IFES systems such as in the following examples:

- Farmers produce wheat and a bioenergy plant operator buys the straw to convert it to useful heat for sale or use on the farm (Bogdanski *et al.*, 2010a).
- In a Chinese district farm model with a central anaerobic digestor plant, the local farmers contribute to the cost of pigs bought by the farm and raise them. In return, they receive cheap biogas and fertilizer, together with yearly dividends from the district farm (Bogdanski *et al.*, 2010a).
- Village farmers bring agricultural residues to a central anaerobic digestion plant and receive electricity in return. Their electricity bill can be positive or negative depending on the amount of residues collected.
- Two innovative business schemes in Bangladesh involve the private sector's need for bio-fertilizers. This has become a driving factor for successful household biogas developments (ISD, 2010). One scheme is the development of a biogas production plant for a dairy cattle leasing organization, whereby the farmers receive funding to purchase a cow and a calf. They milk the cow and provide dung for the digester and for use on organic tea farms and hence can repay the loan. The second scheme, still in its pilot phase, is where a household can refund the loan contracted for its domestic biogas system through the sale of dung and/or biogas slurry, with the possibility to continue selling the slurry and dung for profit once the biogas installation has been paid for.
- "Fee-for-service" schemes such as leasing, energy service companies (ESCOs), or concession schemes also exist for the benefit of farmers.

The World Bank (2008) provided a review of various business schemes suitable for renewable energy investments. While many of these are still in their infancy and therefore their performance is difficult to assess, preliminary analysis showed that no institutional scheme significantly provided better success rates than others (GIZ, 2011).

In summary, renewable energy systems are well understood and the global industry is growing rapidly. Where good renewable energy resources are available, using agricultural land for both food production and energy capture and conversion is feasible. Further along the food-supply chain, food processing plants often have biomass co-products available that are suitable for bioenergy applications. Several co-benefits for landowners, businesses and rural communities result from developing renewable energy systems. Improved energy access in low-GDP countries can result. Integrated food-energy systems are being demonstrated as feasible examples of the landscape approach to energy-smart solutions. However, further analysis is needed to demonstrate the real benefits. Innovative financial schemes are also possible regarding institutional arrangements that can involve several types of public and private partners.

6.5 POLICY OPTIONS

Making the agri-food chain more 'energy-smart' for both small- and large-scale agricultural systems, as well as improving energy access for subsistence farmers, will not happen without strong and long-term supporting policies. Successful examples exist of policy drivers that support agri-businesses throughout the sector, but these will need to be significantly scaled up if a cross-sectoral landscape approach is to be achieved.

---

[24]See for example, www.precisionirrigation.co.nz

Many policies are in place to encourage businesses and householders to improve their efficiency of energy use (WEC, 2010) and to support the deployment of renewable energy projects (REN21, 2012) while other policies have specifically targeted energy access (Practical Action, 2009). Such policies have proved to be largely successful but not always cost effective (IEA, 2008). To date, few international organizations or national and local governments have strongly linked their energy policies with policies that support the agricultural sector and those targeting future food supply security.

National policies relating to food production and security should reflect environmental, social and economic drivers. Such policies can be implemented to emphasize food and energy security as a priority for all countries. Policies on climate change mitigation are mainly of interest for Kyoto Protocol signatories, whereas energy access and social and economic development policies are more of a priority for low-GDP countries. Both are very relevant to the energy and agri-food sectors (IPCC, 2011a).

The European Commission, through its Integrated Policy Platform, reported that the food supply chain was responsible for 20–30% of environmental impacts (external costs) resulting from climate change, ozone, acidification and eutrophication (JRC, 2006). After evaluating European Union consumption trends using various LCAs, the conclusion was that meat and dairy production have the largest environmental impacts in EU-27 countries, together with energy use in buildings and light vehicle road transport. The report also highlighted that these impacts could be reduced by 20% if improvements in agricultural production were made, food wastage was avoided by households, and electricity was saved by conservation and improved efficiency (JRC, 2008). A change of diet to reduce consumption of red meat was recommended but it was realized that "because food and nutrition are strongly rooted in traditions and habits, policy measures aiming at stimulating a change towards healthy diets need to include a combination of different instruments ranging from consumer awareness raising to public procurement activities" (JRC, 2009).

Rather than implement agricultural and food policies in isolation from energy and climate policies, they should be developed in parallel. Supporting energy efficiency improvements and the deployment of renewable energy could be incorporated within existing policies in order to promote win-win situations and capitalize on co-benefits such as reduced pressure on natural resources and the ecosystem that can result from these types of interventions. Where appropriate, this could be in association with other national policies for transport, health, rural development, immigration, innovation and economic growth.

This section gives examples of policies that have been implemented but could be used more widely to support the goals and ambitions to reduce the present high energy dependence of the agri-food sector, particularly on fossil fuels. Ideally, a cross-sectoral approach should be taken, rather than having specific policies for each separate sub-sector as tends to be the current situation.

6.5.1   *Present related policies*

Current policies on energy and agriculture should be scrutinized to assess if they are conducive to the promotion of sustainable energy deployment in the agri-food sector. If necessary, appropriate modifications to existing policy frameworks may be needed in order to align policies to support a transition towards an 'energy-smart' agri-food sector. For example, global fossil fuel subsidies that were around US$ 558 billion in 2008 (IEA, 2010) may need to be amended as might the European Union's common agricultural policies.

In Annex-1 signatory countries to the Kyoto Protocol[25] energy policies related to greenhouse gas mitigation are often linked with meeting Kyoto Protocol targets. Under the Copenhagen

---

[25]These originally comprised mainly OECD and the former Soviet Union, but excluding the USA. However, Canada, Japan and New Zealand have now withdrawn from the second commitment period of the Protocol out to 2020.

Accord in 2009, many countries adopted nationally appropriate mitigation actions (NAMAs) to stabilize their GHG emissions aimed to curb global temperature rise below 2°C (UNFCCC, 2011). However, this initial collective effort only resulted in a trajectory to around 3.5°C (IEA, 2010) and more stringent long-term policy developments are needed.

Providing energy access to the poor is usually the social responsibility of national and regional governments. Relying on the free market approach, as was the policy followed by several high-GDP countries, is not usually considered to be the best approach to provide access to energy services in rural areas of low-GDP countries (IPCC, 2011a). Several initiatives such as the multi-partner *Poor People's Energy Outlook* (Practical Action, 2012) are being carried out to provide a baseline and a practical means to measure energy access for the poorest. However, the Millennium Development Goals (MDGs) make no reference to specific objectives or targets for energy access or for renewable energy. Therefore energy has not been a high priority issue during international and national policy debates despite the common acceptance by multilateral and bilateral agencies, governments, academia and civil society, that energy access and security is critical for sustainable development (IPCC, 2011b) and the move from poverty. This dilemma has now been recognized by the UN General Assembly that designated 2012 as the *International Year of Sustainable Energy for All* (UN General Assembly, 2011), serving as a platform to raise awareness of the relevance of energy to sustainable development. One key goal is to adopt a coordinated global energy strategy in conjunction with consistent and stable national policies in order to bring down the cost of renewable energy technologies, including off-grid systems, for use by the poorest segments of the population living in rural areas.

Policies relating to managing the demand for energy and seeking efficiency improvements along the agri-food system have been implemented by some governments over several decades. These include more general policies that target reducing energy use by the transport, industry and domestic sectors of an economy since each of these contributes to the agri-food chain. Such policies can include the introduction of freight truck fuel economy standards and payload limits, minimum energy performance standards (MEPS) for electric motors, refrigerators, water boilers etc., and energy performance labeling of domestic appliances. Other policies aim to modify current practices by implementing behavioral changes such as vehicle speed restrictions, packaging recycling regulations, and higher charges for landfill disposal of organic wastes. Policies introduced over past decades by some governments exemplify approaches that can be used by others. Lessons can be learned from such experiences by national, regional and local governments of countries beginning to implement similar policies.

6.5.2 *Future policy requirements*

6.5.2.1 *Agriculture*

Agricultural policies that support energy-smart agri-food systems should include environmentally friendly agricultural practices as well as land tenure security. Recent experiences of "land grabbing" have led to the following recommendations for policy developers (De Witt *et al.*, 2009):

- Establish economic development policies upon a series of "higher principles" such as social equity and natural resource sharing, (as has been accomplished by Burkina Faso and Mozambique).
- Develop institutionalized, formal negotiated partnership mechanisms, organized by forms of common interest between private operators and the local rural population, with government authorities acting as 'referees' and guarantors of law enforcement. Based on such "higher principles", this policy approach can be achieved in preference to creating, in parallel, rural developments aimed at large-scale investors whereby governments take land and grant it to commercial investors.
- Ensure there are adequate levels of stakeholder participation throughout the policy process – from design to implementation and monitoring – in order to produce both 'legal'

and 'legitimate'[26] policy measures that are also feasible and acceptable to all relevant stakeholders.

- Link institutional reforms to policy changes where needed.

Agricultural policies should encourage good land management practices since these can contribute positively to productivity and reduce the demand for energy inputs. Conservation agriculture, coupled with maintaining soil health, integrated food-energy systems, growing of drought tolerant crop varieties, precision farming approaches, urea placement, and the use of water-saving management practices (FAO, 2011b) can all lead to improved productivity, reduced energy inputs and associated GHG emissions. Farmers tend to be risk averse and do not like change, so various financial incentives, together with education and capacity building, are usually necessary to encourage change. Existing policy incentives that encourage conventional cultivation, the wasteful use of fertilizers, and excessive water use for irrigation should be discouraged.

Effective water management policies could encourage the introduction of more efficient irrigation systems including precision irrigation techniques that match varying sprinkler application rates to varying soil type and crops, low-head drip irrigation, recycling of waste water and fert-irrigation (adding liquid fertilizers) which could also reduce overall energy inputs. Fertilizer management practices to avoid excessive application of fertilizers can be promoted through training services on recommended dosage rates and precise application methods. Policies that promote financial incentives to use excess fertilizers should be removed. In several low-GDP countries, farm machinery use has been successfully expanded, especially in Asian countries where agricultural pricing policies have been used to support increased uptake (Ashburner and Kienzle, 2011).

To encourage farmers who have become more mechanized and have access to fuel supplies, policy makers should devise incentives to use energy and natural resources more wisely. Innovative approaches could be used, for example through payments made for environmental services and the use of land tenure related-policies that entitle landowners to benefits resulting from any increases in the value of natural capital. Since capital investment is critical for sustainable development, any means that governments can devise in order to provide access to credit, but with minimal transaction costs, should also be considered.

6.5.2.2   *Energy access*

Existing energy policy frameworks in low-GDP countries do not always respond to the energy needs and capacities of the poor, nor to technology transfer and local energy technology R&D (Practical Action, 2009). In some countries, investments in expanding energy access may be limited by the lack of a public grid and government involvement necessary for infrastructure development. The private sector may therefore need to be actively engaged to support the implementation of energy projects that lead to sustainable development. However, since private investors aim to maximize their returns on investment, many would not be attracted to simply providing energy access unless government subsidies or other financial incentives were put in place. Therefore, clear policies on tariffs and risks should be clearly laid out.

Affordability and adaptability aspects, that relate to energy access (see Chapter 2), also need to be considered when developing new policies. In addition, from the social perspective related co-benefits such as security of water supplies, landscape and biodiversity should be considered in any policy decisions.

---

[26]The *'legality'* of a policy characterizes those land rights acquired through some form of government involvement such as using a specific law and related formal administrative procedures and services. On the other hand, the *'legitimacy'* of a legally acquired right is strongly influenced by a set of power relations which may be legitimised by formal processes, and backed or opposed through pressures from influential stakeholders. Customary rights illustrate the difference between legality and legitimacy. They are often much weaker from a legal point of view, but have strong legitimacy because they are rooted in a long-standing social and cultural consensus.

6.5.2.3 *Climate change*

Due to the complexity of the issues, there is no single approach for developing GHG mitigation policies that would also have positive impacts on security of food supplies, in part due to the broad differences that exist between small- and large-scale food supply systems in high- and low-GDP countries. Food security interventions aimed at reducing energy inputs "for" the agri-food sector whilst also supplying renewable energy "from" the sector, have potential to contribute to energy policies aimed at stabilizing GHG emissions. Schneider and Smith (2009) stated that "strategies that realize GHG emission mitigation potentials must become cost-efficient at the farm level either through market price changes or policies". A combined approach to addressing the food/energy/climate nexus can result in multiple benefits, both national and local, but to meet ambitious emission reduction targets will require significant changes in current policies.

Policies will require different methods of engagement depending upon the national context. Strong policies based on fiscal incentives, public finance and stringent regulations are needed to reduce carbon emissions. Using a combination of different measures, including R&D investment, public awareness and education, is often the most successful approach (IEA, 2007). Technology progress coupled with market-led incentives is a common approach but careful analysis is necessary to assess whether staple foods might then rise in price for the more vulnerable groups. If this did occur, significant adverse impacts on people at the lower socio-economic levels could result in political unrest.

6.5.2.4 *Efficient energy use*

Primary producers at present are unlikely to adopt energy savings measures unless significant financial benefits are evident, such as fuel cost savings in capture fishing. Where energy efficiency measures might lead to a decrease in productivity in the short term, it is important to sustain farmers with appropriate incentives until productivity is restored. Policies seeking to improve energy efficiency behind the farm gate should account for synergies and trade-offs with other policies such as water use, health and food safety. National governments could also stimulate investment options including micro-financing of projects that would result in improving the energy use efficiency of primary production, as well as encouraging the use of solar, wind, hydro and biomass resources arising from land use activities, primary production (including residues and wastes from crops, animals and fish as well as energy crops) and food processing (biomass co-products).

There is increasing agreement that the risk of a rebound effect associated with energy efficiency interventions can be reduced if energy efficiency policy instruments, such as standards and regulations, are combined with instruments such as carbon taxes to increase the price of energy products or GHG caps to contain increased demands for energy (Passey and MacGill, 2009).

6.5.2.5 *Renewable energy deployment*

Policies concerning renewable energy deployment in order to support agri-food systems, mainly by project developments on-farm and at food processing plants, are growing. The British Department for Energy and Climate Change (DECC) has recognized that the need to maintain or increase food production is constrained by the target to reduce agricultural GHG emissions. Generation of renewable energy from land-based resources is one solution identified. Consequently, a feed-in-tariff introduced in April 2010 resulted in many farmers embracing renewable energy schemes with a potential 10% return on investment, though this was partly encouraged by commodity prices dropping sharply around that time (SER, 2011).

The EU Biomass Action Plan contends that lack of policies and poor policies are the most important barrier to overcome in RE development since "it is convincingly proven that whenever appropriate policies are implemented, the market reacts positively and develops the necessary structures and operations systems to deliver results" (European Commission, 2005). Policies supporting renewable energy project development are abundant and have recently spread all over the world (REN21, 2012). They relate mainly to:

- promotion of renewable energy markets;
- financial incentives;
- standards, permitting and building codes;
- capacity building through research, education and information dissemination; and
- stakeholder involvement (IPCC, 2011a; Sawin, 2006).

For example, regulations were adopted in 2002 in Thailand to simplify grid connection requirements for small electricity generators up to 1 MW (World Bank, 2011). This and other policies led to the development of integrated sugarcane and rice bio-refineries that produce food, ethanol, heat and electricity along with the recovery of some nutrients from returning process residues to the soil. By 2008, 73 biomass projects using a variety of residues including bagasse and rice husks had been developed with an installed capacity of 1 689 MW$_{electric}$ (IPCC, 2011a).

Biofuel strategies need to be carefully designed to avoid a net increase in overall GHG emissions (DECC, 2010). Policies that support liquid biofuels such as excise tax exemptions, financial support for production and processing, mandating blending levels with gasoline or diesel, or establishing a "Renewable Transport Fuels Obligation" as in the UK, need to avoid increasing the competition for land use with food production. They should also discourage the use of biofuels produced from non-sustainable sources when substitution for gasoline or diesel can actually increase GHG emissions due to the direct and indirect land use changes (often leading to deforestation) linked with energy crop production. Biofuel mandates have also been criticized for inducing global food insecurity (Pimentel *et al.*, 2009) but this is only one of several factors impacting on food prices (Flammini, 2008). Protection of above- and below-ground biomass stocks and soil degradation are now included in British biofuels legislation (GoS, 2011).

A coordinated global energy strategy needs to be adopted in conjunction with consistent and stable national policies to bring down the cost of renewable energy technologies, including off-grid systems, for use by the poorest segments of the population living in rural areas (UN General Assembly, 2011). Recent reviews (IEA, 2008; IPCC, 2011a; REN21, 2011; World Bank, 2011) provide many examples of renewable energy policies in both high- and low-GDP countries from which some interesting lessons can be drawn.

- A flexible approach is necessary and can be tailor-made to suit specific situations. Some policy elements have been shown to be effective and efficient by rapidly increasing renewable energy project deployment more than others, such as feed-in-tariffs for solar PV in many countries (REN21, 2012). However, there is no universal policy recommendation. Having a mix of policy designs and implementation approaches with the flexibility to be adjusted as technologies, markets and other factors mature and evolve, can help overcome several barriers to renewable energy deployment (IPCC, 2011a). At the same time support policies should guarantee enough reliability for investors to respond.
- Policy sequencing is critical for policy effectiveness. For example, legal and regulatory frameworks for resource and land use, electricity grid connection and integration, and/or for the allocation of permits and rights, should be in place before renewable energy policies are introduced. The process of granting consent permits for new project developments should have a rapid response time and not create bottlenecks.
- Transforming the energy sector to be based on low carbon fuels and technologies over the next several decades will require continuing cost minimization of these technologies over this entire period, not only in the short-term. All costs and benefits to society and the environment should be included in the calculation in order to demonstrate and maximize the benefits, despite the complexity it entails.
- Policies that successfully lead to renewable energy scale-up may not necessarily be economically efficient. Examples include subsidies for biofuels and over-priced feed-in-tariffs.

Renewable energy policies can be designed to support local communities since a significant proportion of public and private investments in renewable energy projects flows through to rural regions. The construction and operation of a renewable energy project primarily benefits the landowner or food processing company. However, increased revenue can also benefit the local community, particularly from larger-scale projects. New income opportunities arise from the use of natural and human capital from rural and peri-urban communities which provides the opportunity for local governments to encourage rural development. Some of the new revenue can be used to enhance the quality and quantity of public services that the local municipality provides, and also to attract new businesses to the region. Long-term employment opportunities arise from both the renewable energy project installation located in the region as well as from the stimulation of new businesses, possibly including renewable energy technology and component manufacturing activities and energy service companies. Higher wages for such jobs have favored the development of local skills and the capacity of rural regions to attract skilled workers. For the more remote rural regions, policies can effectively support locally produced renewable energy when this option is cheaper than developing new infrastructure for electricity or gas distribution, or importing energy from outside the region.

6.5.2.6   *Human behavior*

Reducing demand for energy through changing consumer attitudes is another way of being 'energy-smart'. Policies to that effect concern dietary change and reduction in food losses during food preparation and consumption (Section 6.3). Implementing policies to achieve dietary change, for example, by lowering the consumption of animal products, would be difficult, other than by possibly linking them to health benefits. Establishing financial incentives or taxes to discourage people from eating high-fat diets and fast foods can be linked to health related objectives such as to reduce heart disease and obesity levels. Changing people's behavior is not an easy task and powerful marketing and promotion campaigns would be required to raise awareness and gain public acceptance. In considering this challenge, one should bear in mind that similar campaigns have been successfully undertaken to reduce smoking and drink-driving in some countries.

Reducing food waste will help meet GHG emission reduction targets by preventing disposal in landfills. Obligations to reduce organic wastes going to landfills for example as stipulated in the European Landfill Directive, can be an important mechanism to reduce food waste. The UK government, for example, has set a number of initiatives aimed at reducing food waste including the following:

- The *"Waste implementation programme"*, established in 2002 by the Department for Environment, Food and Rural Affairs (DEFRA), set a target of 35% reduction of the 1995 level by 2020.
- The *"War on food waste"* announced in 2008 following the global food price crisis and launched in June 2009, aims to encourage supermarkets, restaurants, schools, public sector bodies and householders to cut down the amount of food thrown away. It proposes eliminating "best-before-date" food labels as they have become confused with "use-by-date" labels. It discourages supermarkets from marketing cheaper food based on larger quantities purchased and suggests creating smaller sizes of food packages.
- Some food waste is inevitable, so the construction of several community-scale anaerobic digestion plants to produce biogas from food wastes is underway.
- The *"Water and resources action programme"* (WRAP[27]) was created in 2000 to help businesses and individuals use resources more efficiently, develop more sustainable products, and reduce wastes including food. In 2007, WRAP initiated the *"Love Food, Hate Waste"*[28] campaign to increase public awareness by encouraging the use of left-over food through producing special

---

[27] More info at www.wrap.org.uk
[28] More info at www.lovefoodhatewaste.com/save_time_and_money

recipes, avoiding preparation of more food than needed, and planning meals and shopping trips in advance. Claims have been made that around 2–3% of potential food losses (valued at ~£ 300M) have been avoided after 2 years of this campaign.

In summary, there is need for an effective policy environment to support the transition of the global agri-food sector to one depending less on fossil fuels and more on renewable energy. A range of energy- and climate-smart policies and measures are available and good examples exist of their successful and cost-effective implementation. They should therefore be ready for replication and scaling up. Policy makers should also consider implementing additional enabling policies to ensure full benefits will accrue. These could involve investments in technology transfer and adaptation, applied RD&D (research, development and demonstration), more energy efficient technologies, fiscal support mechanisms, capacity building, extension services, education and training. A policy environment without allocation of resources for implementation, up-scaling and facilitating the desired smart-energy changes needed on farms, in fisheries, by food processing factories, and during food transport, storage, preparation and cooking may prove to be unsuccessful. Policies that encourage food labeling, dietary change, reducing obesity and avoidance of food losses can all help reduce the energy demand of the agri-food sector at relatively low cost, but social acceptance of such policies could be a barrier.

6.6   ACHIEVING ENERGY-SMART FOOD

This chapter has aimed to put energy into context for the agri-food sectors of both high- and low-GDP countries. It confirmed that future energy supply constraints are a reason why present methods of producing, delivering and consuming food at both small- and large-scales will need to be reassessed. It showed that the way forward involves strategic developments of the agri-food sector at national and local levels in response to these questions:

- What does energy demand entail in terms of future food supplies?
- Where can improvements in energy management be made?
- To what extent can renewable energy substitute for fossil fuel demand?
- What policy and institutional mechanisms will be needed to ensure wide implementation of energy-smart agri-food programs over the long-term?

Significant and sustained efforts at the international, national and local governance levels are necessary to address the food/energy/climate nexus whilst also taking account of future demands for water. The world will need to produce a lot more food by 2050. If prices and supplies of fossil fuels remain stable, and GHG emissions resulting from activities along the food supply chain, can be significantly reduced or even avoided, then this goal could probably be achieved as in the past. In reality, the future security of fossil fuel supplies at affordable prices is uncertain due to both the limited reserves and their propensity to further increase annual GHG emissions which will have to be accounted for. The current status is that energy used throughout the agri-food sector accounts for over 30% of global consumer energy demand largely based on fossil fuels. The sector also produces over 20% of global GHG emissions. The dependence of the agri-food sector on fossil fuels could therefore become a constraint on increasing food production over the next few decades.

Rural development and sustainability goals are also driving the need for a paradigm shift away from the present pathway towards a more sustainable agricultural system. This needs a longer term vision by policymakers but immediate actions based on existing knowledge and technologies are also needed since we cannot afford to wait. Therefore the key question is how best to start on this new energy-smart transition pathway. Continuing under business-as-usual is not an option.

We fail to consume around one third of the food we produce. This is due to harvest and storage losses as well as wastes from producers, retailers and consumers. Such wastage will need to be avoided, new technologies will need to be deployed, innovative multi-stakeholder institutional arrangements will have to be developed, public awareness will need to be increased, and policies aimed at intensifying food production will have to take future energy supplies into account.

Various pathways can lead to increased productivity at both the small- and large-scales of primary production. All of them require significant energy inputs. The challenge is therefore to decouple food prices from fluctuating energy prices by decreasing the energy intensity [MJ/kg] of food produced and using more renewable energy without reducing productivity along the entire agri-food sector.

In theory, the agri-food supply chain could eventually become energy self-sufficient and independent of fossil fuels, but this is a long term prospect. In the short term, the goal of reducing the vulnerability of the sector's dependence on fossil fuel inputs can be met by a combination of energy efficiency measures and renewable energy substitution. This will require considerable investment in RD&D, capacity building, financial support and incentives in order to reduce GHG emissions per unit of food produced (FAO, 2011a).

In addition, access to reliable and affordable energy supplies for many rural, forest and fishing communities in low-GDP countries who are currently without basic energy services can be achieved using a combination of small-scale renewable energy systems together with, at least in the short term, some fossil fuel inputs and more efficient use of traditional biomass. Leapfrogging directly to renewable energy systems should be encouraged where suitable resources exist in order to avoid investment in technologies that will lock users into using fossil fuels for the foreseeable future. Renewable energy carriers, including heat, power and liquid biofuels, can supply useful direct energy inputs for farms, fisheries and processing plants or be sold off-site to gain additional revenue. Much of the renewable energy could come from local resources including wind, solar radiation and small hydro captured from agricultural land, and biomass as a co-product of primary production and food processing enterprises. The potential co-benefits of renewable energy deployment can also result in improvements in people's livelihoods, education, employment prospects, health, rural development and social cohesion. Awareness raising, capacity building and technical field support for installation and maintenance of renewable energy technologies are essential components if projects are to be constructed and operated successfully.

Policies can be employed at various levels in order to develop the resilience of the agri-food sector to possible future energy supply constraints and adaptation to climate change impacts. Rapid deployment of efficient and renewable energy technologies in the sector will require regulatory measures as well as financial incentives and micro-financing to overcome the high up-front capital costs of some technologies and allow for a shift towards more sustainable farming systems.

Deployment of sustainable energy systems throughout the global agri-food sector is a huge undertaking and a multi-stakeholder initiative approach at the international level would help to catalyze this process. An international effort will be essential in order to implement solutions in a non-fragmented and cost effective way. National and local governments will need to consider the development and implementation of policies and measures that support rural development and combine food security with energy security whilst meeting sustainable development and GHG emission reduction targets. Technology transfer and adaptation strategies should be components of the package of interventions.

Progressing towards energy-smart agricultural systems can therefore encourage; improved energy efficiency at all stages along the agri-food supply chain; fossil fuel substitution by continuing the deployment of renewable energy systems; and energy access for all, especially in rural communities.

ACKNOWLEDGEMENTS

The original FAO project report *Energy Smart Food for People and Climate* (FAO, 2011a) on which this chapter was largely based, was led and managed by Olivier Dubois (Climate, Energy and Tenure Division of the Natural Resource Management and Environment Department, FAO). Erika Felix and Anne Bogdanski contributed sections of text and useful review comments were received from Francis Chopin, Theodor Friedrich, Uwe Schneider, Peter Holmgren, Josef Kienzle, Michela Morese, David Muir, Martina Otto, Prof. N H Ravindranath, Jonathan Reeves, Prof. Pete Smith and Peter Steele.

REFERENCES

Arizpe, N., Giampietro, M. & Ramos-Martin, J.: Food security and fossil energy dependence: an international comparison of the use of fossil energy in agriculture (1991–2003). *Crit. Rev. Plant Sci.* 30 (2011), pp. 45–63.

Ashburner, J.E. & Kienzle, J.: Investment in agricultural mechanization in Africa, conclusions and recommendations of a round table meeting of experts. Arusha, Tanzania. June 2009. Agricultural and Food Engineering Technical Report 8, Food and Agricultural Organization of the United Nations, Rome, Italy, 2011.

Azar, C.: Biomass for energy – a dream come true or a nightmare? WIREs Climate Change. John Wiley and Sons Ltd., 2011, www.wires.wiley.com/climatechange (accessed June 2013).

Bailey, A.P., Basford, W.D., Penlington, N., Parka, J.R., Keatinge, J.D.H., Rehman, T., Tranter, R.B. & Yates, C.M.: A comparison of energy use in conventional and integrated arable farming systems in U.K. agriculture. *Ecosyst. Environ.* 97:1–3 (2003), pp. 241–253.

Balmer, M.: Energy poverty and cooking energy requirements: the forgotten issue in South African energy policy? *J. Energy in South Africa* 18:3 (2007), pp. 1–9, http://www.erc.uct.ac.za/jesa/volume18/18-3jesa-balmer.pdf (accessed July 2013).

Bardi, U.: Solar power agriculture: a new paradigm for energy production. *Proceedings 2004 New and Renewable Energy Technology Developments for Sustainable Development* conference, Evora, Portugal June, 2004, http://www.spiritviewranch.com/pdf/Christoph/archive/Solar%20Power%20Agriculture.pdf.

Bernatz, G.: Apples, bananas and oranges; using GIS to determine distance travelled, energy use and emissions from imported fruit. St Mary's University of Minnesota, Winona, USA, 2010, http://www.gis.smumn.edu/GradProjects/BernatzG.pdf (accessed June 2013).

Bhatt, B.P. & Bujarbaruah, K.M.: Eco-energetic analysis of integrated agro-aquaculture models, North eastern Himalayan region, India. *J. Sustain. Agricult.* 55 (2011), pp. 495–510.

Biggs, S. & Justice, S.: Rural development and energy policy; lessons for agricultural mechanization in South Asia. Occasional paper #19, Observer Research Foundation, New Delhi, India, 2011, http://www.observerindia.com/cms/export/orfonline/modules/occasionalpaper/attachments/occ_rural_1296292421217.pdf (accessed June 2013).

Bogdanski, A., Dubois, O., Jamieson, C. & Krell, R.: Making integrated food/energy systems work for people and climate – an overview. Environment and Natural Sources Management working paper 45, Food and Agriculture Organization of the United Nations, Rome, Italy, 2010a, http://www.fao.org/docrep/013/i2044e/i2044e00.htm (accessed June 2013).

Bogdanski, A., Dubois, O. & Chuluunbaatar, D.: Integrated food energy systems –project assessment in China and Vietnam 11–29 October. Climate, Energy and Tenure Division, Food and Agriculture Organization of the United Nations, Rome, Italy, 2010b.

Bomford, M.: Beyond food miles. Post Carbon Institute, 2011, http://www.postcarbon.org/article/273686-beyond-food-miles (accessed June 2013).

Bruisma, J.: The resource book to 2050: by how much do land, water and crop yields need to increase by 2050? FAO expert meeting, How to Feed the World in 2050, Food and Agriculture Organization of the United Nations, Rome, Italy, 2009, www.fao.org/docrep/012/ak542e/ak542e00.htm (accessed June 2013).

Canning, P., Charles, A., Huang, S., Polenske, K. & Waters, A.: Energy use in the U.S. food system. USDA Economic Research Service, Washington, DC, 2010, http://www.ers.usda.gov/AmberWaves/September10/Features/EnergyUse.htm (accessed June 2013).

Cleland, D.: Towards a sustainable cold chain. 1st International Cold Chain Conference, International Institute of Refrigeration, Cambridge, UK, 2010, http://www.iifiir.org/userfiles/image/bookshop/2010-1.jpg (accessed June 2013).

Cuellar, A.D. & Weber, M.E.: Wasted food, wasted energy: The embedded energy in food waste in the United States. *Environ. Sci. Technol.* 44 (20100, pp. 6464–6469.

DARDNI: Heat pumps in agriculture and horticulture. Department of Agriculture and Rural Development, Northern Ireland, UK, 2013, http://www.dardni.gov.uk/ruralni/index/environment/renewables/heatpumps. htm (accessed June 2013).

DECC: 2050 pathways analysis. Department of Energy and Climate Change, UK, 2010, http://www.decc.gov.uk/assets/decc/What%20we%20do/A%20low%20carbon%20UK/2050/216-2050-pathways-analysis-report.pdf (accessed June 2013).

DEFRA: The 2007/2008 agricultural price spikes – causes and policy implications. Global Foods Market Group, a cross-Whitehall group of UK government officials, 2010a, http://cap2020.ieep.eu/assets/2010/1/22/HMT_price_spikes.pdf (accessed June 2013).

DEFRA: Food Statistics Pocketbook 2010. Department for Food, Environment and Rural Affairs, UK, 2010b.

DEFRA: Food Statistics Yearbook. Department for Food, Environment and Rural Affairs, UK, 2011.

De Witt, P., Tanner, C. & Norfolk, S.: Land policy development in an African context – lessons learned from selected experiences. Land Tenure Working Paper 14, Food and Agricultural Organization of the United Nations, Rome, 2009, ftp://ftp.fao.org/docrep/fao/012/ak547e/ak547e00.pdf (accessed June 2013).

El Bassam, N.: Integrated energy farming for rural development and poverty alleviation. In: Resource Management Towards Sustainable Agriculture and Development, Agribios International. Jodhpur, India, 2010, pp. 252–262, http://www.ifeed.org/pdf/Publication_IEF-for-Rural-Development-and-Poverty-Alleviation.pdf (accessed June 2013).

Environment Agency: Anaerobic digestion. Rotherham, UK, 2012, http://www.environment-agency. gov.uk/business/sectors/32601.aspx (accessed June 2013).

European Commission: Biomass Action Plan. Communication from Commission of the European Communities 2005, Brussels, Belgium, 2005, http://ec.europa.eu/energy/renewables/bioenergy/national_biomass_action_plans_en.htm (accessed June 2013).

FAO: Future energy requirements for African agriculture, Chapter 4, Scenarios. Food and Agricultural Organization of the United Nations, Rome, Italy, 1995, http://www.energycommunity.org/documents/ch4_adb.pdf (accessed June 2013).

FAO: How to feed the world in 2050. Food and Agricultural Organization of the United Nations, Rome, Italy, 2009a, http://www.fao.org/fileadmin/templates/wsfs/docs/expert_paper/How_to_Feed_the_World_in_2050.pdf (accessed June 2013).

FAO: Climate change implications for fisheries and aquaculture – overview of current scientific knowledge. FAO Fisheries and Aquaculture Technical Paper 530, Rome, Italy, 2009b, http://www.uba.ar/cambioclimatico/download/i0944e.pdf (accessed June 2013).

FAO: Small scale bioenergy initiatives – brief description and preliminary lessons on livelihood impacts from case studies in Asia, Latin America and Africa. Policy Innovation Systems for Clean Energy Security (PISCES), Practical Action, and FAO, Environment and Natural Resources Management working paper 31, Food and Agricultural Organization of the United Nations, Rome, Italy, 2009c, http://www.fao.org/docrep/011/aj991e/aj991e00.htm (accessed June 2013).

FAO: Energy-smart food for people and climate, Food and Agricultural Organization of the United Nations, Rome, Italy, 2011a, http://www.fao.org/docrep/014/i2454e/i2454e00.pdf (accessed June 2013).

FAO: Save and grow – a policy maker's guide to the sustainable intensification of smallholder crop production. Plant Production and Protection Division, Food and Agricultural Organization of the United Nations, Rome, Italy, 2011b.

FAO: An international consultation on integrated crop-livestock systems for development – the way forward for sustainable production intensification. Integrated Crop Management 13-2010 ISSN 1020-4555, 2011c.

Flammini, A.: Biofuels and the underlying causes of high food prices. 2008, http://www.globalbioenergy. org/fileadmin/user_upload/gbep/docs/BIOENERGY_INFO/0810_Flammini_-_Biofuels_and_the_underlying_causes_of_high_food_prices_GBEP-FAO.pdf (accessed June 2013).

Frater, T.G.: *Energy in New Zealand apple production*. PhD Thesis, Massey University library, Palmerston North, New Zealand, 2011.

Garnett, T.: Cooking up a storm: food, greenhouse gas emissions and our changing climate. Food Climate Research Network, Centre for Environmental Strategy, University of Surrey, Guildford, UK, 2009, http://www.fcrn.org.uk/sites/default/files/CuaS_web.pdf (accessed June 2013).

Geoghegan, T., Anderson, S. & Dixon, B.: Opportunities to achieve poverty reduction and climate change benefits through low-carbon energy access programmes. April 2008, The Ashden Awards for sustainable energy, GVEP International and IIED, 2008, http://www.ashdenawards.org/files/reports/DFID_report.pdf (accessed June 2013).

Giampietro, M.: Energy use in agriculture. In: *Encyclopedia of life sciences*. MacMillan Publishers, Nature Publishing Group, 2002.

GIZ, 2011: Modern energy services for modern agriculture – a review of smallholder farming in developing countries. 2011, http://www.gtz.de/de/dokumente/giz2011-en-energy-services-for-modern-agriculture.pdf (accessed June 2013).

Godfray, C., Beddington, J.R., Crute, I.R., Haddad, L., Lawrence, D., Muir, J.F., Pretty, J., Robinson, S., Thomas, S.M. & Toulmin, C.: Food security: the challenge of feeding 9 billion people. *Science* 327 (2010, pp. 812–818.

GoS: Foresight project on global food and farming futures. Synthesis Report C12: Meeting the challenges of a low-emissions world, UK Government Office for Science, London, UK, 2011, http://www.bis.gov.uk/assets/bispartners/foresight/docs/food-and-farming/synthesis/11-632-c12-meeting-challenges-of-low-emissions-world.pdf (accessed June 2013).

Gustavsson, J., Cederberg, C., Sonesson, U., van Otterdijk, R. & Meybeck, A.: Global food losses and food wastes – extent, causes and prevention. Swedish Institute for Food and Biotechnology and the Rural Infrastructure and Agro-Industries Division, Food and Agricultural Organization of the United Nations, Rome, Italy, 2011, http://www.fao.org/ag/ags/ags-division/publications/publication /en/?dyna_fef%5Buid%5D=74045 (accessed June 2011).

Heinberg, R. & Bomford, M.: The food and farming transition – towards a post-carbon food system. Post Carbon Institute, Sebastopol, CA, 2009, http://www.postcarbon.org/files/PCI-food-and-farming-transition.pdf (accessed June 2013).

Heller, M.C. & Keoleian, G.A.: Life cycle-based sustainability indicators for assessment of the US food system. Center for Sustainable Systems, University of Michigan, Report CSS00-04, 2000, http://css.snre.umich.edu/css_doc/CSS00-04.pdf (accessed June 2013).

Ho, M.-W.: Sustainable agriculture and off-grid renewable energy. Institute of Science in Society (ISIS), London, UK, 2011, http://www.i-sis.org.uk/SustainableAgricultureOffGridRenewableEnergy.php (accessed June 2013).

IEA: Renewable energy heating and cooling. International Energy Agency IEA/OECD, Paris, France, 2007, http://www.iea.org/publications/free_new_Desc.asp?PUBS_ID=1975 (accessed June 2013).

IEA: Deploying renewables – principles for effective policies. International Energy Agency IEA/OECD, Paris, France, 2008, http://www.iea.org/publications/free_new_Desc.asp?PUBS_ID=2046 (accessed June 2013).

IEA: Cities, towns and renewable energy – YIMFY – Yes In My Front Yard. International Energy Agency IEA/OECD, Paris, France, 2009, http://www.iea.org (accessed June 2013).

IEA: World Energy Outlook 2010. International Energy Agency, OECD/IEA, Paris, France, 2010, http://www.iea.org (accessed June 2013).

IEA: World Energy Outlook 2011. International Energy Agency, OECD/IEA, Paris, France, 2011, http://www.iea.org (accessed June 2013).

IMO: Updated study on greenhouse gas emissions from ships 2008–2009. International Maritime Organization, 2008, http://www.unctad.org/sections/wcmu/docs/cimem1p08_en.pdf (accessed June 2013).

IPCC: 4th Assessment Report – Mitigation. Working Group III, Intergovernmental Panel on Climate Change, 2007a, http://www.ipcc-wg3.de/publications/assessment-reports/ar4 (accessed June 2013).

IPCC: 4th Assessment Report – Mitigation, Chapter 9, Agriculture. Working Group III, Intergovernmental Panel on Climate Change, 2007b, http://www.ipcc-wg3.de/publications/assessment-reports/ar4/.files-ar4/Chapter09.pdf (accessed June 2013).

IPCC: 4th Assessment Report – Mitigation, Chapter 7, Industry. Working Group III, Intergovernmental Panel on Climate Change, 2007c, www.ipcc-wg3.de/publications/assessment-reports/ar4/.files-ar4/Chapter07.pdf (accessed June 2013).

IPCC: 4th Assessment Report – Mitigation, Chapter 5, Transport and its infrastructure. Working Group III, Intergovernmental Panel on Climate Change, 2007d, http://www.ipcc.ch/publications_and_data/ar4/wg3/en/ch5.html (accessed June 2013).

IPCC: Special report on renewable energy and climate change mitigation, Chapter 11 Policy, Financing and Implementation. Working Group III, Intergovernmental Panel on Climate Change, 2011a, http://srren.ipcc-wg3.de/report/IPCC_SRREN_Ch11 (accessed June 2013).

IPCC: Special report on renewable energy and climate change mitigation, Chapter 9, Sustainable Development. Working Group III, Intergovernmental Panel on Climate Change, 2011b, http://srren.ipcc-wg3.de/report/IPCC_SRREN_Ch09 (accessed June 2013).

IPCC: Special report on renewable energy and climate change mitigation, Chapter 8, Integration of Renewable Energy into Present and Future Energy Systems. Working Group III, Intergovernmental Panel on Climate Change, 2011c, http://srren.ipcc-wg3.de/report/IPCC_SRREN_Ch08 (accessed June 2013).

IPCC: Special report on renewable energy and climate change mitigation. Working Group III, Inter-governmental Panel on Climate Change, 2011d, http://srren.ipcc-wg3.de/ (accessed June 2013).

IPCC: Special report on renewable energy and climate change mitigation, Chapter 2 Bioenergy. Work-ing Group III, Intergovernmental Panel on Climate Change, 2011e, http://srren.ipcc-wg3.de/report/ IPCC_SRREN_Ch02 (accessed June 2013).

ISD: Increasing access to homestead biogas in Tentulia: pilot project. 2010, http://www.snvworld.org/en/ Documents/Inreasing_access_to_homestead_biogas_in_Tentulia_Bangladesh_2010.pdf (accessed June 2013).

Jianbo, L.: Energy balance and economic benefits of two agro-forestry systems in northern and southern China. *Agricul. Ecosyst. Environ.* 116 (2006), pp. 255–262.

JRC: Environmental impacts of products – analysis of the life cycle environmental impacts related to final consumption of the EU 25. Report EUR 22284 EN, Joint Research Centre of European Commission, 2006, pages, http://ec.europa.eu/environment/ipp/pdf/eipro_report.pdf (accessed June 2013).

JRC: Environmental improvement potentials of meat and dairy products. Report EUR 23491 EN, Joint Research Centre of European Commission, 2008, http://ftp.jrc.es/EURdoc/JRC46650.pdf (accessed June 2013).

JRC: Environmental impacts of diet changes in the EU. Report 23783 EN, Joint Research Centre of European Commission, 2009, http://ftp.jrc.es/EURdoc/JRC50544.pdf (accessed June 2013).

Karekezi, S. & Kithyoma, W.: Cogen for Africa: a project to promote new capacity in several countries. Cogeneration & On-site Power production, November, 2006, http://www.cospp.com/articles/print/ volume-7/issue-6/features/cogen-for-africa-a-project-to-promote-new-capacity-in-several-countries.html (accessed June 2013).

Khakbazan, M., Scott, S.L., Block, H.C, Robins, C.D. & McCaughey, W.P.: Economic effects and energy use efficiency of incorporating alfalfa and fertilizer into grass-based pasture systems. *World Acad. Sci. Eng. Technol.* 49 (2009), pp. 79–84.

LGA: War on waste – food packaging study, Wave 3. Local Government Association, BMRB report 45106324, 2009, http://www.lga.gov.uk/lga/aio/1613930 (accessed June 2013).

Lobell, D.B., Burke, M.B., Tebaldi, C., Mastrandrea, M.D., Falcon, W.P. & Naylor, R.L.: Prioritizing climate change adaptation needs for food security in 2030. *Science* 319:5863 (2008), pp. 607–610.

Miao, Y., Stewart, B.A. & Zhang, F.S.: Long-term experiments for sustainable nutrient management in China – a review. Agronomy for Sustainable Development, 2010, http://www.agronomy-journal. org/index.php?option=com_article&access=doi&doi=10.1051/agro/2010034&Itemid=129 (accessed June 2013).

Moreno, M.M., Lacasta, C., Meco, R. & Moreno, C.: Rainfed crop energy balance of different farming systems and crop rotations in a semi-arid environment: results of a long-term trial. *Soil Tillage Res.* 114:1 (2011), pp. 18–27.

Nassi o Di Nasso, N., Bosco, S., Di Bene, C., Coli, A., Mazzoncini, M. & Bonari, E.: Energy efficiency in long-term Mediterranean cropping systems with different management intensities. *Energy* 36 (2011), pp. 1924–1930.

Nelson, G.C., Robertson, R., Msang, S., Zhu, T., Liao, X. & Jawajar, P.: Greenhouse gas mitigation – issues for Indian agriculture. Discussion paper 00900, International Food Policy Research Institute, Wash-ington DC, 2009, http://www.ifpri.org/sites/default/files/publications/ifpridp00900.pdf (accessed June 2013).

Nguyen, M.L. & Haynes, R.J.: Energy and labor efficiency for three pairs of conventional and alternative mixed cropping (pasture-arable) farms in Canterbury, New Zealand. *Agricult. Ecosyst. Environ.* 52:2–3 (2005), pp. 163–172.

NSCA: Biogas as a road transport fuel. National Society for Clean Air and Environmental Protection, Brighton, UK, 2006.

OECD: Linking renewable energy to rural development. Organisation for Economic and Co-operative Devel-opment, OECD Publishing, Paris, France, 2012, http://www.oecd.org/greengrowth/linkingrenewable energytoruraldevelopment.htm (accessed June 2013).

Passey, R. & MacGill, I.: Energy sales targets: an alternative to white certificate schemes. *Energy Policy* 37 (2009), pp. 2310–2317.

Phong, L.T., De Boer, I.J.M. & Udo, H.M.J.: Life cycle assessment of food production in integrated agricultura-aquaculture systems of the Mekong Delta. *Livestock Sci.* 139 (2011), pp. 80–90.

Pimentel, D., Marklein, A., Toth, M.A., Karpoff, M.N., Oaul, G.S., McCormick, R., Kyriazis, J. & Krueger, T.: Food versus fuels – environmental and economic costs. *Human Ecology*, Springer, 2009, http://www.stopogm.net/sites/stopogm.net/files/foodvsbiofuelspimentel.pdf (accessed June 2013).

Practical Action: Energy poverty – the hidden energy crisis. Practical Action, Rugby, UK, 2009, http://practicalaction.org/energy-advocacy/docs/advocacy/energy_poverty_hidden_crisis.pdf (accessed June 2013).

Practical Action: Poor People's Energy Outlook 2012. Practical Action, Rugby, UK, 2012, http://practicalaction.org/page/15051 (accessed June 2013).

REN21: Renewables 2012 Global Status Report. Renewable Energy for the 21st Century, Paris, France, 2012, http://www.ren21.net/ http://www.ren21.net/Portals/97/documents/GSR/REN21_GSR2012.pdf (accessed June 2013).

Rodriguez, D.: New Holland agriculture's clean energy leader strategy. Climate Action press release, 3 August, 2011, http://www.climateactionprogramme.org/press_releases/new_hollands_clean_energy_leader_strategy (accessed June 2013).

Saunders, C. & Hayes, P.: Air freight transport of fresh fruit and vegetables. Research report 299, Agribusiness and Economic Research Unit, Lincoln University, New Zealand, 2009, http://researcharchive.lincoln.ac.nz/dspace/bitstream/10182/248/1/aeru_rr_299.pdf (accessed June 2013).

Sawin, J.: National policy instruments- policy lessons for the advancement and diffusion of renewable energy technologies around the world. In: D. Aßmann, U. Laumanns & D. Uh (eds): *Renewable energy – a global review of technologies, policies and markets*. Earthscan, London, UK, 2006.

Schneider, U.A. & Smith, P.: Energy intensities and greenhouse gas emissions in global agriculture. *Energy Efficiency* 2 (2009), pp. 195–206.

SER: Farmers flock to invest in renewable energy schemes. *Sustain. Energy Rev.* (5 September 2011), www.internationalsustainableenergy.com/3219/news/farmers-flock-to-invest-in-renewable-energy-schemes/ (accessed June 2013).

Sims, B. & Baudron, F.: Farm power and conservation agriculture. *Landwards* 67:3 (2012), pp. 14–17, Institution of Agricultural Engineers Professional Journal, www.iagre.org. (accessed June 2013).

Sims, R.E.H.: *The brilliance of bioenergy – in business and in practice*. Earthscan Publishers Ltd. London, UK, 2002.

Smil, V.: *Energy in nature and society – general energetic of complex systems*. MIT Press, Cambridge, MA, 2008.

Smil, V.: A skeptic looks at alternative energy. IEEE Spectrum, Institute of Electrical and Electronic Engineers, 2012, http://spectrum.ieee.org/energy/renewables/a-skeptic-looks-at-alternative-energy/0 (accessed June 2013).

Smith, P.D.: *Cities – a guidebook for the urban age*. Bloomsbury, London, 2012, http://www.peterdsmith.com/city/ (accessed June 2013).

SOFIA: State of world fisheries and aquaculture. UN Food and Agricultural Organization, Rome, Italy, 2010, http://www.fao.org/docrep/013/i1820e/i1820e00.htm (accessed June 2013).

Soto, D. (ed): Integrated mariculture: a global review Fisheries and Aquaculture. Technical Paper. No. 529, UN Food and Agricultural Organization, Rome, Italy, 2009, http://www.fao.org/docrep/012/i1092e/i1092e00.htm (accessed June 2013).

Spielman, D.J. & Pandya-Lorch, R.: Proven successes in agricultural development – a technical compendium to "Millions Fed". International Food Policy Research Institute, Washington DC, 2010, http://www.ifpri.org/publication/proven-successes-agricultural-development/ (accessed June 2013).

Steele, P.E.: Agro-mechanization and the information services provided by FAOSTAT. Unpublished internal Report, UN Food and Agricultural Organization, Rome, Italy, 2011.

Thomas, S.A.: White paper: integrated multi-trophic aquaculture. Report of workshop, 14–15 September, 2010, Port Angeles, Washington, DC, 2011, http://www.pacaqua.org/PacAqua_News/wp-content/uploads/2011/05/IMTA-White-Paper-FINAL_14May2011.pdf (accessed June 2013).

Tyedmers, P.H., Watson, R. & Pauly, D.: Fueling global fishing fleets. Royal Swedish Academy of Sciences, Ambio 34:8 (2005), pp. 635–638, http://sres.management.dal.ca/Files/Tyedmers/Fueling_Fleets1.pdf (accessed June 2013).

UNDP: The energy access situation in developing countries – a review focusing on the least developed countries and Sub-Sahara Africa. United Nations Development Programme and World Health Organization, 2009, http://content.undp.org/go/newsroom/publications/environment-energy/www-ee-library/sustainable-energy/undp-who-report-on-energy-access-in-developing-countries-review-of-ldcs—ssas.en (accessed June 2013).

UNEP: Towards a green economy: pathways to sustainable development and poverty eradication – a synthesis for policy makers. United Nations Environment Programme, 2011, http://www.unep.org/greeneconomy (accessed June 2013).

UNFCCC: Compilation of information on nationally appropriate mitigation actions to be implemented by Parties not included in Annex I to the Convention. United Nations Framework Convention on Climate Change, 2011, http://unfccc.int/resource/docs/2011/awglca14/eng/inf01.pdf (accessed June 2013).

UN General Assembly: Promotion of new and renewable sources of energy. Report of Secretary General, 22 August 2011, http://www.un.org/esa/dsd/resources/res_pdfs/ga-66/SG%20report_Promotion_new_renewable_energy.pdf (accessed June 2013).

USEPA: Global mitigation of non-CO_2 GHGs. Environmental Protection Agency report EPA 430-R-06-005, Wahington DC, 2006, http://www.epa.gov/climatechange/economics/downloads/GlobalMitigationFull Report.pdf (accessed June 2013).

Vale, R. & Vale, B.: Time to eat the dog – the real guide to sustainable living. Thames and Hudson, 2010, http://www.newscientist.com/article/mg20427311.600-how-green-is-your-pet.html (accessed June 2013).

Van Iersel, S. & Flammini, A.: Algae-based biofuels – applications and co-products. Environment and Natural Sources Management working paper 44, Food and Agriculture Organization of the United Nations, Rome, Italy, 2010.

Vorley, B.: Small farmers and market modernization, reflect and act. International Institute for Environmental Development, IIED Sustainable Markets Group, July, 2011, http://www.iied.org (accessed June 2013).

Weber, C.L. & Matthews, H.S.: Food-miles and the relative climate impacts of food choices in the United States. *Environ. Sci. Technol.* 42:10 (2008), pp. 3508–3513.

WEC: Energy efficiency – a recipe for success. World Energy Council, 2010, http://www.worldenergy.org/documents/fdeneff_v2.pdf (accessed June 2013).

WEF: Water security – the water-food-energy-climate nexus. World Economic Forum Water Initiative, Island Press, Washington DC, 2011.

White, R.: Carbon governance from a systems perspective: an investigation of food production and consumption. *Proceedings of the European Council for an Energy-Efficient Economy (ECEEE) 2007 Summer Study*, France, 2007.

Williams, A.G., Pell, E., Webb, J., Tribe, E., Evans, D., Moorhouse, E. & Watkiss, P.: Comparative life cycle assessment of food commodities procured for UK consumption through a diversity of supply chains. Final report for DEFRA, Project FO0103, Department for Food, Environment and Rural Affairs, UK, 2007, www.defra.gov.uk (accessed June 2013).

Woods, J., Williams, A., Hughes, J.K., Black, M. & Murphy, R.: Energy and the food system. *Phil. Trans. R. Soc.* B 365 (2010), pp. 2991–3006.

World Bank: RE Toolkit – a resource for renewable energy development, best practices and lessons learnt, technical and economic assessment of off-grid, mini-grid and grid electrification technologies: ESMAP Technical Paper 121/07. Washington DC, 2008, http://go.worldbank.org/IC3FU805H0 (accessed June 2013).

World Bank: Design and performance of policy instruments to promote the development of renewable energy: emerging experience in selected developing countries. Energy and Mining Sector Board, World Bank, Washington DC, 2011, http://www-wds.worldbank.org/servlet/main?menuPK=64187510&pagePK=64193027&piPK=64187937&theSitePK=523679&entityID=000386194_20110718032908 (accessed June 2013).

Woods, J., Williams, A., Hughes J.K., Black, M. & Murphy, R.: Energy and the food system. *Phil. Trans. R. Soc.* B 365 (2010), pp. 2991–3006.

Ziesemer. J.: Energy use in organic food systems. Natural Resources Management and Environment Department, Food and Agricultural Organization of the United Nations, Rome, Italy, 2007, http://www.fao.org/docs/eims/upload/233069/energy-use-oa.pdf (accessed June 2013).

CHAPTER 7

Energy, water and food: exploring links in irrigated cropping systems

Tamara Jackson & Munir A. Hanjra

7.1 INTRODUCTION

The consumption of energy and water are essential for life and form the foundation of agricultural production processes. Concerns over climate change and energy security also drive the need to use these resources more efficiently. The use of these resources and their impact on the environment are interwoven in a complex system, of which both the energy and water components are facing a number of challenges. Long term depletion of natural resources such as coal, oil and natural gas, the issue of global warming and competing demands for water (Graham *et al.*, 2003) indicate the need to consider water and energy factors conjunctively. Energy use of some form is necessary for increasing agricultural productivity and improving food security (Sayin *et al.*, 2005); indeed, intensification of fossil energy use has been associated with an increase in agricultural productivity during recent decades (Conforti and Giampetrio, 1997). Given the global challenge of increasing agricultural productivity to achieve food security, it is likely that the trend of intensifying energy use in agriculture will continue in decades to come. The assessment of water and energy consumption in agricultural production is imperative to identify the sources of waste, and to determine strategies for best allocating scarce resources to enhance crop production.

Many studies have quantified the energy consumption associated with crop production in various countries (Barber, 2004; Canakci *et al.*, 2005; Chamsing *et al.*, 2006; Chaudhary *et al.*, 2006; Erdal *et al.*, 2007; Esungen *et al.*, 2007; Hatirli *et al.*, 2006; Khan *et al.*, 2009; Ozkan *et al.*, 2007; Pimentel *et al.*, 2002; Singh *et al.*, 2007; Tzilivakis *et al.*, 2005). Both Hodges *et al.* (1994) and Lal (2004) found that approximately 23% of direct energy use for crop production in the USA was used for on-farm pumping, indicating that irrigation can be a significant energy cost for primary producers. Energy use for pumping irrigation water can even be far greater in groundwater-based systems than gravity-based surface irrigation. Where a groundwater source is used for irrigation, the use of pressurized micro-irrigation systems can decrease energy consumption where reduced operating pressures and pumping volumes are experienced (Hodges *et al.*, 1994; Jackson *et al.*, 2010; Srivastava *et al.*, 2003).

The inclination towards higher input systems and current levels of energy inefficiency in agricultural systems may be due to traditionally low energy costs. A similar trend has been experienced in regard to water consumption, particularly in developed countries. However, this scenario is changing under emerging conditions, such as the biofuel scenarios in a water perspective (Gerbens-Leenes *et al.*, 2012); the increasing share of bio-energy in world energy supply (Gerbens-Leenes *et al.*, 2009); and carbon pricing and electricity markets (Nelson *et al.*, 2012) where energy costs are becoming a primary consideration for producers, and one of the fastest growing input costs (Chen *et al.*, 2009). This then becomes a driving factor in the need to enhance our understanding of the patterns of energy use in crop production, particularly where irrigated systems are used, so that targeted changes can be made.

Given that water and nutrients are the principal limiting factors for crop production, optimal supply of these farming inputs in cropping systems allows large gains in production (Martin *et al.*, 2006; Molden *et al.*, 2010). However they can also represent large energy inputs into a system;

therefore methods and management practices that allow a reduction in the use of these farming inputs will reduce energy use by agricultural systems.

7.1.1   *Energy in agriculture*

While agriculture is not a major consumer of energy, it is often critically reliant on the supply of fuel and electricity, which means that as an industry it is heavily reliant on energy (Department of Primary Industries and Energy, 1994), and is likely to be influenced by factors outside the industry such as the price and availability of energy (Khan *et al.*, 2009).

Energy consumption in any cropping system can be categorized as either direct or indirect. Direct energy consumption is that energy used from primary or secondary fuel sources such as diesel, oil, gas, solar and electricity. Indirect energy consumption refers to energy that is used to produce equipment and other goods and services that are used in the process of crop production, and as such is the embodied energy in the end product used as an intermediate input into the crop production process. In irrigated agriculture, direct inputs are primarily fuel sources used to operate farm machinery and pumps. In industrialized countries, fuel, lubricants and electricity (where accessible) are the most commonly used direct inputs. These energy sources are used to power tractors, other farm vehicles and irrigation pumps. As previously stated, indirect energy inputs can account for up to two thirds of agricultural energy use. The four most significant categories of indirect energy in terms of quantity are fertilizer, irrigation infrastructure, agrochemicals and machinery manufacture. Other forms of indirect energy inputs, such as human labor, soil improvements and seed are negligible in most cases when compared to these most significant categories.

In an agricultural sense, direct energy consumption is for on-farm practices such as tillage and irrigation pumping; whereas indirect energy consumption is used to produce machinery and other goods and services that are used on-farm (Pimentel, 1992). Direct energy consumption by agriculture is negligible in developed countries, but if indirect energy consumption is included, the figure may more than double; for example, the estimate of energy consumption by agriculture in France nationally is 2%, but when indirect energy consumption is also considered, this figure rises to 5% (Pervanchon *et al.*, 2002). Pimentel (1992) has estimated that direct energy use on farms is just one third of total energy consumption.

Energy is also categorized by source, whether it be commercial or non-commercial. Commercial energy refers to more conventional forms of energy that are available for purchase, such as diesel and electricity and lately solar power. Non-commercial energy mainly refers to renewable forms of energy such as wood, animal manure and crop residues (Stout, 1990). This energy source is widely used in developing countries. Energy is an essential component of any agricultural system, whether the source is human, animal, synthetic or mechanical. Energy consumption in agriculture is also directly related to the development of technology and the level of production from a system (Hatirli *et al.*, 2006). Therefore, the energy-water nexus has important implications for sectoral policies such as crop and irrigation sector as well as broader policies outside agriculture including energy, science and technology, commerce and trade etc.

7.2   THE ENERGY-WATER NEXUS IN CROP PRODUCTION

The water-energy nexus refers to the multiple links between water and energy use. In the context of this chapter, energy is required in the form of farming inputs into crop production as well as that which is required for the sourcing, transport and application of water. Energy use impacts water use, and water use impacts energy use such that every land and water management decision is also an energy management decision. Thus, changes to agricultural production systems may impact on both water and energy demand, since the two are inextricably linked.

Current production patterns have to date been based on cheap energy and almost free water; steep increases in diesel prices since 2008 have been a wakeup call for both producers and consumers. As predicted decades ago by Pimentel *et al.* (1973), when energy and water resources become expensive, significant changes in agriculture take place. Following the oil price shocks of the 1970s, farmers responded by introducing technical and managerial changes that improved energy productivity throughout the next decade, for example by shifting from gasoline to diesel fuel that is more efficient, more effective use of fertilizers, chemicals and irrigation, and continued improvements in plant breeding (Cleveland, 1995). Given current energy prices (and the input use dependent on these prices), as well as an increased awareness of greenhouse gas emissions, major changes in farming practices are likely in the decades ahead. Indeed, energy is becoming a major consideration within all industries, including the agricultural sector. However, it is not clear how technology, investment markets and humans will respond to the emerging energy and water scarcity issues in the future.

We are now experiencing a situation where energy inputs (including fuel, fertilizers and agro-chemicals) and energy prices are soaring, whilst the availability of water for irrigated crop production is declining, prompting a need for change. Current food production systems are under pressure from a myriad of sources: population growth, environmental water demands, increased competition for water from other sectors, physical water scarcity due to regular droughts and a greater demand for biofuels and renewable energy among them. In the broader context, these pressures occur within the complex links between water, energy, food and the environment and have social, economic, ecological and political implications (McCornick *et al.*, 2008).

In terms of irrigated crop production, the water-energy nexus is generally considered in terms of the energy required to extract, transport and apply water for crop growth, as well as the water (and energy) required to produce energy in the form of crops for food, and now, increasingly, biofuels (Gerbens-Leenes *et al.*, 2012). Energy has an important role in groundwater extraction and use (McCornick *et al.*, 2008). In areas where groundwater is a significant source of irrigation water, the levels of energy use can be considerable. This is the case in India, where irrigation accounts for 80% of total water use and 30% of total electricity consumed, due to wide scale use of tubewells as a water source for irrigation. In 1999 there were 18 million tubewells in India, accounting for 56% of irrigation water use (Malik, 2002).

The quality of the water and energy used will also have an impact on the water-energy nexus. For example, the quality of energy and water supply in India (and many other developing countries) is poor; thus, there are different impacts on the water and energy nexus in resource poor countries in comparison to developed countries, where both energy and water supplies are generally reliable, stable and of good quality. This is illustrated by the effects of energy rationing and blackouts and whiteouts that can reduce water use efficiency and cut agricultural production. Coping mechanisms by irrigators where access to water and energy is poor tend to lead to reduced energy and water efficiency and a reduction in agricultural production (Malik, 2002), as well as reluctance to pay for poor quality services. Non-payment leads to further deterioration in service quality and the nexus worsens.

Climate variability and changes to the hydrological cycle can also highlight the intensities related to the water-energy nexus; during drought periods there are fewer water resources and greater demand from end users (Lofman *et al.*, 2002). Where surface water is normally used for irrigation, droughts often reduce the quantity of surface water available, adding pressure to groundwater resources and increasing the energy used as more water is extracted from these sources. Drought and over extraction from groundwater sources also impacts on energy use, as more power is required to extract groundwater as the watertable falls (Lofman *et al.*, 2002) and this could potentially impact in-stream flows in some regions.

7.2.1 *Energy for irrigation*

Irrigation can be considered as both a direct and indirect energy input; direct energy inputs are associated with fuel for extracting, transporting and pressurizing irrigation systems. Pimentel and Pimentel (2008) estimate that 15% of all energy expended for crop production is used to pump irrigation water. Indirect irrigation energy inputs are associated with the energy embodied in irrigation infrastructure and its operation and maintenance. For instance, about 23% of direct energy use for crop production in the USA was used for on-farm pumping (Hodges *et al.*, 1994; Lal, 2004). Similar results were found in the arid zone of India, where irrigation always consumed the largest amount of energy in a farming system, consuming between 33% and 48% of direct energy (Singh *et al.*, 2002). The type of irrigation system used obviously has an impact on the amount of energy consumed, even within pressurized systems, as the energy required for pumping depends on total dynamic head, flow rate and system efficiency (Jackson, 2009).

Energy use for irrigation depends on watertable depth or lift height (Lal, 2004). The source of water used for pressurized systems thus impacts on the amount of energy consumed, since surface water sources and pumping from shallow water tables requires less energy than from deeper water tables. If groundwater is used for irrigation, the use of pressurized irrigation systems can reduce energy consumption due to reduced operating pressure and reduced pumping volumes as higher application efficiency could be achieved, reducing both water and energy demand (Hodges *et al.*, 1994; Jackson *et al.*, 2010; Srivastava *et al.*, 2003). In China, energy use for groundwater pumping was found to be sensitive to changes in pump efficiency (Wang *et al.*, 2012). Thus the hydrologic and biophysical factors impact energy use and its cost in crop production.

The indirect energy costs of installing the various systems must also be taken into account (Lal, 2004), as irrigation water use is associated with large, indirect energy costs in terms of the manufacturing of both irrigation machinery and equipment and water supply infrastructure (Hodges *et al.*, 1994). There is a general trend towards a higher amount of embodied energy as irrigation systems become more sophisticated (Foran, 1998). For instance, pressurized systems are likely to be more costly from an embodied energy perspective as more equipment is required to be manufactured and installed. In addition to this, the construction of the water supply source will incur an extra energy cost. Smerdon and Hiler (1985) estimated that the energy required for the supply of water from surface water and groundwater with a lift of 50 m and 100 m was 0.75, 1.29 and 1.72 GJ/ha per year, respectively. While these costs must be recognized, they will not be considered in this chapter, as it is assumed that these features cannot be easily changed by farmers, as well as the fact that embodied energy requirements will remain approximately the same for a given system.

7.2.1.1   *Factors affecting irrigation energy use*

Energy use in the irrigation sector is influenced by several factors including climatic conditions, season, water management practices, irrigation schedule and system type, proportion of population engaged in agriculture, proportion of cropland irrigated, and government policies on energy pricing and input subsidies. Irrigation energy use is essentially influenced by the quantity of water pumped, the efficiency of the pumping system and the pumping depth. Field testing of irrigation pumps in the USA has found that on average, pumps are operating at below 50% efficiency, far less than the accepted achievable efficiency of 67% (Loftis and Miles, 2004).

Irrigation energy consumption is also affected by climatic conditions, agricultural and water management practices. McChesney *et al.* (1981) identified generally favorable climatic conditions as a factor influencing New Zealand's relatively low energy consumption by their agriculture sector. The use of electricity for irrigation poses problems in terms of electrical load management in relation to climatic conditions; irrigation needs are often highest at peak domestic load times, for example during winter freezes and hot months and afternoons (Hodges *et al.*, 1994). Where possible, steps should be taken to irrigate at off-peak times; this is possible with an automated drip system, which can be programmed to operate and thus use electricity at off-peak times (Srivastava *et al.*, 2003).

In addition to daily and weekly variations in irrigation energy use, which are dependent on the irrigation schedule and system type, there are also seasonal and year-to-year variations (California Electricity Commission, 2001). In California, for instance, where groundwater is pumped to compensate for shortages in surface water supply, there is an increase in energy consumption during drought years.

Ozkan *et al.* (2004) identified the size of the population engaged in the agricultural sector, the amount of arable land and the level of mechanization as being factors which influence energy consumption by the agriculture sector. A large agricultural population would generally correspond with a low level of mechanization, resulting in less energy use. This would generally be representative of the situation in developing countries.

Generally irrigation is a primary user of energy on irrigated farms; therefore at a country level, the proportion of agricultural land irrigated, as well as the different types of irrigation systems used, will impact on the amount of energy used by the agricultural sector. In India, 44% of agricultural land is irrigated, and this accounts for 30% of total electricity consumption at a national level (Malik, 2002). Direct energy use associated with diesel will add to this figure. Aggarwal (1995) showed that significant gains in energy efficiency are possible but adequate water availability is necessary to achieve productive yield. For these reasons, the analysis of energy use in irrigated agriculture is imperative so that areas for potential savings can be identified.

Government policy can also influence energy consumption and energy efficiency in irrigated systems. Indeed, much of the increase in agricultural productivity in the past century was driven by large subsidies of fossil fuels (Cleveland, 1995). Results from a survey in South Asia conducted by the International Water Management Institute (2003) indicate that tube well owners operating their pumps with heavily subsidized electricity ran pumps for between 40 and 250% longer than those buying diesel at market prices. In the province of Balochistan in Pakistan, the cost of pumped water is 54% higher for diesel pump owners than electric pump owners, again due to electricity subsidies (Ahmad, 2005). The result of this government policy has led to an overuse of both energy and water in these areas. A similar situation exists in Syria, where diesel subsidies result in low pumping costs, offering little incentive to farmers to irrigate more efficiently. Modeled results show that diesel price has a negative effect on the efficiency of water application; low pumping costs mean that there is a limited incentive to irrigate more efficiently (Gul *et al.*, 2005). Easing input subsidies on energy intensive items has been cited as a way to improve energy efficiency (McChesney *et al.*, 1981), but the linkages between power pricing and social equity are complex and hotly debated. It could, for instance, lower the income generated by individual farmers from cropping, and reduce agricultural production and raise food prices and social concerns.

7.2.2   *Energy and fertilizer*

Fertilizer is a large source of indirect energy use in agriculture (Hodges *et al.*, 1994; Ozkan *et al.*, 2004; Pimentel, 1992). Up to 40% of energy used in agricultural production in developed countries is devoted to fertilizer production (CAEEDAC, 2000; Hatirli *et al.*, 2006). However, fertilizer production is becoming more energy efficient, particularly that of nitrogen fertilizers, which is traditionally the most energy costly of all fertilizers (Cleveland, 1995; Giampietro *et al.*, 1999; Pimentel, 1992; Wells, 2001).

Fertilizer use also represents an example of factors affected by the water-energy nexus. The use of pressurized systems can lead to indirect energy savings in terms of reduced fertilizer use (Hodges *et al.*, 1994); more efficient water application means that less fertilizer is leached from the rootzone. Fertilizer can also be applied with irrigation water through pressurized systems, a process referred to as fertigation; incorporating the fertilizer into irrigation water reduces the amount of energy used in the system by reducing machinery use. This practice is also undertaken using flood irrigation by some farmers, however the losses from flood fertigation are likely to be higher than when using pressurized systems, as fertilizer is more likely to be leached below the rootzone and is not applied as precisely by flooding.

7.2.3   *Energy and agrochemicals*

While the production of agricultural chemicals can be energy expensive, the relatively small amount of active ingredient used on a per hectare basis reduces their impact on energy inputs. In addition to this, their use is important to the energy outputs of crops due to increased biomass production by crops in the absence of pest and disease (Deike *et al.*, 2008). A German study showed that the net energy output was reduced by 18% while energy intensity (energy inputs per unit of grain equivalent) was increased by 32% where pesticides were not used in an integrated farming system (Deike *et al.*, 2008).

The method of agrochemical application affects energy consumption, with oil formulations and granules being more energy costly than wettable powders (Pimentel, 1992). The development of new technologies such as genetically modified plants that are resistant to certain pests or diseases, as well as the improvement of crop management options such as integrated pest management (van den Berg and Jiggins, 2007) can help to reduce the amount of agricultural chemicals used. This is important from an energy conservation perspective.

7.2.4   Energy for machinery and equipment

The manufacturing and operation of farm machinery and equipment, including pumps for irrigation, is the largest user of commercial energy for agriculture throughout the world (Stout, 1990). Of this, approximately 60% is consumed by operation and 40% by manufacture. Higher levels of mechanization mean that it accounts for a larger share of energy consumption in developed countries in comparison to developing countries. It is important that machinery be correctly selected for its work load, in order to maximize energy efficiency. This includes selecting the correct size of tractor and implements to minimize fuel use for any given situation, as well as maintaining machinery in good working order.

7.2.4.1   Factors affecting input energy use for crop production

Inputs required for agricultural production can be affected by many factors, impacting the level of energy consumption. This is dependent on crop selection, irrigation method and individual management practices, but there are other physical and climatic factors that can affect the level of inputs. The type of farming system and farming practices can change both the total amount of energy consumed as well as the proportions used for various inputs. Simply practicing minimum tillage can reduce energy consumption by reducing the amount of machinery time required for crop production. For example, Singh *et al.* (2008) found that the use of minimum or zero tillage reduced energy input and maximized the energy output-input ratio in rainfed cropping systems studied in India. Savings in energy use due to lower fertilizer input requirements, lower machinery use under no till and reduced irrigation are well accepted benefits of no till systems.

The wide range of agricultural production systems that exist globally must be considered when examining energy consumption. Modern, intensive production methods rely heavily on external energy inputs such as synthetic fertilizers and diesel. In contrast, traditional methods rely more on natural inputs such as animal manure and human labor (Martin *et al.*, 2006). This has significant ramifications for both productivity and the level of impact that the system has on the environment. Purchased inputs are more likely to have higher indirect costs and therefore a higher associated energy cost. Across all systems, the largest inputs of purchased resources are water for irrigation, fertilizer, fuel and labor (Martin *et al.*, 2006).

Organic farming systems are growing in popularity, and while they often have lower yields (outputs) than conventionally grown crops, this is generally coupled with lower inputs, resulting in a higher output-input ratio (Deike *et al.*, 2008). Organic farming systems have a higher proportion of energy used on-site. This is due to increased cultivation due to its substitution for chemical weed control (Wood *et al.*, 2006). Conversely, conventional farming systems are more dependent on indirect energy inputs. However, the adoption of organic farming on a global scale could result in massive yield losses as a result of omitting mineral fertilizers and pesticides. The necessity to ensure global food security will continue to require energy inputs in agricultural production (Deike *et al.*, 2008).

Soil types can impact on energy consumption in a number of ways, as identified by Tzilivakis *et al.* (2005). Cultivation on heavy clay soils requires more energy due to the physical nature of the soil. Sandy soils require far higher amounts of fertilizer and irrigation water due to their poor ability to retain water and nutrients. Weed control is approximately 30% higher on peat soils which are nutrient rich and thus provide conditions suitable for plant growth.

In general, the more closely an agricultural system resembles the original natural ecosystem, the fewer the amount of inputs required for crop production (Pimentel and Pimentel, 2008). Agricultural production carried out in temperate climates without extreme heat or cold and with adequate rainfall is likely to consume less energy. The need for irrigation and frost control is eradicated, reducing energy inputs. However, the very nature of irrigation means that it is typically necessary in arid environments. Therefore an understanding of the patterns of energy consumption for crop production systems can help target the most appropriate areas to effectively reduce energy consumption.

Timeliness of input application (Stout, 1990) allows higher yields from the same amount of energy inputs. Fertilizer and irrigation water applied at critical phenological stages will positively impact on the crop yield in comparison to the same inputs applied at a time when the plant is unable to make productive use of them. Essentially what are required are methods for increasing or maintaining levels of production while minimizing inputs. The use of precision agriculture may help maximize the productivity of crop inputs.

7.3   PATTERNS OF ENERGY CONSUMPTION IN IRRIGATED AGRICULTURE[1]

On-farm energy consumption patterns in irrigated agriculture are highly variable, and depend on many factors. This section uses a case study from two irrigation areas located in southern Australia (NSW and South Australia) to illustrate the variability in energy consumption patterns, the links between energy and water use at the field scale, and some options for reducing energy consumption in irrigated systems. This section builds on a case study that was undertaken between 2007 and 2009 (Jackson *et al.*, 2010).

7.3.1   *Study sites*

The two case study sites are the Coleambally Irrigation Area (CIA) in NSW, and the South East of South Australia (SESA) as shown in Figure 7.1. These regions were selected in order to compare energy and water use for broadacre crops at the field scale in surface water and groundwater dependent irrigation areas, respectively (Jackson *et al.*, 2010).

The CIA is located in southern New South Wales, Australia. The area primarily uses good quality surface water diverted from the Murrumbidgee River to irrigate an area of 79,000 ha, although this area changes based on annual water allocations (for example, 10% in 2007 and 100% in 2012–2013 period). Irrigation water is primarily used for annual crops such as rice, wheat, barley, cotton, maize, canola and soybeans, as well as perennial pasture crops for sheep and cattle. Most irrigators use surface irrigation methods, including lasered contour bays, bed/furrow and some border check. The average farm size is 250 ha, and the irrigated proportion of farm land is close to 100% when water allocations are adequate. The soils common to the CIA are well suited to irrigated crop production, including areas of self-mulching clay that are suitable for rice production, as well as red-brown earths and transitional red-brown earths. The climate of the area ranges from warm temperate to semi-arid conditions, with hot summers and mild winters. Mean annual rainfall is 406 mm, with mean annual potential evapotranspiration of 1797 mm (Jackson *et al.*, 2010), such that irrigation is inevitable for crop production.

In contrast, the region of the (SESA) is almost completely dependent on groundwater for irrigation. The SESA has a total area of 28,120 km^2; of which about 2.8% of the total land area is

---

[1]This section builds on a case study that was undertaken in south eastern and South Australia, and previously published by Jackson *et al.* (2010). The authors wish to thank Elsevier Publishers for granting permission to reuse our work.

Figure 7.1.   Map showing the surface-water irrigation region of Coleambally in NSW, Australia, and the groundwater dependent region of the South East, South Australia.

irrigated (79,000 ha). Irrigation is generally applied to a small proportion of a farm's total area, ranging from 9–23% of the case study farm areas. The availability of good quality shallow groundwater for irrigation allows the production of high value crops such as wine grapes and lucerne for seed production, as well as supporting the meat, dairy and wool industries through pasture production. Over 60% of irrigation systems are pressurized, and there was a shift towards drip and center pivot irrigation systems between 1998/99 and 2003/04 (Econsearch, 2006). Groundwater quality in this region is generally good for most crop production purposes, with salinity of ground water below 1560 μS/cm. Soils in the region include deep sandy soils, loamy and clay soils and significant areas of shallow soils over limestone; in many cases the topsoil barely covers the limestone base (SA Government, 2000). The climate is typified by hot, dry summers and cool, wet winters. The mean annual rainfall is 493 mm, with mean annual potential evapotranspiration of 1,567 mm at Padthaway, in the center of the South East region. Since the 1950s, there has been a general trend of increasing temperatures and decreasing amounts of rainfall (McInnes et al., 2003), increasing the level of dependence on irrigation for crop production (Jackson et al., 2010).

7.3.2   Data requirements

An integrated model of water application and energy consumption at the field scale was developed to describe the relationships between water (source and application rate), energy, irrigation method, climate and soil characteristics. This data was integrated in an accounting model run in Microsoft Excel. To populate the model, field data reported by farmers was used. This data was collected through a comprehensive survey of inputs to crop fields on a seasonal basis, and included a comprehensive data set relating to the farm enterprise and associated socio-economic conditions, as well as pump characteristics and irrigation methods. Information on water use was reported by farmers, who use water meters or dethridge wheels to measure application rates. Data collected for energy inputs was selected based on common categories used in previous studies (Hatirli *et al.*, 2006; Ozkan *et al.*, 2007), and include diesel, electricity, machinery, agrochemicals, fertilizer, soil improvements, seed and human labor. In order to compare the potential effects of converting to other irrigation methods, the model combines operational data reported by farmers with accepted water use data for alternative irrigation methods (flood, center pivot and sub-surface drip). Full details of data and methods are not reported here for brevity and can be found in Jackson *et al.* (2010). Case study farms from each site in CIA and SESA were used to illustrate the potential patterns of water application and energy consumption at the field scale. The most commonly grown crops in the region were included in the analysis, as well as widespread farming practices were considered that ranged from low to high input in terms of water use, fertilizer and agrochemicals.

7.3.3   *Analyzing water application and energy consumption*

7.3.3.1   *Crop water requirements*
For baseline results, the on-farm water application rates reported by farmers were used in the study. For the purposes of simulating the use of alternative irrigation methods, the quantity of water applied was determined separately for each region. In the surface water region (CIA), the values estimated to determine changes to crop water use following the conversion to pressurized irrigation systems follow the method used by Khan *et al.* (2004; 2008). For summer and perennial crops, it is assumed that the use of pressurized irrigation methods would result in a reduction in water application of 10% compared to surface irrigation, while for winter crops (oats and wheat), it was assumed that application would be reduced to 1.1 ML/ha. In the ground water region (SESA) a reduction in water application achieved on the case study farms using pressurized irrigation systems was recorded (SESA Farms 1, 3, 4 and 5). Reported observations show a 50% reduction in water application with the use of center pivot systems compared to flood, and a further saving of 32% where a drip system was used on SESA Farm 1. These reductions in water application have been extrapolated and applied to the other case study farms in the SESA region.

7.3.3.2   *Energy accounting*
In order to quantify the energy consumption from the selected farms in two case study irrigation areas, the method used by Fluck (1992) was selected as a general methodology for performing an energy audit; and a similar method has been used by others (Canakci *et al.*, 2005; Hatirli *et al.*, 2006). First, a boundary is determined around the process to be evaluated; for the purposes of this study, the energy inputs for the on-farm production system are those consumed within the physical boundaries of the irrigation field or farm management unit. The data relates to energy consumption from land preparation to harvest. No post-harvest processes are considered due to data and time constraints. All inputs crossing the 'boundary' are then identified and quantified, with energy values assigned to them. Inputs evaluated include crop inputs and pumping energy consumption as discussed in Jackson *et al.* (2010).

7.3.3.2.1   Crop inputs
All inputs were quantified (e.g. kg/ha or hours of operation/ha) and converted to an energy equivalent using values from current literature; these values and the sources are shown in Table 7.1 (Jackson *et al.*, 2010). The energy requirement for the manufacture and supply of diesel, electricity, machinery, fertilizer and agrochemicals is included as indirect energy within each value given in the table.

7.3.3.2.2   Calculating pumping energy consumption
The energy required for pumping and pressurizing water for irrigation is determined using Equation (7.1), which is an expansion of Bernoulli's equation (Mott, 1994).

$$P\ (\text{kW}) = \frac{(h_A \gamma Q)}{1000 \times pump\ efficiency \times Dr} \tag{7.1}$$

Table 7.1.   Inputs and energy equivalent values and source for inputs assessed in this study.

Input		Unit	Equivalent energy [MJ]	Reference
Human labor		h	2.3	Ozkan et al. (2004); Hatirli et al. (2006)
Fuel	Diesel and diesel oil	L	38.6	Dept. Climate Change (2008)
	Aviation gas	L	33.1	Dept. Climate Change (2008)
Machinery		h	64.8	Hatirli et al. (2006)
Electricity	Electricity	kWh	11.9	Ozkan et al. (2004), Mandal et al. (2002)
Fertilizer	Nitrogen	kg	66.1	Hatirli et al. (2006)
	Phosphorous	kg	12.4	Hatirli et al. (2006)
	Potassium	kg	11.2	Hatirli et al. (2006)
	Sulfur	kg	5.0	Wells (2001)
	Lime	kg	0.6	Wells (2001)
	Manure	tons	303.1	Hatirli et al. (2006), Canakci et al. (2005)
Agro-chemicals	Fungicide	kg/ha	92.0	FAO (2000)
	Herbicide	kg/ha	240.0	FAO (2000)
	Insecticide	kg/ha	200.0	FAO (2000)
Seed	General	kg	14.0	FAO (2000)
	Rice	kg	14.7	Alam et al. (2005)
	Cereals & pulses	kg	25.0	Ozkan et al. (2004)
	Wheat	kg	15.7	Canakci et al. (2005)
	Maize	kg	15.7	Canakci et al. (2005)
	Oil seed	kg	36.0	Ozkan et al. (2004)

Source: Authors based on Jackson et al. (2010).

where P is the rate that energy is added to the fluid in kW, h_A is the total head in meters (the sum of pressure head and elevation head). Pressure head was generally given in kPa or psi and converted to meters. Elevation head is also called suction head. This is the depth from which the water is pumped (m). γ is the specific weight of water (assumed to be 9810 N/m³ at 15°C) and Q is the flow rate (m³/s). *Pump efficiency* is assumed to be 75%, as most large pumps usually operate at efficiencies within the range 74 to 85% in these irrigation systems in Australia (Faour, 2001). The Dr (Engine derating) accounts for efficiency losses between the energy required at the pump shaft and the total energy required. The approximate derating factor for electric motors is 80% and for diesel motors is 75% (Faour, 2001), depending on altitude, temperature and condition of the motor. Actual energy use will be determined by the fuel type used, total pumping hours and the energy added to the fluid (P). Thus, for electric pumps, the total kWh is given by Equation (7.1) multiplied by the total pumping hours. Where diesel pumps are used, the quantities of diesel consumed for pumping is calculated using the assumption that each kWh of engine power consumes 0.25 liters of diesel (Smith, 2004). All results shown correspond to the use of diesel as a fuel source; the patterns of energy consumption using an electric pump are the same as those when using a diesel pump, however the values are slightly higher (13.7%). This is due to higher embodied energy associated with the production of a unit of electricity (from coal in most Australian states) in comparison to a unit of diesel.

7.3.3.2.3   Operating conditions for pressurized systems to estimate conversion effects
Local data from research into alternative irrigation methods (O'Neill et al., 2008) was used to compare the levels of water application and energy consumption for alternative irrigation methods with current practices. The crop types included in the CIA farm case studies include rice, wheat, maize, lucerne (hay and seed production) and oats. Rice is excluded from the modeled results, as it is assumed that it cannot be commercially produced using either center pivot or drip irrigation methods. Table 7.2 shows the assumptions used to determine pumping energy costs in the surface

Table 7.2. Assumptions used to determine pumping energy in the CIA (surface-water) and SESA (ground-water) regions. In the SESA, assumptions are based on modifications of the existing site characteristics, where the * denotes the irrigation method currently in use.

	Case study farm	Irrigation method	Discharge head [m]	Suction head [m]	Flow rate [m³/s]
CIA	All	Centre Pivot	17.2	6	0.09
		Drip	8.27	3	0.04
SESA	Farm 1	F*	0	20	0.09
		CP*	17.5		0.04
		D*	40		0.04
	Farm 2	F*	0	64	0.07
		CP	18.5		0.04
		D	9		0.04
	Farm 3	F	0	20	0.11
		CP*	19.3		0.05
		D	9		0.04
	Farm 4	F	0	42	0.11
		CP*	19.3		0.05
		D	9		0.04
	Farm 5	F*	0	12	0.18
		CP*	5.5		0.04
		D	9		0.04

Source: Authors based on Jackson *et al.* (2010).

and groundwater regions. The impacts of conversion to the pressurized systems are shown in Table 7.3.

7.3.4 *Results and discussion*

7.3.4.1 *Water application and energy consumption: baseline conditions*
Water use and energy consumption are linked, as illustrated by Figure 7.2, which shows a general trend of increasing energy use with increasing water use. This is more pronounced where ground-water is used; for every additional megaliter of water applied, the increase in energy consumption in the groundwater region is more than 2.3 times that in the surface water region. This effect can be both direct and indirect. Where pressurized irrigation or groundwater is used, energy is used directly to operate pumps and equipment. An indirect effect occurs because when a significant investment has been made to fully irrigate a crop to maximize yields, farmers also tend to optimize the use of other inputs such as fertilizers and agrochemicals to ensure high productivity.

7.3.4.1.1 Surface water region
Crop choices and management practices varied among the case study farms. These differences impact on water application and energy consumption; summer and perennial crops with higher water use have higher energy inputs than winter crops. Figures 7.3 and 7.4 illustrate the baseline data for water and energy use for each case study farm. Crop water use is variable, ranging from 2 ML/ha for winter crops such as oats to 15.5 ML/ha for rice. Energy consumption is similarly variable, with a range of 3186–22,049 MJ/ha. Even for the same crop, energy consumption can vary widely; for example, for maize varies from 7975–14,075 MJ/ha; this can be attributed to different agrochemical and fertilizer application rates. These management decisions are made by individual farmers, and can clearly have a significant effect on levels of energy consumption. It is important to understand how and why these decisions are made, and to link energy consumption with the key decision factors, in order to optimize energy inputs.

In this region, baseline energy consumption was mainly attributable to fertilizer and machinery diesel inputs (Fig. 7.5). No energy was used for irrigation operation in this gravity based system.

Table 7.3. The modeled water and energy budgets for conversion from flood to pressurized systems in surface and ground water regions (using a diesel pump), based on key assumptions described under the 'Energy accounting' section of this chapter.

Region	Case study farm	Crop	Flood Irrigation [ML/ha]	Flood Energy consumption [MJ/ha]	Centre Pivot Irrigation [ML/ha]	% change from flood	Centre Pivot Energy consumption [MJ/ha]	% change from flood	Drip Irrigation [ML/ha]	% change from flood	Drip Energy consumption [MJ/ha]	% change from flood
CIA (Surface water)	Farm 1	Lucerne seed	5.0	3186	4.5	−10	8392	163	4.5	−10	7606	139
		Wheat	3.0	5157	1.1	−63	6430	25	1.1	−63	6238	21
	Farm 2	Oats	2.0	8406	1.1	−45	9678	15	1.1	−45	9486	13
	Farm 3	Lucerne hay	11.0	9811	9.9	−10	21264	117	9.9	−10	19534	99
	Farm 4	Maize	8.3	7975	7.5	−10	16651	109	7.5	−10	15340	92
	Farm 5	Maize	8.0	14075	7.2	−10	22404	59	7.2	−10	21146	50
SESA (Ground water)	Farm 1	Lucerne seed	11.0	20325	5.6	−49	16255	−20	3.8	−65	11346	−44
	Farm 2	Pasture	12.0	41759	6.1	−49	36779	−12	4.2	−65	31476	−25
	Farm 3	Clover pasture	11.2	22954	5.7	−49	18316	−20	3.9	−65	13377	−42
	Farm 4	Pasture	7.8	20647	4.0	−49	17392	−16	2.7	−65	13926	−33
	Farm 5	Lucerne seed	12.0	28465	6.0	−50	22907	−20	4.1	−66	17708	−38

Source: Authors based on Jackson et al. (2010).

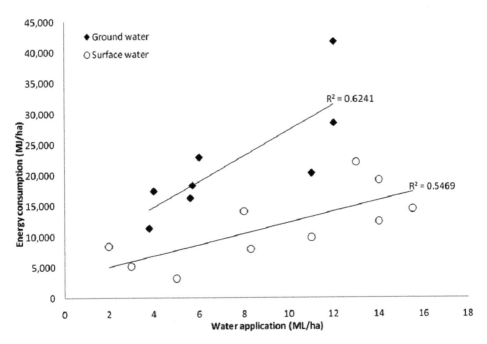

Figure 7.2.   Links between energy use and water application in surface and ground water supplied regions. Dot points relate to the level of water application and total energy consumption for individual farms in the study regions.

7.3.4.1.2   Groundwater region

Pressurized systems reduce water use due to the higher water application efficiencies of these systems. Water use could be approximately halved with the use of center pivot systems compared to flooding, and could be reduced by a further 32% where drip is used on SESA Farm 1 (Fig. 7.6). Energy consumption is very variable, ranging from 11,346 MJ/ha for drip irrigated lucerne seed (which does not require nitrogen fertilizer) to 41,759 MJ/ha for flood irrigated pasture that is part of a dairy enterprise, and hence has high nitrogen application rates (Fig. 7.7). Energy consumption is reduced where pressurized irrigation systems are used.

In the groundwater dependent region, energy inputs are dominated by energy for irrigation pumping (Fig. 7.5); fertilizer for pasture, and chemicals or soil improvements for lucerne is of secondary importance. With the exception of SESA Farm2, which applies high levels of nitrogen fertilizer to achieve high growth rates for pasture used as a dairy feed source, nitrogen fertilizer use is minimal for the production of leguminous perennial pasture crops in this region.

For the purpose of this study, inputs have been separated into several categories including human labor, diesel, aviation fuel, electricity, machinery, fertilizer, agrochemicals, soil improvements and seed, which represent the major types of energy inputs. The distribution of these categories in surface and groundwater regions is shown in Figure 7.5. Understanding the major categories of energy use for a given situation allows targeted improvements to be made where they will be most effective.

7.3.4.2   *Potential energy and water savings using pressurized irrigation systems*

In terms of water application, Table 7.3 shows that there is a potential reduction in water application at the field scale of between 10 and 66%. Pressurized irrigation systems have higher water use efficiency, and thus reduce water application rates. However, while pressurized systems improve water use and water productivity, they also greatly increase energy consumption in surface water

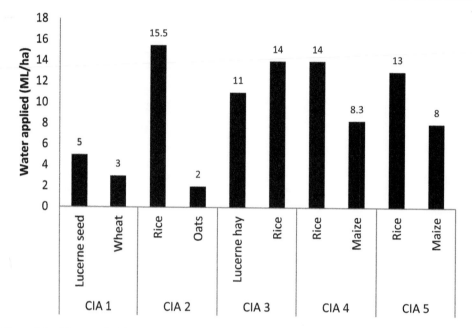

Figure 7.3.   Water applied for each case study farm in the surface water region (the labels CIA1-5 refer to the case study farms in the Coleambally Irrigation Area).

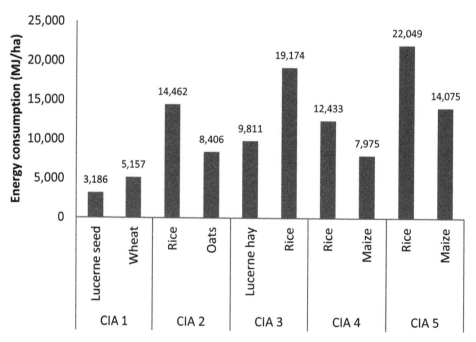

Figure 7.4.   Energy consumption for each case study farm in the surface water region (the labels CIA1-5 refer to the case study farms in the Coleambally Irrigation Area).

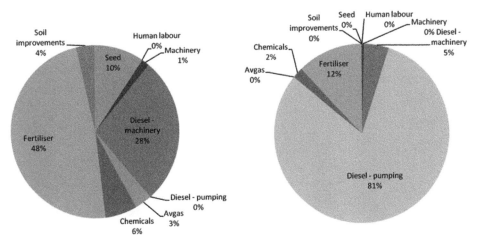

Figure 7.5. Break down of energy use categories for surface water irrigated farms (left) and groundwater irrigated farms (right).

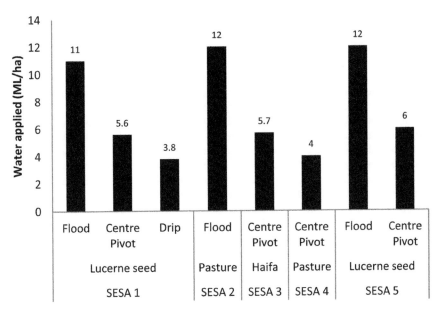

Figure 7.6. Water applied for each case study farm in the groundwater region (the labels SESA1-5 refer to the case study farms in the South East of South Australia).

regions, because when flood irrigation methods are used, energy for gravity irrigation is essentially 'free' in terms of operating energy. For example, when a center pivot is used instead of flood irrigation for lucerne hay production on CIA Farm 3, energy consumption increases by 117%, from 9811 MJ/ha up to 21,264 MJ/ha. The data shows that energy consumption increases more when center pivot systems are used compared to drip systems, because these systems generally require a higher operating pressure and are less efficient, which means that more water volume needs to be pumped in total. This is an important consideration for irrigators in surface water regions who are being encouraged to reduce water use. Given the ever-increasing costs of fuel

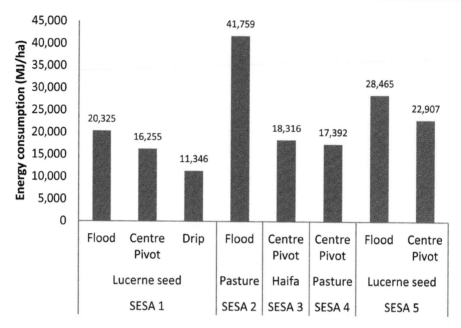

Figure 7.7.   Energy consumption for each case study farm in the groundwater region (using a diesel pump). (The labels SESA1-5 refer to the case study farms in the South East of South Australia).

and energy (diesel and electricity), conversion to a pressurized system would increase farming costs and result in a dependence on these energy sources.

There is a reported reduction in water withdrawals achieved on the case study farms using pressurized irrigation systems (SESA Farms 1, 3, 4 and 5). These reductions in water application were extrapolated to the other case study farms and used to explore the likely impacts of converting to pressurized systems. To be able to quantify the energy consumption, some assumptions were necessary (discharge head, irrigation efficiency, flow rate) for each case study farm using center pivot and drip irrigation methods. These details are described in Table 7.2.

In comparison to the surface water region, the results from this study show that energy consumption can be reduced by 12 and 44% in the groundwater region, indicating that there is potential to reduce energy consumption with the use of pressurized systems. Despite the need for energy when pressurized irrigation systems are installed, total energy consumption is reduced because the amount of water that needs to be pumped is greatly reduced. A similar pattern has been observed in previous studies (Hodges *et al.*, 1994; Srivastava *et al.*, 2003). The variability in results is due not only to the quantity of water pumped, but also to differences in system design and maintenance (irrigation system and pump), as well as operation and management factors. These factors influence water application and subsequently energy consumption (Hamdy *et al.*, 2003; Loftis and Miles, 2004). Higher operating pressures and flow rates increase the quantity of energy consumed (Lal, 2004).

Pressurized irrigation methods are widely adopted in the SESA; more than 60% of irrigation systems are pressurized, with center pivots being the most common. Irrigators know that these systems offer real benefits in terms of reducing water use at the field scale. Given the current and projected upward trends in energy costs and related greenhouse gas (GHG) emissions, it is useful to assess whether the reduction in water use is enough to offset the increased energy demand of these systems.

The amount of change of energy consumption following conversion to a pressurized system varies between the case study farms and crops (Table 7.3). This is because other energy input costs are included in the total energy consumption (e.g. fertilizer, chemicals and machinery).

When energy consumption was modeled, only the change in pumping energy requirements was different to the baseline data (Table 7.3).

7.3.5   *Summary*

This case study illustrates the implications of converting to pressurized irrigation systems in terms of water and energy consumption. Importantly, it demonstrates that the use of pressurized systems in groundwater dependent areas can result in a reduction of both water application and energy consumption. However, in areas where irrigators have a surface water source, the installation of pressurized irrigation systems can greatly increase energy consumption. The results from this case study also highlight the variability of energy consumption patterns, which are influenced by factors such as crop choice, management practices and enterprise type, in addition to the more obvious effects of water source and irrigation method.

It is acknowledged that the small sample size of the data merits further research to confirm the findings presented here; however, in general the results are supported by the literature (Hodges *et al.*, 1994; Mushtaq and Maraseni, 2011; Srivastava *et al.*, 2003), and these conclusions regarding on-farm water and energy use can still be accepted with confidence, given that the case study farms were selected to showcase a range of farming systems in each region. Further, the study results demonstrate water savings at the farm scale and require validation at higher spatial scales.

Along with energy and water impacts, landholders have also identified other benefits associated with pressurized irrigation systems. Yield increases and improved seed quality (the main product) were documented for SESA1. This has also been reported in previous studies undertaken in the CIA in Australia (Khan *et al.*, 2004; 2008; O'Neill *et al.*, 2008) and other countries (Playan and Mateos, 2006). Irrigators in the SESA also reported a reduction in labor use; this was also reported by Khan *et al.* (2004), where 50–80 labor days were saved per year in southern Australia. Additional benefits include reduced weed management costs associated with less agrochemical application and less cultivation.

Whether increased yields and crop quality and a reduction in inputs in conjunction with water savings would outweigh the additional costs associated with increased energy consumption would need to be considered on an individual farm basis, given the variability illustrated by this case study. While it is possible for pressurized irrigation systems to reduce both water and energy consumption simultaneously in targeted areas, any investment in pressurized systems must be carefully considered from both a water and energy perspective, and the resource trade-offs must be identified before decisions are made.

7.4   OPTIONS FOR SUSTAINABLE ENERGY AND WATER MANAGEMENT IN IRRIGATED CROPPING SYSTEMS

There are a range of options for achieving sustainable energy and water management targets at different levels, from field scale and local interventions to policy actions. These include technical interventions and policy-based strategies that can both contribute to optimizing energy and water use; it is essential that use of these resources is considered together in order to promote sustainable irrigated systems.

7.4.1   *Technical interventions*

There is wide variation in the performance of different irrigation methods at the field and farm level in terms of water and energy use (Jackson *et al.*, 2011; Maraseni *et al.*, 2012). The case study in the above section demonstrates that pressurized irrigation systems can reduce water and energy consumption where groundwater is used for irrigation. The use of pressurized systems in such conditions should be encouraged through capacity building efforts for farmers and local irrigation service providers. However, in surface water regions, the use of pressurized irrigation systems should be considered carefully, and perhaps used only where other benefits associated with precise water management are notable, such as the achievement of quality characteristics in wine grape production (Chaves *et al.*, 2007) or other high value crops (Ayars *et al.*, 1999). Instead, attention should be given to designing and/or improving existing surface systems to peak efficiency as a way of improving water use efficiency (Playan and Mateos, 2006). Additionally, where pressurized systems in these regions are associated with increases in energy use, these may be potentially offset by reductions in other inputs such as fertilizer or machinery use.

In addition to converting from gravity-fed to pressurized systems, targeted replacement of older sprinkler systems, that are inefficient in terms of both water and energy use, with efficient pressurized systems (drip or sprinkler), has been shown to reduce both water use and greenhouse gas emissions (Maraseni *et al.*, 2012). Energy consumption for irrigation was reduced by 50% in case studies where hand-shift and roll-line sprinkler irrigation systems were replaced by drip and center pivot irrigation systems respectively (Maraseni *et al.*, 2012).

Where irrigation delivery networks are pressurized, energy consumption can be reduced through appropriate network design, for example using optimum pipe diameter and pump station selection. Management strategies can also be employed, such as grouping individual irrigation operators into sectors that are able to operate during defined periods; the aim of this approach is to maintain the overall head of the systems at its lowest possible level, and to optimize the performance of the pumping units. Such an approach has been found to reduce energy consumption by around 16%, with theoretical savings of up to 22% available (Jimenez-Bello *et al.*, 2011).

In summary, technical interventions must be based on better designed and well maintained irrigation systems, considering both water and energy consumption. Education and training for farmers and irrigation water providers will improve the skill levels and ensure optimal management of on-farm irrigation and wider delivery systems.

7.4.2 *Policy strategies*

The ongoing development of anthropogenic climate change and mitigation policies need to account for links between water consumption, energy use and productivity levels, particularly where energy intensive responses may be promoted in response to water scarcity (Maraseni *et al.*, 2012). For example in China, the 12th Five Year Plan (2011–15) incorporates water conservation measures designed to increase irrigation water use efficiency by 3% and increase grain production by around 13%, in conjunction with targeted reductions in energy and carbon intensity of 16% and 17% respectively (Wang *et al.*, 2012). In this instance, water, energy and carbon emission intensity reduction and grain production targets are in competition, depending on the strategies employed to achieve them. Thus policies aimed at water saving should also be evaluated in terms of energy consumption patterns and other possible benefits (Wang *et al.*, 2012) including greenhouse gas emissions per unit of value added (Randers, 2012). Although Australian farmers are very efficient users of water and energy and achieve some of the highest crop yields in the world, public policy must move beyond the economic incentives and provide social incentives to the farmers to absorb the new cost pressures. For instance, Australian farmers are much more exposed to the risks associated with the introduction of a price on carbon and its impacts on existing energy markets than the average nation within OECD (Nelson *et al.*, 2012).

Current government policy in Australia supports the modernization of irrigation systems, with the aim of generating water savings within the Murray-Darling Basin, where 85% of Australia's irrigation takes place (Department of Sustainability, Environment, Water, Population and Communities, 2013). About AU$ 4 billion has already been committed to upgrade and modernize water and irrigation infrastructure. Under the *On-farm Irrigation Efficiency Program*, another AU$100 million has been allocated in the most recent round of funding (2012). Irrigators are encouraged to modernize on-farm irrigation infrastructure, equally (50:50) sharing water savings between irrigators and the environment. It is reasonable to assume that many irrigators will adopt pressurized irrigation systems under this program, in order to generate significant water savings; this will have far-reaching implications for water and energy consumption, as well as environmental impacts. The energy impacts of these on-farm changes should be considered in conjunction with the proposed water savings.

One policy option for reducing energy use in the irrigation industry could link energy tariff arrangements with the use of Best Management Practice (BMP) for a given crop or irrigation practice. The intention of formally identifying BMPs is to allow the optimal management of resources for economic and environmental sustainability (Boland *et al.*, 2006); payments or policy incentives for energy consumption could be tied to the relevant BMP targets, and achievement of these targets would result in efficiency of resource use in a positive economic environment for irrigators. Given that energy is a major cost for irrigators, and the rapid increases in energy prices being experienced currently, this arrangement would give irrigators an incentive to minimize energy use and therefore costs.

Australian farmers provide invaluable services to the nation and globally in terms of food security and other social benefits (Qureshi *et al.*, 2013) and are critical stakeholders in voluntary climate action. Some critical questions that future research must address from a farmer and industry BMP perspective is that: How much must I reduce my greenhouse gas emissions if I want to do my fair share to contribute towards the national and global effort to keep global warming below the 2°C rise in average temperature (Randers, 2012) under the post Copenhagen climate regime? How does that affect my competitiveness in national and global agri-value chains, and what is the greenhouse gas emission intensity per unit of value added (kg CO_{2-e}/$) in irrigation value chains? These data could be translated into a corporate resolution – say, reduce GHG emissions per $ of value added by 3% per annum – by the irrigation and farming industry for voluntary action, and compulsory publication of the emissions and production data could help the public to gauge BMPs and best-performers when making choices off the shelf.

7.5 CONCLUSIONS

The world's population is reliant on improving agricultural production as a key factor underpinning food security. The agricultural industry is heavily reliant on water and energy resources to maintain and increase levels of crop productivity. In particular, irrigated agriculture is highly productive, but it also uses large volumes of water, and this water consumption is strongly linked to energy use. This chapter highlights the links between water use and energy consumption for crop production, particularly in terms of the energy inputs of on-farm irrigation and other farm inputs such as fertilizer, machinery and agrochemicals. Recent years have seen energy prices skyrocket, which has played a part in rising food prices for consumers. As an industry, it is vital that the agricultural sector understands these links, and targets the most effective areas for efficiency gains, so that production can be maintained while costs are minimized. Producers need both economic and social incentives to help internalize extra costs and adapt to emerging challenges and energy policy changes.

The influence of water source and irrigation method on the energy-water nexus is explained through the use of a case study from Australia. This case study shows the variability of on-farm energy consumption across the two irrigation systems, and illustrates that both environmental and management factors have a significant influence on the level of energy consumption between the farms. The results from the case study farms show that it is possible to reduce water and energy consumption in crop production systems in groundwater dependent regions with the introduction of pressurized systems, while the opposite effect for energy use is true in the surface water regions; the water savings can be achieved, but energy consumption will increase. The water and energy consumption for each case study farm changes for the given situation, indicating that there are many factors that influence energy and water consumption. The variability between individual systems in terms of their design means that energy and emissions may change at different rates to water savings. In addition to this, the results presented are for total energy use related to inputs at the field scale. Energy for irrigation is one part of total energy use, and the proportion of total energy that is attributed to irrigation energy consumption will therefore change with each situation. However, in all cases, energy for pumping groundwater is the largest contributor to energy consumption and greenhouse gas emissions.

Understanding the links between energy and water for crop production, particularly in irrigation areas where this link is intensified, is vital for irrigators, irrigation service providers and policy makers. Any policy aiming to achieve water savings must also account for energy implications; further work is required in order to consider the impacts of the dynamic irrigation environment, in particular the quantifications of savings in water and other inputs when pressurized systems are implemented. Irrigation infrastructure modernization decisions and the water-energy nexus in irrigated agriculture also have implications for rainfed crop production and ecosystems and is a topical issue for further work. Balance must be met between achieving production targets and optimizing use of energy and water inputs. Investments in BMPs and voluntary action by the irrigation and farming industry can help promote more sustainable water-energy use.

ACKNOWLEDGEMENTS

The authors would like to thank a number of reviewers for their comments and feedback which has helped to improve this chapter. We are grateful to colleagues who have contributed informally to the development of this chapter through comments on previous drafts and discussions. Particular thanks go to Elsevier Publishers for granting permission to reuse our work as a case study. We also acknowledge the in-kind and financial support of our home institutions and funding organizations.

REFERENCES

Aggarwal, G.C.: Fertiliser and irrigation management for energy conservation in crop production. *Energy* 20:8 (1995), pp. 771–776.

Ahmad, S.: Issues restricting capping of tubewell subsidy and strategy for introducing the smart subsidy in Balochistan. *Policy Briefings* 2:1 (2006), pp. 1–9.

Ayars, J.E., Phene, C.J., Hutmacher, R.B., Davis, K.R., Schoneman, R.A., Vail, S.S. & Mead, R.M.: Subsurface drip irrigation of row crops: a review of 15 years of research at the Water Management Research Laboratory. *Agric. Water Manag.* 42 (1999), pp. 1–27.

Barber, A.: *Seven case study farms: total energy and carbon indicators for New Zealand arable and outdoor vegetable production.* 2004, AgriLink, Auckland, New Zealand, http://www.agrilink.co.nz/Portals/Agrilink/Files/Arable_Vegetable_Energy_Use_Main_Report.pdf (accessed May 2013).

Boland, A.-M., Bewsell, D. & Kaine, G.: Adoption of sustainable irrigation management practices by stone and pome fruit growers in the Goulburn/Murray valleys, Australia. *Irrig. Sci.* 24 (2006), pp. 137–145.

California Electricity Commission: *Agricultural energy rates in California* 2001. California Energy Commission, prepared CSU Chico Research Foundation, Chico, CA, http://www.energy.ca.gov/reports/2002-05-13_400-01-020.PDF (accessed April 2001).

Canakci, M., Topakci, M., Akinci, I. & Ozmerzi, A.: Energy use pattern of some field crops and vegetable production: case study for Antalya region, Turkey. *Energ. Convers. Manag.* 46:4 (2005), pp. 655–666.

Chamsing, A., Salokhe, V. & Singh, G.: Energy consumption analysis for selected crops in different regions of Thailand. *Agr. Eng. Int.: the CIGR Ejournal* 8 (2006).

Chaudhary, V., Gangwar, B. & Pandey, D.: Auditing of energy use and output of different cropping systems in India. *Agr. Eng. Int.: the CIGR Ejournal* 8 (2006).

Chaves, M.M., Santos, T.P., Souza, C.R., Ortuno, M.F., Rodrigues, M.L., Lopes, C.M., Maroco, J.P. & Pereira, J.S.: Deficit irrigation in grapevine improves water-use efficiency while controlling vigour and production quality. *Ann. App. Biol.* 150 (2007), pp. 237–252.

Chen, G., Baillie, C. & Kupke, P.: Evaluating on-farm energy performance in agriculture. *Austral. J. Multi-Disciplin. Eng.* 7:1 (2009), pp. 55–61.

Cleveland, C.J.: The direct and indirect use of fossil fuels and electricity in USA agriculture, 1910–1990. *Agric. Ecosyst. Environ.* 55 (1995), pp. 111–121.

Conforti, P. & Giampetrio, M.: Fossil energy use in agriculture: an international comparison. *Agric. Ecosyst. Environ.* 65 (1997), pp. 231–243.

Deike, S., Pallutt, B. & Christen, O.: Investigations on the energy efficiency of organic and integrated farming with specific emphasis on pesticide use intensity. *Eur. J. Agron.* 28 (2008), pp. 461–470.

Department of Climate Change: Carbon Pollution Reduction Scheme. Australia's Low Pollution Future. Edited by Department of Climate Change. Canberra, Australia, 2008.

Department of Primary Industries and Energy: Saving energy in agriculture. Canberra, Australia, 1994.

Department of Sustainability, Environment, Water, Population and Communities, Canbberra, Australia, http://www.environment.gov.au/water/policy-programs/srwui/irrigation-efficiency/index.html (accessed June 2013).

EconSearch: *Profile of irrigated activity in the south east catchment region, 2003/04.* EconSearch. Unley, South Australia, 2006.

Erdal, G., Esungun, K., Erdal, H. & Gunduz, O.: Energy use and economical analysis of sugar beet production in Tokat province of Turkey. *Energy* 32 (2007), pp. 35–41.

Esungun, K., Erdal, G., Gunduz, O. & Erdal, H.: An economic analysis and energy use in stake-tomato production in Tokat province of Turkey. *Renew. Energ.* 32 (2007), pp. 1873–1881.

FAO: *The energy and agriculture nexus.* Food and Agriculture Organisation of the United Nations, Rome, Italy, 2000.

Faour, K.: *Winter crop budgets – Southern zone irrigated 2001.* 2001. NSW Agriculture, Yanco, Australia.

Fluck, R.C.: Energy analysis in agricultural systems. Chapter 5 in: R.C. Fluck (ed): *Energy in farm production.* Elsevier, Amsterdam, The Netherlands, 1992, pp. 45–52.

Foran, B.: Working Document 98/13: Looking for opportunities and avoiding obvious potholes: some future influences on Agriculture to 2050. In: Proceedings of *9th Australian Agronomy Conference.* 20–23 July 1998, Wagga Wagga, Australia, 1998.

Gerbens-Leenes, P.W., Hoekstra, A.Y. & van der Meer, T.H.: The water footprint of energy from biomass: A quantitative assessment and consequences of an increasing share of bio-energy in energy supply. *Ecol. Econ.* 68:4 (2009), pp. 1052–1060.

Gerbens-Leenes, P.W., Lienden, A.R.v., Hoekstra, A.Y. & van der Meer, T.H.: Biofuel scenarios in a water perspective: The global blue and green water footprint of road transport in 2030. *Global Environ. Chang.* 22:3 (2012), pp. 764–775.

Giampietro, M., Bukkens, S.G.F. & Pimental, D.: General trends of technological changes in agriculture. *Crit. Rev. Plant Sci.* 18:3 (1999), pp. 261–282.

Graham, P.W. & Williams, D.J.: Optimal technological choices in meeting Australian energy policy goals. *Energ. Econ.* 25 (2003), pp. 691–712.

Gul, A., Rida, F., Aw-Hassan, A. & Buyukalaca, O.: Economic analysis of energy use in groundwater irrigation of dry areas: a case study in Syria. *Appl. Energy* 82:4 (2005), pp. 285–299.

Hamdy, A., Ragab, R. & Scarascia-Mugnozza, E.: Coping with water scarcity: Water saving and increasing water productivity. *Irrig. Drain.* 52 (2003), pp. 3–20.

Hatirli, S.A., Ozkan, B. & Fert, C.: Energy inputs and crop yield relationship in greenhouse tomato production. *Renew. Energ.* 31 (2006), pp. 427–438.

Hodges, A.W., Lynne, G.D., Rahmani, M. & Casey, C.F.: Adoption of energy and water-conserving irrigation technologies in Florida. University of Florida, Gainesville, FL, 1994.

International Water Management Institute: *The energy-irrigation nexus.* Colombo, Sri Lanka, 2003.

Jackson, T.M.: An appraisal of the on-farm energy and water nexus in irrigated agriculture. School of Environmental Sciences, Charles Sturt University, Wagga Wagga, Australia, 2009.

Jackson, T.M., Khan, S. & Hafeez, M.M.: A comparative analysis of water application and energy consumption at the irrigated field level. *Agric. Water Manag.* 97 (2010), pp. 1477–1485.

Jackson, T.M., Hanjra, M.A., Khan, S. & Hafeez, M.M.: Building a climate resilient farm: A risk based approach for understanding water, energy and emissions in irrigated agriculture. *Agric. Sys.* 104:9 (2011), pp. 729–745.

Jimenez-Bello, M.A., Alzamora, F.M., Castel, J.R. & Intrigliolo, D.S.: Validation of a methodology for grouping intakes of pressurized irrigation networks into sectors to minimise energy consumption. *Agric. Water Manag.* 102 (2011), pp. 46–53.

Khan, S., Akbar, S., Abbas, A., Dassanayke, D., Robinson, D., Rana, T., Hirsi, I., Xevi, E., Carmichael, A. & Blackwell, J.: Hydrologic economic ranking of water saving options. Murrumbidgee Valley Water Efficiency Feasibility Project. Pratt Water, Melbourne, Australia. 2004.

Khan, S., Abbas, A., Gabriel, H.F., Rana, T. & Robinson, D.: Hydrologic and economic evaluation of water-saving options in irrigation systems. *Irrig. Drain.* 57:1 (2008), pp. 1–14.

Khan, S., Khan, M.A., Hanjra, M.A. & Mu, J.: Pathways to reduce the environmental footprints of water and energy inputs in food production. *Food Policy* 34:2 (2009), pp. 141–149.

Lal, R.: Carbon emission from farm operations. *Environ. Int.* 30 (2004), pp. 981–990.

Lofman, D., Petersen, M. & Bower, A.: Water, energy and environment nexus: The California experience. *Water Res. Devel.* 18:1 (2002), pp. 73–85.

Loftis, J.C. & Miles, D.L.: Irrigation pumping plant efficiency. Colorado State University Extension Co-operative, Colorado State University, Fort Collins, CO, 2004.

Malik, R.P.S.: Water-energy nexus in resource-poor economies: The Indian experience. *Water Res. Devel.* 18:1 (2002), pp. 47–58.

Mandal, K.G., Saha, K.P., Ghosh, P.K., Hati, K.M. & Bandyopahyay, K.K.: Bioenergy & economic analysis of soybean-based crop production systems in Central India. *Biomass Bioenerg.* 23 (2002), pp. 337–345.

Maraseni, T.N., Mushtaq, S. & Reardon-Smith, K.: Climate change, water security and the need for integrated policy development: the case of on-farm infrastructure investment in the Australian irrigation sector. *Environ. Res. Lett.* 7 (2012), pp. 1–12.

Martin, J.F., Diemont, S.A.W., Powell, E., Stanton, M. & Levy-Tacher, S.: Emergy evaluation of the performance and sustainability of three agricultural systems with different scales and management. *Agric. Ecosyst. Environ.* 115:1–4 (2006), pp. 128–140.

McChesney, I.G., Sharp, B.M.H. & Hayward, J.A.: Energy in New Zealand agriculture: Current use and future trends. *Energ. Agr.* 1 (1981), pp. 141–153.

McCornick, P.G., Awulachew, S.B. & Abebe, M.: Water-food-energy-environment synergies and tradeoffs: major issues and case studies. *Water Policy* 10 (Supplement 1) (2008), pp. 23–36.

McInnes, K.L., Suppiah, R., Whetton, P.H., Hennessy, K.J. & Jones, R.N.: *Climate change in South Australia. Assessment of climate change, impacts and possible adaptation strategies relevant to South Australia.* CSIRO, Aspendale, Victoria, Australia, 2003.

Molden, D., Oweis, T., Steduto, P., Bindraban, P., Hanjra, M.A. & Kijne, J.: Improving agricultural water productivity: Between optimism and caution. *Agric. Water Manag.* 97:4 (2010), pp. 528–535.

Mott, R.L.: *Applied fluid mechanics.* Fourth ed., Prentice-Hall, Inc. Upper Saddle River, NJ, 1994.

Mushtaq, S. & Maraseni, T.N.: Technological change in the Australian irrigation inductry: impications for future resource management and policy development. Waterlines Report, Canberra, ACT, Australia, 2011.

Nelson, T., Kelley, S. & Orton, F.: A literature review of economic studies on carbon pricing and Australian wholesale electricity markets. *Energ. Policy* 49 (2012), pp. 217–224.

O'Neill, C.J., Humphreys, E., Louis, J. & Katupitiya, A.: Maize productivity in southern New South Wales under furrow and pressurised irrigation. *Aust. J. Exp. Agr.* 48 (2008), pp. 285–295.

Ozkan, B., Akcaoz, H. & Fert, C.: Energy input-output analysis in Turkish agriculture. *Renew. Energ.* 29 (2004), pp. 39–51.

Ozkan, B., Fert, C. & Karadeniz, C.F.: Energy and cost analysis for greenhouse and open-field grape production. *Energy* 32 (2007), pp. 1500–1504.

Pervanchon, F., Bockstaller, C. & Girardin, P.: Assessment of energy use in arable farming systems by means of an agro-ecological indicator: the energy indicator. *Agric. Sys.* 72 (2002), pp. 149–172.

Pimentel, D.: Energy inputs in production agriculture. In: R.C. Fluck (ed): *Energy in farm production.* Elsevier, Amsterdam, The Netherlands, 1992, pp. 13–29.

Pimentel, D. & Pimentel, M.: *Food, energy and society.* Third ed., CRC Press, Boca Raton, FL, 2008.

Pimentel, D., Hurd, L.E., Bellotti, A.C., Forster, M.J., Oka, I.N., Sholes, O.D. & Whitman, R.J.: Food production and the energy crisis. *Science* 182 (1973), pp. 443–449.

Pimentel, D., Doughty, R., Carothers, C., Lamberson, S., Bora, N., Lee, K., Hansen, D., Uphoff, N. & Slack, S.: Energy inputs in crop production in developing and developed countries. In: E. Lal (ed): *Food security and environmental quality in the developing world.* CRC Press, Boca Raton, FL, 2002, pp. 129–151.

Playan, E. & Mateos, L.: Modernization and optimization of irrigation systems to increase water productivity. *Agric. Water Manag.* 80 (2006), pp. 100–116.

Qureshi, M.E., Hanjra, M.A. & Ward, J.: Impact of water scarcity in Australia on global food security in an era of climate change. *Food Policy* 38 (2013), pp. 136–145.

Randers, J.: Greenhouse gas emissions per unit of value added ("GEVA") — A corporate guide to voluntary climate action. *Energy. Policy* 48 (2012), pp. 46–55.

Sayin, C., Mencet, M.N. & Ozkan, B.: Assessing of energy policies based on Turkish agriculture: current status and some implications. *Energy. Policy* 33 (2005), pp. 2361–2373.

Singh, H., Mishra, D. & Nahar, N.M.: Energy use pattern in production agriculture of a typical village in arid zone, India – part I. *Energy Convers. Manag.* 43 (2002), pp. 2275–2286.

Singh, H., Singh, A.K., Kushwaha, H.L. & Singh, A.: Energy consumption pattern of wheat production in India. *Energy* 32 (2007), pp. 1848–1854.

Singh, K.P., Prakash, V., Srinivas, K. & Srivastava, A.K.: Effect of tillage management on energy-use efficiency and economics of soybean (*Glycine max*) based cropping systems under the rainfed conditions in north-west Himalayan region. *Soil Till. Res.* 100:1–2 (2008), pp.78–82.

Smerdon, E.T. & Hiler, E.A.: Energy in irrigation in developing countries. In: *Water and water policy in world food supplies*. Texas AM University Press, College Station, TX, 1985, pp. 279–284.

Smith, P.: *Agfact E5.12 Is your diesel pump costing you money?* NSW Agriculture, Tamworth, Australia, 2004.

South Australian Government: Sustainable sevelopment in south east South Australia. Information Path to Sustainable Development of Primary Production, South Australian Government, Adelaide, Australia, 2000.

Srivastava, R.C., Verma, H.C., Mohanty, S. & Pattnaik, S.K.: Investment decision model for drip irrigation. *Irrigation Sci.* 22 (2003), pp. 79–85.

Stout, B.A.: *Handbook of energy for world agriculture*. Elsevier Science Publishers Ltd. Essex, UK, 1990.

Tzilivakis, J., Warner, D.J., May, M., Lewis, K.A. & Jaggard, K.: An assessment of the energy inputs and greenhouse gas emissions in sugar beet (*Beta vulgaris*) production in the UK. *Agric. Sys.* 85 (2005), pp. 101–119.

van den Berg, H. & Jiggins, J.: Investing in farmers – the impacts of farmer field schools in relation to integrated pest management. *World Dev.* 35:4 (2007), pp. 663–686.

Wang, J., Rothausen, S.G.S.A., Conway, D., Zhang, L., Xiong, W., Holman, I.P. & Li, Y.: China's water-energy nexus: greenhouse-gas emissions from groundwater use for agriculture. *Environ. Res. Lett.* 7:1 (2012), pp. 1–10.

Wells, C.: Total energy indicators of agricultural sustainability: Dairy farming case study. Ministry of Agriculture and Forestry, Wellington, New Zealand, 2001.

Wood, R., Lenzen, M., Dey, C. & Lundie, S.: A comparative study of some environmental impacts of conventional and organic farming in Australia. *Agric. Sys.* 89 (2006), pp. 324–348.

CHAPTER 8

Energy use and sustainability of intensive livestock production

Jukka Ahokas, Mari Rajaniemi, Hannu Mikkola, Jüri Frorip, Eugen Kokin,
Jaan Praks, Väino Poikalainen, Imbi Veermäe & Winfried Schäfer

8.1   ENERGY AND LIVESTOCK PRODUCTION

World population was 3 billion in 1960, 7 billion in 2010 and forecasts for 2050 show an estimate of over 9 billion. During the same, time the number of production animals (cattle, chicken and pigs) has changed from 46 billion in 1960 to 221 billion in 2010 (FAO Database, 2013).

An increase in human population means increasing demand of food products. With increased prosperity people prefer animal products over crop products as food. Global meat and milk production is estimated to double from 2000 to 2050. Figure 8.1 shows how the annual main animal production per capita has increased during the last 50 years, except for bovine meat, which has stayed at about the same level (FAO Database, 2013). When animal production per capita has incresed or stayed the same, the total production has increased because world population has increased.

Production has been able to increase due to an increased number of animals and better production efficiency. Animal production always demands more energy than plant production. In plant production the energy ratio (energy in the product/energy used in the production) is from 3 to 15 depending on the type of plant and the intensity of the production. On the other hand, in animal production the energy ratio is usually less than one, meaning that more energy has been used in the production than is embedded in the product itself. Agricultural production has a constant

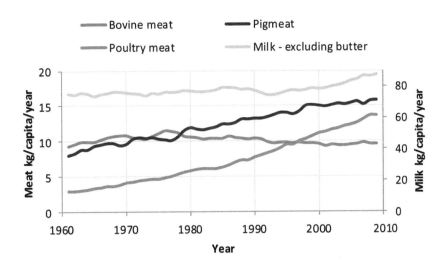

Figure 8.1.   Trends in animal production (FAO Database, 2013).

195

growth of raw material and fossil energy consumption due to the intensification and mechanization of production technologies. Energy efficiency is one of the key factors for developing more
sustainable agricultural practices.

During the last decade, animal production as well as plant production has become energy
dependent. The feed material is produced in the fields with machinery, and artificial fertilizers are
used for plant nutrition. Fertilizers are manufactured with natural gas or coal. Animal production
uses machines, which consume fossil energy.

Capper *et al.* (2009) compared the energy consumption and carbon footprint of dairy production
in the years 1944 and 2007. They concluded that, when calculated per animal, the carbon footprint
was twofold in 2007, but when calculated per produced milk kg, the footprint in 2007 was one
third of that in 1944. The milk production efficiency has improved during these years and reduced
the footprint.

8.1.1 *What is energy*

Energy can be in various forms, kinetic, potential, radiation or thermochemical energy, the latter
is released during burning. Burning can be directly utilized in heating or it can be converted into
mechanical work e.g. in internal combustion engines.

Depending on the type of energy, it is sold and measured in different units. Small consumers
buy oil in liters but crude oil is sold in barrels (oil barrel $= 42$ US gallons $= 159$ liters). Firewood
is sold in cubic meters (m^3), electricity in kWh for small consumers, and in MWh for large
consumers. The electricity consumption on a national scale is given in TWh or PWh. For energy
analysis different amounts of energy must be converted into the same unit. The basic energy SI
unit is Joule (J, $1\,J = 1\,N\,m = 1\,kg\,m^2/s^2$). However, Joules are not used in trading. On farms kWh
is a more common unit because it is used in electricity consumption meters ($1\,kWh = 3.6\,MJ$ or
$1\,MJ = 0.278\,kWh$). In agricultural production energy is bought into the farm in different forms
(solid, liquid and gas fuel, electricity), which also affects the variety of units.

Energy contents of materials are determined with bomb calorimeters. The bomb calorimeter
yields the maximum energy (heating value), which can be released from matter. Three different
heating values can be determined:

- Higher heating value (*HHV*, gross calorific value, upper heating value): When hydrocarbon
  fuel is burned the flue gases include vapor. The energy released in the condensation of vapor
  is included in the higher heating value.
- Lower heating value (*LHV*, net calorific value): The condensation heat of flue gas vapor is not
  included in the lower heating value.
- Gross heating value: Many materials such as biomass contain water in addition to dry matter.
  This has to be taken into account when the heating value of wet material is defined. Only the
  dry matter part of the fuel mass is burnable and the other part (water) is vaporized. Gross
  heating value is calculated from the lower heating value of fuel.

Table 8.1 shows typical lower heating values for different materials. When the material includes
water the gross heating value H_g can be calculated with Equation (8.1), where H_{LHV} is the lower
heating value of material [MJ/kg] and w is the moisture content of the material [as a decimal
number]:

$$H_g = H_{LHV}(1 - w) - 2.443 \cdot w \tag{8.1}$$

Example. Dried wood logs have a moisture content of about 15%. The gross heating value is
19 MJ/kg $\times$ (1 − 0.15) − 2.443 MJ/kg $\times$ 0.15 = 15.8 MJ/kg.

Animals cannot utilize the whole energy content of the feed. Figure 8.2 shows an example
of energy utilization in milk production. Part of the feed is indigestible depending on its composition. From the digestible energy, part is metabolizable. Metabolizable energy is the part of

Table 8.1. Typical dry matter lower heating values.

Material	Lower heating value [MJ/kg]	Lower heating value [kWh/kg]	Bulk density [kg/m³]	Lower heating value [kWh/m³]
Crops	20	5.6	700	3890
Straw	19	5.3	100	530
Rape seed	37	10.3	600	6170
Wood	19	5.3	400	1410
Diesel oil	43	11.9	830	9910
Ethanol	27	7.5	790	5930

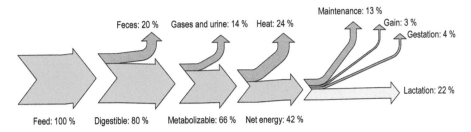

Figure 8.2. Example of energy utilization of feed in milk production (Dairy Note).

digestible energy that is not lost to gases (primarily methane) and urine. Animals also produce heat, and part of the metabolizable energy is used for heat production. When the energy of the heat production is subtracted from metabolizable energy, the result is net energy. Net energy is used for maintenance, gain as well as gestation and lactation. Depending on the production and the production circumstances the shares of energy vary in great extent.

8.1.2   *Energy consumption and emissions*

Energy consumption in animal production can be divided into four main categories: production management, purchase of feed material, animal housing, and livestock machinery. These categories and their subcategories are shown in Figure 8.3. The manager of the farm decides the type of production, machinery, buildings, and cattle feeding and breeding strategy. Management also has effect on the other categories. In feed production the choice is between own feed production and purchased material. When energy use is considered, own feed material consumes less energy than bought feed material. Regarding animal housing and microclimate, animal welfare has to be accounted for but without high energy losses. This means temperature, humidity, gas concentration, and illumination control according to the animal's needs. Work in the cattle house is done with machines, and the type of machinery and how it is used has an effect on energy use.

The management itself does not use much energy. Therefore the energy consumption is normally divided into three parts: energy needed for feed material, energy needed to run livestock machinery and energy needed for the building. Feed material production is the most energy demanding part. In a cold climate, heating is needed in pork and broiler production and this may consume large amounts of energy. Cows tolerate low temperatures better and normally heating is not needed. In milk production, electricity is used to run the production machinery and milking system. Electricity is needed in milk production somewhat more than in pork or broiler production.

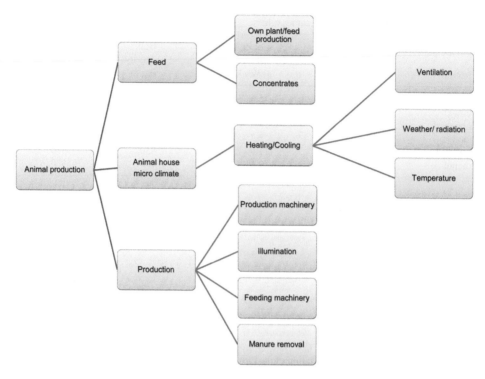

Figure 8.3.   Animal production system.

8.1.3   *Direct and indirect energy*

The energy used for cattle production can be divided into direct and indirect energy, depending on how it comes to the farm. The allocation of the two energy parts is not always clear. Normally fuel and electricity are considered direct energy use because the energy is bought directly to the farm. Goods, machinery, chemicals and maintenance also need energy for production and transport. This is indirect energy, sometimes also called embodied or embedded energy. Some goods such as wooden construction materials could be used as fuel at the end of their lifetime. The caloric part of the goods is then called embedded or embodied energy, because it is not used primarily as fuel.

The major indirect energy in agricultural plant production is the energy used in chemicals, mainly in fertilizers. In animal production it is in the feed material of the animals. Additionally, farmers buy machines and construct buildings. The energy needed in the manufacturing, transportation and maintenance of these goods is allocated to indirect energy usage. The indirect energy can be grouped as follows:

- Living inputs for plant and animal production: e.g. seeds, seedlings, calves, piglets, chicks.
- Agrochemicals and minerals for production: e.g. mineral fertilizers, crop protection agents, seed dressing agents, drugs, cleaning materials.
- Feed for animal production: e.g. feed from outside the farm includes concentrates, additives and forages. Feed produced inside the farm boundary is categorized into the output of crop production e.g. crop, silage and hay.
- Investments: farm machinery, tools, buildings used by the farm processes, drainage, office equipment.

It should be clearly indicated if indirect energy is also included in the energy analysis. It is quite typical that energy needed to manufacture chemicals (fertilizers) is included in agricultural energy analysis but not, for instance, the manufacturing energy of machines or buildings.

Direct energy in animal production normally comprises the following items:

- Electricity used in the production. This means, for instance, lights, ventilation fans, milking system, manure removal, feeding machines, and water pumps.
- Diesel oil for diesel operated machinery (feeding, manure removal).
- Fuels for heating the buildings and water (electricity, burning oil, gas, biofuels).

8.1.4 *Efficiency*

Energy usage is never 100% efficient. The usage of energy also needs some energy and the conversion of energy from one form to another is not without loss. Efficiency tells how much one can exploit from the heating value of the material. This can be calculated either from the power or energy of the process. For example, in burning, the heating furnaces typically have an efficiency of 80–90% and, for instance, internal combustion engine efficiencies are at best about 50%. Efficiency η can be calculated with Equation (8.2), where P_{out} or E_{out} are the power or energy produced in the process and, correspondingly, P_{in} or E_{in} are the power or energy needed to run the process:

$$\eta = \frac{P_{out}}{P_{in}} = \frac{E_{out}}{E_{in}} \qquad (8.2)$$

Example: A wood burning furnace uses 10 kg of wood of a 15% moisture content and the furnace produces 126 MJ of energy. The energy content of this wood is according to Equation (8.1) $19 \cdot (1 - 0.15) - 2.443 \cdot 0.15 = 15.8 \, MJ/kg$. The wood mass has $10 \, kg \cdot 15.8 \, MJ/kg = 158 \, MJ$ energy if all the wood energy could be utilized. The efficiency η in this case is $126/158 = 0.8$.

8.1.5 *Energy analysis*

8.1.5.1 *Methodology of energy analysis*
There are no standards for energy analysis in agricultural production. Thus, numerous approaches render incomparable results. Nearly all theoretical and methodological issues such as a commonly accepted definition of system boundaries remain unresolved (Jones, 1988). However, a significant step forward from taxonomy to scientific methodology was done by Odum and his scholars, who developed an energy accounting method (Odum, 1996). Although this methodology is employed by scientists worldwide, it has not yet been utilized in common environmental decision making. One reason may be that thermodynamic laws are difficult to integrate into transdisciplinary research.

Scientifically sound methods based on thermodynamic laws focus rather in sectorial energy accounting (e.g. Bastianoni *et al.*, 2009; Brown and Ulgiati, 2004; Rydberg and Haden, 2006; Ugidwe & Bakshi, 2007; Ukidwe & Bakshi, 2007) but are also employed in energy accounting for animal production (e.g. Alfaro-Arguello *et al.*, 2010; Castellini *et al.*, 2006; Dong, 2012; Ro'tolo *et al.*, 2007).

The present mainstream methodology is the life cycle assessment (ISO 14040), a part of the ISO 14000 environmental management standard family. These guidelines provide practical tools for industrial production companies. Nevertheless, many agricultural scientists employ LCA in animal production for environmental accounting using spreadsheet calculations. The results mainly cover direct fossil energy input and the proportion of fossil fuels embedded in feed, machinery and buildings as well as their impact on environmental parameters (e.g. Cederberg and Mattsson, 2000; Grönroos *et al.*, 2006; Kraatz, 2012; Meul *et al.*, 2007; Refsgaard *et al.*, 1998; Thomasson *et al.*, 2008).

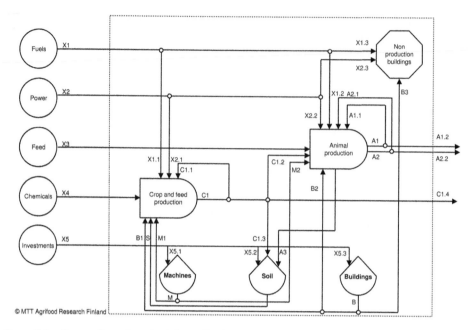

Figure 8.4.   Energy flow of a mixed farm. X = input, C = crop production output, A = animal production output, M = embedded energy in machines, S = embedded energy in soil nutrients,
B = embedded energy in buildings.

At the farm level, a simplified energy analysis should be used to enable farmers to perform
the analysis by themselves. Figure 8.4 shows an example of energy flows on a farm. The whole
farm is bordered with a dashed line and for the whole farm analysis only the energy flows into
or out of the farm are taken into account. Farms normally produce more than one product and, in
livestock production, the farm may also have crop production and mass and energy flow between
them. Figure 8.4 also shows these flows and also the direct and indirect energy flows within the
system boundary. Figure 8.5 shows an example of a farm energy analysis result.

In the energy analysis of animal production the physical farm boundary can be used as a system
boundary. This approach offers many advantages:

- mirrors reality, agricultural products are produced on farms
- embraces the interactions between crop production, animal production, soil, and environment
- is applicable as a system boundary for most energy analysis methods
- clearly shows the environmental sustainability criteria, less nonrenewable input and fewer
  emissions to water body and atmosphere across the system boundary increase sustainability
  and vice versa
- may apply to any farm whether it is a beef or broiler production farm, or a mixed dairy farm.
- may account for the time factor (e.g. crop rotation, weather conditions)
- does not limit the chain of embedded energy calculation for products across the system
  boundary.

However, there are some weaknesses or possibilities for abuse of the model. For example, inputs
rendered from external contractors like tillage or combined harvesting may bias the real energy
consumption if the energy analysis will account for only the fuel consumed by the contractor's
machinery during the work inside the farm boundary. As a result the farm energy balance favors
outsourcing services.

The definition of the farm boundary as a system boundary does not yet solve the other methodological problems mentioned above. If we follow the present mainstream methodology then we have to decide the fossil energy input embedded in each input item. For fossil fuels and electric power country-specific figures are available from literature.

The area inside the dashed lines shows the system boundary. The arrows show the energy flow from input groups to processes, consumers and tanks or stocks of the farm, the branches of energy flow within the farm marked by a dot, and the energy output. Emissions are not shown. The symbols are described by Odum (1996) and generally applied by scientists of the International Society for the Advancement of Energy Research (ISAER).

Investments into machinery ($X5.1$), buildings ($X5.3$), and soil ($X5.2$) such as land clearing or drainage are usually depreciated over a long period. The energy used for their provision is after procurement (= after passing the farm gate) stored in a tank and tapped during the depreciation period by the production processes in the farm.

The energy of seed input ($C1.1$) is considered as the energy proportion of crop production output $C1$ used for propagation. In the case where seed is procured outside the farm, it can be considered as an additional input. $C1.2$ stands for the energy of feed, $C1.3$ for the energy of crop residues like straw or roots of legumes, and $C1.4$ for the energy of crops leaving the farm.

The energy flow for animal production is shown in a similar way. In a dairy farm, $A1$ may be the energy of milk, $A1.1$ stands for the milk consumed within the farm boundary (e.g. used as feed for calves) and $A1.2$ for the energy of sold milk. Correspondingly, $A2$ may be allocated to the energy embedded in animal bodies (dead or alive) or animal products, $A2.1$ to the energy in young animals replacing the old generation, $A2.2$ to the energy in animals leaving the farm, and $A3$ means the manure applied to the soil. Farms that procure young animals from outside, such as pork and poultry, require an additional external input.

Part of the machinery may be used for both crop and animal production, and their use is usually taken into consideration via the embedded energy of their production and maintenance before they leave the farm as scrap. The same principle is employed for buildings, in crop production, e.g., dryer, machine shed, and grain store, in animal sheds, silos, and feed storage buildings.

Buildings not used directly for production but located inside the farm boundary such as living quarters, workshops, and forgeries are considered consumers of energy.

However, as mentioned before, there are no standards for how to make the accounting results comparable. In addition to the energy flow methodology (Fig. 8.4), Sankey diagrams are a common method to illustrate energy flows in a process. An example of an energy flow diagram can be found in Figure 8.5. Both methods need dedicated software for the analysis.

If energy ratios and specific energy consumption of farms are calculated and compared, then the items in Table 8.2 must be addressed.

In energy analysis production energy ratio and specific energy consumption are the two commonly calculated numbers.

8.1.5.2 *Energy ratio*
In agricultural production the energy efficiency of the production is defined with an energy ratio, N_e (Equation (8.3)). E_o is energy output of the production (gross heating value) and E_i is the energy input of the production excluding sun energy.

$$N_e = \frac{E_o}{E_i} \tag{8.3}$$

Example. Annually, the farm uses 21 GJ/ha (input) and the yield is 53 GJ/ha (output). The energy ratio is $N_e = 53/21 = 2.5$.

Energy ratios can be calculated in many ways. Typically, in crop production all the direct energy consumptions are included, as well as fertilizers as indirect energy consumption. The indirect energy of machinery and buildings is often neglected. The analyzer should always explain what

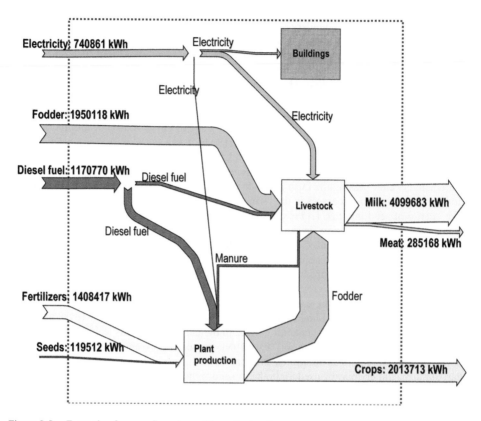

Figure 8.5.   Example of energy flows in a milk producing farm.

has been included in the analysis. Table 8.3 gives some examples of energy ratios in agricultural production. These numbers are calculated on the farm level, so that in cattle production the inputs are not feed material heating values but energy used to produce them.

Energy ratio is also the efficiency of agricultural production. Based on efficiency definition (Equation (8.2)), it cannot be over one. It is not possible to get more energy from a system than what is the input energy. In agricultural production the energy from the sun is neglected and, therefore, energy ratios of agriculture obtain values over one.

8.1.5.3   Specific energy consumption

To different productions a proper meter is needed. Specific energy shows how much energy is used to produce one kilogram of a product, Equation (8.4). N_s is the specific energy consumption, E_i the energy used in the production and y is the yield:

$$N_s = \frac{E_i}{y} \tag{8.4}$$

Example. The farm uses annually 21 GJ/ha (input) and the yield is 3500 kg/ha (output). The specific energy consumption is $N_s = 21$ GJ/ha/3500 kg/ha = 6 MJ/kg.

Specific energy can be used in comparisons of different farms. The lower the number the more energy efficient is the production. Sometimes energy use per animal (or animal place) per year

Table 8.2. Problematic points in farm energy analysis.

Item	Method
Inputs	Gross heating values should be used in energy flow calculations. Because farms are compared, the energy needed for fuel or electricity preparation and transportation is not taken into account. For fertilizers and chemicals, and average production energy figures should be used.
Outputs	Milk if possible should have protein and fat correction, carcass weight should be used for meat and normal merchandize moisture content should be used for crops and dry matter for hay material. In energy ratio calculations the gross heating values are used and in specific energy analysis the weight is used.
Living inputs, seeds	Farmers can use their own seed or they can buy seeds outside. If the farmer uses their own seed then they loose the heat value of the seed when not selling it. In addition they loose the energy needed for seed preparation. To get comparable figures the energy should be calculated with the gross heating value + energy needed in seed cultivation, preparation and transport. The other possibility is that the amount of bought seed is subtracted from the yield and the energy used for seed preparation is calculated for this amount.
Living inputs, calves, chickens and piglets	There are two possibilities for the calculations. First possibility is to include only the weight gained at the farm in the analysis. In this case the incoming weight must be subtracted from the outgoing weight. Then the farm energy consumption is allocated to the animal weight gain during the breeding. The other possibility is to add the energy for nursing to the farm energy consumption. With this methodology the farms can be compared whether they bring up the animals on the farm or buy them.
Feed material and concentrates	In comparison of farm production effectiveness the energy used for the feed material and concentrates production and transport should be known or average figures should be used.
Manure	Manure produced on the farm and used on their fields decreases the fertilizer need. This decrease should be converted to energy gain with energy amount deceased in the fertilizer manufacturing.
Renewable fuels	Renewable energy does not cause GHG emissions but for instance when fossil energy is compensated with biofuels the energy consumption does not change. In energy analysis the gross heating value of the bioenergy should be used.
Investments	Normally can be neglected. It must however remembered that machines and buildings need energy for manufacturing, transport and maintenance.
Contractor work	Contractor work must be in line with investments, if investments are not taken into account then only contractor work fuel consumption is included.

Table 8.3. Typical energy ratios in agricultural production on farms.

Production	Typical energy ratio
Crop	3–5
Forage	5–8
Milk	0.5–1.0
Pork	0.4–0.9
Broiler	0.1–0.4

is used. With this it is easier to estimate and compare farms because only the number of animals, not the produced amount is needed in the analysis.

The same kind of methodology should be used in the analysis. For instance, the amount of produced material should be clearly defined. It is normal that in crop production the weight is given in the merchandize moisture level (13–14%), in hay production the dry matter yield is recommended, and in meat production the carcass weight should be used. In milk production the amount of fat and protein corrected milk (FPCM) is recommended to be used.

8.1.5.4   Types of energy analysis
There are two ways to perform the energy analysis, top-down or bottom-up. While the top-down analysis is focused on system boundary and processes on farm within the boundary, the bottom-up analysis works on process and factor levels.

8.1.5.4.1   Bottom-up analysis
Bottom-up analysis starts from the factor of each process. Each process may be broken down in subprocesses (Figs. 8.4 and 8.5). For example, dairy production may be split into the process of milk production and the process of nursing calves and young cattle, the crop production process may include cash crops and feed crops. Subprocesses of feed crop production are e.g. pasture, silage production, and hay production. Each process requires different work steps or tasks requiring specific factor input. The bottom-up analysis is usually made from data in two tables: one that allocates the factor to a certain task (e.g. fuel consumption for plowing) and another that allocates the tasks to a process or subprocess (e.g. plowing for wheat production).

It is obvious that the bottom-up approach requires a lot of data to be collected or calculated before an energy balance on a process or farm is possible. Often certain factors are neglected or difficult to allocate. As a result the total energy input of all production steps is often lower than the energy consumed by the farm.

Especially in milk production this approach faces some methodological problems, for instance how to allocate the embedded fossil energy input in relation to the output in terms of milk, meat, manure, and breed. An example using Figure 8.4 may illustrate the problem. Assume that we wish to calculate the energy input to produce one liter of milk on a dairy farm, where crop production is limited to feed production. Then we first consider the animal production process and sum up the energy of all inputs. Energy consumption should be measured at every stage. This means the measurement of fuel and electricity consumption and bookkeeping of the work done and allocation of measured energy to each category. In the analysis the points shown in Table 8.2 should be addressed.

Now the energy input per kg milk may be calculated by dividing the energy input by the yield of the milk sold. However, meat is a co-product of milk production using the same energy input. Because it is difficult to allocate the input between milk and meat, one solution is to subtract the energy output of meat from the input.

This simple example shows why the range of specific energy numbers is so wide in the bottom-up analysis because there are no calculation standards available.

8.1.5.4.2   Top-down analysis
The great advantage of the top-down approach is that the total energy input figures are available from the bookkeeping figures of the farm. The energy efficiency of the farm is easy to calculate directly from the bookkeeping data and the conversion factors of the indirect energy input. The overall efficiency of the farm model given in Figure 8.4 is:

$$N_e = \frac{A1.2 + A2.2 + C1.4}{X1 + X2 + X3 + X4 + X5} \tag{8.5}$$

Similarly, the specific energy consumption per ha or the electric power consumption per kg milk is easy to calculate. However, tracing the energy consumption from the total input down to each process and each task of each process is as difficult as in the bottom-up approach. The problem is to allocate the inputs to the processes and tasks. In case of fuels and electricity, this is possible only by measuring the consumption on the particular spot.

The top-down approach immediately shows which of the external inputs offers the greatest energy-saving potential. The conclusion is that animal production is most efficient when the external input is minimized and feed and fuel are produced inside the farm boundary by harnessing renewable energy resources. This also explains why, for hundreds of years, the agricultural sector has been the energy producing sector worldwide, farms having produced more energy than they consumed.

The top-down approach indicates that animal production is always linked to crop production. Crop production is limited by the available area of agricultural soil. Consequently the use of specific energy figures in animal production becomes questionable, if the required land area is neglected. The bottom-up approach often neglects this fact focusing mainly on optimizing the metabolic efficiency of animal production. Nutrient recycling by manure fertilization instead of import of chemicals is rediscovered as an effective measure to save energy in animal production as well.

8.2   LIVESTOCK PRODUCTION SUSTAINABILITY

8.2.1   *Sustainability*

Bauman and Capper (2011) define a sustainable agricultural system as follows: "A sustainable agriculture system includes an economic dimension represented by an industry that is productive, efficient and profitable; an environmental dimension characterized by an industry that makes the most effective use of resources; maintains air and water quality and preserves wildlife habitat and rural landscape; and a social dimension demonstrated by an industry that cares for and takes into consideration the community, employees and animal welfare". Because of its shortage as a resource and influence on economics and environment, the use of energy is an important factor in the consideration of sustainable agricultural practices.

To develop sustainable milk production practices, a comparison of different management systems is essential. However, in milk (and meat) production energy consumption varies widely due to the choice of analytical methods, the included and excluded parameters and the allocation of production. Also the results can be very different depending on system boundaries and used energy conversion factors.

To get reliable results, foodstuff energy conversion factors need to be studied more comprehensively. To estimate the uncertainty associated with energy coefficients, use of fossil energy was measured by Vigne *et al.* (2012) for dairy farming systems in two French territories: Poitou-Charentes (PC) in mainland France and Reunion Island (RI) (a tropical island in the Indian Ocean). Comparative uncertainty and sensitivity analyses were performed. Fossil energy use for milk in PC was 4.6 MJ/L and in RI 7.9 MJ/L. Uncertainly analysis performed via 30,000 sets of randomly selected energy coefficients estimated 95% confidence intervals from 3.6 to 5.0 and 5.8 to 8.2 MJ/L, respectively. This also highlights the need for a standardized and properly defined method for calculations of energy coefficients, especially regionally adapted ones, to accurately estimate fossil energy use in agricultural systems throughout the world. Energy inputs in milk production from different articles are given in Table 8.4.

In a comprehensive survey of 150 dairy farms throughout New Zealand, the energy usage ranged from 0.9 to 5.6 MJ/kg per liter of milk, indicating great variability in dairy farm energy consumption. An average dairy farm consumed 1.84 MJ/kg of energy (Wells, 2001). In 2006 Hartman and Sims found the average total energy input in New Zealand to be 3.9 MJ/kg (range 3.0–5.4 MJ/kg), inputs of irrigated farms being higher. In the regions where the use of concentrates is higher than that used in New Zealand, the energy consumption per 1 kg of milk tends to be larger.

The use of energy is lower in organic systems than in conventional systems due to feeding strategies and cultivation practices in plant production. The difference between conventional and organic systems has been demonstrated e.g. by Dalgaard (2003) in Denmark, Grönroos *et al.* (2006) in Finland, Bos *et al.* (2007) in The Netherlands.

During 2009–2011 fossil energy consumption was measured in three Estonian uninsulated loose cowsheds for 1775, 523 and 596 dairy cows with milk production 8079, 9540 and 10,066 kg/year, respectively. This input energy is presented in Figure 8.6.

Table 8.4.   Energy inputs in milk production.

References	Energy input, [MJ/kg milk]	Remarks
Refsgaard et al. (1998)	3.3	Denmark, conventional farming
Refsgaard et al. (1998)	2.1	Denmark, organic farming
Ceberberg and Mattsson (2000)	3.5	conventional farming
Ceberberg and Mattsson (2000)	2.5	organic farming
Dalgaard (2003)	3.1	Denmark, conventional
Dalgaard (2003)	2.8	Denmark, organic
Wells (2001)	1.8	New Zealand, range 0.9–5.6
Hartman and Sims (2006)	3.9	New Zealand, range 3.0–5.4
Grönroos et al. (2006)	6.4	Finland, conventional
Grönroos et al. (2006)	4.4	Finland, organic
Bos et al. (2007)	4.9	Netherlands, conventional, range 4.3–5.5
Bos et al. (2007)	4.0	Netherlands, organic, range 3.6–4.5
Smil (2008)	5–7	
Kraatz and Berg (2009)	3.5	Germany (ECM)
Mikkola and Ahokas (2009)	1.6	Finland, feed production energy consumption
Mikkola and Ahokas (2009)	3.2	Finland, energy consumption of feed production and housing
Neuman (2009)	0.4–1.2	Only direct energy (electricity and diesel oil) use included
Ahokas (2013), ed.	4.0–6.1	Finland and Estonia
Woods et al. (2010)	2.7	England and Wales
Vigne et al. (2012)	4.6	France, Poitou-Charentes
Vigne et al. (2012)	7.9	France, Reunion Island

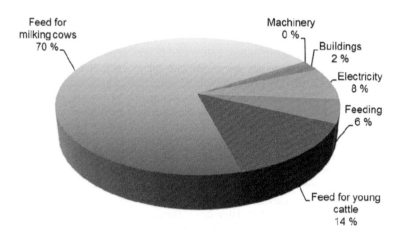

Figure 8.6.   Example of energy use on a dairy farm.

The average fossil energy consumption during the three years was 5.1 (4.8–5.5) MJ/L. Feed for dairy cattle (70.2%) and young stock (14.0%) was the dominant term in the usage of energy; 13.4% of the energy input made up direct energy. The energy input of livestock buildings (1.9%) and machinery (0.5%) had only a slight influence on the energy intensity.

Because feed production for dairy cattle requires most of the total energy input, it is important to find possibilities to produce feed in an economic and sustainable way. To decrease energy costs in dairy farm renewable energy sources (biogas, wind and solar energy) may be considered. Renewable energy itself does not decrease energy input but it reduces CO_2 emissions.

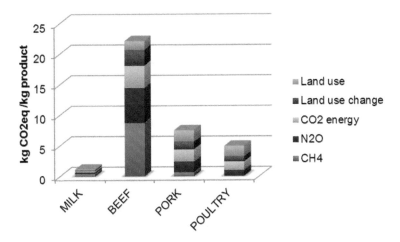

Figure 8.7.　Livestock production GHG emissions (OECD/FAO, 2011).

8.2.2　CO_2 – equivalents

The global warming potential (GWP) of the production is calculated with carbon dioxide equivalent $CO_{2\text{-eq}}$. Usually the methodology of IPCC is utilized (IPCC, 1996). There are a number of greenhouse gases (GHG) which are seen to contribute to climate change, such as carbon dioxide (CO_2), methane (CH_4) and nitrous oxide (N_2O). The GWP of each gas is defined as its warming influence relative to that of carbon dioxide (100 year time period): $CO_2 = 1$, $CH_4 = 21:1$, $N_2O = 310:1$. The coefficients may vary and they depend on the yearly IPCC reports. In agricultural production the emissions from the soil are included. Nitrogen and nitrogen fertilizer in the soil will produce N_2O gases. Because the coefficient of N_2O is high a small emission will introduce a high CO_2 equivalent. The main sources of methane emission from agriculture are the rumen of ruminants and animal manure.

When biofuels are burned the direct carbon dioxide emissions are neglected because plants will use it from the atmosphere. Burning of biofuels also produces other gases besides CO_2 and their GWP should be included in the analyses.

Emission terminology also lacks standardization. Carbon footprint in many reports means $CO_{2\text{-eq}}$ and GHG emissions are the same as $CO_{2\text{-eq}}$.

8.2.3　Livestock GHG emissions

The share of livestock to GHG emissions in the EU countries has been reported to be 9 to 13% of the total agricultural GHG emissions. From these, 23% is from methane, 24% from nitrous oxide, 21% from energy use and 29% from land use. Figure 8.7 shows how the emissions are divided between various categories and productions. The highest emissions come from beef production and the lowest from milk production (OECD/FAO, 2011).

Emissions of dairy production including milk production, processing, and transportation are 2.7–4.0% of the total global GHG emissions. The higher value also includes meat production relative to the keeping of dairy cattle while the lower value includes only dairy production. Regional variation is high, from 1.3 to 7.5 kg $CO_{2\text{-eq}}$/kg milk. From these, methane's emission is the main source, namely 52%. Nitrous oxides come next with a 27–38% share (FAO, 2010).

Methane emissions are mainly from the feed material of cows while nitrous oxide emissions are related to the soil and manure and fertilizer use in plant production. They are the two largest emission sources in dairy and related meat production. Reduction of them has the largest effect on

GHG emissions. From methane emission 5–20% comes from manure and by biogas production it could be reduced (FAO, 2010).

In pork production, the GHG emission in European conditions varies from 2.6 to 8.6 $CO_{2\text{-eq}}$ per kg carcass. Part of the variation is due to different analysis methods. The main part of the emissions comes from the feed material; its share can be three quarters of the whole GHG emission. The study included also storage and food preparation at home, which can be 20% of the emission (Kingston et al., 2009).

Lundshøj-Dalgaard (2007) found that from the GHG emissions, 72% came from nitrous oxide, 17% from methane and the rest from carbon dioxide. From nitrous oxide emissions, 43% originated from denitrification of nitrate leached from the field, and 44% from denitrification of N fertilizer (artificial and slurry). The main factor in GHG emissions is to lower the nitrogen emissions.

The GHG emission in broiler production varies from 1.4 to 5.1 kg $CO_{2\text{-eq}}$ per kg carcass (da Silva et al., 2010; Nielsen et al., 2011; Pelletier, 2008). The emissions vary because of different climatic and production conditions and also because of differences in analysis methods. According to Nielsen et al. (2011) the GHG emissions in broiler production in Denmark was 2.31 kg $CO_{2\text{-eq}}$ per kg carcass weight. The hatch egg production was 13.5%, the broiler production on farm 76.4% and the slaughterhouse share was 10.1%. The distribution of GHG emission was such that 91% came from feed production, 4% from use of electricity, 2% from gas and 3% from other sources. The heating of the broiler houses in the Nielsen et al. (2011) analysis was done with renewable energy which does not have GHG emissions. Nitrous oxides had a 53.8% share of GHG emissions; carbon dioxide had a 42.8% share and methane a 3.5% share.

GHG emissions of cattle production can be reduced:

- For cows the feed material has an effect on methane production, which is the main GHG source. Diet of the cow affects the emissions.
- Biogas production reduces the manure methane emissions considerably.
- Covered and cold manure have lower emissions because methane and ammonia evaporation is lower.
- In pork and broiler production feed material production has the largest effect on GHG emissions. In feed material production the use of fertilizer consumes a large amount of energy and nitrous oxide emissions are also high. Good manure utilization will decrease fertilizer use and emissions.
- Improvement in digestibility of the feed decreases emissions because a smaller field area is needed for feed production.

8.3   ENERGY CONSUMPTION IN LIVESTOCK PRODUCTION

In this section the direct energy use and from indirect energy feed energy use of the livestock production is considered. In practice, this means electricity and fuel used in the production and production houses and energy needed for fertilizer manufacturing.

When more detailed values of energy consumption are needed, the consumption of individual parts has to be known. Normally this procedure is used in the bottom-up analysis.

8.3.1   *Feed material production*

Of the total energy inputs feed (concentrates, crop, forage or grass) is the dominant term in energy use (Fig. 8.8, 8.9 and 8.10). For milking cows its share is more than 70–80% and if the feed needed for young animals is also included the share is over 80%. Electricity and energy needed for feeding is the second largest contribution.

The energy needed for feed production is shown in Table 8.5. Usually grass and hay production need less energy than crop production. This is due to higher biomass yield and perennial plants.

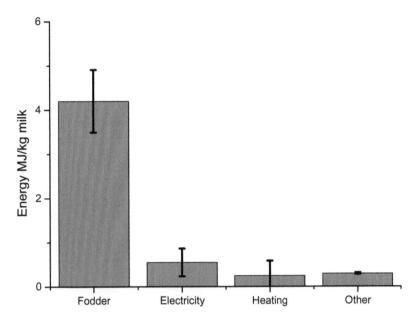

Figure 8.8. Energy consumptions and deviation in milk production during farm follow-up.

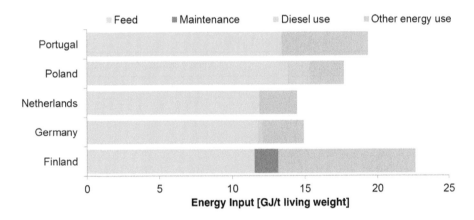

Figure 8.9. Energy consumption in pork production in five European countries (AGREE, 2012).

8.3.1.1 *Crop production*

Figure 8.11 shows an example how the energy consumption is divided in barley production (Ahokas, 2013). The largest input is chemicals, the second largest agricultural field works (tillage, sowing, spraying and harvesting 33%) and as the last, drying. Energy used for drying depends on the moisture content of harvest and if very wet grain is harvested drying consumption will increase. Under favorable climate conditions drying is not needed but grain cooling may instead be required. The major part of chemicals consists of fertilizers.

Energy consumption in crop production can be reduced:

- If possible, grass production should be favored because of the lesser amount of energy needed.
- For conservation of crop drying should be avoided if possible. Other methods like airtight silos could be used.

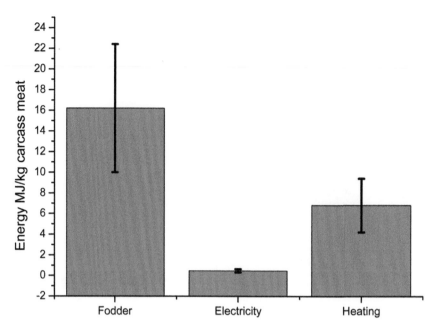

Figure 8.10.   Share of heating, electricity and feed in the energy input of one broiler farm.

Table 8.5.   Energy use in feed production (Ahokas, 2013; Mikkola and Ahokas, 2009; Sainz, 2003).

Feed	Energy consumption [MJ/kg]
Barley	3.8–4.3
Hay	2.8
Corn gluten concentrate	12.6
Corn grain	5.0
Corn silage	2.3
Grass silage	1.8
Pasture	0.7
Oat	2.8
Salt and minerals	0.4
Soy husk (milled)	5.8
Wheat	4.0

- A good nutrient circulation system should be arranged on the farm and manure used as effectively as possible.
- With nitrogen binding plants and proper plant rotation, the amount of fertilizer can be reduced and energy saved.

8.3.1.2   *Grass and hay production*

In grass and hay production, energy efficiency is much better than in crop production. Hay plants are perennials and they can be grown on the same field for several years, which reduces energy needed for tillage. Also nitrogen binding plants (clover) can be used in the plant mixture, to reduce need for fertilizer. Biomass yield is high because in crop production straw is not utilized and in hay production straw is utilized. The energy needed for grass and hay production is shown in Table 8.5.

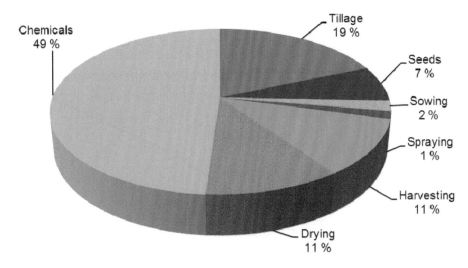

Figure 8.11.   Energy consumption in barley production (3900 kWh/ha total consumption, Ahokas 2013).

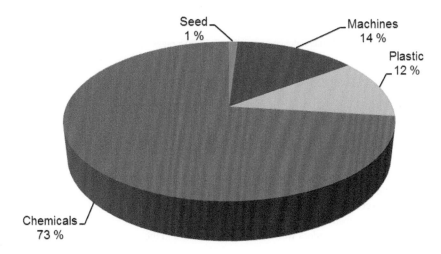

Figure 8.12.   Energy use share in silage production (15.5 MJ/ha).

Figure 8.12 shows how the energy use is divided in silage production when plastic wrapping for bales is used. The main part (73%) also in this case comes from chemicals (fertilizers).

Energy consumption in grass and hay production can be reduced:

• With nitrogen binding plants and proper plant rotation the amount of fertilizers can be reduced and energy saved.

8.3.1.3   *Concentrate production*

Using concentrate always means higher energy consumption in the feeding of animals. Concentrates must be manufactured and transported and this needs energy. However, on many farms the use of concentrate is modest or low and its share in the feed material is low. Therefore, energy used in the concentrate production is not always high. On the other hand, if concentrates form a larger part of the feed material, they have to be included in the analysis.

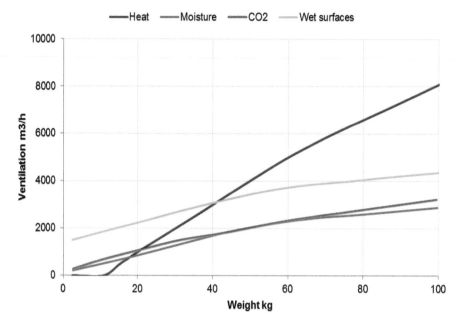

Figure 8.13.   Ventilation rate dependence on weight of pigs in a piggery. Ambient temperature is $-8°C$ and humidity 50%. Curves for heat, moisture, CO_2 and wet surfaces show how much ventilation is needed to remove the excess from the building (Mannfors and Hautala, 2011)

8.3.2   Ventilation

Ventilation is needed in cattle houses for the following reasons:

- Animal welfare, ventilation improves the microclimate of the cattle house:
  o cooling the cattle house
  o removing excess humidity from the building
  o removing gases from the building
- Human welfare: cattle tenders work in the cattle house and the microclimate should be healthy for the workers.
- Building welfare: high moisture content and harmful gases produce rust and mold in the structures.

Animals in the cattle house produce heat, moisture, carbon dioxide and other gases such as ammonia and methane. For a good microclimate in the house, ventilation is needed to remove the harmful emissions and bring fresh air into the building. Water is used for washing and as drinking water, and part of this is also vaporized and needs to be removed from the building. Which from these three matters dictate the ventilation rate depends on animal density, weight and ambient weather.

Figure 8.13 shows an example from a piggery. When pigs are small, the needed ventilation rate is low but when they grow the rate must be increased. This is natural because larger animals produce more heat and moisture. During the growth of the pigs the ventilation criteria changes. In the beginning the wet surfaces (washing water) define the ventilation rate but when the pigs have grown, heat removal is more important. The criteria also change when ambient temperature changes. In hot weather heat removal is the most important but in cold weather the CO_2 removal becomes more important. This means that ventilation cannot have a constant rate, but must be managed according to the changing conditions. Depending on the situation, it could be temperature, humidity or CO_2 content, which defines the air quality control.

The energy needed for ventilation can be calculated from Equation (8.6), where q_v is the air volume flow, Δp the pressure difference between the fan inlet and outlet, η the combined efficiency of the fan and motor running the fan and t is the time the ventilation is running:

$$E = \frac{q_v \cdot \Delta p}{\eta} \cdot t \qquad (8.6)$$

Example. One cow requires a ventilation rate between 75 and 300 m^3/h. If the average rate is 150 m^3/h (0.042 m^3/s), the pressure difference is 30 Pa and the efficiency of electrical motor is 85 % and the fan 50%. The combined efficiency to run the fan is $0.85 \times 0.5 = 0.425$. There are 8760 hours in a year and then the energy needed to run the fan is:

$$E = \frac{0.042 \text{ m}^3\text{s}^{-1} \times 30 \text{ Pa}}{0.425} \times 8760 = 26 \text{ kWh}$$

For animal welfare and energy use it is important that the ventilation rate is according to need. If the animal house needs heating during cold season, this means that also the ventilated air must be heated and the energy need will increase.

Instead of forced ventilation, natural ventilation can be used. It is used typically in cold or semi-cold cow houses and the ventilation rate is controlled with the adjustment of the openings and outlets. The micro climate in these buildings is, in most cases, good because of a sufficient ventilation rate. During the hot seasons by having doors, windows and other openings fully open the wind will remove heat effectively and the microclimate in the building will be good. Cows tolerate low temperatures better than, for instance, pigs or chicks. Therefore, heating is not needed in cow houses.

In hot circumstances when the animals are in danger of having heat stress shades for the animals have to be provided and the air speed can be increased to cool the animals. If this is not sufficient, evaporative cooling can be used. Coarse or fine water spray can be used and the evaporation of water will lower the temperature and help the animals to dissipate heat (DairyCo.; Smith *et al.*, 1998; Wiersma and Short, 1983).

Energy consumption in ventilation can be reduced:

- Natural ventilation needs less energy and saves energy compared to forced ventilation.
- Ventilation rate should be adjusted to the needs of the production to offer a healthy microclimate and to keep energy consumption low. Ventilation control changes depending on temperatures, number of animals and animal weight; the control system should be able to function with changing criteria.

8.3.3 *Illumination*

Lighting is an animal welfare question but also proper lighting is needed for human workers. When animals are kept in building lighting is needed but also appropriate period is needed for rest without artificial lighting. Illumination affects safety, animal growth, fertility and production. For instance with milking cows proper lighting can increase yield 5 to 16% (Crill *et al.*, 2002).

In different production and growing periods illumination needs are different. Illumination can be achieved with natural lighting or with artificial lights when natural lighting is not sufficient or available during late evenings or early mornings.

In lighting the trend is to replace incandescent lightbulbs with more energy efficient lamps. The electric power of incandescent bulbs was also commonly used to indicate their lighting capacity. The proper SI unit for lighting capacity is luminous flux (lm = lumen), which should be used. This is a measure of the total amount of visible light emitted by a bulb. How much light is needed on a

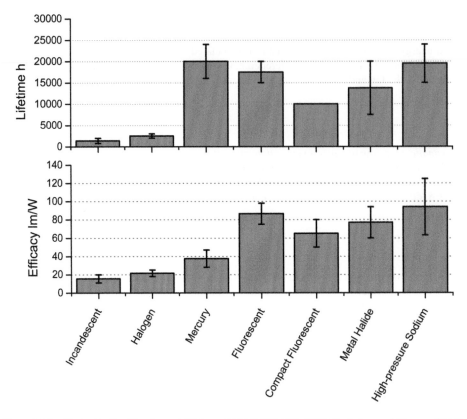

Figure 8.14.   Typical bulb efficacies and lifetimes (ASAE EP344.3, 2010, Gustafson and Morgan, 2004).

Table 8.6.   Typical illumination values for animal houses (ASAE EP344.3, 2010).

Building	Recommended illumination [lx]
Cowshed	150
Milking parlor	100–200
	500 near udders
Broilers	5–30
Pigs	>100
Office	500

surface depends on luminance, which is measured with lux (1 lux = 1 lm/m^2). Lumen describes the light power of the bulb and lux shows the light power that illuminates a surface. Besides the bulb, lux depends on the distance to the bulb, its shade and how large the solid angle of the light is. Luminous efficacy of the bulb can be calculated by dividing the bulb lumen value by the power consumption of the bulb. For instance, if a bulb produces 850 lm light power by taking 15 W of electrical power, the luminous efficacy is 850/15 = 57 lm/W. The higher this value, the more energy efficient the bulb is. Figure 8.14 shows typical bulb efficacies. Economically, the price, energy use and lifetime of the bulb should be utilized in choosing the right bulb type.

Recommendations for the illumination of cattle houses are available. Table 8.6 shows typical values for illumination (ASAE EP344.3 2010).

Table 8.7.   Critical and optimal temperatures [°C] in animal production (MMM RMO C2.2).

Animal	Lower critical temperature	Higher critical temperature	Optimal temperature
Milking cows	−25 ... −15	23 ... 27	5 ... 15
Calves	0 ... 10	30	15 ... 25
Beef cattle	−35 ... −15	25 ... 30	−10 ... 15
Pigs	7 ... 15	25 ... 27	15 ... 22
Piglet	25	34	30 ... 32

Although the energy demand of a single bulb is low, the number of lamps is high and they can be on continuously making the energy use high. For instance in milk production, lights can consume 10–30% of the total electricity used (Hörndahl, 2008; Ludington and Johnson, 2003). To save energy the lights should be used only when needed and with special lighting programs. In addition places which are seldom used could have automatic light switching systems. Control of lighting can be arranged with three different factors: movement, time or ambient light intensity (DairyCo). The incandescent bulbs should be changed to more energy efficient bulbs, for instance fluorescent bulbs, LED bulbs or HID (high density discharge) bulbs.

Good natural lighting can be achieved with 10–15% roof area (DairyCo). Depending on the time of year the length of daylight period varies and especially in northern and southern parts of the hemisphere artificial lights are needed during the dark periods of the year.

Energy consumption in illumination can be reduced:

- With energy saving bulbs energy consumption can be decreased
- Lights should be used only when needed and continuous light use should be avoided
- Daylight should be utilized as much as possible to decrease the artificial lighting and intensity
- Lighting programs, obscuring and automatic switching systems save energy

8.3.4   Heating of animal houses

In cold climate cattle houses must be heated for good animal welfare. Different animals require different temperatures. Table 8.7 shows the critical and optimal temperatures for some animals. When the temperature is critical the production is lower. In low temperatures animals need more energy (feed) to retain their body temperature while high temperatures produce heat stress for the animals. The temperature range changes during the animal's growth and, in the beginning, higher temperatures are required.

Heat is lost through the structures of the building generally as conduction and with the ventilation air as forced convection. The structures can also gain heat by radiation from the sun. Figure 8.15 shows an example of a piggery heat balance as a function of varying ambient temperature. The heat production of 2400 pigs is about 300 kW. In this example the animal heat production is in balance with building heat losses at −14°C and extra heating is needed if the ambient temperature decreases from this, otherwise the inside temperature will decrease. The main heat losses result from the ventilation losses. Ventilation is needed for a good microclimate in the building. With heat exchangers energy could be returned to the building and additional heating would not be necessary.

8.3.4.1   Heat conduction

In conduction, heat flows through floors, walls and ceiling from higher temperature to lower temperature (Fig. 8.16). During cold ambient weather heat flows from inside to outside and during hot weather the direction of the heat flow is opposite.

The heat power which flows through the structures with conduction can be calculated with Equation (8.7). In the equation λ is the thermal conductivity of the material of the structure,

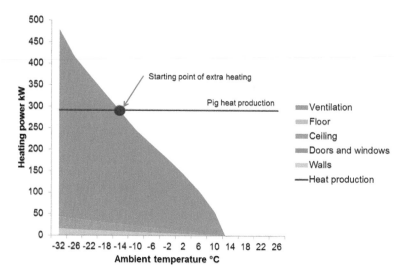

Figure 8.15.   Example of heat losses and pig heat production (2400 pigs of weight of 60 kg/pig, walls have a U value of 0.4 W/m²/K and ceilings of 0.3 W/m²/K) (Mannfors and Hautala, 2011).

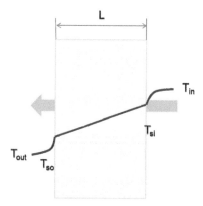

Figure 8.16.   Heat conduction through structures (L is wall thickness, T_{in} is indoor temperatre, T_{si} is inside wall surface temparuture, T_{so} is outside wall surface temperature and T_{out} is outdoor temperature).

A is the cross section area, ΔT is the temperature difference of the surfaces across the structure ($T_{si} - T_{so}$) and L is the thickness of the structure.

$$P = \lambda A \frac{\Delta T}{L} \qquad (8.7)$$

Thermal conductivities of some materials are given in Table 8.8.

The wall structures are not made from only one material but they consist of several materials. In these cases conduction is calculated by the total heat transfer coefficient (U-value) of the wall. The U-value for each wall layer i is calculated with Equation (8.8), where L is the thickness of the material:

$$U_i = \frac{\lambda_i}{L_i} \qquad (8.8)$$

Table 8.8. Material thermal conductivities.

Material	Thermal conductivity [W/m/K]
Air	0.024
Glass wool	0.045–0.050
Rock wool	0.040–0.070
Saw dust	0.08–0.12
Brick	0.4–0.9
Wood	0.14
Concrete	1.7

Thermal conductivity of the walls depends on their structure. The insulating material and its thickness mostly affect thermal conductivity. Heat can also flow through walls by convection if walls do not have windshield characteristics. Moisture can also destroy heat insulation if it penetrates the material and condensates. The total U value of the wall can be calculated as:

$$\frac{1}{U} = \frac{1}{U_1} + \frac{1}{U_2} + \cdots + \frac{1}{U_n} \tag{8.9}$$

Example. The wall consists of a 150 mm thick fibreglass layer ($\lambda = 0.05\,W/m/K$) and a 22 mm thick wooden board ($\lambda = 0.14\,W/m/K$) on both sides. If the wall area is 50 m^2 what is the heat conduction flow through the wall when the outside wall temperature is $-15°C$ and inside wall temperature is 15°C (temperature difference is 30 K).

U-values for the layers are: outside board $U_1 = 0.14/0.022 = 6.36\,W/m^2/K$, insulation $U_2 = 0.05/0.15 = 0.33\,W/m^2/K$ and inside board $U_3 = U_1$. The total U value is: $1/U = 1/6.36 + 1/0.33 + 1/6.36 = 3.34 \Rightarrow U = 0.30\,W/m^2/K$. Heat power flowing through the wall is: $P = 0.30\,W/m^2/K \times 50\,m^2 \times 30\,K = 0.45\,kW.$

The wall surface temperatures (T_{si}, T_{so}) are not the same as indoor and outdoor temperatures (T_{in}, T_{out}). There is an air boundary layer on the surfaces and this has also an insulating effect. The U-values of the boundary layers depend on radiation, air speed (wind) at the surface and surface properties. For outside walls the U-value of the boundary layer is in the range of 20–50 W/m^2/K and for inside walls 7–12 W/m^2/K (Henderson et al., 1997; Vuorelainen, 1977). If the walls have an effective heat insulation, then the effect of boundary layers is minor but with poor heat insulation this has to be taken into account in U-value calculations (Equation (8.9)).

Walls normally have windows and doors. The U-values of the openings are different from those of the wall. Typically their U-values are 2–3 W/m^2/K. Normally their heat losses are calculated separately or a mean U-value is calculated for the walls and openings.

The temperature in the building varies in the vertical direction so that on the ceiling the temperature is warmer than on the floor. The recommendations for the insulation of the building take this into account and the insulation demands are higher for ceilings than for the walls.

At floor level, the temperature under the floor is warmer during cold weather than the outside temperature. Therefore, heat losses through floors are lower than for other parts of the building. The heat losses at the floor are largest at the outside perimeter. There the heat flows from the floor to the soil and from the soil to the air. Floor insulation is therefore under the floor and mainly near the outside walls. Heat losses through the floor are difficult to calculate precisely because the soil temperature under the floor is not known.

There are recommendations for U-values in cold climates and quite commonly, outer walls should have a U value of 0.4 W/m^2/K and ceilings of 0.3 W/m^2/K (MMM RMO C2.2).

8.3.4.2   Heat losses by ventilation

If the production buildings are heated with heaters, ventilation produces heat losses when warm air flows out and cold air streams in. The sensible heat loss by ventilation can be calculated with Equation (8.10), where c_i is the specific heat capacity of air [1.0 kJ/kg/K], q_v the air volume flow, ρ_i the air density and ΔT is the difference between inside and outside temperatures:

$$P = c_i \cdot q_v \cdot \rho_i \cdot \Delta T \qquad\qquad (8.10)$$

Example. One milking cow requires a minimum ventilation rate of $55\,m^3/h$. If the inside temperature is $+12°C$ and outside temperature $-20°C$, how much energy is needed to warm the ventilated air for one cow?

   Air density is about $1.2\,kg\,m^{-3}$ at $12°C$, temperature difference is, $\Delta T = (12 - (-20)) = 32\,K$. $P = 1.0\,kJ/kg/K \times 55\,m^3/3600\,s \times 1.2\,kg\,m^3 \times 32\,K = 0.6\,kW$. One cow produces about this amount of sensible heat, so the cow's own heat production is sufficient to warm the ventilated air needed. The building also has heat losses through the structures, which means that to retain the temperature additional heating is required if the inside temperature is wished to be at $+12°C$.

Low air flow means low energy consumption but this leads to poor air quality. A well-functioning ventilation system should ensure a good microclimate for the animals without additional air flow. As mentioned in Section 8.3.2, the ventilation rate needed depends on circumstances.

Heat exchangers can be used to recover the energy of the exhaust air. This will decrease energy use or, with proper exchangers, no additional heating is required and the heat the animals produce is sufficient for the heating of the building. This is quite easy to achieve in moderate weather conditions because with proper insulation the conductive heat losses are low.

However, there are problems with heat exchangers because the incoming and exhaust air must meet. This means additional piping, or the air flows must be close to each other. The air in a cattle house is very humid and dusty. These can block or freeze the exchanger and the system needs maintenance.

Cows have better thermoregulation than pigs or poultry. This means that they can be kept in cold or semi-cold cow houses without problems when it is taken care that drinking water or manure is not frozen. Semi-cold buildings have only some insulation so that the heat produced by the cows is enough to keep the temperature above zero. Cow houses do not need heat insulation or only a minor insulation is used and no heat recovery system is either needed.

If the farm has bioenergy another possibility is to use it for heating. This does not decrease energy use but for the farmer it can be more economical and it also reduces CO_2 emissions because biofuels are carbon neutral fuels.

Energy consumption in heating animal houses can be reduced:

- When animal houses need heating heat losses follow. These can be reduced with proper heat insulation.
- Cows can tolerate lower temperatures better than pigs or poultry. Cow houses can be cold or semi-cold buildings.
- In ventilation large amount of heat is lost. With good ventilation control losses can be kept low and heat recovery systems could be used to save energy. Ventilation control changes depending on temperatures, number of animals and animal weight, the control system should be able to function with changing criteria.

8.3.5   Energy use follow-up

In order to improve the energy efficiency on a farm, the farmer must follow the farm energy consumption and compare it to the values of similar farms. A simple analysis (top-down) can be

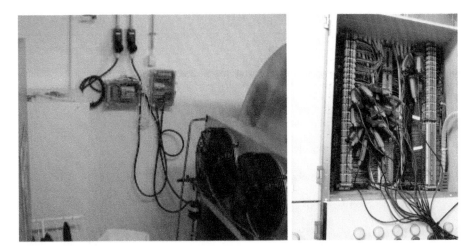

Figure 8.17.   Examples of electricity measurements.

done by calculating the annual energy use. This gives a coarse estimate of the energy efficiency but does not show the energy leakages of the production. For this a detailed analysis (bottom-up) is needed, in which the energy consumptions of different machinery must be measured. However, this is a demanding procedure because electrical assemblies are fixed and a measuring system must be introduced. Only qualified persons can do such changes. Figure 8.17 shows an example of measurements. The example on the left of Figure 8.17 has an energy meter between the plugs of the machine and electrical grid. In the right, current clamps are used in an electrical center.

The assemblies of energy measurement systems are costly and it would be convenient if the machinery would be equipped with embedded energy meters.

Figure 8.5 shows an example of energy flows on a milk farm. Part of the plant production is used as a feed for the cows and part of the crop is sold out. Livestock produces manure, which partly compensates for the fertilizer need.

Figure 8.5 shows only the energy flows of the farm. To compare the consumption, specific values should be used. The consumption should be calculated per produced product. These numbers are compared with other farms and if they are not of the same magnitude, the reason should be found.

Energy use follow-up on farms:

- For farm energy analysis the farmer should follow the energy consumption and compare it to average values.
- During electricity installations and new equipment, purchase of an energy meter acquisition should be considered.

8.4   ENERGY USE AND SAVING IN LIVESTOCK PRODUCTION

8.4.1   *Energy consumption in livestock production*

The animal production is a poor converter of energy because it is based on a double energy transformation. First, solar energy and soil nutrients are converted into biomass by green plants. When the plants are fed to animals, a major share of energy intake is used in body metabolism and only a small portion is used to produce meat and milk (Heilig, 1993) (Fig. 8.2).

Fossil energy is a major input in livestock production systems, used mainly for the production, transport, storage and processing of feed. Depending on location (climate), season of the year

Table 8.9.　Livestock production intensities (FAO, 2011).

Scale of producer	Overall input intensity	Human labor units	Animal power use	Fossil fuel dependence	Capital availability	Major food markets	Energy intensity
Subsistence level	Low	1–2	Common	Zero	Micro-finance	Own use	Low
Small family unit	Low	2–3	Possible	Low/medium	Limited	Local, fresh, process, own use	Low to high
	High	2–3	Rarely	Medium/high	Limited	Local fresh/ process/own use	Low to high
Small business	Low	3–10	Rarely	Medium/high	Medium	Local/ regional/export	Low to high
	High	3–10	Never	High	Medium	Local/ regional/export	Low to high
Large corporate	High	10–50	Never	High	Good	Regional process/export	Low to high

Table 8.10.　Grain and forage inputs per kilogram of animal product produced, and fossil energy inputs required to produce the same amount of animal protein (Pimentel, 2004).

Livestock	Grain [kg]	Forage [kg]	E input/E protein
Lamb	21	30	57:1
Beef cattle	13	30	40:1
Eggs	11	–	39:1
Grass-fed beef cattle	–	200	20:1
Swine	5.9	–	14:1
Dairy (milk)	0.7	1	14:1
Turkeys	3.8	–	10:1
Broilers	2.3	–	4:1

and building facilities, energy is also needed for thermal control (cooling, heating or ventilation) and for collection and treatment of animal waste.

Large differences in energy consumption exist between countries, livestock species and types of production system. In the developing world, fossil fuels are seldom used. For example, bullocks are used for transport, farmlands grazed cattle, goats and sheep do not require fuel. FAO (2009) divides livestock production systems into grazing, mixed farming and industrial (or 'landless systems'). Industrial systems include intensive beef cattle, pigs and poultry, fed on feeds that are purchased from outside the farm. According to production intensity, livestock enterprises can also be categorized as in Table 8.9.

It must be remembered that farms are run mostly by economical and work capacity reasons and at the moment energy savings are not very significant. However, energy questions will become more important because of the limited fossil energy resources.

Feed is the dominant term in energy use whether it is used as concentrates, conserved forage or grazed grass. The major fossil energy inputs required to produce grain and forage for animals includes fertilizers, farm machinery, fuel, irrigation, and pesticides. The energy inputs vary according to the particular grain or forage being grown and fed to livestock. Forage can be fed to ruminant animals because of their ability to convert the forage cellulose into digestible nutrients through microbial fermentation. Grain and forage inputs per kilogram of animal product produced, and fossil energy inputs required to produce the same amount of animal protein (Pimentel 2004) are presented in Table 8.10.

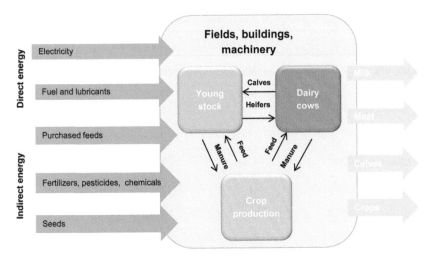

Figure 8.18.   Principal description of the energy flows in a milk production farm with milk, meat and crop production.

8.4.2   *Energy consumption in milk production*

8.4.2.1   *Milk production system*
A principal description of energy flows in a milk production system is given in Figure 8.18.

Direct energy inputs are fuel and lubricants used in feed processing and for energizing of machinery. The electrical energy is used for milking, milk cooling, water heating and pumping, lighting, ventilation, heating, electrical fencing, manure handling, and office and personnel working environments. 26% of fossil energy input is direct energy (electricity and fuels) Woods *et al.* (2010). Indirect energy is embedded in the products used on the farm, mainly in feed (71%) and building (3%) (see also Section 8.4.2.2).

The different farming systems for dairy cows, which have an important role in the development of more sustainable farming practices, depend on the agro-ecological and socio-technical situation. In practice, the boundaries between different types of farming systems may be indistinct. There are three main farming systems (Norton *et al.*, 2006):

1. Mixed livestock farming:
    Mixed farming is an integration of crops and livestock production. Mixed farming is common in traditional agriculture. It helps to manage risk, makes efficient use of labor and land, and helps maintain soil fertility.
2. Intensive production systems:
    The main principle of intensive systems is a high and certain return per unit of input. The intensification in milk production is possible through technical achievements:
    * Genetic improvement to increase the milk production per cow and the contents of valuable constituents in the milk;
    * An intensive use of fertilizers to increase the production of farm feeds;
    * An almost full-scale mechanization, advanced building design and automated control to reach labor-saving operations.
    Most of the milk produced in Europe derives from intensive farming. The majority of those farms use loose housing with cubicle cowsheds. The average herd size is increasing with the reduction in the number of farms. Intensive dairy farming systems are land-based, but the area of land is not sufficient to feed the cows. Therefore complementary feed is needed. Often the cows are kept inside throughout the year (zero-grazing management). Intensive livestock production systems are characterized by high fossil energy utilization.

Table 8.11.   Energy consumption in milk production.

Task	[kJ/kg milk]	Reference
Lighting	44–91	Hörndahl (2008)[1]
	99	Neuman (2009)
	16–82	Ludington and Johnson (2003)[1]
Milking and milk cooling	154–229	Hörndahl (2008)[1]
	80–177	Ahokas (2013), ed.
	173	Neuman (2009)
	58–231	Ludington and Johnson (2003)[1]
Manure removal	9–16	Hörndahl (2008)[1]
	20–50	Ahokas (2013), ed.
	41	Neuman (2009)
	2–14	Ludington and Johnson (2003)[1]
Ventilation	0–32	Hörndahl (2008)[1]
	25	Ahokas (2013), ed.
	66	Neuman (2009)
	10–107	Ludington and Johnson (2003)[1]
Feeding	53–261	Hörndahl (2008)[1]
	147	Neuman (2009)
	5–99	Ludington and Johnson (2003)[1]
Misc.	35–90	Hörndahl (2008)[1]
	73–140	Ahokas (2013), ed.
	28	Neuman (2009)

[1] Calculated from the source with the assumption of 9000 kg annual milk yield.

3. Extensive production systems:
   In extensive systems the return per unit of input is relatively low. Extensive systems are commonly based on livestock grazing.

Organic dairy farming is a land-based and value-based method, described by four principles (EFSA, 2009; Norton, 2006):

1. To sustain and enhance the health of soil, plant, animals and humans.
2. To be based on living ecological systems and cycles.
3. To be built on relationships of fairness to the common environment and life opportunities.
4. To be managed in a manner to protect the health and well-being of current and future generations and the environment.

8.4.2.2 *Energy used in milk production*

According to several studies the energy needed to produce one kilogram of milk varies between 0.4–7.9 MJ/kg milk (Table 8.11). The differences are partly due to the analysis method. For instance, some analyses include fodder production, some only the direct energy consumed on the farm. Energy inputs from different sources are collected in Table 8.4. Figure 8.6 shows an example of milk energy inputs (Ahokas, 2013). In this example, the largest portion of the energy (80%) is used in fodder production and second largest portion comes from electricity use. Woods *et al.* (2010) give a 71% share of the total energy consumption for feed material.

In dairy buildings usually only the milking parlor and offices are heated and the heat consumption is low. The variation in energy consumptions is large, which also means good possibilities to improve energy efficiency of production. Results of detailed consumption figures in dairy production are given in Table 8.11. The values vary due to differences in machinery, production systems and geographic place.

8.4.2.3   *Feed production and feed material*

The feed for dairy cattle is mainly based on grass (silage) and concentrates produced from cereals. In summer time the pasturing is often used. Embedded energy of some feed ingredients for calculating fuel use in livestock is given in Table 8.4.

Depending on the animals' diet the impact of the feed production can vary because the process to produce concentrates is more energy consuming than to produce fodder. Extensive grazing systems for ruminants tend to have lower energy inputs than intensive livestock systems. Pasture has the lowest energy demand (0.84 MJ/kg of dry matter) because machines are used only for fertilization and cultivation operations (Kraatz and Berg, 2009).

The process in livestock production requiring the most energy is the production and processing of concentrate. Therefore, the feeding efficiency could be improved by reducing the amount of concentrate in feed. Feeding efficiency can also be increased by efficient management in feed production (see Section 8.3.1); transport expenses decrease when the forage and crops are produced on dairy farm. By Kraatz (2012) increasing milk yields lead to a decrease of the energy intensity in dairy farming. However, this effect diminishes with yields above 8000 kg milk/cow/year because of higher inputs of concentrate in the diet.

Dairy cattle are kept for milk and meat production, on-farm energy inputs (including indirect energy for feed, buildings and equipment) per unit of cattle food product are 5–7 MJ/L of fresh milk and 80–2100 MJ/kg of beef meat (Smil, 2008). Ingested feed energy allocation is 85.6% to milk and 14.4% to meat (IDF 2010).

Fossil energy requirements for veal production are much higher than for beef. Direct fossil energy use is higher due to the more sophisticated housing systems and indirect fossil energy use for calves' feed production is high, (e.g. drying wet substances to obtain skim milk powder). Brand and Melman (1992) calculated the fossil energy requirements for The Netherlands, it being 27.83 MJ for 1 kg milk substitute.

8.4.2.4   *Use of direct energy*

Electricity is a type of energy on which many different technological processes rely. Milk production is not an exception. According to different researchers and to on-site measurements, the electricity share is 7% of the total farm energy input. In research done is Estonia the electricity share was 7.7%.

Figure 8.19 shows an example of electricity use on a dairy farm. The most dominating part on this farm is the miscellaneous electricity use (others), which includes lights in the building and in the yard, office equipment and electrical heaters used during cold period. As for the production, milking and milk handling forms the largest portion. The remaining (heating, ventilation, manure removal and water pumping) are about the same magnitude. The numbers and shares vary considerably between farms, locations and different production managements.

Figure 8.20 shows an example of the variations in specific and total electricity consumption during one year in an Estonian cowshed. The peak consumption comes during winter time because of the cold weather. Then the heating, manure handling, and milk production require the highest amount of electricity. The lowest total consumption is in summertime. Specific energy consumption depends also on milk production and for this reason their peaks differ from the total consumption peaks.

8.4.2.5   *Milking and milk cooling*

Milking machines are operated with partial vacuum by removing air from the milking machine. At the teats the vacuum should be constant and it should remain constant even if some of the clusters are kicked off. This means that the vacuum pump capacity must be higher than the capacity needed in a normal milking operation. The normal operation of vacuum pumps means working at full capacity when switched on. The vacuum is managed with a vacuum regulator, which controls the airflow through the system.

For energy saving variable speed motors are used which change their speed according to the air need. This can save energy up to 40–50% in a milking machine (Dunn *et al.*, 2010). The energy

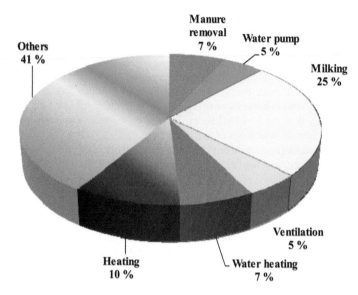

Figure 8.19.   Example of electricity use on a dairy farm in Estonia.

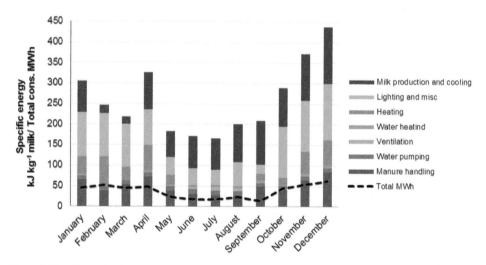

Figure 8.20.   Example of a measured monthly specific [kJ/kg] milk and total [MWh] electricity consumption in an Estonian cowshed in 2011 (596 cows).

consumption of a milking pump is 30–100 kJ/kg milk. For one cow the savings in electricity can be 540 MJ (150 kWh).

The temperature of milk after milking is 35–37°C. To maintain high milk quality, including low bacteria counts, the raw milk temperature needs to be quickly lowered to around 3–4°C. The specific heat capacity of milk is 4.19 kJ/kg/K. When one liter of milk is cooled by 30°C, the energy released by the milk is 126 kJ. If the annual milk production of a cow is 9000 kg, then 1.1 MJ energy could be recovered from the milk at maximum. The cooling system also produces heat. This heat could also be utilized on the farm.

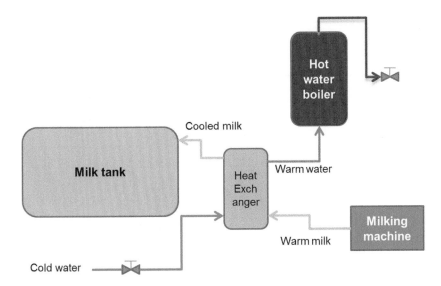

Figure 8.21.   Milk pre-cooler system.

The typical refrigeration systems consist of a milk tank, a motor-driven compressor unit, an evaporator and a condenser unit. The energy amount consumed by the cooling system varies depending on the number of cows (the daily amount of milk) and the temperature of the milk tank room.

The easiest way to utilize the cooling energy is to conduct the hot air from the cooling system to heating of the building. The other possibilities are to heat water with milk pre-cooling or heat recovery systems. The warm water can be used as drinking water for the cows or it can be used for heating the hot water of the cattle house or farm.

In cold and semi-cold cow houses drinking water and manure removal systems must not be frozen. The heat from cooling milk could be used for these purposes. For example, it could be used in a floor warming circuit. In addition the milking parlor must be warm and the heat released from cooling could be used for heating it.

In milk pre-cooling, a heat exchanger is assembled in the milk pipes before the milk tank. This will pre-cool the milk and pre-heat the hot water. The milk cooling system will work at lower power when the milk is cooled by some degrees and also the milk in the milk tank will have better quality because its temperature does not change so much as without pre-cooling. This system is especially suitable for robot milking where milk production is rather constant.

Figure 8.21 shows the layout of milk pre-cooler. A heat exchanger is in the milk line before the milk tank. According to Karlsson *et al.* (2012) when the water flow is equivalent to the milk flow the milk can be cooled to 17°C. This system can utilize the heat of the milk only when there is a need for hot or warm water during the milking. It is also possible to use cold water for precooling of milk without heat utilization. Then the water is only pumped through the heat exchanger without further use. This will still reduce the energy consumption of the milk cooling system. If the heated water is used for washing, its temperature must be increased at least to 65°C for hygienic reasons.

Figure 8.22 shows a heat recovery system integrated in the milk cooling system. The system utilizes the heat of the milk cooling media. The system produces warm water (50–55°C) which must be heated in the hot water boiler to a higher temperature. If the warm water tank is full of warm water the heat exchanger cannot cool the cooling media and the condenser fans start to operate. This system can utilize two thirds of the energy in the cooling media (Karlsson *et al.*, 2012).

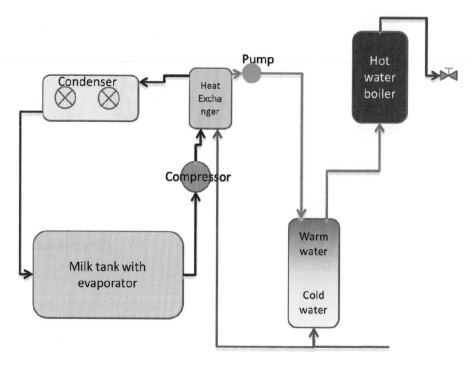

Figure 8.22.   Example of heat recovery system in milk cooling system.

8.4.2.6   *Lighting*

The appropriate lighting can improve productivity and safety on a dairy farm. Factors that contribute to the increase in milk production include the spectrum of light and the length of the lighting period. The last influences the energy consumption as well as lighting efficiency.

Lighting requirements on a dairy farm can be divided into three categories. The first category is visually intensive task lighting, which requires the highest lighting level (Ludington *et al.*, 2004). Areas that benefit from this type of lighting include milking parlors, equipment washing, equipment maintenance and repair areas, offices, maternity, veterinary treatment areas and utility rooms. The second category includes lighting for holding areas, feeding areas, animal sorting and observation and general clean-up. These areas and tasks require high to moderate lighting levels. Finally, low to moderate lighting levels are adequate for general lighting for livestock resting areas, passageway lighting, general room lighting and indoor and outdoor security lighting.

Results of electrical measuring show that in winter in Estonia, when the daylight time is much shorter than in summer, the amount of energy used for lighting is significantly higher (Fig. 8.23).

8.4.2.7   *Ventilation*

If natural ventilation is used in the cattle house the ventilation system needs only a small amount of energy for adjustments and control. In forced ventilation systems the fans are running almost continuously and the ventilation should be adjusted according to the microclimate inside the building. In forced ventilation systems heat exchangers can be used to recover some of the outgoing heat energy.

Proper ventilation is needed in dairy barns throughout the year to help maintain animal health and productivity, the barn's structural integrity, milk quality and a comfortable work environment for the laborers. The air inside a barn becomes warm and humid as cows continuously produce heat and moisture. An efficient ventilation system brings fresh air that is cooler and drier into

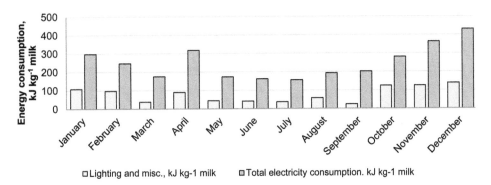

Figure 8.23.   Example of monthly electrical consumption of lighting during year 2011 in an Estonian cowshed (596 cows).

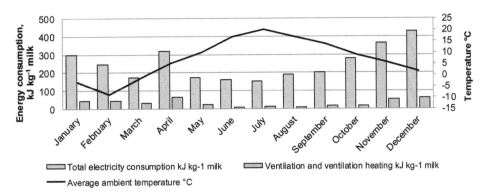

Figure 8.24.   Example of farm ventilation and heating electrical consumption by months during year 2011 in an Estonian cowshed (596 cows).

the barn, mixes and circulates it, and exhausts and dilutes the moisture, dust, manure gases and odors, and airborne contaminants.

Cows that are exposed to fresh air are less likely to develop heat stress, respiratory ailments, mastitis, reproductive problems and other diseases that could decrease milk production.

There are two types of ventilation: natural and mechanical (forced). Natural ventilation uses the least amount of energy but, similarly to mechanical ventilation; it requires the openings in the walls or roof for air exchange, the ability to control ventilation rates, flexibility to provide the cows a comfortable environment throughout the year and good barn construction. Barns that are in the path of prevailing winds and designed with adjustable side and end walls can take advantage of breezes.

Mechanically ventilated barns require air inlets, outlets and fans. Each inlet targets a certain area to provide uniform control of moisture and gases. Ventilation systems are often designed with air inlets and exhaust fans placed on opposite ends of a building. In addition to these, in the barns with high ceilings, the air mixing fans with vertical shafts are installed under the ceiling.

The results of electrical energy measurements in an Estonian dairy barn (Fig. 8.24) show that the ventilation systems use more electricity in winter. During winter the incoming air has to be heated to sustain a comfortable temperature in the barn. In the summer the optimal humidity and temperature in the dairy barn is ensured by natural ventilation. In un-insulated barns it is most effectively achieved by adjustable side walls.

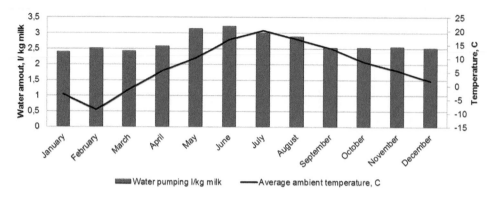

Figure 8.25.   Example of the monthly water consumption by months during year 2011 in an Estonian cowshed (596 cows).

8.4.2.8   *Water pumping and hot water*

Water demand on a dairy farm can be divided into two parts: the drinking water for farm animals and water for technologic needs (udder, milking equipment, materials, buildings and other washing).

Fresh water is one of the most important factors in the animals' diet, because it is involved in all life processes. If the water loss in an animal's body is more than 20% of its weight, the animal cannot survive. Therefore, it is critical that the animal always has access to fresh drinking water.

Cattle get their drinking water from three sources: mostly from fresh water sources, from feeds and a small amount from metabolism within the body. Calves get all of the needed drinking water from milk or milk substitute until a certain age. In their first week, calves need approximately 1 kg of water a day, when they reach the age of 4 weeks the need is 2.5 kg a day.

Milking cows need 3–5 kg of fresh water for every kg of dry food. The amount of water needed for milk production also must be added to this water usage. The amount of water required will rise if the amount of dry food is raised or the cattle room temperature rises. The amount will also increase if the amount of salt, proteins and soda is higher in the food.

Technological water is used for milk hygiene and cleaning. It is also used for:

- Milk plate coolers;
- Water sprayers;
- Slurry flush systems;
- Irrigation;
- Domestic use.

The amount of water used in technological processes depends on the technologies that are used on the farm. Figure 8.25 shows an example of dairy farm water consumption. In the summer the use of water is higher than in winter.

8.4.2.9   *Bringing up young cattle*

Milk production sustainability has achieved by replacement of cows. However, the energy costs of maintaining breeding stock may be considerable. By Kraatz and Berg (2009), the cumulative energy demand for the growing period of a heifer (0–25 months) is 13 GJ per heifer including pasture in the diet and 16 GJ per heifer with indoor stock keeping. 70% of the cumulative energy demand is used by the feed supply. On an Estonian dairy farm, 15% of the input energy was feed energy for young stock, whereas energy for raising a heifer varied between 15 and 22 GJ.

With lower lifetime of dairy cows, the energy intensity for replacement increases and gains a stronger influence on the energy intensity of the dairy farming process. Rising milk yields are mainly accompanied by increasing replacement rates. The influence of rising milk yields outweighs the effect of increasing replacement rates at least up to annual milk yields of 10,000 kg per cow (Kraatz, 2012).

Energy can be saved in milk production:

- Feed material production is the most energy demanding part of the production. In feed production nitrogen fertilizer is the dominant factor. With proper manure handling the nutrients can be partly restored to the fields. Cultivation of nitrogen binding plants also reduces need for nitrogen.
- Depending on the production type, machinery and circumstances, the direct energy consumption varies greatly. Energy can be saved by:
  - using natural ventilation
  - using heat recovery systems in milk cooling
  - controlling lighting so that only the necessary luminance is used
- Heifer production consumes energy, and prolonging cows' lifespan, energy can be saved in the upbringing of new cattle.

8.4.3 *Energy consumption in pork production*

8.4.3.1 *Pork production*

Pork meat is the most popular meat for human nutrition (OECD/FAO, 2012). Pork production is increasing worldwide but the growth rate of poultry production is faster than that of pork and poultry meat is expected to exceed the production of pork meat during 2018–2020. China is the leading pork producer in the world (FAO, 2013). Production was 40,000 tons in 2000 and it increased by 28% (11,000 tons) in the period of 2000–2010 (van der Sluis, 2012). In addition to China, other Asian countries, the USA and many European countries (FAO, 2013) are on the list of top twenty pork producing countries. The list tells that pork is produced in varying climatic and economic conditions.

The size of pig farms is growing. In China, 48% of the pigs were produced on farms of less than 100 pigs per year in 2010 but the number of these farms decreased by 9% in the period 2009–2010. Similar structural change is also going on in Europe (EUROSTAT, 2013). Small farms do not have possibilities to invest in production technology and young people are no longer interested in raising pigs (van der Sluis, 2012). Due to urbanization self-sufficiency animal husbandry is no longer possible. Animal products have to be produced on larger farms with less labor force.

However, it is worth thinking whether small-scale pork production would be more energy efficient than large-scale production. In small-scale production there are good possibilities to utilize biowaste from the kitchen and garden for pig feeding. This is an important aspect because feed was the biggest single energy input of swine production in Lammers' (2009) research. Furthermore, nitrogen is the major energy input in crop production (Mikkola and Ahokas, 2009) and therefore Honeyman (2013) has concluded nitrogen management to be the key tool in reducing the use of nonrenewable energy in pork production. This means that nitrogen recycling should be promoted and ammonia emissions should be minimized in pig manure management.

Besides feed, ventilation is another key energy consumer. Ventilation is needed to guarantee animal welfare and reasonable working conditions for workers, and also to reduce the stress of moisture on constructions of buildings. In a cold climate heating is needed to keep the temperature above critical limits (e.g. Videncenter for Svinproduktion, 2011).

Small-scale pork production is labor-intensive which makes it economically unviable. Small-scale production is possible only in rural areas but pork meat is also demanded in cities. Therefore, large-scale pork production is needed to supply pork to urban consumers. Small-scale pork producers are a heterogenic group of producers and no data is available from energy consumption in this group. For these reasons this examination concentrates on intensive production in larger (more than 100 pig places) professional pork meat producing units.

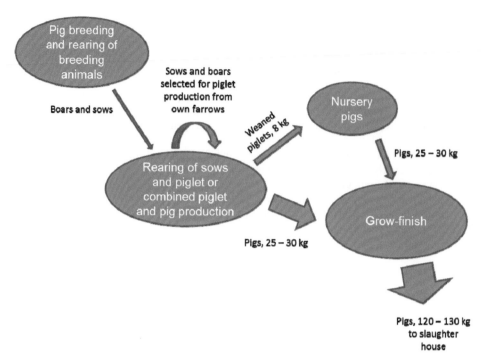

Figure 8.26.   Stages of pork meat production.

Pork production can be separated into three–four phases (Fig. 8.26). Pig breeding is continuous work which ensures that pork producers constantly get new and more favorable genotypes for production. Demands from consumers, authorities and farmers define principles of pig breeding. High feed efficiency is an important target from the point of view of energy efficiency.

Only part of the sows and boars used for piglet production come from breeding stations or farms specialized for pig breeding. Most of the sows are selected from piglets produced for meat production. Careful bookkeeping and monitoring of sows and farrows is used as criteria for selection. Sows should produce piglets which meet the high requirements of well-being or health, feed efficiency, meat quality, and growth speed.

Piglet production demands special expertise and, for this reason, piglet production and housing of finishing pigs have increasingly separated to different farms. Growing of weaned piglets from the weight of 8 kg to 25–30 kg can also take place on a separate farm. In the most specialized pork production systems, breeding sows, housing bred sows, and farrowing are also separated into different farms. This is so called satellite pork production. Specialization gives the possibility to optimize all stages of the production chain and helps to deepen know-how on farms focusing on a narrow sector of production. On the other hand, specialization means that pigs are moved from one place to another many times. It can increase risk of animal diseases and stresses animals. Farms involved in highly specialized production chains are more dependent on each other than farms which take care of all or most of the production actions themselves.

A study of Neuman (2009) takes into account the impact of different production systems on energy consumption but other energy studies made on pork production have focused mainly on the production of finishing pigs. Lammers (2009) has studied two types of facilities—conventional confinement and hoop barn-based—within farrow-to-finish pig production systems scaled to produce either 5200 or 15,600 market pigs annually, but Lammers (2009) has not studied the impact of locating the actions of the production to different sites and its impact on energy consumption.

Table 8.12. Energy consumption in pork production.

Task	[MJ/kg meat]	References
Lighting	0.04–0.4	Seifert *et al.* (2009)[1]
	0.03–0.14	Neuman (2008)[1]
Manure removal	0.04–0.2	Seifert *et al.* (2009)[1]
	0.01–0.05	Neuman (2008)[1]
Ventilation	0.7–2.5	Seifert *et al.* (2009)[1]
	0.2–1.0	Neuman (2008)[1]
Feeding	0.04	Seifert *et al.* (2009)[1]
	0.2–0.8	Neuman (2008)[1]
Heating	1.8–6.7	Seifert *et al.* (2009)[1]
	2.2	Ahokas (2013)
	0.1–0.3	Neuman (2008)[1]
Electricity[2]	1.1	Seifert *et al.* (2009)[1]
	2.2	Ahokas (2013)
	1.1–17.6	Neiber and Neser (2010)
	0.5–2.2	Neuman (2008)[1]

[1] Calculated from data with the assumption of 100 kg meat and three batches in a year.
[2] Total electricity consumption (lighting + ventilation + manure removal + feeding).

Producing pigs using hoop barns for grow-finish and gestation required less embodied energy than using conventional confinement facilities. Hoop barn-based pig production required similar quantities of total operating energy as conventional facilities. Increasing the scale of production from 5200 to 15,600 market pigs annually lowered the embodied energy and the use of nonrenewable energy. Hoop barn-based production scaled to produce 5200 market pigs annually required similar amounts of embodied energy compared to conventional confinement facilities scaled to produce 15,600 market pigs annually (Lammers, 2009).

8.4.3.2 *Pork production energy consumption*
When the consumption is divided into different categories, Table 8.12 is obtained. Heating is the major energy input and next comes ventilation. The other inputs are low compared to these. There is large variation in the figures and the references concern only a part of the global production. For instance heating is not needed in warm and hot climates.

Neuman (2008) studied direct energy consumption of 83 Swedish farms out of which 17 farms were producing mainly or only piglets and 14 farms were growing mainly or only finishing pigs. Energy inputs recorded were: electricity for feeding, diesel oil for feeding, ventilation, electricity for manure removal, diesel oil for manure removal, lighting, electricity for heating, oil for heating, biofuel for heating, other electricity and other diesel oil consumption.

Farms producing piglets had from 60 to 600 sows (on average 241) and they produced totally some 100 000 piglets per year. Energy consumption per piglet was 95 MJ on farms producing weaned piglets, which were transported to a nursery after weaning (satellite system). When the piglet production and the nursery were combined, energy consumption was 172 MJ per piglet, and when all three stages were combined (piglet production, nursery, and growing of finishing pigs) the energy consumption was 165 MJ per piglet. Energy for transporting piglets was not taken into account. Neuman (2009) is careful in doing conclusions because there was a large variation between farms and the number of studied farms was low. However, the results indicate that specialized piglet production would have better energy efficiency than combined production. The major inputs of direct energy were heating 51%, ventilation 19%, lighting 14%, and feeding 10%. Pig feed was not included because only direct energy inputs were considered (Neuman, 2008).

Energy consumption for growing finishing pigs was 83 MJ per pig on farms which had specialized for growing finishing pigs. On farms producing piglets and finishing pigs energy consumption was 112 MJ per pig. Specialization indicated lower energy consumption also in growing finishing pigs. The major inputs of direct energy in finishing pigs were ventilation 42%, energy for feeding 33%, heating 14 %, and lighting 6%. Increasing farm size indicated lower energy consumption both in piglet and in pork meat production (Neuman, 2008).

Energy consumption in pork production has been studied in the currently ongoing AGREE project (Agricultural Energy Efficiency), Figure 8.9. Feed was found to be the major energy input also in this study. Differences between countries result from differences in the class "other energy consumption" which includes energy consumption e.g. for heating and ventilation. Energy consumption for these inputs depends on climate conditions and also on animal welfare requirements (AGREE, 2012).

8.4.3.3   *Feed production and feed material*

Feed was found to be the highest energy input in the study of Lammers (2009) and in the AGREE (2012) project. In the study of Mikkola and Ahokas (2009) it was the second highest energy input. There is sense to focus on efforts to increase energy efficiency on the major energy inputs because even a small saving from a big energy input is probably more than a big saving from a small energy input. Maize, wheat, and barley are popular components in pig feed and energy saving efforts should be focused on cultivation and processing of these crops. As Lammers (2009) states, nitrogen management is essential and it means that in all stages of manure handling ammonia emissions should be kept at minimum.

By-products from the food industry (dried distillers grains with solubles, whey, mash, yeast stock) can be used locally for pig feeding. The feeding menu of pigs can vary widely depending on the availability of different feed components. Use of these components increases energy efficiency in pork production if there is no other use for them or if energy demanding processing would be needed to refine them to other products than pig feed. Use of by-products may also improve energy efficiency and economy of the main product (i.e. malt, cheese).

In countries where crop is normally dried before storage and use for animal feed, energy could be saved by using the silage method for grain conservation instead of drying. If the moisture content at harvesting is 18% and after drying 14%, up to 250 MJ/ton energy could be saved. Energy efficiency of drying can also be improved. Heat insulation of hot metal surfaces of grain dryers cuts 10–15% of energy consumption and increasing the temperature of drying air from 80°C to over 100°C cuts 5–10% more. If ambient air or solar heated drying can be used instead of hot air drying, energy efficiency increases.

Energy can be saved in pork production:

- Feed is the main energy consumption in pork production. By improving feed production energy efficiency increases.
- Heat recovery from ventilation air could be used in heating of the piggery. Then additional heating is not needed or only for a smaller extent.
- Ventilation rate changes with changing circumstances (pig weight, humidity, temperature, gases). Control of the ventilation rate should work according to microclimate of the building (temperature, humidity, gases). Ventilation rate, which is adjusted according to animal needs guarantees a good microclimate and low energy consumption.

8.4.4   *Energy consumption in broiler production*

8.4.4.1   *Broiler production*

The poultry sector has changed significantly during the last decades. Nowadays it is one of the fastest growing meat production sectors worldwide. OECD/FAO has predicted that poultry meat production will overtake pork production in size by the year 2021. Production has increased

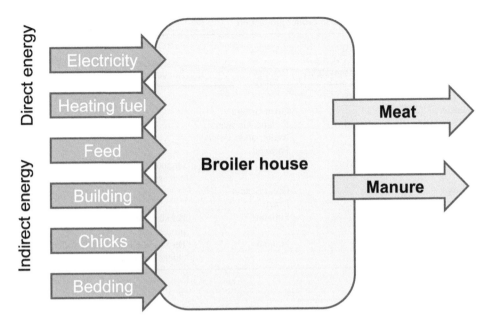

Figure 8.27.   Energy inputs and outputs of broiler house.

fastest in developing countries (OECD/FAO, 2012). At the moment, most of the poultry meat is produced in the United States of America, China and Brazil (Faostat, 2011).

Broilers are grown in different production circumstances around the world. The modern day commercial-scale broiler production (poultry meat production) is a very specialized and integrated production sector. Big units of tens of thousands of birds are typical in a commercial-scale broiler production. Five or six broiler flocks are grown a year (Lindley and Whitaker, 1996). Short production cycles are typical. Modern day broilers reach about 2.5 kg live weight during the 42 day growing period (Gutiérrez del Álamo Oms, 2009). However, the growing period varies on different farms, typically from about five to seven weeks.

Broiler production is a very energy intensive production sector. Energy ratio of broiler (output/input) production is low (e.g., Barber *et al.*, 1989; Heidari *et al.*, 2011; Qotbi *et al.*, 2011). Hence, much more energy is needed to grow a broiler than what is received from the meat (and manure). Barber *et al.* (1989) reported that energy is used over three times more to produce a broiler than what is received from the meat and manure. Figure 8.27 shows energy inputs and outputs of a broiler house. During the growing period, energy is used directly as electricity for ventilation, lighting and feeding and fuel for heating and manure handling. In many countries, heating is the largest direct energy input in broiler houses. Most of the total input energy is used indirectly to produce, process, and transport broiler feed. Feed is also the largest economical cost in broiler production.

8.4.4.2 *Energy consumption in broiler production*
Energy consumption varies on different broiler farms around the world due to a different climate, housing type, size of houses, management practices and devices. Broilers are grown in different regions and different environmental circumstances. In colder climates, the share of heating energy of the total energy consumption is larger than in warmer climates.

The energy input of broiler production (broiler meat production) varies from 10 to 51 MJ per carcass weight (Table 8.13). It is challenging to compare the total energy input of broiler production because of different analyzing methods used in different studies. Sometimes there is

Table 8.13.   Energy consumption in broiler production.

References	Energy input [MJ/kg of carcass weight]	Country	Remarks
Smil (2008)	25–35	–	
Woods et al. (2010)	17	Great Britain	
AGREE (2012)	10–18	Finland, Germany, Netherlands, Poland, Portugal	
Seguin et al. (2011)	48–51	France	Organic production, figure converted to carcass weight from living weight
Williams et al. (2008)	16	Great Britain	
	12	Brazil	
Katajajuuri et al. (2006)	17	Finland	Per sliced broiler fillet, only broiler production
	21	Finland	Per sliced broiler fillet, broiler and chicken production

also a lack of relevant information on production systems or reliable data of energy inputs. Energy input is also presented in many different ways, for example, per live/market weight (Baughman and Parkhurst, 1977; Liang et al., 2009), per bird (Heidari et al., 2011; Hörndahl, 2008), per target weight (Hörndahl, 2008), per kg (Heidari et al., 2011), per carcass weight (Katajajuuri et al., 2006; Rajaniemi and Ahokas, 2012), per flock (Hörndahl, 2008; Rajaniemi and Ahokas, 2012) or per m² (e.g., Bokkers et al., 2010). However, previous studies (e.g., AGREE, 2012; Atilgan and Koknaroglu, 2006; Barber et al., 1989; Baughman and Parkhurst, 1977; Katajajuuri et al., 2006; Woods et al., 2010) show that feed is the largest energy input in broiler production.

Nowadays commercial-scale broiler production is very intensive and has fast cycle production, but broilers are also grown in more extensive systems in free range (pasture poultry). Energy consumption can vary also between farms and broiler houses within the country. Energy consumption varies due to different housing systems, management practices, devices, and local climate. Figure 8.10 shows an example of the share of the total energy consumption in one broiler house (1600 m²) in Finland. In this case the total energy consumption includes heating, electricity and feed. Energy input of feed was, on average, 69% of the total energy consumption. The energy input of heating is probably higher in Finland than in many other countries due to the colder climate.

Data of electricity and heating energy consumption in broiler houses is shown in Table 8.14. Hördahl (2008) measured the direct energy consumption of two broiler houses (100,000 birds per house, target weight of bird 1.5 kg) in Sweden. One house was measured in detail and in the other house only the total electricity consumption was measured. Energy consumption included data from five broiler flocks. Heating energy was clearly the largest direct energy input. Liang et al. (2009) collected direct energy consumption data from four commercial-size (12 m × 121 m) broiler houses in northwest Arkansas. Data was collected for seventeen years. For the first fifteen years broilers were grown in open-curtain houses and after that period houses were converted to enclosed, solid wall systems. More electricity was consumed in an enclosed house than in an open curtain house. Ventilation and lighting were the largest electricity consumers comprising 87% of the total electricity consumption (Liang et al., 2009). Baughman and Parkhurst (1977) monitored energy consumption of two broiler houses (one insulated and fan ventilated and the other uninsulated, with drop curtain side walls) in north Carolina. Energy was monitored during the summer and winter periods (two trials). Result of the study was that the environmentally controlled house (insulated and fan ventilated) consumed less energy to produce the same net

Table 8.14.   Electricity and heating energy consumption in broiler houses.

Item	Consumption [MJ/kg live weight]	Reference
Electricity	0.32	Hörndahl (2008)[1]
	0.37	Liang *et al.* (2009)
	0.45–0.70 (conventional house)	Baughman and Parkhurst,
	1.30–1.53 (environmental house)	(1977)
Heating energy	1.8	Hörndahl (2008)[1]
	4.0, range 1.3–6.8	Rajaniemi and Ahokas (2012)
	1.5	Liang *et al.* (2009)
	1.5–9.2 (conventional house)	Baughman and Parkhurst,
	0.9–4.9 (environmental house)	(1977)

[1] Electricity consumption was presented per bird. Converted per kg of live weight using 1.5 target weight.

weight than the conventional house (uninsulated with drop curtain side walls). Broiler production is probably more energy efficient nowadays than almost 40 years ago. Broiler breeding has improved weight gain, which has reduced growing period.

8.4.4.3 *Lighting*
The energy consumption of lighting depends on the lighting program, dimming of lights, housing type and energy efficiency of bulbs. Continuous or near continuous lighting (1 h dark and 23 h light) is commonly used through the growing period in many countries. This system is assumed to maximize feed intake and growth. However, when the length of the light period is increased, mortality and leg problems also increase. Nowadays, increasing interest for birds' welfare and risen energy prices have increased the use of different lighting programs or a shorter light period. (Lewis and Morris, 2006) Continuous or almost continuous lighting is no longer allowed in broiler houses in the European Union. The EU Council's welfare directive for meat chicken (43/2007) includes minimum rules for lighting in broiler houses (European Union, 2007). The directive includes minimum stipulation in light intensity, day length and length of dark period. Different lighting programs are still used around the world. Length of dark and light cycles varies in different lighting programs.

Length of dark and light cycles has a direct effect on the energy consumption of lighting. Figure 8.28 shows an example of lighting program for a flock at one Finnish broiler farm (Rajaniemi and Ahokas, 2012). The growing period was, on average, 38 days and during this time birds reached 1.8 kg carcass weight (carcass weight is 73.25% of live weight). The lighting program of the farm follows the EU council rules of lighting. Dimmable fluorescent tube lights (36 W) were used for lighting in the case farm. Continuous lighting is used for the first two days of the growing period to enable the one-day old chicks to find the feeding and drinking places. From three days of age, light intensity is decreased stepwise. Every 24 hours cycle includes at least 6 hours' dark period. Length of the dark period directly influences the energy consumption, but also dimming of lights significantly decreased the energy consumption of the broiler house. The electricity consumption of lighting was on average 0.01 kWh (0.04 MJ) per carcass weight and about 500 kWh (1800 MJ) per flock. It was largest at the beginning of the growing period. Differences in energy consumption between flocks were minor. (Rajaniemi and Ahokas, 2012) Liang *et al.* (2009) also reported about a 500 kWh electricity consumption per flock in same size building using dimmable fluorescent lights. The electricity consumption data of lighting included only two broiler batches in both cases. Hörndahl (2008) reported a much higher electricity consumption of lighting, 9790 kWh per flock and 100,000 birds. Electricity consumption per target weight of the birds is still high (0.065 kWh per kilogram of live weight).

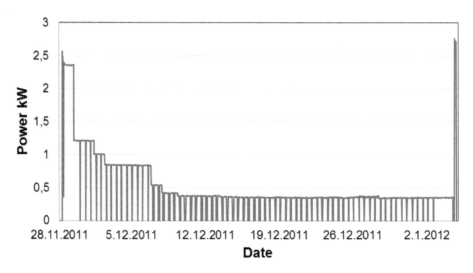

Figure 8.28.   Power of lights in a broiler house (Rajaniemi and Ahokas, 2012).

Type of light has a large effect on the electricity consumption of lighting. Energy efficient bulbs save energy and money. Only about 10% of incandescent light power is converted to visible light (Lewis and Morris, 2006). Incandescent bulbs also attract insects and get dirty. The rest of the light power is emitted as heat (Lewis and Morris, 2006). A more energy efficient lighting solution in broiler houses is fluorescent tube lights or newer, fast developing LED lights (light emitting diode). In addition the lifetime of LED lights is many times longer than the lifetime of incandescent or fluorescent lights. Liang *et al.* (2009) noticed that energy consumption of broiler house lighting decreased from 60 to 75% by replacing incandescent lights with dimmable fluorescent lights and cold cathode fluorescent lights. For example, the Commission of European Community has given a regulation (244/2009) which defines the requirements for energy efficiency of lamps. In practice, this means that production and importation of incandescent lamps is prohibited in the EU (European Union, 2009).

Electricity consumption of lighting is typically lower in broiler houses with curtain sidewalls than in totally enclosed broiler houses (e.g., Liang *et al.*, 2009; Tabler, 2007). The reason for the lower electricity consumption of lighting in curtain sidewall houses is the partial use of natural light.

8.4.4.4   *Ventilation*

Ventilation is usually the biggest electricity consumer in a broiler house (e.g., Liang *et al.*, 2009; Rajaniemi and Ahokas, 2012) if mechanical ventilation is used. However, natural ventilation is used in some broiler houses. This decreases electricity consumption of ventilation compared to the farms which use only mechanical ventilation. Energy consumption depends on the ventilation system, energy efficiency of fans, maintenance practices and accessories of fans. According to Ford *et al.* (2001) maintenance practices and fan accessories have the largest effect on energy efficiency and performance of fans. Electricity is used more in warm and hot seasons than in colder seasons. Electricity consumption is also smaller in the beginning of the growing period, when the ventilation demand of birds is smaller. Hörndahl (2008) reported a 3340 kWh electricity consumption per 100,000 birds in Sweden. This means about 0.08 MJ per kg of live weight. Rajaniemi and Ahokas (2012) obtained similar results in Finland. Electricity consumption of ventilation was 0.09 MJ per kg of live weight. Electricity consumption is usually lower in open curtain systems than in enclosed systems. In enclosed houses additional electricity is needed especially during the summertime, when additional ventilation is needed.

There is a wide variety in the energy efficiency of fans. MSWP (1999) reported that energy efficiency of fans (in pressure of 25 Pa) varied from 14.1 to 31.6 m³/h/W for a 91 cm diameter fan and from 7.82 to 36.36 m³/h/W for a 122 cm diameter fan. Energy efficiency of fans has a direct influence on electricity consumption. If the least energy efficient fans are used in a broiler house compared to the most energy efficient fans, more electricity is needed to change the same amount of air.

Fan accessories (e.g. shutters, guards) and air ducts increase pressure drop and reduce air flow and energy efficiency of ventilation, whereas discharge cones and well-designed housing can increase energy efficiency (15% or more) (Ford *et al.*, 2001). Shutters are commonly used in tunnel ventilated houses to prevent backflow through the fans when not running (Simmons and Lott, 1997) and guards are used mainly for safety reasons (Ford *et al.*, 2001). Guards reduce energy efficiency and airflow less than 5%. Shutters get dirty quickly and need cleaning at regular intervals. Accumulated dust causes the shutters to work poorly. Air flow decreases and more fans are needed to do the same job (Czarick and Lacy, 1993). Simmons and Lott (1997) reported a 10.9% decrease in airflow when popular sized fan (91.4 cm) was used with clean shutters. After a 42 day growing period, 0.454 kg of dust and dirt was accumulated on a fan and its shutter causing a 16.3% reduction in airflow. This reduction is even larger if fans are not cleaned between growing periods. Ford *et al.* (2001) reported that shutters reduce airflow and efficiency of fans from 10 to 25% and dirty shutters even 40%. Also fan propellers should be cleaned because dirt will decrease also their performance. Also loosen or worn belts can cause slippage resulting in decrease of fan performance and efficiency (Ford *et al.*, 2001; MWPS-32, 1990).

8.4.4.5 *Heating*

Broilers achieve best growth when ambient temperature is in the thermoneutral zone. Hence, temperature of the broiler house has an important effect on birds' feed consumption and growth. Especially in hot climates, the temperature of a broiler house could be a problem because a high ambient temperature decreases the feed intake and weight gain. This causes a declination in the feed conversion ratio (Donkoh, 1989) The broiler house is usually heated to 32°C when day-old chicks arrive. Temperature is decreased stepwise into 21°C when birds reach market weight. For this reason, heating energy consumption is higher mainly in the beginning of the growing period when heating demand is higher (e.g. Liang *et al.*, 2009) and the birds also produce less heat.

Heating is the largest direct energy input in broiler houses in many countries (e.g. Barber *et al.*, 1989; Hörndahl, 2008; Rajaniemi and Ahokas, 2012). Variation in the heating energy consumption between houses depends on climate, type of house, insulation and size of house. Heating energy consumption is naturally higher during colder seasons than warmer seasons when the difference between inside and outside temperature is higher. Heat losses through the structures and ventilation also affect heating energy consumption. Poor insulation increases heat losses.

Figure 8.29 shows the cumulative energy consumption of seven broiler flocks in Finland. Heating energy consumption is significantly higher in colder seasons than in warmer seasons. It was on average 4.0 MJ per kg of live weight (varies from 1.3 to 6.8 MJ) (Rajaniemi and Ahokas, 2012). Table 8.14 shows that heating energy consumption varies in different studies. Energy consumption in different countries can vary depending on weather conditions.

Previous studies (e.g., Liang *et al.*, 2009, Xin *et al.*, 1993) showed that the best way to save heating energy in a broiler house is to minimize heat losses. Large amount of heat is lost through the exhaust air. Heat losses through ventilation air depend on temperature differences between outside and indoor temperature. Heat dissipation is challenging especially during the colder seasons. It is possible to decrease heat losses by using heat recovery. This system usually increases electricity consumption, but at the same time it decreases heating energy use. However, dust and condensation (freezing) are usually problems in this system (Bucklin *et al.*, 1992). It is possible to reduce the heat fluctuation through the walls, ceiling and floor by using good insulation. Energy efficiency of heating also depends on the heating system.

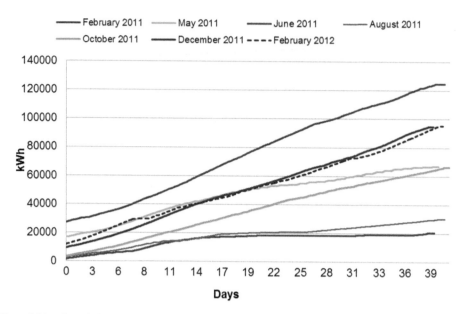

Figure 8.29.   Cumulative energy consumption of seven broiler flocks in Finland. Energy consumption includes pre-heating and heating during the growing period (Rajaniemi and Ahokas, 2012).

8.4.4.6   *Feed and feeding*

Feed is the biggest energy input and expenditure on broiler farms. Previous studies (e.g., AGREE, 2012; Atilgan and Koknaroglu, 2006; Barber *et al.*, 1989; Baughman and Parkhurst, 1977; Katajajuuri *et al.*, 2006; Woods *et al.*, 2010) showed that the feed's share of total energy input in broiler production varies in different countries. Differences in analyzing methods and production circumstances have an effect on the share of energy input of feed.

A large amount of energy is used to produce, process and transport the broiler feed. The processing and transportation energy input of feed is usually rather small compared to the energy input of grain production. Most of the feed production energy is consumed indirectly as agrochemicals (mainly nitrogen fertilizers), but a large amount of energy is also used as fuel for threshing, sloughing and for grain drying or cooling (e.g. Mikkola, 2012). Corn and wheat are commonly used broiler feed around the world (Leeson and Summer, 2005; Sulhattin, 2003). Wheat is mainly used as poultry feed in the European Union and corn in Brazil, China and United States of America (Gerber *et al.*, 2007). The energy consumption of grain production (per kg of grain) varies in different countries due to different climates, yield level and production intensity. Piringer and Steinberg (2006) reported that previous studies showed the energy consumption of wheat production varies from 3.1 to 4.9 MJ per kg of wheat in the US Meyer-Aurich *et al.* (2012) calculated the energy consumption of wheat production in Germany. Energy consumption was 2.5 MJ per kg of wheat. Kraatzt *et al.* (2008) reported energy consumption of corn production 2.7 MJ per kg of corn in Germany and from 1.5 to 1.8 MJ per kg of corn in Wisconsin. Artificial drying is the main reason for higher energy consumption of corn production in Germany.

Feed is known to be the largest energy input in broiler production, but how efficient feed users are broilers themselves? Pimentel (1980) reported already over thirty years ago that broilers are one of the most efficient production animals to convert feed energy into meat protein. Modern day broilers achieve market weight faster than a few decades ago. Leeson and Summers (2005) reported that over the past 20 years, the annual increase in broiler body weight at the age of 42 days is about 25 g. This growth has been even faster during the last few years (Leeson and Summers, 2005). Broilers use feed for growth and maintenance. Leeson and Summers (2005) noticed that better feed efficiency of broilers is a simple consequence of reduced need of maintenance energy. Probably the use of more environmentally controlled (precise temperature and ventilation control) houses are one reason. Stocking density is one major factor which affects feed intake and growth. If the stocking density is high the birds eat less.

Efficiency of feed utilization is an important economical aspect in broiler production. It has also an effect on energy efficiency when different animals are compared. Feed efficiency is usually calculated by dividing the body weight by feed intake. Broilers' feed conversion ratio (feed intake/body weight of broiler) was about 2.2 in early 1960 and now it is 1.75. (Leeson and Summers, 2005) Hence, less feed is needed nowadays to produce broiler than about 50 years ago. Mikkola *et al.* (2002) reported a 3.4 kg feed consumption per broiler and water consumption of about 6.5 liters per bird during the 37 day growing period. Feddes *et al.* (2002) reported approximately 3 kg feed consumption per bird and water consumption of 54 L per bird during the 38 and 42 day growing periods. Feed conversation ratio was about 1.7. Previous studies showed that water consumption is almost twice as much than feed consumption.

Energy can be saved in broiler production:

- Feed is the main energy input in broiler production. By improving feed production, energy efficiency and also broiler production efficiency increases.
- Ventilation heat recovery could be used in heating of a broiler house.
- Good insulation decreases demand for heating energy.
- Energy efficient lights and fans use less electricity than less energy efficient choices.
- Regular maintenance and cleaning of devices save energy.
- Ventilation rate changes with the changing circumstances (chicken weight, humidity, temperature, gases). The ventilation rate control should work according to microclimate of the building (temperature, humidity, gases). Ventilation rate, which is adjusted according to the animal needs, guarantees a good microclimate and low energy consumption.
- Optimal growing circumstances assure the best feed efficiency of broilers.

8.5 CONCLUSIONS

Energy consumption in livestock production can be reduced by fine tuning the production in an energy efficient way. At the moment, economics and work load mostly dictate the production system. The economical influence on the energy consumption will be higher in the future, meaning that energy will be one key element in the economy of the farm.

The energy analysis methodologies need standardization. There are too many unsolved questions on how to handle different production systems and what is included and what is excluded from the analysis.

To save energy the farmers need to have energy bookkeeping and do energy analysis. The energy consumption of machines is not however easy to follow because in most cases energy consumption is measured by larger units, not by individual machines. In future livestock and other agricultural machinery could have an option for individual energy metering or during electricity assemblies also individual energy metering is taken into account.

In intensive animal production most energy is used in the fodder production and there mostly for fertilizer (nitrogen) manufacturing. With proper manure handling the nutrients can be partly recycled to the fields. Cultivation of nitrogen binding plants also reduces need for nitrogen.

Cows can tolerate cold weather, which means that heating of cow houses is unnecessary. In cold climates, it must be remembered that manure and drinking water must not be frozen.

Animals, which do not tolerate cold weather, must have warm houses. This means in many cases heated houses. Most of the energy in these houses is however lost through ventilation. With heat exchangers this energy could be gained and extra heating is not always necessary.

Livestock machinery also consumes energy. The energy efficiencies of this machinery can vary considerably and energy can be saved with proper control and production systems.

Livestock GHG emissions are originate from the animal (methane and ammonia), from the usage of nitrogen as fertilizer (nitrous oxides) and from direct energy use (carbon dioxide). These emissions will increase because more food is needed to feed the increasing population. Animal emissions can be reduced with change of feed composition and with the proper handling of manure. Nitrous oxide emissions need more research to be solved. Direct carbon dioxide emissions can be reduced by reducing energy use or using biofuels. It must be noted that, at the moment, emissions do not have any influence on farm economy.

REFERENCES

AGREE: State of the art on energy efficiency in agriculture. Country data on energy consumption in different agro production sectors in the European countries. 2012, http://www.agree.aua.gr/index-4.html (accessed February 2013).

Ahokas, J. (ed): Maatilojen energiankäyttö. Enpos-hankkeen tulokset. (In Finnish, Energy use on farms. Results of ENPOS-project), Department of Agricultural Sciences, Publications 15. Univerisity of Helsinki, Helsinki, Finland, 2013.

Alfaro-Arguello, R., Diemont, S.A.W., Ferguson, B.G., Martin, J.F., Nahed-Toral, J., Álvarez-Solís, J. D. & Ruíz, R.P.: Steps toward sustainable ranching: An emergy evaluation of conventional and holistic management in Chiapas, Mexico. *Agric.Sys.* 103 (2010), pp. 639–646.

ASAE: EP344.3. Lighting systems for agricultural facilities. American Society of Agricultural and Biological Engineers, ASABE Standards ASAE EP344.3 JAN2005, R2010, St. Joseph, MI, USA.

Atilgan, A. & Koknaroglu, H.: Cultural energy analysis on broiler reared in different capacity poultry houses. *Ital. J. Anim. Sci.* (2006), pp. 8393–8400.

Barber, E.M., Classen, H.L. & Thacker, P.A.: Energy use in the production and housing of poultry and swine – an overview. *Can. J. Anim. Sci.* 69 (1989), pp. 7–21.

Bastianoni, S., Pulselli, R. & Pulselli, F.: Models of withdrawing renewable and non-renewable resources based on Odom's energy systems theory and Daly's quasi-sustainability principle. *Ecol. Mod.* 220 (2009), pp. 1926–1930.

Baughman, G.R. and Parkhurst, C.R.: Energy consumption in broiler production. *Transact. ASAE* 20:02 (1977), pp. 341–344.

Bauman, D. E., & Capper, J. L.: Future challenges and opportunities in animal nutrition. *Proc. Southwest Nutr. Management Conf.*, 2011, pp. 70–84.

Bokkers, E.A.M., van Zanten, H.H.E. & van den Brand, H.: Field study on effects of a heat exchanger on broiler performance, energy use, and calculated carbon dioxide emission at commercial broiler farms, and the experiences of farmers using a heat exchanger. *Poultry Sci.* 89 (2010), pp. 2743–2750.

Bos, J.F.F.P., de Haan, J.J., Sukkel, W. & Schils, R.L.M.: Comparing energy use and greenhouse gas emissions in organic and conventional farming systems in The Netherlands. *3rd QLIF Congress*, Hohenheim, Germany, 2007, http://orgprints.org/9961/1/Bos-etal-2007-EnergyGreenhouse.pdf (accessed February 2013).

Brand, R.A. & Melman, A.G.: Energie inhoudnormen van de veehouderij; deel 2 proceskaarten, (Energy values of inputs of animal production). TNO, Instituut voor milieu- en energietechnologie, Apeldoorn, The Netherlands. 1993, http://www.fao.org/WAIRDOCS/LEAD/X6111E/x6111e05.htm (accessed February 2013).

Brown, M. & Ulgiati, S.: Emergy analysis and environmental accounting. In: C. Cleveland (ed): *Encyclopedia of Energy*. Academic Press, Elsevier, Oxford, UK, 2004, pp. 329–354.

Bucklin, R.A., Nääs, I.A. & Panagakis, P.B.: Energy use in animal production. In: R.C. Fluck: Energy in farm production. *Energy in World Agriculture*, 6. 1992, pp. 257–266.

Capper, J.L., Cady, R.A. & Bauman, D.E.: The environmental impact of dairy production: 1944 compared with 2007. *J. Anim. Sci.* 87 (2009), pp. 2160–2167.

Castellini, C., Bastianoni, S., Granai, C., Dal Bosco, A. & Brunetti, M.: Sustainability of poultry production using the emergy approach: Comparison of conventional and organic rearing systems. *Agric. Ecosy. Env.* 114 (2006), pp. 343–350.

Cederberg, C. & Mattson, B.: Life cycle assessment of milk production – a comparision of conventional and organic farming. *J. Clean. Product.* 8 (2000), pp. 49–60.

Crill, R., Hanchar, J., Gooch, C. & Richards, S.: Net present value economic analysis model for adoption of photoperiod manipulation in lactating cow barns. Cornell University, Department of Animal Science, 2002, http://www.ansci.cornell.edu/pdfs/photoperiod.pdf (accessed July 2013).

Czarick, M. & Lacy, M.: Exhaust fans and shutter opening devices. *Poultry Housing Tips* 5:3 (1993), The University of Georgia Cooperative Extension Service, http://www.poultryventilation.com/sites/default/files/tips/1993/vol5n3.pdf (accessed August 2013).

DairyCo: Dairy housing – a best practice guide. 2012, DairyCo, a division of the Agriculture and Horticulture Development Board, Warwickshire, UK, http://www.dairyco.org.uk/resources-library/technical-information/buildings/dairy-housing-a-best-practice-guide/ (accessed July 2013).

DairyNote: Energy partitioning. AgroMedia International Inc, Calgary Canada, 1997, http://www.agromedia.ca/ADM_Articles/content/f1r1e1.pdf (accessed February 2013).

Dalgaard, T.: On-farm fossil energy use. *IFOAM Ecology and Farming*, January–April (2003), p. 9, http://www.okoforsk.dk/funktion/vidsyn/energi/efjan03.pdf (accessed 2 February 2013).

da Silva V.P., van der Werf, H. & Soares S.R.: LCA of French and Brazilian broiler poultry production scenarios. *XIIIth European Poultry Conference*, 23–27 August 2010, Tours, France, 2010.

Dong, X.; Yang, W.; Ulgiati, S. Yan, M. & Zhang, X.: The impact of human activities on natural capital and ecosystem services of natural pastures in North Xinjiang, China. *Ecol. Modelling* 225 (2012), pp. 28–39.

Donkoh, A.: Ambient temperature: a factor affecting performance and physiological response of broiler chickens. *Int. J. Biometeorol.* 33 (1989), pp, 259–265.

Dunn, P., Butler, G., Bilsborrow, P., Brough, D. & Quinn, P.: Energy + efficiency, renewable energy and energy efficiency options for UK dairy farms. Newcastle University, 2010, http://www.morrisons.co.uk/Global/0_FarmingPage/Energy%20Efficiency%20Options%20for%20UK%20Dairy%20Farms.pdf (accessed 12 February 2013).

EFSA: Scientific report on the effects of farming systems on dairy cow welfare and disease. Report of the Panel of Animal Health and Welfare. Annex to the *EFSA Journal* (2009) 1143, p. 284, http://www.efsa.europa.eu/en/efsajournal/doc/1143r.pdf (accessed February 2013).

European Union: Council Directive 2007/43/EC of 28 June 2007 laying down minimum rules for the protection of chickens kept for meat production. *Official Journal of the European Union* (2007), Brusseles, Belgium.

European Union: European Commission Regulation (EC) No 244/2009 of 18 March 2009 implementing Directive 2005/32/EC of the European Parliament and of the Council with regard to ecodesign requirements for non-directional household lamps. *Official Journal of the European Union* (2009), Brusseles, Belgium.

EUROSTAT: Pig farming statistics. FAO, Rome, Italy, undated, avialable at http://epp.eurostat.ec.europa.eu/statistics_explained/index.php/Pig_farming_statistics#Production_of_pigmeat (accessed January 2013).

FAO: The State of Food and Agriculture. Livestock in the balance (2009). The state of food and agriculture. FAO, Rome, Italy, 2009, http://www.fao.org/docrep/012/i0680e/i0680e00.htm (accessed January 2013).

FAO: Greenhouse gas emissions from the dairy sector. FAO, Rome, Italy, 2010, http://www.fao.org/docrep/012/k7930e/k7930e00.pdf (accessed 4 January 2013).

FAO: 'Energy-smart' food for people and climate. FAO Issue paper. FAO, Rome, Italy, 2011, http://www.fao.org/docrep/014/i2454e/i2454e00.pdf (accessed January 2013).

FAO: Production. Livestock primary. FAO, Rome, Italy, 2013, http://faostat3.fao.org/home/index.html (accessed January 2013).

FAO Database 2013: avilable at http://faostat3.fao.org (accessed January 2013).

FAOSTAT: Food and agricultural commodities production, top production-indigenous chicken meat 2011. FAO, Rome, Italy, 2011, http://faostat.fao.org/site/339/default.aspx (accessed January 2013).

Feddes, J.J.R., Emmanuel, E.J. & Zuidhof, M.J.: Broiler performance, body weight variance, feed and water intake and carcass quality at different stocking densities. *Poultry Sci.* 81 (2002), pp. 774–779.

Ford, S.E., Riskowski, G.L., Christianson, L.L. & Funk, T.L.: Agricultural ventilation fans, performance and efficiencies. BessLab, Dept. of Agricultural Engineering, University of Illinois, Urbana-Champaign, IL, 2001.

Gerber, P., Opio, C. & Steinfeld, H.: Poultry production and the environment – a review. 2007, http://www.fao.org/ag/againfo/home/events/bangkok2007/docs/part2/2_2.pdf (accessed January 2013).

Grönroos, J., Seppälä, J., Voutilainen, P., Seuri, P., & Koikkalainen, K.: Energy usage in conventional and organic milk and rye bread production in Finland. *Agri. Ecosyst. Environ.* 177 (2006), pp. 109–118.

Gustafson, R.J. & Morgan, M.T.: *Fundamentals for electricity for agriculture.* 3rd edition. American Society of Agricultural Engineers, St Joseph, MI, USA, 2004.

Gutiérrez del Álamo Oms, Á.: *Factors affecting wheat nutritional value for broiler chickens.* 2009. PhD Thesis, Wageningen University, Wageningen, The Netherlands, http://edepot.wur.nl/4262 (accessed 9 January 2013).

Hartman, K. & Sims, R.E.H.: Saving energy on the dairy farm makes good sense. *Proceedings of the 4th Dairy3 Conference,* Hamilton New Zealand. Centre for Professional Development and Conferences, Massey University, Palmerston North, New Zeland, 2006, pp. 11–22.

Heidari, M.D., Tabatabaeifar, A., Omid, M, & Akram, A.: An investigation on the effects of number of chicks and ventilation system type on energy efficiency in yazd broiler farms. *Concrete for a Sustainable Agriculture. 7th International Symposium,* September 18–21 2011, Québec, Canada, pp. 142–147.

Heilig. G.K.: Lifestyles and energy use in human food chains. Working paper WP-93-14. IlASA International Institute for Applied Systems Analysis, A-2361, Laxenburg, Austria, 1993, http://www.iiasa.ac.at/publication/more_WP-93-014.php (accessed January 2013).

Henderson, S.M., Perry, R.L. & Young J.H.: Principles of process engineering. 4th ed. American Society of Agricultural Engineers, St Joseph, MI, 1997.

Honeyman, M.: Sizing up pork production's energy use. 2013, http://www.swineweb.com/sizing-up-pork-production%E2%80%99s-energy-use/ (accessed January 2013).

Hördahl, T.: Energy use in farm buildings – a study of 16 farms with different enterprises. Swedish University of Agricultural Sciences, Faculty of Landscape Planning, Horticulture and Agricultural Science, Report 2008:8, Alnarp, Sweden, 2008.

IDF: A common carbon footprint approach for dairy. The IDF guide to standard lifecycle assesment methodology for the dairy sector. Bulletin of the International Dairy Federation 445/2010. Interantional Dairy Federation, Brussels, Belgium, 2010.

ISO 14040: Life cycle assessment. Principles and framework. International Organization for Standardization, Geneva, Switzerland, 2006.

IPCC: IPCC Guidelines for national greenhouse gas inventories 1996. Intergovernmental Panel on Climate Change, Geneva, Switzerland, 1996, http://www.ipcc-nggip.iges.or.jp/public/gl/invs1.html (accessed January 2013).

Jones, M.R.: Analysis of the use of energy in agriculture – approaches and problems. *Agric. Syst.* 29:4 (1989), pp. 339–355.

Karlsson, E., Hörndahl, T., Pettersson, O. & Nordman, R.: Energiåtervinning från mjölkkylning (In Swedish, Energy recover from milk cooling). Rapport 401, Landtbruk & Industri, JTI – Institutet för Jordbruks-och Miljö Teknik, Uppsala, Sweden, 2012.

Katajajuuri, J.M., Grönroos J., Usva, K., Virtanen, Y., Sipilä, I., Venäläinen, E., Kurppa, S., Tanskanen, R., Mattila, T. & Virtanen, H.: Environmental impacts and improvement options of sliced broiler fillet production. MTT. *Maa-ja elintarviketalous* 90 (2006), http://www.mtt.fi/met/pdf/met90.pdf (in Finnish), (accessed January 2013).

Kingston, C., Meyhoff-Fry, J. & Aumonier S.: Scoping life cycle assessment of pork production. Environmental resources management AHDBS, Final report. Environmental Resources Management Limited, 2009, http://www.pigprogress.net/PageFiles/20858/001_boerderij-download-PP5978D01.pdf (accessed August 2013).

Kraatz, S.: Energy intensity in livestock operations – Modelling of dairy farming systems in Germany. *Agric. Syst.* 110 (2012), pp. 90–106.

Kraatz, S., & Berg, W.E.: Energy efficiency in raising livestock at the example of dairy farming. *ASABE Annual International Meeting,* Grand Sierra Resort and Casino Reno, Nevada, 21–24 June 2009, ASABE Meeting Presentation Paper Number: 096715, 2009.

Lammers, P.: *Energy and nutrient cycling in pig production systems.* PhD Thesis Iowa State University, Ames, IA, 2009, http://lib.dr.iastate.edu/cgi/viewcontent.cgi?article=1609&context=etd (accessed January 2013).

Leeson, S. & Summers J.D.: *Commercial poultry nutrition.* 3rd edition, Nottingham University Press, Nottingham, UK, 2005.

Lewis, P. & Morris, P.: Poultry lighting the theory and practice. Northcot, Hampshire, UK, 2006.

Liang, Y., Tabler. G.T., Watkins, S.E., Xin, H. & Berry, I.L.: Energy use analysis of open-curtain vs. totally enclosed broiler houses in northwest Arkansas. *Appl. Eng. Agr.* 25:4 (2009), pp. 577–584.

Lindley, J.A. & Whitaker J.H.: *Agricultural buildings and structures.* Revised edition, American Society of Agricultural Engineers, St Joseph, MI, 1996.

Ludington, D. & Johnson, E.: Dairy farm energy audit summary report. New York State Energy Research and Development Authority, New York, 2003.

Lundshøj-Dalgaard, R.: The environmental impact of pork production from a life cycle perspective. University of Aarhus Faculty of Agricultural Sciences, Aarhus, Denmark, 2007.

Mannfors, B. & Hautala, M.: Eläinten hyvinvointiin perustuva tuotantorakennusten mikroilmasto: Ilmanvaihtoon ja lämpötilaan liittyvät suositukset. Ilmanvaihtolaskurit. (Animal welfare and livestock building microclimate: Ventilation and temperature recommendations. Ventilation calculators). *Department of Agricultural Sciences. Publication 6*, Maataloustiteiden laitos, Helsingin yliopisto. 2011, (in Finnish), avialable at http://www.helsinki.fi/maataloustieteet/tutkimus/agtek/proj/karva/index.html (accessed July 2013).

Meul, M., Nevens, F., Reheul, D. & Hofman, G.: Energy use efficiency of specialised dairy, arable and pig farms in Flanders. *Agric. Ecosys. Env.* 119 (2007), pp. 135–144.

Meyer-Aurich, A., Ziegler, T.H. Jubaer, H., Scholz, L. & Dalgaard, T.: Implications of energy efficiency measures in wheat production. *International Conference of Agricultural Engineering CIGR-Ageng* 2012, 8–12 July 2012, Valencia, Spain, 2012.

Mikkola, H.: Field bioenergy production in Finland. *Department of Agricultural Sciences. Publication 10*, Helsinki, Finland, 2012 (in Finnish).

Mikkola, H.J. & Ahokas, J.: Energy ratios in Finnish agricultural production *Agri. Food Sci.* 18 (2009), pp. 332–346.

Mikkola, H., Puumala, M., Kallioniemi, M., Grönroos, J., Nikander, A., & Holma, M.: Best available techniques in livestock farmin in Finland. *The Finnish Environment* 564, 2002, http://www.ymparisto.fi/download.asp?contentid=4582 (in Finnish, accessed January 2013).

MMM RMO C2.2: Maatalouden tuotantorakennusten lämpöhuolto ja huoneilmasto (In Finnish, Agricultural production buildings, heat production and micro climate). Ministry of Agriculture and Forestry, Helsinki, Finland.

MWPS-32: *Mechanical ventilation systems for livestock housing.* First edition, Iowa State University, Ames, IA. MidWest Plan Service, 1990.

Neiber, J. & Neser, S.: Energy consumption and energy saving potentials in piglet production. *Landtechnik* 65:6 (2010), pp. 421–425.

Neuman, L.: Kartläggning av energianvändning på lantbruket 2008. (In Swedish, Survey of energy is in agriculture 2008). LRF KONSULT AB, Stockholm, Sweden, 2008.

Nielsen, N.I., Jørgensen, M. & Bahndorff, S.: Greenhouse gas emission from the Danish broiler production estimated via LCA methodology. Knowlede Center for Agriculture, Aarhus, Denmark, 2011.

Norton, G.W., Alwang, J. & Masters, W.A.: *Economics of agricultural development world food: systems and resource use.* Routledge, Oxford, UK, 2006.

Odum, H.: *Environmental accounting, emergy and decision making.* John Wiley, NY, USA 1996.

OECD/FAO: OECD-FAO Agricultural Outlook 2011–2020, OECD Publishing and FAO, 2011, http://dx.doi.org/10.1787/agr_outlook-2011-en (accessed August 2013).

OECD/FAO: OECD-FAO Agricultural Outlook 2012–2021. OECD publishing and FAO, 2012, http://dx.doi.org/10.1787/agr_outlook-2012-en. (accessed August 2013).

Pelletier, N.: Environmental performance in the US broiler poultry sector: lifecycle energy use and greenhouse gas, ozonedepleting, acidifying ande utrophying emissions. *Agr. Syst.* 98 (2008), pp. 67–73.

Pimentel, D.: *Handbook of energy utilization in agriculture.* CRC Press, Boca Raton, FL, 1980.

Pimentel, D.: Livestock production and energy use. In. R. Matsumura (ed): *Encyclopedia of energy.* Elsevier, San Diego, CA. 2004, pp. 671–676.

Piringer, G. & Steinberg, L.J.: Reevaluation of energy use in wheat production in the United States. *J. Indus. Ecol.* 10:1–2, pp. 149–167.

Qotbi, A.A.A., Najafi, S., Ahmadauli, O., Rahmatnejad, E. & Abbasinezhad, M.: Investigation of poultry housing capacity on energy efficiency of broiler chickens production in tropical areas. *Afr. J. Biotechnol.* 10:69 (2011), pp. 15,662–15,666.

Rajaniemi, M. & Ahokas, J.: Energy consumption in broiler production. *International Conference of Agricultural Engineering, CIGR-AgEng2012*, Valencia, Spain, 2012, http://cigr.ageng2012.org/images/fotosg/tabla_137_C1277.pdf (accessed January 2013).

Refsgaard, K., Halberg, N. & Kristensen, E.S.: Energy utilization in crop and dairy production in organic and conventional livestock production systems. *Agric. Syst.* 57:4 (1998), pp. 599–630.

Ro'tolo, G.C., Rydberg, T. Lieblein, G. & Francis, C.: Emergy evaluation of grazing cattle in Argentina's pampas. *Agric. Ecosys. Env.* 119 (2007), pp. 383–395.

Rydberg, T., Haden, A.: Emergy evaluations of Denmark and Danish agriculture: Assessing the influence of changing resource availability on the organization of agriculture and society. *Agric. Ecosys. Envir.* 117 (2006), pp. 145–158.

Sainz, R.D.: Fossil fuel component. Framework for calculating fossil fuel use in livestock systems. 2003, University of California, Davis, CA, USA, http://www.fao.org/WAIRDOCS/LEAD/X6100E/x6100e00.htm#Contents (accessed February 2013).

Seguin, F., van der Werf, H., Bouvarel, I. & Pottiez, E.: Environmental analysis of organic broiler production in France and improvement options. *LCM 2011 Conference*, Berlin, 2011.

Seifert, C., Wietzke, D. & Fritzsche, S.: Energy for heating and ventilation in pig production on farms. *Landtechnik* 6, 2009.

Simmons, J.D. & Lott, B.D.: Reduction of poultry ventilation fan output due to shutters. *Appl. Eng. Agr.* 13:5 (1997), pp. 671–673.

Smil, V.: *Energy in nature and society- general energetic of complex systems*. MIT Press, Cambridge, MA, 2008.

Smith, J., Harner, J., Dunham, D., Stevenson, J., Shirley, J., Stokka, G. & Meyer, M.: Coping with summer weather. Dairy Management Strategies to Control Heat Stress. Kansas State University, Agricultural Experiment Station and Cooperative Extension Service, 1998, http://www.ksre.ksu.edu/bookstore/pubs/MF2319.pdf (accessed July 2013).

Sulhattin, Y.: Performance, gut size and ileal digesta viscosity of broiler chickens fed with awhole wheat added diet and the diets with different wheat particle sizes. *Int. J. Poultry Sci.* 2:1 (2003), pp. 75–82.

Tabler, G.T.: Applied broiler research farm report: electricity usage before and after renovation. *Avian Advice Newsletter* 9:2 (2007), http://www.thepoultrysite.com/articles/847/applied-broiler-research-farm-report-electricity-usage-before-and-after-renovation (accessed August 2013).

Thomassen, M.A., van Calker, K.J., Smits, M.C.J., Iepema, G.L. & de Boer, I.J.M.: Life cycle assessment of conventional and organic milk production in The Netherlands. *Agric. Syst. 96* (2008), pp. 95–107.

Ukidwe, N.U. & Bakshi, B.R.: Industrial and ecological cumulative exergy consumption of the United States via the 1997 input–output benchmark model. *Energy* 32 (2007), pp. 1560–1592.

van der Sluis, W.: China aims for higher pork production. *Pig Progress* 27 June 2012, 2012.

Videncenter for Svinproduktion: Viden/Stalde/Staldklima/Staldtemperatur. Danish Agriculture & Food Council, Pig Research Centre 2011, http://vsp.lf.dk/Viden/Stalde/Staldklima/Staldtemperatur.aspx (in Danish), (accessed February 2013).

Vigne, M., Martin, O., Faverdin, O. & Peyraud J.L.: Comparative uncertainly analysis of energy coefficients in energy analysis of dairy farms from two French territories. *J. Clean. Product.* 37 (2012), pp. 185–191.

Vuorelainen O.: LVI-tekniikka. Lämmöntarve ja lämmöneristys. (In Finnish, HVAC technology. Heat demand and insulation.) Otapaino Espoo, Finland 1977.

Wells, C.: Total energy indicators of agricultural sustainability: dairy farming case study. Ministry of Agriculture and Forestry, Wellington New Zealand, 2001.

Wiersma, F. & Short, T.: Evaporative cooling. In: M. Hellickson & J. Walker (eds): Ventilation of agricultural structures. *ASAE Monograph* 6, pp. 103–118, American Society of Agricultural Engineers, 1983.

Williams, A.G., Pell, E., Webb, J., Tribe, E., Evans, D., Moorhouse, E. & Watkiss, P.: Final Report for Defra Project FO0103, Comparative Life Cycle Assessment of Food Commodities Procured for UK Consumption through a Diversity of Supply Chains. Department for Environment Food & Rural Affairs, UK, 2008.

Woods, J., Williams, A., Hughes, J.K., Black, M. & Murphy, R.: Energy and the food system. *Phil. Trans. R. Soc.* B 365 (2010), pp. 2991–3006.

Xin, H., Berry, I.L., Barton, T.L. & Tabler, G.T.: Sidewall effects on energy use in broiler houses. *J. Appl. Poultry Res.* 2 (1993), pp. 176–183.

CHAPTER 9

Diesel engine as prime power for agriculture: emissions reduction for sustainable mechanization

Xinqun Gui

9.1 DIESEL ENGINE AS PRIME POWER FOR AGRICULTURE

The world population surpassed seven billion people in 2012. For the first time, in 2012, 50% of the world population live in urban areas, accelerating a trend of population migration from rural to urban areas. In 1900, farmers represented nearly 40% of the labor force in the USA. Today, that number has plunged to well below 2%. In the meantime, energy intensity in agriculture has steadily decreased since World War II (Davidson). Remarkably, the world has adequate food supply to feed ever-increasing world population with less farm labor force. According to the USDA Economic Research Service, the level of the USA farm output in 2009 was 170% above its level in 1948, due in part to mechanization (http:www.ers.usda.gov). It is no wonder that the US National Academy of Engineering has identified agricultural mechanization as one of the greatest engineering achievements in the 20th century (http://www.greatachievements.org/). Center to agricultural mechanization is the internal combustion engine, first introduced to farm tractors in 1902, which provides prime power for tractors, combines and other forms of farm equipment. There are few people, if any, who even remember horse-drawn carriage anymore.

  In the 1930s, the diesel engine was introduced to farm machines to replace gasoline fueled internal combustion engines. Diesel engines provided more power at lower cost than gasoline engines. Today, diesel engines are the standard for almost all agricultural machines, with the exception of small utility equipment. The fuel efficiency of farm machines has increased steadily over the years, as demonstrated in Figure 9.1, which shows relative fuel consumption since 1960 for a class of row crop tractors.

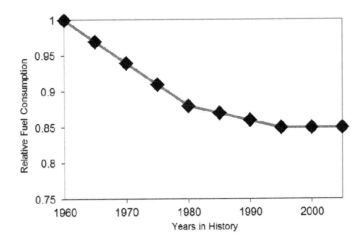

Figure 9.1.   Fuel efficiency improvements of row crop tractors over the years.

The government of the USA enacted Clean Air legislation for the automotive industry in the 1960s. The clean air legislation started to apply to truck and bus industry in the late 1980s and non-road mobile equipment (farm and construction machines) industry in the 1990s. Criteria for vehicle or engine exhaust gas emissions were established for nitric oxide (NO_x), hydrocarbon (HC), carbon monoxide (CO), and particulate matter (PM). These emissions have been widely referred to as criteria pollutants. Following the lead of the USA, emissions regulations have been enacted across the globe. Billions of dollars have been invested in research, engineering development, and manufacturing by the auto and engine industry to reduce criteria pollutants. As a result, we now breathe cleaner air, and live in an overall healthier environment. This chapter focuses on emissions reduction for non-road mobile equipment. In Section 9.2, we highlight global emissions criteria and test procedures for non-road engines and compare them with on-road requirements. In Sections 9.3 and 9.4, we describe critical components and system integration technologies as building blocks for meeting emissions criteria. In Section 9.5, we describe engine development through emissions tiers. We briefly discuss the use of biofuels in Section 9.6. Finally, we conclude with a summary and perspectives. With this focus on emissions, we will limit our discussion of other product attributes, such as reliability, not because they are not important (they are, in fact, critical) but because thorough discussion of those attributes requires lengthy papers on their own.

9.2   GLOBAL NON-ROAD EMISSIONS REGULATIONS

Non-road diesel engine emissions are defined by power category and by tiers that were implemented over time. Tier 1 requirements were first implemented for engines between 130 kW (175 hp) and 560 kW in 1996, with a NO_x limit of 9.2 g/kWh and a particulate matter (PM) limit of 0.54 g/kWh. Tier 1 requirements for other power ranges were implemented between 1997 and 2000, with appropriate limits at different power levels. Figure 9.2 shows a summary of criteria emissions requirements since Tier 1 by power category and by tiers over the time frame from 2001 through 2015 by the US EPA. Over this period, the NO_x limit was reduced from 9.2 to 0.4 g/kWh and the particulate matter limit was reduced from 0.54 to 0.02 g/kWh for most power categories. In addition to the limitations on NO_x and PM, hydrocarbon (HC), particularly non-methane hydrocarbon (NMHC) and carbon monoxide (CO), are also regulated. Figure 9.3 shows the same information in graphical form for a visual perspective. When Tier 4 implementation is complete by 2015, non-road diesel engines will produce near-zero criteria pollutants.

The tiers of regulations have provisions for averaging, banking, and trading emission credits in the USA, commonly referred to as ABT credits, which can result in complex product strategies unique to each manufacturer.

In addition to the 8-mode test of previous tiers, the Tier 4 standard has two added requirements: one is the not-to-exceed (NTE) zone, and the other is the non-road transient test cycle NRTC). The NTE zone is defined by the torque curve, the 100% engine speed, the 15% engine speed, the 30% power line, and the 30% torque line. Figure 9.4 shows an example of a NTE zone. Within the NTE zone, the peak exhaust emissions cannot exceed the cycle emissions times a multiplier, which is 1.5 if the exhaust NO_x emission is below 2.5 g/kWh and 1.25 if the exhaust NO_x is above 2.5 g/kWh over the NRTC.

The non-road transient test cycle (NRTC) is defined by a time sequence of engine speed and torque, as shown in Figure 9.5. The engine must be run in accordance with this time sequence for certification testing. The NRTC must be executed twice: the first run is when engine temperature has been at an equilibrium of 19.5°C (67°F), which is referred to as the cold cycle, and the second run is 20 minutes after the completion of the cold cycle. This second run is referred to as the hot cycle. The reported emission is the weighted sum of 5% from the cold cycle and 95% from the hot cycle for the USA, and 10% and 90% respectively for Europe.

The change from steady-state 8-mode testing to non-road transient testing is quite significant in itself. A fully functional transient test cell that meets EPA certification requirements costs multiple million US dollars to build, requires several months if not longer to commission, and

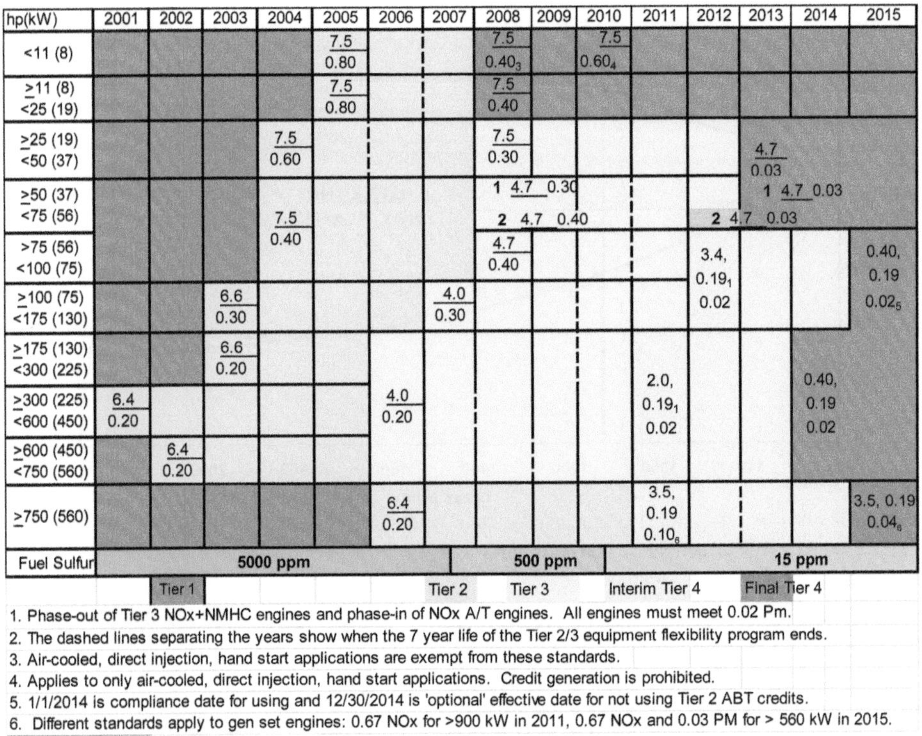

hp(kW)	2001	2002	2003	2004	2005	2006	2007	2008	2009	2010	2011	2012	2013	2014	2015
<11 (8)					7.5 / 0.80			7.5 / 0.40[3]		7.5 / 0.60[4]					
≥11 (8) <25 (19)					7.5 / 0.80			7.5 / 0.40							
≥25 (19) <50 (37)				7.5 / 0.60				7.5 / 0.30					4.7 / 0.03		
≥50 (37) <75 (56)								1 4.7 0.30 / 2 4.7 0.40					1 4.7 0.03 / 2 4.7 0.03		
>75 (56) <100 (75)				7.5 / 0.40				4.7 / 0.40				3.4, 0.19[1] / 0.02			0.40, 0.19 / 0.02[5]
≥100 (75) <175 (130)			6.6 / 0.30			4.0 / 0.30									
≥175 (130) <300 (225)			6.6 / 0.20								2.0, 0.19[1] / 0.02			0.40, 0.19 / 0.02	
≥300 (225) <600 (450)	6.4 / 0.20				4.0 / 0.20										
≥600 (450) <750 (560)		6.4 / 0.20													
≥750 (560)					6.4 / 0.20						3.5, 0.19 / 0.10[6]				3.5, 0.19 / 0.04[6]
Fuel Sulfur	5000 ppm							500 ppm			15 ppm				

Tier 1 | Tier 2 | Tier 3 | Interim Tier 4 | Final Tier 4

1. Phase-out of Tier 3 NOx+NMHC engines and phase-in of NOx A/T engines. All engines must meet 0.02 Pm.
2. The dashed lines separating the years show when the 7 year life of the Tier 2/3 equipment flexibility program ends.
3. Air-cooled, direct injection, hand start applications are exempt from these standards.
4. Applies to only air-cooled, direct injection, hand start applications. Credit generation is prohibited.
5. 1/1/2014 is compliance date for using and 12/30/2014 is 'optional' effective date for not using Tier 2 ABT credits.
6. Different standards apply to gen set engines: 0.67 NOx for >900 kW in 2011, 0.67 NOx and 0.03 PM for > 560 kW in 2015.

Figure 9.2.   US EPA emissions requirement for non-road diesel engines by power category (40 CFR Part 89 and Part 1039 (CFR, 2005).

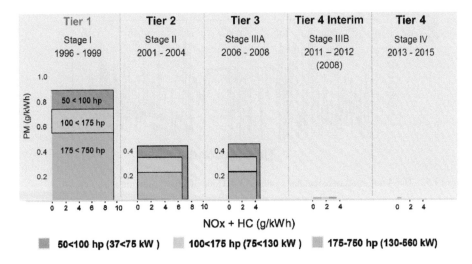

Figure 9.3.   Non-road emissions requirements for diesel engines by tiers.

must be operated and maintained by skilled staff. Details of test procedure and facility guidelines are available in the US Code of Federal Regulations, Title 40 (CFR, 2005).

Figure 9.6 shows the NRTC points overlaid on an engine torque-speed map in comparison with the 8-mode test of the previous tiers. When viewed on the torque-speed map, several clusters of test points can be seen. These clusters represent typical operating duty cycles of non-road

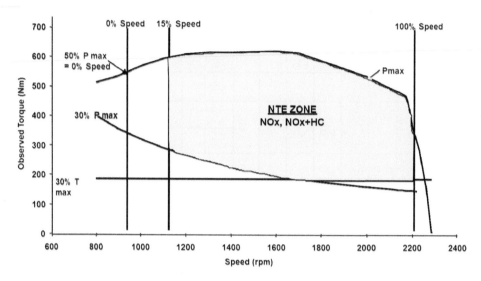

Figure 9.4.   Not-to-exceed (NTE) area definition.

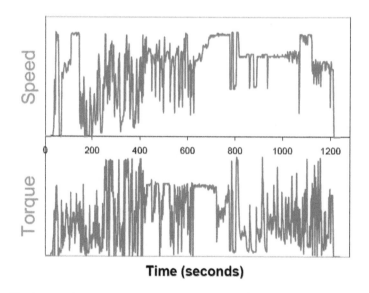

Figure 9.5.   Tier 4 non-road transient test cycle per US Code of Federal Regulations, Title 40.

machine types, such as row crop tractors, combines, loaders, excavators, backhoes, and crawlers. Therefore, NRTC can be viewed as the sum of non-road machine operating profiles. Note that the 8-mode test is still required, and not replaced by NRTC, for Tier 4 engine emissions certification.

The emission standards discussed above must be met over the entire useful life of the engine. The US EPA requires application of deterioration factors (*DF*s) to all engines under regulation. The *DF* is a factor applied to the emission certification test data to represent emissions at the end of the useful life of the engine. The engine useful life and the in-use testing period, as defined by the EPA for emission testing purposes, are listed in Table 9.1 for different engine categories. These requirements remain the same for Tier 4 engines.

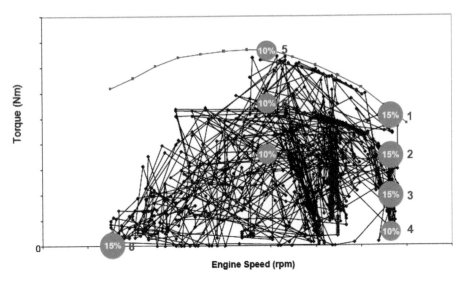

Figure 9.6. Non-road transient test cycle (NRTC) comparison with 8-mode test.

Table 9.1. Engine useful life requirements.

		Useful life		In-use testing period	
Power rating	Rated engine speed	Hours	Years	Hours	Years
<19 kW	All	3000	5	2250	4
19–37 kW	Constant speed ≥ 3000 rpm	3000	5	2250	4
	All others	5000	7	3750	5
>37 kW	All	8000	10	6000	7

Criteria emissions reductions are accompanied by fuel sulfur reduction. There are two motivating factors to require sulfur reduction in parallel with criteria emissions reduction. One is that sulfur becomes sulfate coming out of the engine exhaust and is counted as particulate emission. The second is that removal of sulfur enables some emissions reduction technologies to be adopted. These technologies include exhaust gas recirculation, NO_x sensors, and aftertreatment systems. For tiers 1 through 3, the sulfur content in non-road diesel fuels was not limited by environmental regulations. The oil industry specification was 0.5% (wt., max), with an average in-use sulfur level of about 0.3% or 3000 ppm. To enable sulfur-sensitive control technologies in Tier 4 engines, the EPA required that sulfur content in non-road diesel fuels be 0.05% or 500 ppm effective June 2007, and 0.0015% or 15 ppm (ultra-low sulfur diesel) effective June 2010.

Engine emissions regulations started in the USA, and many nations now regulate engine emissions. For the non-road equipment sector, the EU refers to non-road emission regulation by stages and is largely harmonized with the emission tiers of the USA, including NRTC. The exception is the NTE, which is yet to be fully defined at this time. Brazil will adopt the equivalent of ECE Stage IIIa in 2015 for construction equipment and 2017 for agriculture equipment. Russia has adopted ECE stage II and encourages Stage IIIa. India adopted Bharat Stage II (Tier 1) in 2005 and implemented Bharat Stage IIIA (Tier 3) between 2010 and 2011. China implemented National Stage I (Tier 1) in 2007 and National Stage II (Tier 2) in 2010 and has drafted National Stage III (Tier 3) with possible implementation date of October 2015. For a single-source overview of worldwide diesel emissions requirements including non-road engines, refer to www.dieselnet.com.

On-road diesel emissions regulation typically leads the non-road regulation by three to four years in the USA. It is often natural to compare non-road emissions with the on-road requirement,

Table 9.2. Comparison of representative test cycles worldwide.

	Non-Road Transient Cycle (NRTC)	Heavy-Duty Federal Transient Procedure (FTP, USA on-road)	European Transient Cycle (ETC, European on-road)
Average torque	39.3%	24%	36.7%
Average speed	67.7%	41.5%	50.9%
Test repeat	2	3	One test only
Cold cycle weighting	5% USA, 10% Europe	1/6	None

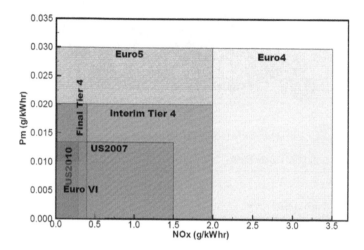

Figure 9.7. Comparison of on-road and non-road emissions requirements.

but they are not directly comparable. For example, fuel sulfur levels are different, although they are converging. Useful life requirements are defined by operating hours for non-road and by driving distance for highway vehicles, and most notably, the test cycles are different. Table 9.2 summarizes key differences in representative test cycles. The NRTC has substantially higher cycle engine speed and torque than both the Federal Transient Procedure (FTP) and the European Transient Cycle (ETC). These differences result in differences in engine combustion signature, as well as exhaust gas temperature and flow. Engine components such as turbochargers and aftertreatment devices behave differently. Electronic engine controls must adapt to these differences and behaviors. Therefore, any attempt to compare non-road and on-road diesel engine emissions can only be an approximation.

With these qualifications, Figure 9.7 gives comparison between non-road and on-road diesel emissions requirements. In this comparison, non-road interim Tier 4 falls between EPA 2007 and the European on-highway requirement of Euro 5, and non-road final Tier 4 is comparable to EPA 2010 and the European on-highway requirement of Euro 6. As will be discussed later, engine technology choices are similar between non-road interim Tier 4 and EPA 2007 or Euro 5, and between non-road final Tier 4 and EPA 2010 or Euro 6.

9.3   BUILDING BLOCKS OF DIESEL ENGINES

9.3.1   *Combustion system*

Rudolph Diesel invented the diesel thermodynamic cycle in 1897. In the ideal diesel cycle, combustion occurs at constant pressure. Before emission standards were regulated, the study of diesel

Temperatures

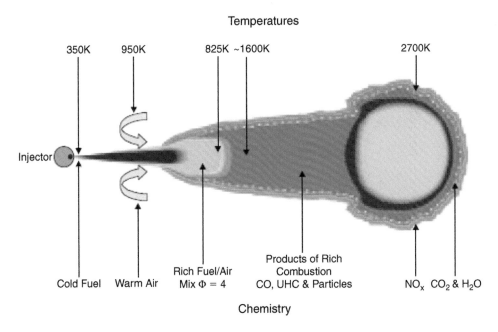

Figure 9.8.   Diesel combustion process (source: Dec, 1996; 2005).

combustion process focused entirely on thermodynamic efficiency, a subject of many college textbooks. Emissions regulations required engineers and scientists to optimize thermodynamic efficiency while criteria pollutants are minimized. The characteristics of the diesel combustion process in advanced diesel engines are shown in Figure 9.8, which was a result of collaborative study by the diesel engine industry and scientists at Sandia National Laboratory during the 1990s. The key attribute of the process is the presence of a cold fuel jet into which hot air is entrained. The fuel air mixture enters a diffusion flame sheath where it is further heated to approximately 825K and reactions are initiated, which consumes all the local oxygen and releases about 15% of the total heat of combustion. This jet supplies the interior of the plume with products of rich combustion that contain mostly CO and partially burned HC. Surrounding the burning plume, a thin diffusion flame is formed where complete oxidation occurs, yielding the products of complete combustion. In this diffusion flame, fuel fragments and particulates are converted to CO_2 and water vapor. Temperatures at the diffusion flame interface are high, providing an ideal environment for NO_x formation reactions. Thus, the process can be thought of as a cold liquid jet, entrained with warm air, supplying the reactants for a rich premixed reaction that feeds the interior of the plume. John Dec provides a comprehensive review of recent advances in diesel engine combustion (Dec, 1997; 2005). As Dec explains, ignition needs to occur near top dead center (TDC) to achieve high efficiency and the fuel-air mixture needs to be lean or diluted and premixed to obtain low emissions. A lean combustion mixture can be achieved through turbocharging, which also improves fuel efficiency. Dilution is typically accomplished through use of cooled exhaust gas recirculation (cEGR).

Above understanding of the diesel combustion process, NO_x and particulate formation are helpful for engine design practitioners to effect engine out emissions. Figure 9.9 shows intake oxygen effect on engine out NO_x, which is the combined effect of turbocharging, cooled EGR and other parameters, for a John Deere 9 L engine. EGR rate has the strongest effect and therefore is an effective means to reduce engine out NO_x. Figure 9.10 shows an example of the effect of intake manifold temperature on engine out NO_x, which can be effected by charge air-cooling. Most modern heavy-duty diesel engines, in both on- and off-road applications, are equipped with

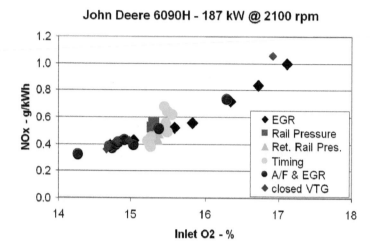

Figure 9.9.   Inlet oxygen content effect on engine out NO$_x$ through parametric study.

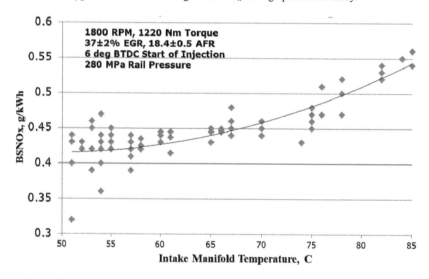

Figure 9.10.   Intake manifold temperature effect on engine out NO$_x$ on a John Deere 6090 engine.

turbocharging, cooled EGR, and charge air-cooling. They are effective tools that engineers can use to abate emissions from the source.

9.3.2   Electronic engine control system

Diesel engines were once governed by mechanical flyballs to achieve speed control. Stringent emissions requirements made it necessary to use electronic control systems. The electronic control system consists of an electronic control unit (ECU), sensors and actuators, control and diagnostics algorithms, and calibration data. Figure 9.11 illustrates a system level schematic of a Tier 4 engine control system. If a traditional diesel engine is considered the base engine, then a modern heavy-duty diesel engine consists of the base engine, the exhaust aftertreatment system, and the electronic engine control system. In addition to accomplishing the tasks of a mechanical flyball governor, the electronic control system manages many more tasks in a modern nonroad diesel engine.

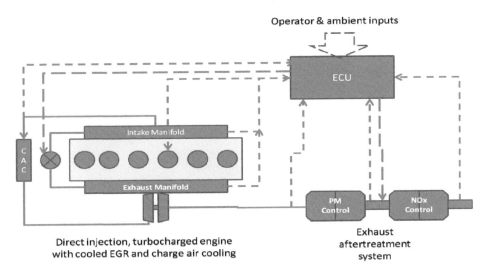

Figure 9.11.   System level schematic of a Tier 4 engine control system.

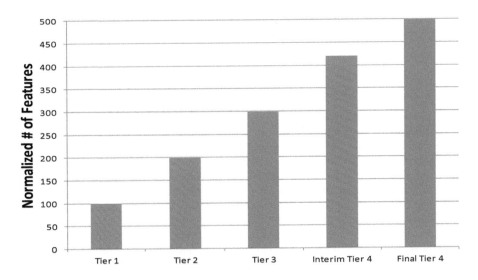

Figure 9.12.   Growth of normalized number engine control features.

It effects the combustion process, based on engine operating conditions, through simultaneous manipulation of fuel injection pressure and timing, the turbocharger boost level, and the EGR rate. It manages the exhaust aftertreatment system through control of exhaust gas temperature and dynamic urea dosing. It compensates engine performance based on operating environments such as altitude and ambient temperatures, and it provides real-time diagnostics of engine and component health. Figure 9.12 shows the growth of electronic control features from Tier 1 to final Tier 4. A feature in this case refers to an independently identifiable function within the engine control software, similar to a component in a mechanical system.

It is necessary to control injection, EGR, and air systems. Additional sensors and controls continue to be developed to improve engine operation. At one time, it was sufficient that an engine produced the desired power without producing excessive emissions. Now, the engine must

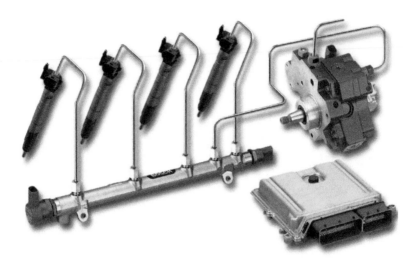

Figure 9.13.   Elements of a common-rail injection system (source: www.alexdiesel.com).

work with the aftertreatment system to provide exhaust gas temperatures and constituents that can be accommodated.

9.3.3   Fuel injection system

The fuel injection system has been a part of the engine since the invention of the diesel cycle. Most diesel engines used to have pump-line-nozzle injection systems in which one pumping element for each cylinder was contained in a block and actuated by a camshaft gear driven from the engine's crankshaft, with individual lines from each pumping element to each injection nozzle. Mid-range diesel engines progressed to use rotary injection pumps. This system shared the pumping elements, and a distributor head connected each injection line to the pumping elements at the appropriate time. Larger engines adopted unit injectors, which are individual piston pumps combined with a nozzle. Unit injectors can be mechanically driven by a camshaft, either directly or through rocker arms and pushrods, or they can be hydraulically driven. The injection system that is currently gaining the most popularity is the common-rail injection system (Fig. 9.13). In this system, a high-pressure fuel pump is driven from the engine's crankshaft in a controlled manner to maintain the desired fuel pressure in a pressure vessel, which is called a "rail". The electronically controlled injectors are connected to this rail with high-pressure lines and inject fuel on command. This provides great flexibility, since the rail pressure can be controlled to desired value independent of engine speed and torque, and multiple injections can be commanded to occur at desired times within each engine cycle. As electronic common-rail injection systems have been developed, the rail pressure and number of injections have been increased and the minimum separation between injections has been decreased. Consequently, new diesel engines are now typically designed with common-rail injection systems. Rail pressures greater than 200 MPa are now available, and 300 MPa capability is being pursued by fuel injection equipment manufacturers. Figure 9.14 illustrates the effect of injection pressure on fuel efficiency and engine out soot. When injection pressure is increased from 200 MPa to 240 MPa, fuel consumption and engine out soot are both reduced substantially. The magnitude of fuel injection pressure effect increases with reducing engine out NO_x levels. It is often believed that high injection pressure is the enabler for successive realization of emissions tiers. In the 1980s, injection pressure of 100 MPa was considered very high. Today, it is common to find injection pressure above 200 MPa on heavy-duty diesel engines.

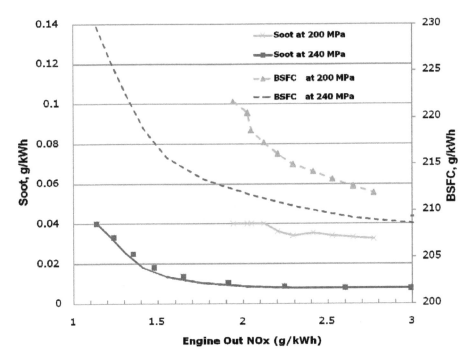

Figure 9.14. Fuel injection pressure of fuel efficiency.

9.3.4 *Turbocharching*

Turbocharging has been long considered as a means to increase engine efficiency and boost power density. A turbocharger consists of a turbine wheel in the exhaust stream that drives a compressor wheel in the intake air stream. The compressor pressurizes the air, allowing the engine to induct more air and thereby burn more fuel and produce more power. Turbocharging is almost a requirement for modern low-emission engines, and cooling of the air compressed by the turbocharger is common. With turbocharging, a turbine wastegate (a valve that opens to allow exhaust to bypass the turbine wheel) can be used to limit the boost at high speeds. A more recent advance is to use a variable-geometry turbocharger (VGT) to control the boost across the operating range. There are two common types of VGT: the swing vane (Fig. 9.15) and the sliding vane. The moving mechanical parts change the velocity and direction of the flow entering the turbine wheel in order to change the work extracted from the exhaust.

Figure 9.16 shows a double layer passage VGT. There are two passages in the turbine housing. Exhaust gas enters the turbine through the inner passage at low engine speed and load, as shown by the red arrow, and through both inner and outer passages at high engine speed and load, as shown by the red and blue arrows. The opening of the outer passage is controlled by a simple actuator, typical of a wastegate actuator.

Turbocharging increases air flow and allows more fuel flow to increase engine power density. In addition, fuel efficiency is improved because the engine can be smaller and have less friction, and because the power needed to compress the cool air at the inlet is obtained from expanding the hot exhaust gases, which is an efficient process. The extra air provided by turbocharging reduces smoke emissions and allows diluent, i.e., EGR (discussed below), to be added to the cylinder while maintaining adequate air in the cylinder.

In recent years, staged turbocharging, using both a low-pressure and a high-pressure turbocharger, has been used to increase engine output, low speed torque, and altitude capability.

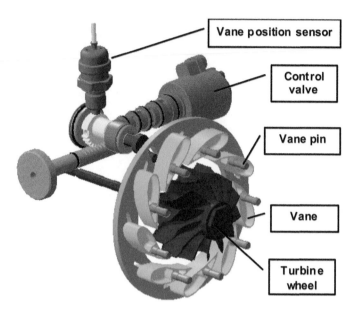

Vane position sensor

Control valve

Vane pin

Vane

Turbine wheel

Figure 9.15.   Swing vane type variable geometry turbocharger (courtesy of Honeywell).

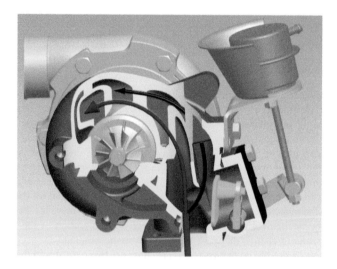

Figure 9.16.   Double layer passage VGT (courtesy of Kangyue Turbocharger, Ltd).

Figure 9.17 shows an example of an air system on a Tier 4 engine, which consists of two turbochargers and cooled EGR. Future concepts include power augmentation of the turbocharger using electrical, mechanical, or hydraulic turbocompounding. Supercharging using a belt- or gear-driven blower remains a remote possibility due to the high power requirements of the supercharger.

As diesel engines have developed over the years, there have been some clearly discernible design trends, although the recent emphasis on very low emission levels has effected some of these trends.

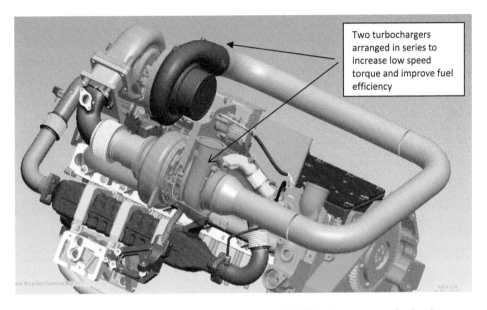

Two turbochargers arranged in series to increase low speed torque and improve fuel efficiency

Figure 9.17.   Air system on a Tier 4 engine consisting of cooled EGR and two-stage turbocharging.

One clear trend has been increased power output and increased cylinder pressure. Technology in materials and design has allowed these increases. Especially notable has been the change from aluminum to steel pistons on larger engines. Another important change has been increased boost from turbochargers and staged turbocharging to provide more air and thereby allow power density to increase. As a consequence of this technology, diesel engines have become lighter and smaller while still maintaining the same power output.

9.3.5   *Exhaust gas recirculation*

Exhaust gas recirculation (EGR) has become an important addition to low-emission diesel engines. This concept involves taking a portion of the exhaust gas and using it as part of the fresh intake air in order to provide inert mass in the cylinder to reduce peak flame temperatures, and thereby reduce NO_x emissions. There are many ways to add EGR to a diesel engine. The simplest is internal EGR, where some of the exhaust is retained in the cylinder or leaked into the fresh air charge while it is being inducted. However, internal EGR is not cooled, and therefore the NO_x reduction is relatively small. External EGR allows the recirculated exhaust gas to be cooled as it flows through a pipe, or more aggressively through a heat exchanger that utilizes engine coolant or cooling air. Cooled EGR reduces engine out NO_x in proportion to its mass flow rate (Fig. 9.18), and larger amounts of EGR can be used without displacing as much air. Therefore, cooled EGR has become a popular method of NO_x control. Recent studies have also found that EGR can also improve brake thermal efficiency (Koeberlein, 2012), contrary to earlier believe that EGR may hinder fuel efficiency (Fig. 9.19).

On a turbocharged engine, the EGR can flow from the turbine inlet to the intake manifold, which is commonly referred to as high-pressure EGR or short-route EGR. Alternatively, the EGR can flow from the turbine outlet to the compressor inlet, which is called a low-pressure EGR or long-route EGR. Both systems require enough pressure difference to drive the EGR through the piping and the EGR cooler. However, in a low-pressure system, the EGR needs to be cleaned of particulate matter to avoid fouling the compressor and charge air cooler. Currently, high-pressure EGR is far more common on heavy-duty applications.

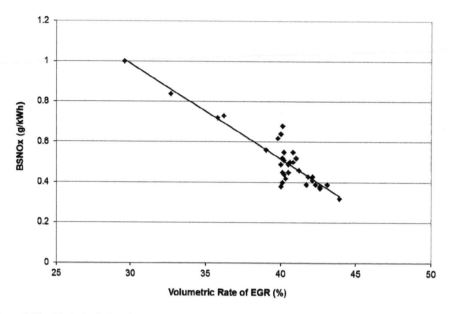

Figure 9.18.   Typical relations between engine out NO_x and volumetric flow rate of EGR.

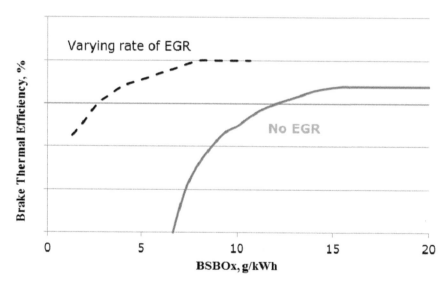

Figure 9.19.   Relations between thermal efficiency, engine out NO_x, and EGR rate (source: Koeberlein, 2012).

Careful design of the EGR system is necessary to properly cool, measure, and control the EGR over the life of the engine. This can be a challenge because the EGR gas contains particulate matter, which can be sticky, so the components must accommodate some accumulation of material without malfunction. The design must also recognize that water condensation can occur in the EGR system, and that this water will be corrosive due to the presence of dissolved sulfuric and nitric acids. Of course, condensed water will freeze at low ambient temperatures.

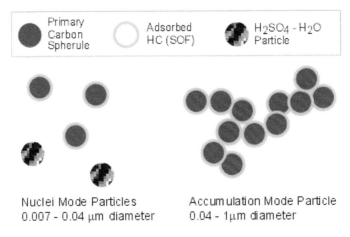

Nuclei Mode Particles
0.007 - 0.04 μm diameter

Accumulation Mode Particle
0.04 - 1μm diameter

Figure 9.20. Representative PM composition and particle size distribution.

9.4 AFTERTREATMENT TECHNOLOGIES

9.4.1 *Particulate matter and NO$_x$*

Before we discuss aftertreatment component technologies, it may be fitting that we first define particulate matter (PM) and nitrogen oxide or NO$_x$. PM is composed of carbonaceous soot particles, soluble organic fraction (SOF), and sulfates formed by diesel engine combustion. The volatile content is highly dependent on engine combustion conditions. Sulfates are formed by combustion of sulfur present in diesel fuel and lube oil. A representative PM composition and particle size distribution are shown in Figure 9.20.

Nitrogen oxide (NO) and nitrogen dioxide (NO$_2$) are formed during high-temperature diesel combustion from reactions of N$_2$ and O$_2$. Typically, when exhaust gas recirculation (EGR) is increased, peak flame temperature and engine out NO$_x$ are reduced. At the same time, engine out PM quantity is increased. This behavior is known as the NO$_x$-PM trade-off.

9.4.2 *Exhaust filtration*

Exhaust particulate filters are divided into two categories: full-flow filters (or wall-flow filters) in which all exhaust gases flow through the filtration media and filtration efficiencies are above 85%, and partial-flow filters in which only part of the exhaust gases flow through the filtration media and filtration efficiencies are typically less than 50%. If a soot layer is established on a full-flow filter, the filtration efficiency stays above 95% because the soot layer is also a filtration media. Inorganic material formed from the combustion of lube oil also accumulates on filters as ash. An example of a full-flow filter is shown in Figure 9.21.

9.4.3 *Regeneration types*

Particulates trapped on filters can be oxidized (cleaned) by NO$_2$ and O$_2$. This process is called regeneration. Passive regeneration refers to particulate oxidation by nitrogen dioxide (NO$_2$), and active regeneration refers to particulate oxidation by O$_2$, as shown in Figure 9.22. Passive regeneration occurs at normal engine operation above 300°C without any special engine control intervention. Engine out NO$_x$ is predominately NO, so most of the NO$_2$ is formed by the diesel oxidation catalyst. Active regeneration requires inlet gas temperature to be above 550°C. Oxygen levels in diesel exhaust are typically above 5%. Therefore, the O$_2$-based active soot burn rate is primarily limited by exhaust temperatures.

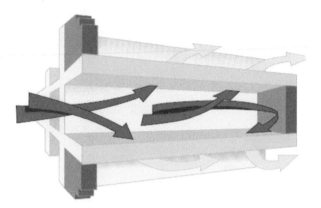

Figure 9.21.   Full-flow filter example: alternatively plugged cordierite diesel particulate filter.

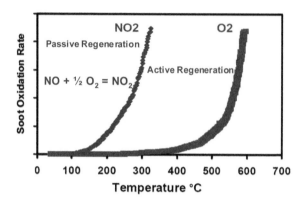

Figure 9.22.   Passive and active DPF regeneration temperature window.

Passive soot oxidation reaction:

$$C + 2NO_2 \rightarrow CO_2 + 2NO$$

Active soot oxidation reaction:

$$C + O_2 \rightarrow CO_2$$

9.4.4   *Active regeneration technologies*

Due to the high PM reduction efficiency (>85%) required to meet Tier 4 non-road regulations, some manufacturers have adopted the use of full-flow diesel particulate filters (DPF) with active regeneration. Active regeneration requires a means of raising the exhaust gas temperature above 550°C. Two primary heating technologies for active regeneration are exhaust diesel fuel burners and diesel oxidation catalysts (DOC). The diesel fuel burner can be integrated with either the engine or the diesel particulate filter. A potential advantage of a burner is its ability to regenerate the DPF at idle even with cold ambient temperatures. Disadvantages are design, control, and installation complexity due to the combustion air supply, flame stability, and temperature uniformity. Regeneration with a DOC requires a controlled amount of hydrocarbon (HC) to be injected upstream of the DOC. In order to reduce heat loss, the DOC is preferably integrated with the DPF. Advantages of a DOC system are enhanced passive regeneration and robustness with exhaust flow and O_2 levels (A/F ratio).

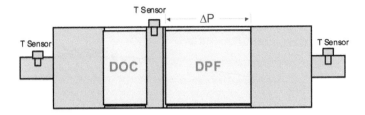

Figure 9.23.   Integrated DOC|DPF with associated sensors.

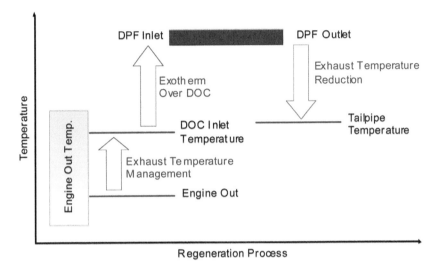

Figure 9.24.   Overall exhaust temperature control schematic.

To meet interim Tier 4 or Stage IIIb non-road emission requirements, some manufacturers, such as John Deere, have chosen to adopt cooled EGR, an advanced combustion system, flexible air and fuel systems, and an electronic engine control unit (ECU). The engine out NO_x is reduced with a cooled EGR and combustion optimization to meet the interim T4 regulations, and PM is reduced with a DPF. An integrated DOC|DPF with three temperature sensors and a differential pressure sensor, as shown in Figure 9.23, is installed on the engine. The engine is capable of achieving DOC light-off temperatures at low speed and light load by combustion modification for exhaust heating. Fuel is introduced in front of the DOC and oxidized over the DOC to raise exhaust temperature above 550°C. The hot exhaust leaving the DPF is diluted with ambient air by an exhaust diffuser, so the exit gas temperature is reduced to acceptable levels. The overall exhaust gas temperature control concept is shown in Figure 9.24.

9.4.5   *Diesel oxidation catalyst (DOC)*

The primary function of the DOC is to oxidize hydrocarbon. DOC substrates are typically flow-through monoliths, as shown in Figure 9.21, made from cordierite material. Cordierite has a very low coefficient of thermal expansion (CTE of 6×10^{-7}°C^{-1}) and great thermal stability (>1200°C). A catalytic coating (washcoat) containing precious metals, such as platinum (Pt) and palladium, (Pd) is deposited on the substrate internal walls to maximize exhaust gas contact. In the case of active regeneration, hydrocarbon fuel is injected upstream of the DOC and is oxidized into CO_2 and water by the DOC. The fuel energy released heats the DOC and raises the exhaust

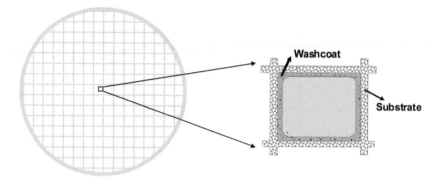

Figure 9.25.   DOC with catalytic coating deposited on a monolithic cordierite substrate.

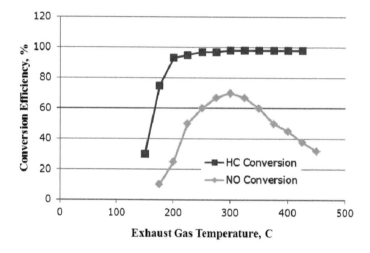

Figure 9.26.   Typical DOC HC conversion light off curve and NO to NO_2 conversion efficiencies.

gas temperature. The DOC requires a minimal exhaust temperature at its inlet to light-off and be active, as shown in Figure 9.26. Above the DOC light-off temperature, the DOC stays hot and hydrocarbon is oxidized at greater than 90% efficiencies, which ensures very little hydrocarbon slippage. High efficiency and robustness with varying O_2 levels and flow rates enable precise DOC outlet temperature control.

Under normal operating conditions, the DOC cleans up CO, HC, and SOF of PM emitted from the engine. The DOC also oxidizes NO to NO_2 to promote passive soot oxidation. Optimal NO_2 conversion efficiency occurs around 300°C, as shown in Figure 9.22, where the kinetic reaction rate is fast and NO_2 yield is not limited by thermodynamics.

9.4.6   Diesel particulate filter (DPF)

The primary purpose of a DPF is to trap particulate matter (PM). The two most common material choices are silicon carbide (SiC) and cordierite, as shown in Figure 9.27. Key DPF material properties are summarized in Table 9.3. SiC filters allow a higher soot loading limit of up to ~8 g/L, compared to the 5 g/L capacity of cordierite filters. However, due to high thermal expansion, SiC filters must be segmented and assembled with cement boundaries, while this design provision is

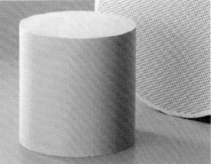

Figure 9.27.   Segmented SiC DPF (top) and cordierite DPF (bottom).

Table 9.3.   Key material properties of different diesel particulate filters.

Porous material	Melting temperature [°C]	CTE [$\times 10^{-7}$°C^{-1}]	Intrinsic density	Specific heat at 500°C [J/g/°C]	Thermal conductivity at 500°C [W mK^{-1}]
Cordierite	1460	6	2.51	1.11	1
Aluminum titantate (TiAl$_2$O$_5$)	1600	10	3.4	1.06	1
alpha-SiC	2400	45	3.23	1.12	20
Si-bonded alpha-SiC	1400	43	3.19	1.12	10

not required for cordierite DPF. Both cordierite and SiC filters can be further catalyzed to enhance passive regeneration and reduce HC and CO emission during active soot burn.

9.4.7   Catalyst canning

In most nonroad applications, the ceramic DOC and DPF are canned into an integrated converter, as shown in Figure 9.18, and packaged onto a vehicle with connecting pipes, brackets, isolators, flex elements, etc. An integrated DOC|DPF converter is preferred over separated DOC and DPF elements due to better space utilization and reduced heat loss. In a typical design, the DOC or DPF is wrapped with a ceramic mat and secured into a stainless steel casing under pressure, as shown in Figure 9.28. Friction force immobilizes the DOC and DPF. In general, the retention force through friction must exceed the combined g force from vibrations and the pushing force from the exhaust gas. The ceramic mat further serves as an exhaust gas seal and thermal insulation, so the converter skin temperature and heat loss remain low during regeneration. The inlet cone geometry is critical to guarantee uniform exhaust flow entering the DOC and DPF. Additional design provisions include sensor installation, flanges, and clamps so that the DPF can be removed for ash cleaning when necessary.

9.4.8   Exhaust fuel dosing system

Fuel required for DPF active regeneration can either be introduced by a late post injection during the power stroke or by an external exhaust fuel doser. If fuel is introduced by an engine late post injection, the fuel is fully evaporated and is well mixed with exhaust gas. The DOC|DPF can therefore be close-coupled with the turbocharger to minimize heat loss. Since fuel is injected late in the combustion cycle, frequent post injections increase the potential of oil dilution with fuel,

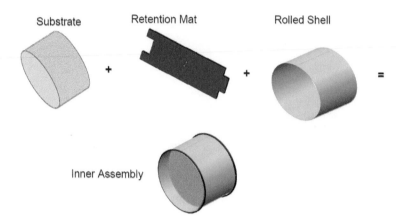

Figure 9.28.   DOC and DPF canning.

which can lead to shorter oil change intervals. Careful engine control calibrations are necessary to minimize such potential. An external fuel dosing system can also be used, in which case, sufficient mixing distance is required to ensure fuel evaporation and uniform mixing before the fuel-carrying exhaust gas enters the DOC.

9.4.9   *Aftertreatment system integration and controls*

Active DPF regenerations require sophisticated controls to manage exhaust temperature and the soot burn rate. The soot oxidation reaction with oxygen is highly exothermic. Heat released from soot combustion will increase the DPF temperature and further accelerate the reaction rate and increase thermal stress (Boger *et al.*, 2008).

To prevent device-damaging runaway regenerations, the soot loading on the DPF must be limited. These "safe" soot levels are highly dependent on DPF material, geometry, exhaust flow rate, DPF and temperatures. Key DPF control features such as DOC outlet temperature control, soot-loading prediction, and active regeneration control are discussed in more detail below.

9.4.9.1   *DOC outlet temperature control*
A metered amount of diesel fuel is introduced to achieve a desired DOC outlet temperature based on the DOC inlet temperature and desired DPF regeneration temperature. Fuel energy is released to heat exhaust gas and DOC before entering the DPF, which follows the energy balance shown in Equation (9.1). Uniform fuel mixing ensures a uniform DOC outlet and DPF inlet temperature radial distribution. Computational fluid dynamics (CFD) simulations are commonly used to study the fuel evaporation rate and fuel uniformity. One such example for a straight pipe at exhaust gas temperatures of 330°C is shown in Figure 9.29:

$$d(\Delta T)/dt = (M_{fuel} * LHV * DOC_{eff}$$
$$- M_{fuel} * Cp_{exhaust} * \Delta T$$
$$- M_{exhaust} * Cp_{exhaust} * \Delta T - \Delta H_{loss})$$
$$\div (M_{DOC} * Cp_{DOC}) \qquad (9.1)$$

where ΔT is the temperature rise, M_{fuel} is the amount of fuel injected, LHV is the fuel low heating value, DOC_{eff} is the DOC energy conversion efficiency, $Cp_{exhaust}$ is the specific heat capacity of the exhaust gas, ΔH_{loss} is the heat loss to ambient, M_{DOC} is the diesel oxidation catalyst mass, and Cp_{DOC} is the specific heat capacity of the diesel oxidation catalyst.

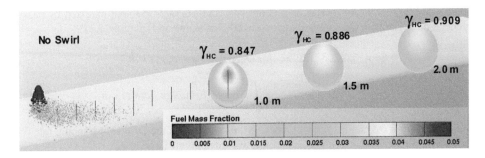

Figure 9.29.   CFD analysis of diesel fuel evaporation and mixing uniformity.

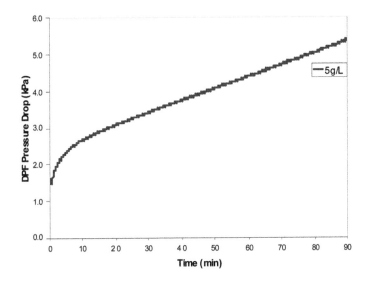

Figure 9.30.   Steady-state soot loading vs. measured differential filter pressure drop.

9.4.9.2   *Soot loading prediction*

As soot accumulates, the pressure drop (ΔP) across the DPF will increase. Soot prediction models typically use a flow-adjusted pressure drop as the primary signal to estimate soot loading. Redundant soot prediction models based on soot mass and operational hours are typically used to complement the ΔP model. As an example, at steady-state engine operation conditions, ΔP increases linearly with DPF soot loading after a soot layer is established on the filter, as shown in Figure 9.30. The ΔP over a DPF has three components, as shown in Equation (9.2), and a detailed physical model has been well described by Konstandopoulos *et al.* (2002) and Boger *et al.* (2008):

$$\Delta P = \Delta P_{filter} + \Delta P_{Forchiemer} + \Delta P_{expansion/contraction} \qquad (9.2)$$

where ΔP_{filter} is the pressure drop of the filter wall/soot layer, $\Delta P_{Forchiemer}$ is the pressure drop from the Forchiemer effect, and $\Delta P_{expansion/contraction}$ is the pressure drop due to exhaust contraction and expansion as exhaust gas enters and exits the DPF.

As ash accumulates on the DPF, the ΔP model needs to be corrected for the ash contribution.

The soot mass model integrates engine out soot and subtracts the soot oxidized by passive regeneration to estimate the soot loading on the DPF, as shown in Equation (9.3):

$$\text{Soot mass on DPF} = \int EO_{soot}(t) * \eta_{DPF\_trapping} * dt$$

$$- \int Passive\_oxidation\_rate(t) * dt \tag{9.3}$$

where $EO_{soot}(t)$ is the engine out soot rate (g/h), $\eta_{DPF\_trapping}$ is the DEF soot-trapping rate (%), and $Passive\_oxidation\_rate(t)$ is the soot passive oxidation rate (g/h).

9.4.9.3   Active regeneration control

The soot burn rate is a function of exhaust temperature, O_2 level, exhaust flow rate, soot loading, and other parameters. The energy balance equation is:

$$d(\Delta T)/dt = (\Delta H_{C+O2} + M_{exhaust}$$

$$* Cp_{exhaust} * \Delta T - \Delta H_{loss})$$

$$\div (M_{DPF} * Cp_{DPF}) \tag{9.4}$$

where ΔH_{C+O2} is the heat release from soot burn, $M_{exhaust}$ is the exhaust mass flow, $Cp_{exhaust}$ is the exhaust gas specific heat capacity, ΔT is the exhaust temperature rise over DPF, ΔH_{loss} is the heat loss to ambient, M_{DPF} is the DPF mass, and Cp_{DPF} is the DPF specific heat capacity.

In most cases, heat loss to ambient is relatively small compared to the total energy introduced by fuel injection. An example of heat loss (combined from converter and pipe) at different exhaust inlet exhaust gas temperatures based on a free convection model is shown in Figure 9.31. Heat loss to ambient increases with DOC inlet temperature and exhaust flow rate:

$$\Delta H_{loss} = \Delta H_{conduction} + \Delta H_{radiation} + \Delta H_{convection} \tag{9.5}$$

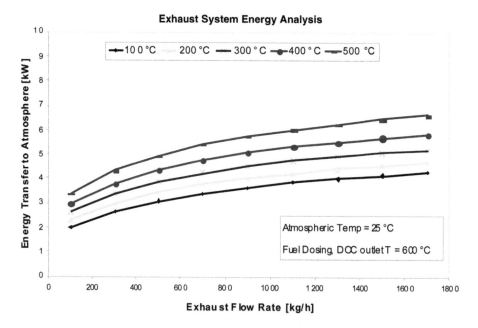

Figure 9.31.   Simulation of heat loss from DOC|DPF during at different regeneration conditions.

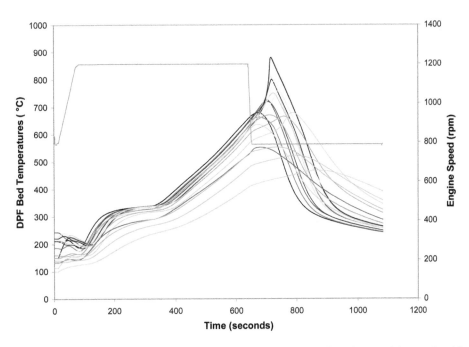

Figure 9.32. DPF temperature distribution during active regeneration with engine speed drops to low idle.

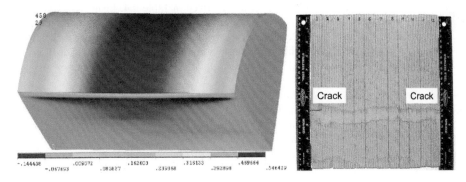

Figure 9.33. FEA analysis shows high stress area around DPF perimeter near midpoint (left) and picture of failed part (right).

Ceramic DPFs are susceptible to thermal shock. A soot-loaded DPF has a high propensity to cracking if the engine suddenly transitions to idle operation after initiation of DPF regeneration, often referred to as drop to idle. The soot burn will continue along the DPF flow axis with insufficient exhaust flow to dissipate the heat. During drop to idle, the soot burn becomes highly non-uniform, local high temperature is created toward the end of the DPF, and the large temperature gradient induces thermal stress. A thermocouple-instrumented DPF is often used to measure the temperature gradient and soot limit during drop to idle. The DPF temperature distribution (as in Fig. 9.32) is fed into FEA models to predict stress and compare the prediction with measured material properties. In this example, the high-stress area predicted by the FEA (Fig. 9.33) agrees well with the cracking detected in post mortem analysis.

Table 9.4.   Trade-offs of different de-NO$_x$ technologies.

	Lean NO$_x$ Catalyst (LNC)	Lean NO$_x$ Trap (LNT)	Selective catalytic reduction (SCR)
NO$_x$ efficiency	~30%	~80%	~95%
Reductant	Diesel fuel	Diesel fuel	NH$_3$ (e.g., from urea)
Advantages	Simple	No extra fluid	Fuel efficiency
Disadvantage	Low efficiency, HC slip, and fuel penalty	Sulfur poisoning, durability risk, and fuel penalty	Complex fluid storage/delivery, and freeze and thaw

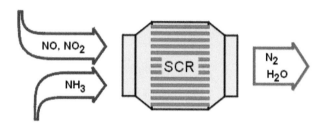

Figure 9.34.   SCR converter and NO$_x$ and NH$_3$ reaction.

9.4.10   *Diesel engine NO$_x$ aftertreatment technologies*

Depending on engine out NO$_x$ levels, several NO$_x$ aftertreatment technologies could be considered. Trade-offs of these technologies are summarized in Table 9.4. Lean NO$_x$ Traps (LNT) are viable for light-duty applications and selective catalytic reduction (SCR) is the technology of choice for almost all heavy-duty applications due to its proven longevity and fuel efficiency advantages. SCR has been in use on heavy duty trucks in Europe since 2008 and in the USA since 2010. It has been or is being adopted for non-road applications as of this writing.

9.4.10.1   *Selective catalytic reduction (SCR)*

SCR is based on the selectivity of ammonia (NH$_3$) reaction with NO$_x$ in a temperature window of 150°C to 500°C, as shown in Figure 9.34. Unlike a HC lean NO$_x$ catalyst, NH$_3$ prefers to react with NO or NO$_2$ over O$_2$. A NH$_3$ reaction with O$_2$ becomes profound only above 500°C. Typical and fast SCR reactions are shown below.

  Standard SCR reaction:

$$4NO + 4NH_3 + O_2 \rightarrow 4N_2 + 6H_2O$$

  Fast SCR reaction:

$$4NH_3 + 2NO_2 + 2NO \rightarrow 4N_2 + 6H_2O$$

Currently, heavy duty applications have selected a high-purity 32.5% urea/water solution called diesel exhaust fluid (DEF), which is called AdBlue in Europe, as the media to deliver NH$_3$. DEF thermally decomposes into NH$_3$ and isocyanic acid (HNCO), which further hydrolyzes into NH$_3$ and CO$_2$, as shown below.

  Thermolysis:

$$(NH_2)_2CO \rightarrow NH_3 + HNCO$$

  Hydrolysis:

$$HNCO + H_2O \rightarrow NH_3 + CO_2$$

Successful implementation of SCR to achieve high NO$_x$ conversion requires NH$_3$ to be uniformly mixed with exhaust gases. Non-uniform distribution results in NH$_3$ and NO$_x$ slip. Three

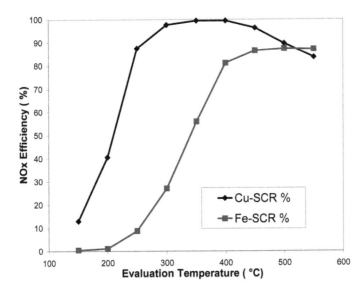

Figure 9.35. NO_x conversion efficiency over temperature with 1:1 NH_3/NO ratio.

types of SCR catalyst technologies exist in the market today. V-SCR (V_2O_5-TiO_2-WO_3) is commonly extruded into monoliths, although it can be coated on a cordierite substrate. Fe-zeolite and Cu-zeolite SCR catalysts with much higher thermal stability than V-SCR are preferred in an aftertreatment system with an actively regenerating DPF. The NO_x reduction temperature window of Fe-SCR and Cu-SCR with 500 ppm NO and 500 ppm NH_3 is shown in Figure 9.35. In the absence of NO_2, Cu-SCR gives much greater NO_x removal than Fe-SCR. If a DOC is used and a favorable NO_2/NO_x ratio is established before the SCR, the performance difference of Fe-SCR and Cu-SCR diminishes.

DEF solution freezes at −11°C. Therefore, the DEF tank must be designed with an active heating function for use in northern environments. The DEF storage tank is typically thawed by engine coolant due to the high energy requirement, while DEF transfer lines and the DEF pump are commonly thawed by electrical power. The DEF solution decomposes above 65°C, which creates additional vehicle integration, storage, and transportation challenges in hot environments. The onboard delivery system of DEF includes a supply pump and an injector. Pressurized DEF is metered and injected into the exhaust stream as a fine spray. Injection at low exhaust temperature below 180°C is avoided since DEF cannot be fully decomposed into NH_3 and the catalyst is not active. The cost of DEF erodes part of the fuel economy benefit brought about by the SCR technology.

A combination of lean NO_x trap (LNT) and SCR has also been studied in which NH_3 is produced during LNT regeneration and is temporarily stored on a downstream SCR. This combined system eliminates the need for complex onboard urea storage and delivery system (Ngan, E. *et al.*, 2011; McCarthy *et al.*, 2011). The combined LNT-SCR system marginally improves the total NO_x conversion efficiency of the LNT by the complementary NO_x reduction on SCR (Bremm *et al.*, 2008; Dou *et al.*, 2002; Hu *et al.*, 2006; Johnson, 2008; Kharas and Bailey, 1998; Lambert *et al.*, 2004).

9.5 MEETING DIESEL EMISSIONS THROUGH TIERS

9.5.1 *Tier 3 and earlier engines*

The first US federal standards for new off-road engines were adopted in 1994. These were called Tier 1 standards and were phased-in from 1996 to 2000 for engines over 37 kW. Engines made

Table 9.5. Tier 3 engine technologies.

	Category 1 Low tech intensity	Category 2 Medium tech intensity	Category 3 High tech intensity
Two-valve cylinder head	Yes	Maybe	–
Four-valve cylinder head	Maybe	Maybe	Yes
Mechanical fuel injection	Yes	–	–
Common-rail fuel injection	–	Yes	Yes
Standard turbocharger	Yes	–	–
Wastegate turbocharger	–	Yes	–
Variable-geometry turbocharger	–	–	Yes
Charge air cooling	Maybe	Yes	Yes
Cooled exhaust gas recirculation	–	–	Yes
Full electronic control	–	Yes	Yes, sophisticated

before the Tier 1 standards are referred to as Tier 0 engines. Tier 2 standards were phased-in from 2001 to 2006, and Tier 3 standards were applied from 2006 to 2008. These first three phases of emission control required improvement on fuel injection and air systems. While mechanical fuel injection pumps were still viable, many engines adopted high-pressure common rail systems. Fuel injection pressure increased from below 100 MPa to as high as 180 MPa during this period. Some manufacturers adopted cooled EGR technology for Tier 3 which, although more complex and expensive than non-EGR engines, were more fuel-efficient. Most manufacturers chose to avoid exhaust aftertreatment.

Tier 3 engines on the market can be grouped into three categories: those that have low technology intensity, those that have medium technology intensity, and those that have high technology intensity. Table 9.5 summarizes key differences among these categories. The John Deere PowerTech M engine, for example, falls into category 1, the PowerTech E engine falls into category 2, and the PowerTech Plus engine, the most sophisticated of the three, is an example of a category 3 engines.

Engines with higher technology intensity typically cost more and are more fuel-efficient. By way of an example, Figure 9.36 shows a John Deere Tier 3 PowerTech Plus engine that features cooled exhaust gas recirculation (cEGR), a pivot vane type variable-geometry turbocharger, a high-pressure common-rail fuel injection system, and a swirl-assisted combustion system. The cEGR is taken from the exhaust manifold, cooled through a gas-to-water heat exchanger, and electronically controlled by a poppet-type valve. The heat exchanger (EGR cooler) is strapped to the engine. The EGR cooler works in a very harsh environment. It is subject to engine vibration that can cause mechanical cracking and fatigue. It is intended to cool the exhaust gas, which contains soot particles, hydrocarbon, and nitric oxide that can cause fouling inside the cooler. It is also subject to high temperatures that can induce thermal stress, and thermal cycles that can cause thermal fatigue. These are challenges for EGR cooler design and verification. Computational fluid dynamics (CFD) analysis can help understand flow and temperature distribution, and finite element analysis can help optimize the structure, vibration, and thermal design. Fundamental understanding and elimination of EGR cooler fouling is more of an art than a science.

9.5.2 *Meeting US EPA Tier 4*

US EPA Tier 4 standards include interim and final Tier 4 standards, which are equivalent to Stage IIIB and Stage IV in Europe. They are being phased-in from 2008 to 2015. In comparison with the previous Tier 3 standards, they require over 90% reduction in nitric oxide and particulate matter. not-to-exceed (NTE) limits and nonroad transient cycle (NRTC) tests add additional dimensions to the Tier 4 standards. Either or both particulate matter and NO_x exhaust aftertreatments are

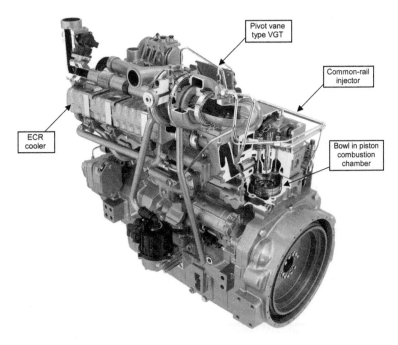

Figure 9.36. John Deere Tier 3 PowerTech Plus engine (courtesy of John Deere).

required for engines above 56 kW, although research is underway to avoid or reduce the increased cost and complexity, especially for smaller engines. This section describes alternatives for meeting interim Tier 4 and final Tier 4 standards.

Figure 9.37 depicts a full set of a final Tier 4 solution. Using a Tier 3 engine discussed above with common rail fuel injection system, turbocharging, and cooled EGR as base, the final Tier 4 solution has added two stage turbocharging for power density and low speed torque, DOC and DPF for particulate control, and SCR for NO_x abatement. To fulfill Interim Tier 4 requirements, a subset of a final Tier 4 solution is necessary. Either DOC and DPF or SCR alone can take a Tier 3 engine to meet interim Tier 4, and both solutions are available in the market place for nonroad mobile machinery (NRMM) applications including agricultural and construction equipment.

In the case of the interim Tier 4 solution without SCR, particulate requirement of 0.02 g/kWh is accomplished by the high filtration efficiency of DPF with engine out particulate levels of less than 0.1 g/kWh and NO_x requirement of 2 g/kWh is achieved through controlled application of cooled EGR at a rate of about 20% of the total engine through flow. In the case of the interim Tier 4 solution without DPF, particulate standard is reached through combustion optimization at an engine out NO_x of greater than 5 g/kWh and NO_x requirement is met by using a SCR with a conversion efficiency in the range of 70%. However, it is possible to meet the interim Tier 4 target without the use of any aftertreatment through other means for power categories in the 56 to 130 kW range.

To meet final Tier 4 standard in the 56 to 560 kW power range, there are multiple solutions that are technically feasible. It is possible to meet final Tier 4 using high rate of cooled EGR, however, there are two main drawbacks for this solution. One is poor fuel economy even at injection pressure of greater than 300 MPa and the other is compromised transient response rate. As a result, we likely will not see this solution in the market place anytime soon. It is also possible to meet final Tier 4 by utilizing high NO_x conversion efficiency of SCR without use of cooled EGR. Final Tier 4 products have been displayed at trade shows although they are not in the market yet as of this writing. These trade displays indicate that most final Tier 4 engines will

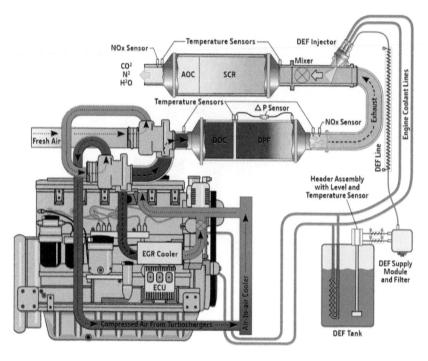

Figure 9.37.   Schematic illustration of final Tier 4 engine (Courtesy of John Deere).

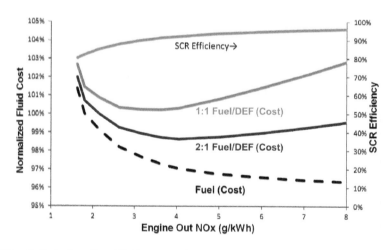

Figure 9.38.   Optimization of total fluid consumption.

use a range of technologies including both cooled EGR and SCR for NO_x control. Some engine manufacturers will use a DPF for particulate reduction and some will not. Figure 9.38 shows an example illustrating one aspect of system optimization for a particular product application. For applications that stress optimal fluid (fuel plus DEF) consumption, one may choose to apply both cooled EGR and SCR, as in production class construction machines. For applications with low total life cycle operating hours, one may opt for minimizing product cost by avoiding cooled EGR, as in utility farm tractors.

9.6   BIOFUEL FOR MODERN DIESEL ENGINES

Today, it is common to find that biodiesel blends are used in nonroad mobile machines including agricultural equipment. Biodiesel blends up to 5% by volume in North America and 7% in Europe are routinely used in agriculture machines without impacting machine performance or durability. Biodiesel producers have adopted quality standards to ensure quality of the biodiesel fuel. Engine manufacturers have improved designs to the engine diesel fuel supply system and the combustion process to enable such biodiesel blends. At higher biodiesel blends, however, additional care by the user is needed to ensure performance and durability. As an example, up to 20% of biodiesel blends can be used in John Deere engines with little compromise on performance if keep-clean additive is used with the biodiesel blend. The additive is necessary at a high level of biodiesel blend because biodiesel blends contain oxygen, which have higher propensity to cause coking on injector nozzles, resulting in the degradation of engine power and service life.

9.7   SUMMARY AND PERSPECTIVES

Agriculture machines, powered by diesel engines, have helped to improve agricultural productivity dramatically over the course of more than a century. Agricultural production, brought about in part by mechanization, has more than kept pace with the global population increase and urbanization, while the farm labor force in developed countries and emerging economies alike has been steadily declining. Diesel engines, used as the prime power source for agriculture machines, have become nearly zero emissions as a result of clean air regulations. While becoming clean, these engines continue to gain fuel efficiency, contrary to some early expectations of a better environment only at the expense of energy efficiency. Nebraska tractor test records show that some production tractors, such as the John Deere 8000 series with Tier 3 engines, have the best fuel economy ever. We shall remain optimistic that fuel economy will hold its place through engineering innovation and total vehicle system integration.

Nonroad engines are becoming more complex throughout the emissions tiers. Tier 4 engines will have added hardware components, such as aftertreatment, and ever more sophisticated control software. Tier 4 powertrain engineering skills consist of design engineers and engineers with specialization in thermal and fluid science, materials science, chemistry and chemical engineering, electrical and electronics engineering, and embedded software engineering. World-class products depend on world-class talent in all of these areas.

Looking beyond Tier 4, greenhouse gas reduction and on-board diagnostics will likely serve as motivators for product improvement for environmental sustainability. Perhaps that will be another 20-year or longer journey. We are proud that agricultural mechanization, powered by the internal combustion engine, was one of the greatest engineering achievements of the 20th century. We shall be equally proud to have contributed to feeding a world with 50% more population while using fewer resources when we will look back at the technological innovation of the modern diesel engine at the end of this century.

ACKNOWLEDGEMENTS

The author sincerely thanks the many Deere colleagues for technical data and helpful discussions.

REFERENCES

Boger, T., Rose, D., Tilgner, I.C. & Heibel, A.K.: Regeneration strategies for an enhanced thermal management of oxide diesel particulate filters. SAE 2008-01-0328, Warrendale, PA, 2008.
Bremm, S., Pfeifer, M., Leyrer, J., Mueller, W., Kurze, S., Paule, M., Keppeler, B. & Vent, G.: Bluetec emission control system the U.S. Tier 2 bin 5 legislation. SAE 2008-01-1184, Warrendale, PA, 2008.

CFR: CFR 40 CFR Parts 85, 86, 89, 90, 91, 92, 94, 1039, 1048, 1051, 1065, and 1068: Test procedures for testing highway and non-road engines and omnibus technical amendments. *Federal Register* 70 (2005), pp. 40, 419–440, 468.

Dec, J.: A conceptual model of DI diesel combustion based on laser-sheet imaging. *SAE Transaction Paper* 970873, 1997.

Dec, J.: Advanced compression ignition engines – understanding the in-cylinder processes. *Proceedings of the Combustion Institute* 32, 2009, pp. 2727–2742.

Dou, D., Pauly, T., Price, K, & Salyers, L.: A systematic investigation of parameters affecting diesel NOx adsorber catalyst performance. *DEER Conference*, San Diego, 2002.

Hu, H., Reuter, J., Yan, J. & McCarthy, J. Jr.: Advanced NOx aftertreatment system for commercial vehicles. SAE 2006-01-3552, Warrendale, PA, 2006.

Johnson, T.V.: Diesel emission control technology in review. SAE 2008-01-0069, Warrendale, PA, 2008.

Kharas, K.C.C. & Bailey, O.H.: Improvements in intimately coupled diesel hydrocarbon adsorber/lean NOx catalysis leading to durable Euro 3 performance. SAE 982603, Warrendale, PA, 1998.

Koeberlein, D.: Cummins Super Truck Program: Technology and system level demonstration of highly efficient and clean diesel powered Class 8 Trucks. May 17, 2012.

Konstandopoulos, A.G., Skaperdas, E. & Masoudi, M.: Microstructural properties of soot deposits in diesel particulate traps. SAE 2002-01-1015, Warrendale, Pa.: SAE, 2002.

Lambert, C.K., Hammerle, R.H., McGill, R.N. & Khair, M.K.: Technical advantages of urea SCR for light-duty and heavy-duty diesel vehicle applications. SAE 2004-01-1292, Warrendale, PA, 2004.

McCarthy Jr, J., Yue, Y., Mahakul, B., Gui, X., Yang H., Ngan, E. & Price, K.: Meeting nonroad final Tier 4 emissions on a 4045 John Deere engine using a fuel reformer and LNT system with an optional SCR showing transparent vehicle operation, vehicle packaging and compliance to end-of-life emissions. *SAE Int. J. Engines* 4:3 (2011), pp. 2699–2717, doi:10.4271/2011-01-2206.

Ngan, E., Wetzel, P., McCarthy Jr, J., Yue, Y. & Mahakul, B.: Final Tier 4 emission solution using an aftertreatment system with a fuel reformer, LNT, DPF and optional SCR. *SAE Technical Paper* 2011-01-2197, 2011, doi:10.4271/2011-01-2197.

ABBREVIATIONS

BSFC = brake specific fuel consumption
cEGR = cooled exhaust gas recirculation
DOC = diesel oxidation catalyst
DEF = diesel emissions fluid
DPF = diesel particulate filter
GHG = greenhouse gases
HC = hydrocarbon
LNC = lean NO_x catalyst
LNT = lean NO_x trap
NRTC = non-road transient test cycle
NTE = not to exceed area of the torque-speed map
OBD = on-board diagnostics
SCR = selective catalytic reduction, commonly refers to the urea-based system

Section 3
Biofuels

CHAPTER 10

Biofuels from microalgae

Malcolm R. Brown & Susan I. Blackburn

10.1 INTRODUCTION

10.1.1 *Introduction to biofuels*

World energy use has been predicted to increase five-fold by the end of the century (Huesemann, 2006). While today, about 80% of the energy demand is met from fossil fuels, their stocks are rapidly depleting and production costs are rising. Further, the increasing concern over the environmental impact of burning fossil fuels and its link to climate change has necessitated the development and use of alternative energy sources that are sustainable and renewable. Hydro, solar, wind and biomass are being exploited to reduce the use of fossil fuels, though the use of combustible biomass resources, i.e. biofuels, will also play an important role contributing to fossil fuel replacement. The advantage of biofuels is that they enable energy to be chemically stored, and derivatives can be used within existing engines and transportation infrastructure by blending with conventional fuels, e.g. petroleum diesel (Amaro *et al.*, 2011). However, some consider that the light blending at current levels (e.g. E10) is actually a fossil fuel enhancer, and does little to encourage competitive substitution of biofuels for fossil fuels (Batten, 2008).

Biofuels, specifically biodiesel and ethanol have been commercially produced from the processing of crop plants (e.g. soy, palm, corn, beet, sugarcane) for the past few decades. These are the so-called first-generation biofuels. Although their production has rapidly increased (ca. 30–40% p.a.), they still only contribute about 1% to the overall production of liquid fuels (Williams and Laurens, 2010). However, their production is giving rise to increasing scrutiny by the public and non-government organizations. The most common concern is that as their production increases, so does the competition with agriculture for valuable crop land used for food production (Batten and O'Connell, 2007).

The second generation biofuels (also termed advanced biofuels) include ethanol and biodiesel produced from the non-edible portions of novel starch, oil and sugar crops (e.g. *Jatropha*, cassava or *Miscanthus*), and ethanol from lignocellulosic feedstock (e.g. straw, wood, grass) (Amaro *et al.*, 2011). These biofuels are considered more sustainable, though there remain some technical and economic challenges that need to be overcome before they can be produced profitably at a commercial scale (Jones and Mayfield, 2012). Other second generation biofuels may be derived from microalgae, macroalgae (seaweeds) and microbes, though some authors refer to these as third generation biofuels (Ghasemi *et al.*, 2012). Biofuels obtained from microalgae are the focus of discussion in this chapter.

10.1.2 *History of investigation of biofuels from microalgae*

Microalgal biomass has been exploited by humans as food for thousands of years. However, the culturing of microalgae only commenced mid-way during the last century (Borowitzka, 1999). Over the ensuing decades, culture systems have developed either as outdoor systems (e.g. ponds) or indoors as semi-enclosed systems (cylinders, polythene bags) in scales of hundreds to many thousands of liters. The initial applications were as human food or food supplements, or as aquaculture feeds (Borowitzka, 1999). During the 1970s, mass production of microalgae was investigated as a source of lipid or oil for biofuel through several programs, particularly the US$ 25 M Aquatic Species Program (ASP) which ran from 1978 to the mid-1990s (Sheehan *et al.*, 1998). Its major conclusion was that microalgal biofuels were a promising alternative to conventional biofuels, but at that time were not cost-effective. Nevertheless the research made important advances to understanding the biology, selection and screening of strains, metabolic engineering and the mass culture of selected strains (Williams and Laurens, 2010).

During the last decade, the interest in microalgae as a potential feedstock for biofuels has escalated. More scientific papers were published on the topic from 2007–2010 than the preceding 25 years (Konur, 2011). In the early 2000s, there were only a few companies investigating biofuels from microalgae; in 2010 the number was estimated to be 75, with more than a hundred others pursuing efforts at various levels (OILGAE, 2011). Examples of the major companies directly investigating biofuel production, or providing services to the developing industry are shown in Table 10.1.

10.1.3   *Potential advantages of microalgae as biofuel feedstock*

Microalgae possess a range of positive attributes that gives them a potential advantage over the other aforementioned biofuel feedstock that has accelerated the interest in their commercial exploitation. These include their ability to:

- Photosynthesise with higher activity and with high growth rates, which may, under certain conditions, exceed those of terrestrial crops by approximately tenfold.
- Grow in culture systems located on non-arable land, thus avoiding direct competition with conventional food crops.
- Use freshwater resources efficiently. Many species may be cultured in brackish, saline or hypersaline seawater. Freshwater species need lower rates of freshwater renewal than terrestrial crops require as irrigation water.
- Utilize nutrients (e.g. P, N) from a variety of wastewater sources, thus providing the advantage of wastewater bioremediation and nutrient recycling.
- Survive and proliferate over a wide range of environmental conditions, due to the large number of species/strains with high levels of genetic diversity.
- Accumulate energy-rich oils, which may surpass 50% of their cell dry weight (DW).
- Produce valuable products (e.g. pigments, nutraceuticals, etc.), which significantly improves the economics of producing low-value biofuel from the same microalgae.
- Utilize CO_2 from flue gases produced by power plants and other industrial sources, thus utilizing and potentially reducing greenhouse gas (GHG) emissions.
- Provide multiple pathways to the production of several types of biofuels.

10.1.4   *Overview of the production of biofuel from microalgae*

A schematic of the "microalgae to biofuel pipeline" is shown in Figure 10.1. The first step is the selection of the species or strains of microalgae. Laboratory-scale screening experiments are often undertaken to identify strains with rapid growth rate and high oil (lipid) content. However, the technical feasibility and cost-efficiency of many downstream processes may be influenced by the strain chosen. Such factors include the ability of the strain(s) to grow at scale at the available sites, and the ease to which its biomass may be harvested, extracted, and converted to fuel. The quality of the fuel (e.g. biodiesel) that might be produced will also be determined by strain selection, since microalgae may differ significantly in the fatty acid profiles of their constituent oils. Most commonly, microalgae are grown photoautotrophically, i.e. using CO_2 as a carbon source and the energy provided by sunlight or artificial illumination, in either open ponds or within closed systems called photobioreactors (PBRs). The addition of inorganic nutrients or fertilizers to these systems facilitates growth to much higher biomass densities than would normally occur in natural environments. Alternatively, some strains can be grown heterotrophically, i.e. using agriculture

Table 10.1. Examples of prominent companies investigating the commercialization of microalgal biofuels.

Company (location) – website	Comments – technology, products, scale
Algae.Tec Ltd (Atlanta, GA, USA & Perth, WA, Australia) – algaetec.com.au	Supplier of modular high-yield algae-to-biofuels bioreactor systems. Aims to use waste CO_2 and sunlight to produce biodiesel and jet fuel.
Algenol Biofuel (Bonita Springs, FL, USA) – www.algenolbiofuels.com	Outdoor photobioreactors, ethanol from cyanobacteria. Pilot-scale biorefinery on 36 acre site (Fort Meyers). Demonstrated productivity of 30,000 L ethanol per acre per year.
Aquaflow (Nelson, New Zealand) – www.aquaflowgroup.com	Utilizes microalgae from municipal waste water, and biomass from other waste streams to produce gasoline, jet fuel and biodiesel. At demonstration facility stage.
Aurora Algae, Inc. (Hayward, CA, USA) – www.aurorainc.com	Pond production. Biorefinery approach where producing biodiesel, and other co-products including omega-3 oils and feeds. Facilities in Florida, California, Mexico and Australia
Bioalgene (Seattle, WA, USA) – www.bioalgene.com	Pond production, utilizing flue gas emissions from power plants, and cow manure. Pathways developed to produce aviation fuel, biodiesel and co-products. Pre-pilot scale.
Bionavitas, Inc. (Redmond, WA, USA) – www.bionavitas.com	Company claims to have developed technology that enables better light utilization by microalgae for environmental remediation, manufacturing health and nutraceutical products, and biofuels.
Biofuel Systems, SL (Alicante, Spain) – www.biopetroleo.com	Upright cylindrical photobioreactors, pilot scale. Production of green crude. Utilized CO_2 from flue gas from adjacent cement works.
Bodega Algae, LLC (Boston, MA, USA) – www.bodegaalgae.com	Produces closed continuous-flow photobioreactors (PBRs) for use in the production of biofuel. Systems are modular and stackable, allowing it to be co-located on the premises of industrial plants.
Cellana, Inc. (La Jolla, CA, USA) – www.cellana.com	Coupled production using closed system PBRs and ponds. Six-acre Kona demonstration facility on Kona, Hawaii. Multiple products: biofuels, nutraceuticals, feeds and cosmetics.
EWBiofuel and ReEnergy (Gujarat, India) – www.ewbiofuel.com	Utilizes Jatropha for biodiesel; aiming to incorporate microalgae and becoming the first Asian company to commercially produce biodiesel from microalgae.
Ingrepo (Borculo, Netherlands) – www.ingrepro.nl	Pond production. Largest industrial microalgal production in Europe.
Neste Oil (Helsinki, Finland) – www.nesteoil.com/	Worlds largest producer of renewable diesel, utilizing feedstock including microalgae.
Parabel, Inc. (Melbourne, FL, USA) – www.parabel.com/	Licenses technology and provides management support for production and processing of microalgae to develop biofuels, feed and food. Focus is on Asia, Latin America and African regions.
Sapphire Energy, Inc. (San Diego, CA, USA) – www.sapphireenergy.com	Pond production; green-crude. 300-acre site for raceway ponds and biorefinery under construction in New Mexico. Targeting production of over 5 M gallons of crude per year at site.
Seambiotic Ltd. (Tel Aviv, Israel) – www.seambiotic.com	Pond production on a 0.4 acre site, Ashkelon, Israel. Pilot scale, utilizing flue gas from Israeli Electric Company. Multiple products: biofuels, omega-3 oils, animal feeds.
Solazyme, Inc. (Sth. San Francisco, CA, USA) – www.solazyme.com	Heterotrophic production in 128,000 L fermentors. Produce biodiesel and jet fuel, and other co-products. Pilot-scale demonstrated, commercial-scale facilities under construction (Moema, Brazil).
Solix Biofuels, Inc. (Fort Collins, CO, USA) – www.solixbiofuels.com	Provide culture systems through their 4000 L floating photobioreactors and other technological solutions for biofuel production. Demonstration facility in Colorado.
Synthetic Genomics, Inc. (La Jolla, CA, USA) – www.syntheticgenomics.com	Designing metabolic pathways for the production of next generation biofuels and biochemical from a variety of feedstocks, including microalgae.

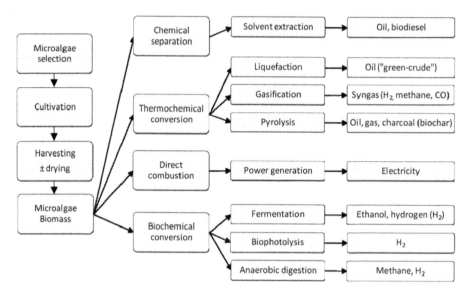

Figure 10.1.   Overview of processes within the microalgae-to-bioenergy pipeline. Conversion of microalgal biomass may proceed along one or several of the processing options (based on Huesemann and Benemann, 2009).

waste or by-products as a carbon (C) source and not requiring photosynthesis. Once grown at scale, the microalgae are harvested using techniques such as flocculation, filtration or centrifugation. The primary purpose of this step is to remove most, or all of the water (a drying step may be required) to provide a concentrated biomass that can then be processed further, leading to the production of microalgal fuels. Depending on the path chosen, these include ethanol, crude oil (= "green crude"), biodiesel, methane, hydrogen and syngas (a mixture of the latter and carbon monoxide). Biomass may also be directly combusted to yield energy.

10.1.5   *Current status of commercial microalgal biofuel production and future prospects*

The technical feasibility of some of the individual steps or processes in the microalgae to biofuel pipeline have been shown at a pilot or commercial scale, but there are major hurdles that still need be overcome before biofuels from microalgae could be regarded as cost-effective and able to compete with other biofuel feedstock. Currently there is no commercial-scale production of biofuel from microalgae. A key challenge is to ensure that more energy is obtained in the final microalgal biofuel products, than is needed to generate it. High costs and energy requirement for producing biomass, especially mixing, harvesting and de-watering are major obstacles. Significant cost-reduction could be effected by increasing the scale of production by about tenfold, developing improved strains with higher oil (or extractable energy) productivity, and technological advances that reduce harvesting and energy-extraction costs (Hall and Benemann, 2011). Genetic and engineering approaches may have a role here to improve strains. The use of nutrients from wastewater, and CO_2 from industrial flue effluent may also assist in reducing the input costs and therefore production costs. Finally, it will be essential to adopt a biorefinery approach whereby every component of the biomass is used to extract energy and other valuable products, thus improving the cost-efficiency of process.

Several companies are operating at pre-pilot or pilot scales, with targets to scale up to commercial production over the next five to ten years. Current production costs for microalgae-derived biofuels are estimated to be more than double that of fossil based fuels or non-algal biofuels, but industry is projecting that further technical advances will make microalgal biofuels cost-competitive in the near future. Ongoing investment will be needed to drive this innovation. It is unrealistic to expect that microalgae-based biofuels could ever completely substitute for current crude oil production – land requirements alone could equate to 2–40% of the United States land area (van Beilen, 2010). However, a successful industry could replace 1–2% of current global oil use (Hall and Benemann, 2011). While this number seems small, even 1% of the global oil supply is a vast amount and would contribute to a lowering of GHG emissions and improving energy security.

10.2   GENERAL PROPERTIES OF MICROALGAE

10.2.1   *Taxonomy and general characteristics*

Microalgae are unicellular micro-organisms found as single cells or colonies, ranging in size from one to several hundred microns. As photosynthetic organisms, they can convert solar to chemical energy and are regarded as more efficient than higher land plants due to their simpler cell structure. Because they grow in aqueous environments, microalgae have better accessibility to water, dissolved CO_2 and nutrients. Under optimal conditions they may double their biomass every 1 to 2 days (Sheehan *et al.*, 1998).

Microalgae may be eukaryotic or prokaryotic, i.e. characterized by the presence or absence, respectively, of a membrane-bound nucleus. Eukaryotic microalgae are divided into nine taxonomic groups, i.e. Bacillariophyceae (diatoms), Chlorophyceae (green algae), Phaeophyceae (brown algae), Chrysophyceae (golden algae), Xanthophyceae (yellow-green algae), Prasinophyceae, Rhodophyceae (red algae), Dinophyceae (dinoflagellates) and Eustigmatophyceae (Van Den Hoek *et al.*, 1995). Cyanophyceae, commonly known as blue green algae or cyanobacteria are prokaryotic, and are strictly a form of bacteria. However because of their similarities to eukaryotic microalgae and their photosynthetic ability, in this review they are considered together with the latter under the general term of microalgae. The number of described species has been estimated as 27 to 47 thousand (Sharma and Rai, 2011). Hence, microalgae represent a tremendous source of biodiversity, and as such, survive and proliferate in a range of climatic and aqueous environments, from freshwater to marine to hypersaline (Jeffrey *et al.*, 1991).

Most microalgae only grow photoautotrophically, though some can grow heterotrophically. Photoautotrophic species use CO_2, and/or bicarbonate as a carbon source and light as a source of energy. Heterotrophic microalgae use only organic compounds for growth. Mixotrophic species can grow either phototrophically or heterotrophically, depending on the culture environment. This chapter will mainly focus on microalgae grown photoautotrophically, with occasional reference to heterotrophs.

10.2.2   *Sourcing and maintaining microalgae species or strains*

Several laboratory techniques have traditionally been used for isolating individual cells from natural samples, including separation using micropipettes, serial dilution cultures, successive plating on agar media and gravity/centrifugation techniques (Andersen and Kawachi, 2005). These techniques may be supplemented with use of antibiotics, as it may be important for maintenance to render cultures axenic. During the last few decades, automated technologies like flow cytometry with fluorescence-activated cell sorting have been applied for isolating small cells, which can then be sorted into multi-well plates for establishing new microalgal cultures (Thi Thai Yen *et al.*, 2011).

Once isolations are established, microalgae are grown in enriched culture media. The choice of medium will depend on whether the microalga comes from a fresh, brackish, or a marine environment. Media typically contain essential macronutrients such as nitrogen, phosphorus and silicon, as well as major ions (magnesium, calcium, chloride, sulfate), micronutrients (iron, zinc, cobalt, cadmium and molybdenum) and vitamins (e.g. B1, B12 and biotin) (Harrison and Berges, 2005; Watanabe, 2005). Stock cultures are typically kept in cultures tubes (e.g. 10–20 mL) or small glass flasks (125 to 500 mL) stoppered with cotton wool or packed cellulose

Table 10.2.   Major culture collections supplying living microalgae cultures, for commercial or research purposes. Information on other collections is available from the World Federation of Culture Collections (http://www.wfcc.info).

Culture collection/supplier and website	Description
The Culture Collection of Algae – UTEX web.biosci.utexas.edu/utex/	Approx. 3000 strains of marine and freshwater microalgae; representatives from extreme environments
Culture Collection of Algae and Protozoa – CCAP www.ccap.ac.uk	Approx. 2500 strains of marine and freshwater microalgae and protozoa
SAG Collection – University of Göttingen www.uni-goettingen.de/en/184982.html	Approx. 2300 strains (1400 species); mainly freshwater
Microbial Culture Collection at the National Institute of Environmental Studies – NIES mcc.nies.go.jp	Over 2300 strains (700 species) of marine and freshwater microalgae and protozoa
Provasoli-Guillard National Center for Marine Algae and Microbiota (NCMA) https://ncma.bigelow.org/	Over 2000 strains of marine microalgae.
Australian National Algae Culture Collection – ANACC www.csiro.au/ANACC	Over 1000 strains (300 species); marine and freshwater; cold-water to tropical strains; majority Australian isolates
ACOI Coimbria Collection of Algae acoi.ci.uc.pt	Over 4000 strains (1000 species); freshwater

plugs, and maintained within constant environment rooms. Stock cultures are used only to provide inocula for starter cultures when required, and are sub-cultured every 2–6 weeks by aseptically transferring a small volume to a sterilized flask with culture medium (e.g. 1–2 mL inoculum to 150 mL).

Culture collections (Table 10.2) provide strains that are usually well characterized for growth with physiological, biochemical and genetic information, highlighting their value as repositories of global microalgal biodiversity. Collections can provide a variety of different species or strains that may be utilized for research, and industrial applications including biofuels, nutraceutical, pharmaceuticals, food crops and aquaculture feeds.

10.2.3   Chemical profiles of microalgae

10.2.3.1   Proximate composition

The initial products of CO_2 fixation by microalgae during photosynthesis are carbohydrates. Further metabolic processes convert the carbon into proteins, lipid and nucleic acids. Approximately ninety percent of the carbon within the cell is bound up in these macromolecules. The precise ratios of these and the non-organic fraction (i.e. mineral components, or ash) can vary substantially between microalgae species. Under standard growing conditions (i.e. where light or nutrients are not limited), microalgae typically contain 25 to 50% of DW as protein, 10 to 30% as lipid, 5 to 30% as carbohydrate and 5 to 40% as ash (Volkman and Brown, 2006). Cryptophytes are generally rich in protein (40 to 60% DW), whereas diatoms have more ash (20 to 40% DW) (McCausland et al., 1999). Cyanobacteria contain less lipid (usually 5 to 15% of DW) (De Oliveira et al., 1999) than the eukaryotic microalgae. Nucleic acids may contribute between 1 to 8% of the DW, and crude fiber may account for up to 10% of total carbohydrate in some species (Volkman and Brown, 2006).

When microalgae become nutrient limited (stationary phase), growth slows and this is usually accompanied by significant compositional changes. Depending on the species, limiting nutrient, and other factors, increases in lipid, carbohydrate, or both, may occur at the expense of protein (Volkman and Brown, 2006). Light (intensity and photoperiod), temperature, salinity, and pH can also influence the proximate composition though not in a consistent way across all species (Sharma *et al.*, 2012).

10.2.3.2 *Qualitative aspects of proximate composition – amino acids and sugars*

The biochemical profiles of microalgae can directly affect extraction and processing efficiency of biomass, and define the chemical and physical properties of the final fuel. As well, the composition may influence the economic value of the residue remaining after fuel extraction, for exploitation as a co-product (e.g. animal feeds, cosmetics, etc.).

Most of the amino acids, i.e. 88 to 97%, within microalgae are incorporated as protein. The amino acid composition of the protein fraction is very similar across different species (Brown, 1991). Aspartate and glutamate are in highest concentrations (8 to 12% of total amino acids); cystine, methionine, tryptophan and histidine are in lowest (typically 0.5 to 3%) and values of other amino acids range from 4 to 8%. These profiles represent a balanced composition, comparing favorably to other marine proteins (Brown, 1991).

Carbohydrates in microalgae are distributed between the polymeric (polysaccharide) and simple sugars, i.e. the mono-, di- and oligo-saccharides. Most polysaccharide is readily-hydrolysable, which typically accounts for 80 to 95% of carbohydrate (Brown, 1991). Microalgae can also have up to 13% DW as non-hydrolysable fiber (Mišurcová *et al.*, 2010). Of this, some species may have chitin, cellulose matrix, or pectic polysaccharides associated with cell wall structures (Vesk *et al.*, 1990), though, unlike many terrestrial crops, microalgae are deficient in lignin.

Glucose is the predominant sugar within microalgal polysaccharide, ranging from 20 to 90% of total sugars (Brown, 1991). Galactose (1 to 20% of sugars) and mannose (2 to 46%) are also common, with fucose, arabinose, rhamnose, ribose and xylose found in varying percentages (0 to 17%). These observations are consistent with glucans (polysaccharides rich in glucose) being the major polysaccharide in microalgae. The principal glucans in chlorophytes, prasino-phytes and rhodophytes are amylase and/or amylopectin; in cyanobacteria, glycogen and in the diatoms, chrysolaminarin (Beattie *et al.*, 1961; Williams and Laurens, 2010). In diatoms, structural polysaccharides contribute to the cell wall matrix, i.e. glucuronomannans and polysaccharides with widely varying proportions of galactose, glucose, mannose, xylose, fucose and rhamnose.

10.2.3.3 *Lipid class and fatty acids*

Lipids act as energy reserves and also have a structural role within the cell. Neutral lipids fulfill the role of energy reserves, and the major form is triacylglycerol (TAG), i.e. esters derived from glycerol and three fatty acids. The polar lipids are most commonly found as membrane components and are typically comprised of a glycerol moiety bonded to two fatty acids, with the third position esterified to a polar molecule. In microalgae the major polar lipids are the phospholipids (phosphatidylcholine, phosphatidylethanolamine and phosphatidylglycerol) and glycolipids (monogalactosyldiacylglycerol, digalactosyldiacyl-glycerol and sulfoquinovosyldia-cylglycerol) (Volkman and Brown, 2006). Fatty acids may contain no double bonds (saturated fatty acids; SFA), or have one (mono-unsaturated; MUFA) or more (poly-unsaturated; PUFA) double bonds.

Under nutrient-replete conditions, microalgae synthesize fatty acids primarily for incorporation into polar lipids, which may constitute 70 to 90% of total lipids representing 5 to 20% of their DW (Volkman and Brown, 2006). As cells transition and then reach nutrient limitation, or other associated stresses leading to reduced cell division, many microalgae regulate their lipid biosynthesis towards the production and accumulation of TAG. Hence, in stationary phase TAG can constitute between 20 and 80% of total lipid, and 5 to 40% of cell DW (Bigogno *et al.*, 2002; Sharma *et al.*, 2012; Tornabene *et al.*, 1983).

Microalgal lipids may also include free fatty acids, wax esters, sterols, aliphatic alcohols, hydrocarbons, prenyl derivatives and chlorophylls, though usually these are all in low proportions

(Hu *et al.*, 2008; Volkman *et al.*, 1989). One exception is the chlorophyte *Botryococcus braunii*. Strains of this species have been investigated for several decades as a source of fuel, because of their ability to produce large amounts of very long chain (C_{25} to C_{40}), unusual hydrocarbons. Values of up to 75% of DW have been recorded, albeit only after a prolonged stationary phase – hence with low growth and reduced lipid productivity. Several reviews have been published specifically on *Botryococcus* chemistry, biology and potential as biofuel feedstock (e.g. Banerjee *et al.*, 2002). To date, this potentially exceptional hydrocarbon producing microalga has defied efficient growth production at scale.

With respect to fatty acid distribution, most of the PUFA are present within polar lipids, whereas TAG usually has higher proportions of SFA and MUFA (Volkman and Brown, 2006). However, there are exceptions to this. The oleaginous chlorophyte *Parietochloris incisa* accumulated TAG during stationary phase of which more than half was PUFA, and predominantly arachidonic acid (20:4n-6; ARA) (Bigogno *et al.*, 2002).

The fatty acids profiles of many microalgae have been examined, focusing on their link to lipid quality in applications such as biofuels, aquaculture, nutraceutical and microalgal physiology (Brown *et al.*, 1996; Knothe, 2011; Spolaore *et al.*, 2006; Volkman *et al.*, 1989). Profiles of representative microalgae are compared in Table 10.3, together with those of some terrestrial plants that are being examined as biofuel feedstock, i.e. *Jatropha* and palm. The dominant SFA in microalgae is generally palmitic acid (16:0), whereas 18:1 (usually oleic acid, 18:1n-9), while often in significant percentages, is less than that seen in oil of terrestrial plants. Of the other MUFA, microalgae may also contain significant percentages of palmitoleic acid 16:1n-7.

PUFA typically represents $\geq$40% of total fatty acids in microalgae, moreover many strains contain high amounts of long chain ($C \geq 20$) PUFA (LC-PUFA), which are not produced in crop plants (Table 10.3). Of the LC-PUFAs, arachidonic acid (20:4n-6; ARA), eicosapentaenoic (20:5n-3; EPA) and docasohexaenoic acid (22:6n-3; DHA) are of interest for aquaculture because they are essential for marine fish (Volkman and Brown, 2006). EPA and DHA as components of fish oil (obtained by the fish in their diet with microalgae being the primary source) also provide health benefits in humans with a rapidly increasing global market and daily intake recommendations (Ruxton *et al.*, 2007). Diatoms, cryptomonads, eustigmatophytes, rhodophytes and some prymnesiophytes (e.g. *Pavlova* spp.) are good sources of EPA. Dinoflagellates have high DHA, whereas many cryptomonads and some prymnesiophytes (e.g. *Isochrysis* sp. T.ISO) have moderate amounts (Table 10.3 and Volkman and Brown, 2006). Prasinophytes can contain significant proportions EPA or DHA. Eustigmatophytes, rhodophytes and diatoms contain small amounts of ARA (up to 6% of fatty acids). Chlorophytes, though usually lacking the LC-PUFA, typically have high percentages of C18 PUFA (Volkman and Brown, 2006).

The diversity of fatty acid profiles in microalgae both within and between different taxonomic classes may in part be linked to their responses to environmental conditions, but may be underpinned and governed by intrinsic species (or strain) differences in enzyme expression within fatty acid biosynthetic pathways. Compared to land plants, critical knowledge regarding the regulatory and structural genes associated with microalgal fatty acid biosynthesis is lacking, though this is an area of active investigation.

10.2.3.4   *Other chemical components within microalgae of commercial interest*

Apart from the value of microalgae biomass based on their compositional profiles for applications like biofuel feedstock, and feeds, secondary components may also be of commercial significance. Microalgae, either as intact living cells, or processed biomass, are important sources of essential trace nutrients such as vitamins and minerals in aquaculture, and as health food supplements, e.g. dried *Spirulina* and *Chlorella* (Spolaore *et al.*, 2006).

The biodiversity of microalgae is also reflected by their chemical diversity. Products that are currently commercialized or have commercial potential include: pigments for feed additives or health supplements, PUFAs for the nutraceutical market, certain polysaccharides used in the food industry, stable isotopes (e.g. fatty acids, amino acids) for diagnostics, extracts used in cosmetic

Table 10.3. Fatty acid composition (% of total fatty acids) of some microalgae from different taxonomic classes that are being investigated for biofuel potential, and their comparison to profiles of oil from crop plants. Microalgae were grown either phototrophically or heterotrophically.

Fatty acid	Phototrophic							Heterotrophic		Terrestrial crop oil	
	A. pla [CYN]	T. ps [DIA]	T. sue [PRA]	N. oc [EU]	ISO [PRM]	D. ter [CHL]	C. sor [CHL]	C. sor [CHL]	C. coh [DN]	Jatropha	Palm
Saturated fatty acids											
14:0 (myristic)		14.3	0.6	5.4	16.6	0.2			14.9		
16:0 (palmitic)	42.3	11.2	20.3	22.7	11.5	14.7	12.8	13.7	18.7	14.2	40.3
18:0 (stearic)	0.95	0.7	0.9	0.4	0.2	0.4			9.1	6.9	3.1
Other saturates	8.6	1.0	0.5	0.8	0.3	0.1					
Total saturates	47.8	26.2	21.8	28.5	28.3	15.3	12.8	13.7	44.9	21.1	43.3
Monounsaturated fatty acids											
16:1	1.0	18.7	2.7	20.6	2.9	2.9	3.6	4.6		1.4	
18:1	2.0	0.6	12.4	5.9	17.6	2.3	16.1	14.4	0.5	43.1	43.4
Other monounsaturates	2.9	0.3	1.6	0.1	0.1		2.5	2.2			
Total monounsaturates	5.9	19.2	15.1	26.5	20.5	5.2	22.2	21.2	0.5	44.5	43.4
Polyunsaturated fatty acids											
16:2	2.4	7.2	1.1	0.7	1.3	0.7	5.7	5.0			
16:3		12.7	6.5			4.2					
16:4		2.3	13.7			21.0					
18:2	16.2	0.4	13.8	1.7	4.5	4.8	14.5	14.1		34.4	13.2
18:3	20.1	0.3	11.8	0.6	6.1	46.2	26.1	26.4			
18:4(n-3) (stearidonic)		5.3	8.4		24.2	1.0					
20:4 (arachidonic)		0.6	1.8	6.7							
20:5(n-3) (eicosapentaenoic)		19.3	4.3	33.4	0.6						
22:6(n-3) (docosahexaenoic)		3.9			9.0				54.3		
Other polyunsaturates					3.3					0.3	
Total polyunsaturates	39.6	52.6	59.5	43.0	45.7	77.9	46.3	45.5	54.3	34.4	13.2
Others fatty acids	3.0	1.3	1.5	1.0	1.9	1.5					
Reference[2]	[1]	[2]	[2]	[3]	[3]	[2]	[4]	[4]	[5]	[6]	[6]

[1] Species codes: *A. pla* = *Arthrospira (Spirulina) platensis*; *T. ps* = *Thalassiosira pseudonana*; *T. sue* = *Tetraselmis suecica*; *N. oc* = *Nannochloropsis oculata*; *ISO* = *Isochrysis* sp. (T.ISO); *D. ter* = *Dunaliella tertiolecta*; *C. coh* = *Crypthecodinium cohnii*; *C. sor* = *Chlorella sorokiniana*. Microalgae class codes: CYN = cyanobacteria; DIA = diatom; PRA = prasinophyceae; EU = eustigmatophyceae; PRM = pymnesiophyceae; CHL = chlorophyceae; DN = dinoflagellate.
[2] References: [1] Ötleş and Pire (2001); [2] Volkman et al. (1989); [3] Dunstan et al. (1993); [4] Zheng et al. (2012); [5] Jiang and Chen (2000); [6] Sarin et al. (2007).

industries, and variety of bioactive molecules (e.g. microbicides, toxins, enzyme inhibitors, anti-tumor) (Spolaore *et al.*, 2006). Two examples follow:

a) Pigments: β-carotene is produced by the chlorophyte *Dunaliella salina*, at concentrations up to 14% of DW (Borowitzka and Borowitzka, 1988). The major commercial producer is BASF, Australia who produces the microalgae in hyper-saline ponds. β-carotene can attract prices up to US$ 3000/kg (Spolaore *et al.*, 2006). Astaxanthin, another highly-valued carotenoid pigment, is produced by the chlorophyte *Haematococcus pluvialis*, where in stationary phase concentrations can reach 1.5 to 3% of DW. Astaxanthin's major market is as a feed ingredient (flesh colorant) for the salmon industry, and its value is estimated at US$ 200 M per annum with an average price of US$ 2500/kg (Hejazi and Wijffels, 2004). For both pigments, synthetic forms are also available, though the microalgal derived products command premium prices (Spolaore *et al.*, 2006).

b) PUFA: The dinoflagellate *Crypthecodinium cohnii* grown heterotrophically contains over 40% lipid, with a DHA content in excess of 30% of total fatty acids (Ratledge, 2004). Martek Biosciences Corporation (Columbia, MD, USA) are producing oil from *C. cohnii*, which is included as a feed additive to infant milk formula product of various manufacturers, and distributed to over 60 countries. Martek's oil production of DHA-rich oil from *C. cohnii* was estimated to be 240 tons in 2003 (Ratledge, 2004).

10.3   SELECTION OF STRAINS AS CANDIDATES FOR BIOFUEL FEEDSTOCK

10.3.1   *Growth rates and environmental tolerances from small-scale cultures*

There are a number of methods to convert microalgal biomass to energy, and multiple products are potentially obtainable (Fig. 10.1). Where the main product sought is crude microalgal oil ("green crude") or biodiesel, the first step is to identify strains with high oil production and rapid division rates. Much of the early-stage screening of strains has been undertaken in small-scale cultures (e.g. 200–2000 mL flasks) whereupon strains with suitable properties are successively scaled up in volume for more rigorous characterization including experimentation to optimize oil production (Rodolfi *et al.*, 2009).

Strains for screening may be existing isolates from culture collections (see Section 10.2.2), and/or new strains isolated from the same or similar biogeographical regions where scaled-up production might occur. The rationale for investigating local isolates is that they may be better adapted to prevailing environmental conditions and display better growth characteristics. Culture parameters that are typically examined in screening trials are temperature, light (photoperiod and intensity), salinity and nutrients (inorganic and C-source). Strains must not only have favorable growth and production rates under specified, controlled conditions, but also importantly be able to tolerate environmental fluctuations that are likely to be encountered during production-scale growth.

Screening may also target strains that are capable of tolerating high concentration of CO_2 (10 to 40%) in industrial flue gases, e.g. *Chlorococcum littorale*, *Scenedesmus obliquus* and *Chlorella kessleri* (Wang *et al.*, 2008) and references quoted therein). Another example is the screening of strains that can be grown successfully using municipal wastewater, with a view to integrating microalgae cultivation for biodiesel feedstock with wastewater treatment (Li, Zhou *et al.*, 2011). Establishment of whether strains are photoautotrophic, heterotrophic or mixotrophic is undertaken by screening cultures grown in the presence or absence of organic substrates, with or without light.

10.3.2   *Screening for total lipid, and fatty acid quality*

Some several hundred oil-rich (oleaginous) species or strains have been evaluated under laboratory and/or outdoor culture conditions. Hu *et al.* (2008) examined $\approx$100 oleaginous strains, with the

Table 10.4.   Lipid content of some oleaginous microalgae being investigated for biofuel production.

Microalgae species	Lipid content [% DW]	Reference
Attheya septentrionalis	24–45	Knuckey *et al.* (2002)
Botryococcus braunii	25–75	Chisti (2007)
Chlorella vulgaris	25–53	Mujtaba *et al.* (2012)
Dunaliella primolecta	17–37	Thomas *et al.* (1984c)
Extubocellulus spinifera	18–33	Knuckey *et al.* (2002)
Isochrysis sp. (T.ISO)	21–43	Thomas *et al.* (1984a)
Nannochloropsis sp.	30–69	Bondioli *et al.* (2012)
Neochloris oleabundans	7–40	Li *et al.* (2008)
Phaeodactylum tricornutum	20–30	Thomas *et al.* (1984b)
Tetraselmis suecica	22–29	Bondioli *et al.* (2012)
Thalassiosira pseudonana	21–31	Brown *et al.* (1996)

most common group represented being the chlorophytes, followed by diatoms. They attributed the prevalence of chlorophytes as due to their occurrence in diverse habitats, their ease of isolation, and fast growth rates compared with many other species. The average lipid content from these eukarytotic groups were similar, i.e. averaging 23 to 27% of DW under normal growth, and 38 to 46% of DW under stress conditions. While noting that many cyanobacteria had also been screened for lipid production, their content was less (i.e. average 10% of DW under stress conditions) compared to the eukaryotic microalgae. The range of oil or lipid content found in some of the oleaginous species that have been investigated for biofuel are shown in Table 10.4.

The methods for lipid analysis which have commonly been used in screening require extraction of samples with organic solvents, then evaporating the solvent to yield an oil which is weighed, i.e. gravimetric analysis. These methods are precise, and coupled with gas chromatography can yield important information on the quality of the oil. However, they are costly and time-consuming, and in some cases a less precise method would be acceptable for screening if it could provide a much higher analysis throughput. Recently, flow cytometry with fluorescence-activated cell sorting has been applied to cells pre-treated with fluorescent, lipid-staining dyes such as Nile Red and BODIPY. As cells are sorted fluorescence is measured and, generally for most species, the relative fluorescence provides an estimation of the lipid content (Thi Thai Yen *et al.*, 2011). The technology allows for high throughput-screening and is being used to identify and isolate high-lipid microalgal strains as candidates for further assessment as biofuel feedstock. The technique is not without limitations however, as some strains take up the dye unevenly and results may vary with culture conditions (Ratha and Prasanna, 2012). Near infra-red reflectance spectroscopy is also being investigated as a rapid technique for screening the lipid content in microalgae, and it has shown to be accurate for estimation in dried microalgal biomass, but yet to be demonstrated in live cultures (Laurens and Wolfrum, 2011).

The fatty acid profiles of microalgal oils are major determinants in the qualitative properties of derived biodiesel (Knothe, 2011). Saturated fatty acids (SFA) are more favorable for combustion-related characteristics (exhaust emissions and cetane number, a measure of the ignition quality of the fuel) and oxidative stability, whereas unsaturated fats provide better cold flow properties and kinematic viscosity. Compared to other feedstock commonly used for biodiesel (e.g. *Jatropha* and palm oil), microalgal oils often have high proportions of SFA and PUFA (Table 10.4). As a result, such microalgal oils are likely to produce biodiesel with poor cold flow properties and poor oxidative stability, although other properties, e.g. cetane number, may be favorable (Knothe, 2011). Hence, species such as *Nannochloropsis*, *Isochrysis* and *Schizochytrium* that are considered having potential for biofuel production based on lipid production may prove unsuitable for biodiesel based on their fatty acid profile (Knothe, 2011). Other species, such as *Trichosporon capitatutum* or *Chlorella protothecoides* (both can contain more than 65% of 18:1n-9, oleic acid) have more favorable profiles (Gao *et al.*, 2010; Wu *et al.*, 2011).

Therefore, based on their composition, many microalgal oils are unlikely to comply with European biodiesels standards, e.g. EN14214, which set out physical and chemical properties that biodiesel must match or exceed to be acceptable as a diesel substitute (Knothe, 2011). However, apart from strain selection, several strategies may improve the quality of biodiesel derived from microalgae. Fatty acid composition of microalgae can be manipulated within limits by growth conditions, e.g. the cetane number should increase under conditions that promote the accumulation of TAG rich in SFA. Chemical strategies could include the use of alcohols other than methanol in the transesterification process (ethyl and isopropyl esters have superior low-temperature properties), hydrogenation to increase saturation, catalytic cracking, and the use of additives to augment biodiesel properties (Georgianna and Mayfield, 2012). Using transgenic approaches, it may be feasible to insert genes into microalgae leading to increases in the production of 12:0 and 14:0 (optimal chain-length for biodiesel) and/or fatty acids of even shorter chain length suited for production of gasoline or jet fuel (Radakovits *et al.*, 2010).

10.3.3   *Other strain selection criteria*

Apart from lipid quantity and quality, other information from laboratory screening influences selection of potential biofuel candidates. These include:

- The ability of strains to compete with other microalgae and bacteria, and to resist predation. For example, the salinity-tolerant *Dunaliella salina* can successfully outcompete other species in hypersaline environments (Borowitzka, 1999);
- Content of high-value products or other non-lipid components that may be converted to biofuel. Some chlorophytes (e.g. *Chlorella* spp.) and prasinophytes (e.g. *Tetraselmis* spp.) may accumulate up to 50% DW starch, which can be efficiently converted to ethanol using industrial fermentation (Hirano *et al.*, 1997; Thomas *et al.*, 1984c);
- Strains that may directly produce non-lipid fuels, such as ethanol and hydrogen. For example, the chlorophyte *Chlamydomonas rheinhardtii* can produce both products under different selected culture conditions (Hemschemeier *et al.*, 2008); and
- The suitability of strains to potential downstream harvesting techniques, e.g. flocculation (Knuckey *et al.*, 2006). Cell morphology and dimensions, fragility and surface chemistry will influence this process.

10.4   SCALING UP PRODUCTION OF MICROALGAE BIOMASS

Beyond laboratory-scale screening, further assessment is required at scales more likely to reflect actual commercial production conditions. This section looks at the general considerations of microalgal mass culture, the growth systems used (e.g. ponds versus PBRs), and the production rates that may be obtained using varying system scenarios.

10.4.1   *General considerations*

Microalgae can be cultivated from laboratory-scale to industrial scale (e.g. for biofuels or other applications) using aqueous media containing growth nutrients. Cultured microalgae, in essence, behave rather like natural microalgal blooms that form when there is good growth conditions (e.g. light, temperature, nutrient supply), but in controlled conditions and growth systems. Typical growth and proliferation of a batch culture is characterized by a lag phase where there are no increase in cell numbers followed by an exponential or logarithmic (log) phase of growth until some growth factor becomes limiting (e.g. light or one or more nutrients) at which stage the growth pattern changes to a phase of declining relative growth and then stationary phase, followed by a death phase. To optimize lipid yield for biofuel production the term two-stage culture is often cited, where the first stage is the logarithmic phase during which biomass density is established through cell proliferation, and the second stage occurs during late log to stationary phase during synchronous growth where lipid accumulates (Suali and Sarbatly, 2012). The duration of a culture may be extended by using semi-continuous or continuous culturing. In such cultures, fresh medium (nutrients and water) are supplied either semi-continuously or continuously, along with removal of a matching amount of the culture, e.g. 5 to 20% of the culture volume daily (Zhang and Richmond, 2003). This prevents the culture becoming nutrient-limited, avoiding the stationary and death phases.

For phototrophic cultures, factors that influence the growth and composition of microalgae include light (intensity, quality and photoperiod), nutrient concentrations, temperature, salinity, pH and mixing (Schenk *et al.*, 2008 and references quoted therein). The response of different microalgae may differ significantly under varying environments, i.e. conditions that are optimal for one species may not necessarily be so for another, even for closely related species. While these factors are similarly important in mass-culture production, the added complexity is that several of these environmental factors may be subject to little, or limited control. This is especially the case for outdoor cultivation systems, where temperature and light fluctuations are unavoidable and significantly impact on production rates. Therefore, strains chosen must not only have good production characteristics under optimal growing conditions, but also be able to tolerate or adapt to the climatic conditions experienced at the growing location.

10.4.1.1 *Light and temperature*

Light, together with nutrients are considered the key factors that may control the growth of microalgae. Light intensities at which maximum photosynthesis and growth rates are observed vary between microalgal species and groups. Maximum growth rates of cyanobacteria, diatoms and chlorophytes occurred at values of 40, 50 and 210 μmol photon/m^2/s, respectively (Richardson *et al.*, 1983). For some *Chlorella* strains, optimum light intensities may be in excess of 300 μmol photon/m^2/s (Pulz and Scheibenbogen, 1998). At intensities beyond this light saturation point, further increases can actually reduce growth rates. This phenomenon, i.e. photo-inhibition, results from generally reversible damage to the photosynthetic apparatus due to excessive light (Chisti, 2007).

The light saturation levels are generally much lower than typical midday outdoor light intensities in equatorial regions (i.e. 2000 to 2300 μmol photon/m^2/s). A large amount of light energy may be wasted as heat or fluorescence near the culture surface, whereas cells deeper in the culture unit may be light limited because of the attenuation of light. This reduces the overall solar efficiency of the system. One solution is to reduce the antenna pigment content of microalgae by engineering other biological strategies, so that each photosystem receives only the light it needs, instead of an excess (Stephens *et al.*, 2010). Despite ongoing research in this area, no improved strain is yet available that has proven to have higher productivity outside the laboratory (Huesemann and Benemann, 2009).

Also, rapid mixing cycles, e.g. at levels down to a millisecond time scale, can move cells between light and dark regions in the culture unit. As well as reducing photoinhibition, rapid mixing can also improve the efficiency of mass transfer.

The irradiance a single cell experiences in a mixed system changes continuously in relation to the exterior photic zone, the depth of which is determined by the degree of mutual shading which, in turn, is a function of cell density. What is also important therefore is the average irradiance a cell receives, and how this relates to saturation levels (Richmond, 1992). Another key point is that there exists a strong interaction between light intensity and temperature, on photosynthesis and cell growth, i.e. the maximum utilization of light energy can only be realized when temperature is optimal (Richmond, 1992). More discussion on microalgal cell physiology and growth in relation to temperature-light interactions are provided by Richmond (1992) and Williams and Laurens (2010).

A temperature of 15°C or above is considered essential for sustained microalgal production, with optimal growth usually within the range 17 to 28°C depending on species (Jeffrey *et al.*, 1991; Moheimani and Borowitzka, 2006; Thompson *et al.*, 1992). Microalgae with broad temperature tolerances, i.e. with good growth between 10 and 30°C include *Tetraselmis* spp., *Dunaliella tertiolecta, Nannochloropsis oculata, Chaetoceros calcitrans* and *Scenedesmus* sp. (Jeffrey *et al.*, 1991; Li, Hu, *et al.*, 2011). Temperate strains exhibiting good growth between 10 and 20°C, but unable to sustain growth at higher temperatures include *Thalassiosira pseudonana* and *Skeletonema costatum* (Jeffrey *et al.*, 1991). *Chlorella* species generally show temperature optima between 20 and 30°C (e.g. *C. vulgaris, C. protothecoides*), though *C. sorokiana* can grow at a temperature of 40°C (Bechet *et al.*, 2013; Kessler, 1985). Few microalgae can grow at such elevated temperatures, another example includes *Scenedesmus almeriensis*, which grows optimally at 35°C and can also grow at a temperature of 40°C (Sánchez *et al.*, 2008).

10.4.1.2 *Inorganic nutrients*

The main inorganic nutrient requirements for microalgae are nitrogen (N), and to lesser degrees, phosphorus (P), potassium and (for diatoms) silicon. While some cyanobacteria (e.g. *Anabaena* spp., Moreno *et al.*, 2003) can fix N directly from the atmosphere, most microalgae require added soluble N to the culture media in the form of nitrate, ammonia or urea. Within culture media, N is added typically at a rate of 25 mg N/L for phototropic production (i.e. $f/2$ formulation according to Guillard and Ryther, 1962), though inclusions of up to 1.5 g N/L have been used both in phototropic and heterotrophic systems (Converti *et al.*, 2009; Shi *et al.*, 1999). Once cell growth slows, cultivation through to N-limitation or starvation stages can significantly enhance lipid content (Suali and Sarbatly, 2012; Volkman and Brown, 2006). P is required at about one-sixteenth the concentration of N based on cell compositional profiles, but is often added in excess because not all added P is bioavailable. Silica is typically added at 15 mg Si/L, for diatom culture (Guillard and Ryther, 1962).

Agricultural-grade, industrial fertilizers are used for commercial-scale production (Gonzalez-Rodriguez and Maestrini, 1984). These provide an inexpensive growth media, but there is a concern that fertilizer production is dependent on fossil fuels, and this must be included in life-cycle assessments. P is a non-renewable resource and at current extraction rates, global commercial phosphate rock reserves could be depleted within 50 years (Borowitzka and Moheimani, 2013). This puts microalgal production for biofuel in direct competition with food crop production, which relies on phosphate fertilizers. This makes it critical to use P efficiently and to recycle nutrients wherever possible. Along with CO_2, N and P are seen as the dominant constraints in scaling up autotrophic microalgal cultivation for biofuel production (Pate *et al.*, 2011).

P and N from wastewater may also be utilized for microalgal culture (Li, Zhou, *et al.*, 2011), but growth is inconsistent because of compositional variation of the wastewater and generally high concentrations of ammonia (Borowitzka and Moheimani, 2013). There is developing interest in this approach, and it has been argued that microalgal cultivation together with wastewater treatment is the most feasible and sustainable pathway to commercial biofuel production from microalgal biomass in the short term (Batten *et al.*, 2013; Benemann, 2009). Cost savings on the requirements for chemical remediation and minimization of fresh water requirements for biomass production are the primary drivers for producing microalgal biomass as part of a wastewater treatment process.

10.4.1.3 CO_2

Aeration of cultures with CO_2-enriched air is essential to obtain the high productivities for almost all phototrophic microalgae, and also aids in mixing and pH buffering. Each kg of microalgae produced fixes about 1.8 kg of CO_2 – but because of inefficiencies higher amounts e.g. at least 5 kg CO_2 for open ponds, may be required (Petkov *et al.*, 2012). The use of industrial flue gases from power plants, typically containing 10 to 20% v/v CO_2, is potentially a cheap and viable source of CO_2 for microalgal production, and its use for this purpose has been proposed as one way of capturing anthropogenically produced CO_2 (Salih, 2011). However, flue gas may need to be purged of soot, dust, products of incomplete combustion and SO_2 (Petkov *et al.*, 2012). There may be an added cost in the distribution of the CO_2 enriched gas to the microalgal production pond. High concentrations of CO_2, such as those feeding in directly from flue gas may inhibit microalgal growth in which case dilution with air may be necessary. Alternatively, strains tolerant to high concentrations ($\geq$14%) of CO_2, e.g. strains of *Dunaliella tertiolecta*, *Nannochloris* sp., *Tetraselmis* sp., *Chlamydomonas* sp., *Scenedesmus* sp. and *Chlorella* sp. – could be chosen as culture candidates (Salih, 2011).

10.4.1.4 *Land and water*

The largest commercial microalgae production operation (BASF; at their sites in Australia) has raceway pond area in excess of 750 ha, and there is a view that plants producing microalgae for biofuel may need to be of a similar or larger scale (Borowitzka and Moheimani, 2013). Nevertheless, compared to the other types of biofuels produced from biomass such as palm sugar cane or canola, the land footprint of microalgae systems would be significantly less, and, importantly, the land does need to be suitable for conventional food crops. Physical characteristics, such as topography and soil, could also restrict the land available for outdoor production. Topography would be a limiting factor for pond systems because the construction and operation requires a flat terrain, e.g. <5% slope (Darzins *et al.*, 2010). Soil porosity and permeability properties influence the construction costs and design of pond systems, by virtue of the need for pond lining or sealing. Social, legal, economic and political factors may also influence land availability.

Regions with high annual average sunshine should be selected to yield optimal growth, though according to Darzins *et al.* (2010) radiation levels of 1500 kWh/m^2/year are adequate, which means most of the earth's land could be suitable for production. Mean monthly minimum temperatures in excess of 15°C are important for economically feasible, year-round production. Rates of evaporation and rainfall are also important. The site should be located as close as possible to key production inputs such as CO_2, nutrients and water to reduce costs of supply. Various modeling approaches are available and being applied to identify potential geographic regions for microalgal biomass production, based on collected geospatial data, e.g. CO_2 sources, wastewater location with nutrients concentrations, land topography and solar resources (Batten *et al.*, 2011).

Microalgae mass cultures require large volumes of water, e.g. between 40 and 80 kL per ton of microalgal biomass from pond production (Borowitzka and Moheimani, 2013). Freshwater sources are often limited in areas otherwise suited for microalgal production in criteria such as temperature and insolation, and their use at scale may require the appropriation of irrigation water from other agricultural uses. By extension, it is unlikely that freshwater microalgae can make a significant contribution to commercial production of biofuels, and species able to grow in saline water, preferably over a broad salinity range, should receive greater focus (Sheehan *et al.*, 1998). Marginal land in coastal regions with access to seawater and/or brackish water may offer the best opportunities for microalgal biofuel production, from practical, sustainability and economic reasons.

Waste water, particularly from municipal sewage plants, is another possible water resource and has the advantage of presenting a cheap and sustainable source of N and P. Some consider that waste nutrient and water resources from human, animal and industrial wastes are most appropriate for the generation of microalgal biomass for renewable energy production in the short term (Batten *et al.*, 2011). By reducing the reliance on external sources of water and nutrients, the sustainability and overall costs of such facilities for microalgal growth and oil production can improve significantly. However another opinion is that microalgal biomass that could be produced at these facilities may be inadequate to justify infrastructure investment for oil extraction and liquid fuel production, and conversion of the biomass by anaerobic digestion or into animal feeds may be more attractive (Darzins *et al.*, 2010).

10.4.2 *Pond systems*

Open ponds include naturally occurring ponds as well as low technology constructions that are low cost and relatively easy to build and operate. Pond systems vary in size, shape, material used for construction, method of mixing, and inclination. Overall, other than natural ponds or lakes, there are three major types: (i) inclined systems where mixing is achieved through pumping and gravity flow, (ii) circular ponds with agitation provided by a rotating arm, and (iii) raceway ponds constructed as an endless loop, in which the culture is circulated by paddle wheels (Tredici, 2004).

Figure 10.2.   Examples of raceway ponds: a) Experimental-scale pond raceways and suspended polythene
PBRs at Sapphire Energy's R&D Facility at Las Cruces, NM, USA. b) Pilot-scale raceways
at Sapphire Energy's commercial demonstration facility, Columbus, NM. Photos provided by
Sapphire Energy, Inc.

Inclined systems have promise because of highly turbulent flow and very thin culture layers
(e.g. potentially <1 cm), thus promoting high cell densities and a higher surface area to volume
ratio compared with other open systems (Tredici, 2004). Their major disadvantages include high
rates of evaporation and CO_2 desorption, and a large energy requirement for continuous pumping.
Although circular ponds also require high energy input for mixing, they are widely used in Asia for
Chlorella mass production, also in aquaculture for producing microalgae for larval feed (Brennan
and Owende, 2010).

Raceway ponds are preferred for many operations because of their ease of construction and
operation, and the vast experience that exists on their application (Borowitzka, 1999). They are
typically made of closed loop, oval shaped circulation channels (Figs.10.2a,b) between 0.2 and
0.5 m deep. Channels are built in concrete or compacted earth, and may be lined in plastic (Chisti,
2007). Paddlewheels or water jets are used for flow and mixing, keeping the microalgae suspended
in the water and regularly circulating cells back to the surface for exposure to sunlight. Generally
the ponds are operated in a continuous manner, with nutrients and CO_2 being constantly fed to
the ponds, while microalgae-containing culture is removed at the other end.

Compared to closed photobioreactor systems (PBRs; see next section), open pond or raceways
have many advantages, but also disadvantages (see Table 10.5). Raceways have a significantly
greater requirement for land area, but since non-arable land may be used, production does not

Table 10.5. Comparing the advantages and disadvantages of open (pond) versus closed (photobioreactor systems; PBRs) for the phototrophic production of microalgae (modified from Pulz, 2001, and Ratha and Prasanna, 2012).

Parameter	Open ponds	PBRs
Land requirement	High; flat land	Low for PBR itself
Water loss	Very high	Low
CO_2 loss	High, influenced by pond depth	Low
Oxygen concentration	Typically low due to continuous spontaneous outgassing	Gas exchange system required to prevent O_2 build-up and photoxidation, photoinhibition
Temperature	Can be highly variable, some control by pond depth	Cooling may be required (e.g. spraying PRB or water
Weather dependence	High (light, temperate, rainfall, wind)	Medium (light, cooling required)
Cleaning	Limited required	Required; wall-fouling reduces light intensity
Shear	Low due to gentle mixing	Usually high – turbulent flows needed for mixing and pumping through as exchange devices
Contamination (competing species, or grazers)	High – limiting the species that can be grown	Low
Biomass concentrations	Low; between 0.1 and 0.5 g/L	High; between 1 and 8 g/L
Biomass quality	Variable	Reproducible
Production flexibility	Only few species possible; difficult to change	High, easily changeable
Process control/ reproducibility	Limited (flow rate, mixing, temperature by changing depth)	Possible
Production start-up	6–8 weeks	2–4 weeks
Capital investment	Low-medium	High–very high
Operating costs	Low (paddle wheel, CO_2 and nutrient addition, water replacement)	High (CO_2 and nutrient addition, pH control, O_2 removal, increased maintenance)
Harvesting costs	High (large volumes); variable by species	Lower due to higher concentration of biomass

necessarily compete with existing agricultural crops. Raceways also have lower energy input requirements, cleaning and maintenance is easier, and therefore have the potential to yield larger net energy production (Rodolfi *et al.*, 2009). Limitations of open ponds include poor light utilization by the cells, high evaporative losses and loss of CO_2. Temperature variations due to diurnal cycles and season are difficult to control in ponds and reduce productivity. Contamination by other microalgae, predators and other fast growing heterotrophs has also restricted the production of microalgae in open systems to only those organisms that can grow under extreme conditions. *Arthrospira* (*Spirulina*) (grows under high alkalinity), *Dunaliella salina* (adaptable to hypersaline conditions) and *Chlorella* (adaptable to high nutrient concentrations) are microalgae that have been grown commercially utilizing their competitive growth advantage under extreme environments (Borowitzka, 1999). Specifically relating to biofuels production, raceways are one of the main growth technologies being assessed at the pilot-scale by several companies investigating biofuels from microalgae (Table 10.1 and Fig. 10.2b).

Figure 10.3.   The "biofence" photobioreactor system. Photo provided by the National Research Council of Canada.

10.4.3   *Photobioreactors (PBRs)*

Photobioreactors (PBRs) are culture systems where the principle is to reduce the light path (e.g. to between 2 and 10 cm *c.f.* raceways, 20 to 50 cm) thus increasing the amount of light available to each cell. PBRs are well mixed to optimize light utilization, and to enhance gas exchange (Borowitzka, 1999). PBRs may be installed outdoors to receive solar illumination, or indoors and artificially illuminated. Tredici (2004) has defined PBRs as culture systems in which >90% of light does not fall directly on the culture surface, but has to pass through the transparent wall of the unit to reach cells. Another characteristic of PBRs is that they do not allow, or at least strongly limit, the direct exchange of gases or contaminants (microorganisms, atmospheric particles, etc.) between the culture and the atmosphere and therefore are usually considered to be closed systems.

PBR technologies have been well reviewed, and a variety of types have been described (Eriksen, 2008; Pulz, 2001; Tredici, 2004). Three of the most common designs are tubular, flat plate and column PBRs. Tubular PBRs (e.g. Fig. 10.3) consist of an array of plastic or glass tubes, that can be aligned vertically, horizontally, inclined, or as a helix (Eriksen, 2008; Tredici, 2004). Cultures are recirculated with a pump or airlift system, the latter providing mixing as well as enabling CO_2 and O_2 exchange. Tubular PBRs are limited in design to the length of tubing, which is dependent on the potential rates of CO_2 depletion, O_2 accumulation and the variation of pH within the systems. Therefore, individual tubular PBR units cannot be scaled up indefinitely, so large-scale production plants are based on combining units. The two largest production plants utilizing PBRs base their production on tubular PBRs, i.e. the 700 m^3 plant in Klötze, Germany and the 25 m^3 reactors at Mera Pharmaceuticals, Hawaii (Eriksen, 2008). Not all species are suited for culture in tubular PBRs, i.e. delicate species may be damaged by the circulation systems, whereas other "sticky" species may adhere to tubing and provide excessive fouling. Species that have been successfully grown in tubular PBRs include *Chlorella* spp., *Nannochloropsis* sp., *Phaeodactylum tricornutum*, *Arthrospira platensis*, *Tetraselmis* spp. and *Isochrysis galbana* (Borowitzka, 1999; Tredici, 2004).

Flat plate systems have been developed and assessed following research in the late 1980s by scientists is Italy, Israel and Germany (Tredici, 2004; Zmora and Richmond, 2004). An advantage

Figure 10.4.   The bubble-column reactor, from Greenfuels site. Image courtesy of http://www.flickr.com/
photos/jurvetson/58591531/.

of this technology is that elevated flat panels can be easily re-oriented to provide the optimal
exposure to sunlight, thus increasing the efficiency of light conversion and the areal productivity. This technology can support very high densities of microalgal cells, e.g. >80 g DW/L for
Chlorococcum littorale (Hu *et al.*, 1998). Other species successfully grown in flat plate PBRs
include *Nannochloropsis* sp., *Arthrospira (Spirulina) platensis*, *Chlorella* sp., *Isochrysis galbana*
and *Chaetoceros muelleri* (Tredici, 2004; Zhang and Richmond, 2003). Flat plat PBRs are considered suitable for mass cultivation of microalgae due to low accumulation of dissolved O_2 and
a high photosynthetic efficiency when compared to tubular systems (Richmond, 2000).

Column PBRs are installed or hung vertically, and aerated from below (e.g. a bubble-column
reactor; Fig. 10.4). The PBRs are illuminated through the outer transparent walls, or internally
(Eriksen, 2008; Rodolfi *et al.*, 2009). Construction may be rigid (e.g. glass, glass-fiber, poly-
carbonate) or flexible (polyethylene sleeves), and units range from 5 to 50 cm in diameter, with
volumes from 100 to 1000 L. Annular systems are a variation of the design. Polythene sleeves (Fig.
10.2a) offer a cheap, disposable option, that may be operated independently, or interconnected
for continuous or semi-continuous addition of culture media and inoculum as well as harvesting,
providing a high-degree of automation. These systems are used commonly within the aquaculture
industry, e.g. the SeaCAPS system of 40 × 500 L bags (Seasalter Shellfish (Whitstable) Ltd, UK).
Column PBRs are characterized by the most efficient mixing, the highest mass transfer rates and
more controllable growth conditions (Eriksen, 2008).

The proliferation of pilot-scale production facilities using PBRs compared to open ponds could
be ascribed to more rigorous parameter control and higher production rates, and therefore potentially a higher production of biofuel and co-products (Brennan and Owende, 2010). An added
advantage resulting from the higher cell densities they produce is that the biomass is easier and
more-cost effective to collect for further down-stream processing. Nevertheless, there remain
some significant disadvantages of closed PBRs compared to open ponds (Table 10.5), particularly the higher construction and operational costs. Another main limitation with production
in narrow pathlength PBRs are the maintenance of necessary turbulent flows in long lengths
of tubing and the possibility of photo-inhibition due to the oxygen accumulation (Williams and
Laurens, 2010).

10.4.4 *Fermentation systems*

Fermentation technology is much more mature and established at an industrial scale compared with phototrophic production systems. The major advantages of this technology are greater process control and a cost reduction due to no lighting requirements, and associated higher cell densities (e.g. 15 to 50 g/L DW for *Chlorella* spp.) in large-scale systems (up to 11,000 L) (Li et al., 2007; Xiong et al., 2008). The number of microalgae produced heterotrophically remains very small. Technology development has in general not focused on the particular requirements of microalgae and only a few microalgae have been shown to grow effectively under heterotrophic conditions. These include species that can produce bioproducts of interest, such as the chlorophytes *Chlorella*, *Dunaliella* and *Haematococcus,* the diatom *Nitzschia* and the dinoflagellate *Crypthecodinium cohnii* (Perez-Garcia et al., 2011). Though set up costs are minimal and light is not required, the system may effectively use more energy than phototrophically-grown microalgae because the process cycle requires the initial production of organic carbon through photosynthesis (Brennan and Owende, 2010). In most experimental studies to date, refined carbon sources (e.g. glucose, acetate, glycerol) have been utilized, but given the inherent high cost of feedstock the production costs could be further reduced if low-value, sustainable alternatives (e.g. sweet sorghum or Jerusalem artichoke) were used (Gao et al., 2010).

10.4.5 *Hybrid growth systems*

Various multiphase cultivation strategies have been considered. Richmond (1987) suggested combining a tubular reactor connected to an open raceway in order to maximize biomass production by optimizing environmental variables e.g. using the raceway cultures in the hot part of the day and using the photobioreactors for high productivities during the periods of the day with lower environmental stress. Multiphase cultivation strategies can be devised that ensure maximum production of biomass in one stage and maximum induction and accumulation of desired products in the other. This concept has been successfully applied to outdoor cultures of *Dunaliella* for β-carotene production as well as to photobioreactor cultures of *Haematococcus* for astaxanthin production (Hu, 2004).

Recently, a two stage heterotrophic and phototrophic culture strategy for microalgal biomass and lipid production was demonstrated experimentally (Zheng et al., 2012). Here, cultures of *Chlorella sorokiniana* were grown heterotrophically on a substrate of food-waste hydrolysate, then the harvested biomass was used as inocula for phototrophic growth in 40 L ponds, and cultured to achieve a final density of 0.8 g DW/L.

10.4.6 *Productivities of microalgae growth systems*

Because of their more efficient utilization of light energy and greater degree of process control, higher biomass concentrations and growth rates are achievable in closed PBRs compared to open pond systems. Table 10.6 presents data on the yield of microalgal biomass and lipid, and also the productivity, using different culture systems. Productivity is expressed on a volumetric and/or aerial basis; both measures provide useful metrics though aerial productivity is the more widely used and relevant to economic modeling of the microalgal production process, especially for outdoor systems. Additionally, it should be noted that most of the data are for experimental-scale systems that may serve as an indicator of potential commercial-scale productivity, but further confirmation is needed.

For phototrophic production, maximum or average biomass concentrations generally fall in the range of 0.1 to 0.5 g/L for open ponds and 1 to 8 g/L for PBRs. Data in Table 10.6 and from other studies (Brennan and Owende, 2010; Eriksen, 2008) also show that aerial productivities are generally comparable, e.g. ranging from 10 to 70 g/m²/day for raceways compared to 10 to 50 g/m²/day for PBRs, though more typically values that have been sustained over extended periods fall in the range 20 to 30 g/m²/day for both technologies (Darzins et al., 2010).

Table 10.6. Biomass and lipid productivity data for microalgal production comparing phototropic (flasks, ponds, photobioreactor systems) and heterotrophic culture systems.

Culture scale/ Microalgae species	Vol. [L]	Biomass [g/L]	Lipid [mg/L]	Vol. productivity [mg/L/day]		Aerial productivity [g/m²/day]		Reference[1]
				Biomass	Lipid	Biomass	Lipid	
Flasks or carboys:								
Isochrysis sp.	10	0.26–0.86	62–290		6.4–21	6–19	1.4–4.4	[1]
Porphyridium cruentum	0.25			370	35			[2]
Tetraselmis spp.	0.25			280–320	27–43			[2]
Tetraselmis sp.	10	0.61–1.2	85–310		19–23	33–45	3.9–4.8	[1]
Nannochloropsis spp.	0.25			170–210	38–61			[2]
Neochloris oleabundans	1			310–630	38–133			[3]
Open ponds, raceways:								
Tetraselmis suecica	5000	0.1–1.1		370		40		[4]
Anabaena sp.	100–2000	0.1–0.3		90–190		9–19		[5]
Nannochloropsis sp.	?			90	26			[6]
Neochloris oleabundans	?			90	26			[6]
Spirulina maxima	?			210	9			[6]
Pleurochrysis carterae	150			12–230	4–76	2.3–47.7	0.8–15.9	[7]
Photobioreactors:								
Chaetoceros muelleri	300	1.5		150		15.4		[8]
Chlorella vulgaris	1	1.3–1.9	290–850	130–180	29–78			[9]
Isochrysis sp. (T.ISO)	300	1.3		130		13.3		[8]
Isochrysis galbana	50	1.2–4.2		280–320				[10]
Nannochloropsis sp.	590					9.9	6.5	[11]
Nannochloropsis sp.	440			140–270		6.7–14.2		[12]
Nannochloropsis sp.	120			600–1450				[13]
Tetraselmis suecica	16					7.6	1.7	[11]
Heterotrophic:								
Chlorella vulgaris	1			80–150	27–35			[14]
Chlorella protothecoides	5	51.2	25800	7300	3700			[15]
Chlorella protothecoides	5	15.5	7100	2020	930			[16]
Chlorella protothecoides	11000	14.2	6300	1850	820			[16]

References[1]: [1] Huerlimann et al. (2010); [2] Rodolfi et al. (2009); [3] Li et al. (2008); [4] Laws et al. (1986); [5] Moreno et al. (2003); [6] Gouveia and Oliveira (2009); [7] Moheimani and Borowitzka (2007); [8] Zhang and Richmond (2003); [9] Mujtaba (2012); [10] Molina-Grima et al. (1994); [11] Bondioli et al. (2012); [12] Cheng-Wu et al. (2001); [13] Zittelli et al. (2000); [14] Liang et al. (2009); [15] Xiong et al. (2008); [16] Li et al. (2007).

Some studies have shown that both technologies have similar productivity if subjected to the same environmental conditions (Weissman *et al.*, 1988). However, greater process control can improve the productivity of PBRs, as heavy rain or cold weather severely disturbs open ponds (Moheimani and Borowitzka, 2006).

Aerial productivity data must be used with caution for PBRs because values will depend on the system orientation. Vertically-oriented PBRs can produce high and misleading productivities (Darzins *et al.*, 2010). Vertical PBRs modules must be separated far enough apart to avoid shading, so that the basic limitation on productivity remains the same for both PBRs and open ponds. Volumetric productivities are typically 5 times or greater in PBRs (e.g. 0.3 to 4 g/L/day) compared to open pond or raceway systems (0.05 to 0.3 g/L/day) because of their higher surface area to volume ratios (SA:V). Heterotrophic production, by virtue of its independence from light and CO_2 to sustain growth, is characterized by significantly higher volumetric production rates (e.g. 6.3 to 26 g/L/day; Table 10.6) than either of the phototrophic growth technologies.

10.4.7   *Improving productivity through technical and biological approaches*

10.4.7.1   *Culture system design*
Culture unit design may profoundly influence microalgal production. PBRs with narrower light paths can reach significantly higher cell densities before light may become limiting, compared to pond systems. However, for a given culture volume, the latter will be more thermally stable to ambient temperature fluctuations. The culture unit design will also influence the level of turbulence associated with mixing, which in turn will effect in-culture light climate, gas and nutrient transfer and level of hydrodynamic stress, all of which may impact of productivity (Grobbelaar, 2010).

In general, pond technology is based on simple structures that have been well studied over many years, so the potential for technical innovation here to significantly improve productivity is limited. One possible exception is thin-layer (e.g. 6 mm depth) open systems that are characterized by high cell density and productivity (Doucha *et al.*, 2005), though these systems are not easily scalable to mass culture and there are issues with culture stability. There is scope for incremental improvements in productivities through innovation in PBR designs; although more impact on cost-efficiency could be gained from a significant lowering of the production costs of the systems. One of the more successful designs reported was a triangular "3DMS reactor" developed by GreenFuel Technologies Corpn., that combined the principle of a bubble column with mixing through in-built static mixers in an external "downcomer". This unit exhibited an average productivity over several weeks of 98 g/m²/day (Pulz, 2007), which is a similar productivity to the maximum theoretical value for microalgal productions calculated by J.C. Weissman, of 100 g/m²/day (Schenk *et al.*, 2008). However, it was not established if these productivities could be sustained over longer periods, and the company eventually folded in 2009 citing high system productions costs as a key contributing factor (Kanellos, 2009a).

10.4.7.2   *Ecological approaches*
To achieve productivities in commercial ponds that match those based on laboratory predictions, it is necessary to deal with culture invasions by contaminants, predators and competitors. Instead of trying to exclude other organisms, insights based on ecology and species interactions, may provide alternative strategies to manage these systems.

Based on the principle that biological systems tend to increase in complexity over time, likewise microalgal monocultures in ponds will increase in complexity through biota invasion. Kazamia *et al.* (2012) proposes that monocultures of microalgae lack the stability for industrial scale-up, and instead microalgae should be produced in a "synthetic community" with carefully selected co-habitants, which increases overall productivity. Other microalgae with complimentary nutrient requirements could be co-inoculated to improve nutrient utilization. Microalgal-bacterial inter-actions may also be positive, e.g. bacteria can excrete vitamin B_{12}, which may be required for growth by certain microalgae. For example, co-inoculation of presumed bacterial contaminants isolated from a culture of *Chlorella ellipsoidea* with the same microalga, resulted in 0.5–3 times increase in growth compared to un-inoculated cultures of *C. ellipsoidea* (Park *et al.*, 2007).

Top-down bio-manipulation strategies may also improve productivity. For example, the introduction of a gape-limited zooplankton will favor the growth of larger microalgae, thus eliminating smaller-sized competitors. An alternative concept is to introduce one or more species of zooplanktivorous fish, to reduce the microalgal losses associated with zooplankton grazers (Smith *et al.*, 2010).

10.4.7.3 *Breeding and genetic engineering*

Breeding systems are well described for the model chlorophyte species *Chlamydomonas*, but not for others (Georgianna and Mayfield, 2012). In part, the reason may be a lack of mating compatible isolates, or the greater emphasis in most biofuel research on applying molecular and transgenic technologies instead of breeding. Nevertheless, despite the technical challenges with breeding microalgae, their short sexual cycles and rapid growth rates (*c.f.* plant crops) are an advantage, and the technology does represent a pathway to strain improvement that warrants further investigation. Moreover, conventional breeding methods need to become well established for the full potential of genetic engineering to be realized in some microalgal species, e.g. diatoms (Radakovits *et al.*, 2010).

The potential for genetic engineering to yield strains expressing modified fatty acid profiles to improve biodiesel quality was discussed in section 10.3.2. From sequencing technologies, the understanding of microalgal genomes has improved rapidly, particularly with genes involved in metabolic responses, indicating the potential for increasing lipid yield by metabolic engineering (Georgianna and Mayfield, 2012). The inhibition of starch synthesis in genetically engineered mutants of *Chlamydomonas reinhardtii* has led to an overproduction of lipid through redirecting metabolism (Li *et al.*, 2010). A further engineering strategy could be to modify the photosynthetic apparatus to adjust light-harvesting pigments so as to minimize the light saturation effect (Hall and Benemann, 2011). Whether microalgal mutants grow faster than their wild-type progenitor is yet to be established, and there is further research required to establish whether growth rates specifically can be improved through genetic engineering. However, the technology could eventually enable microalgae to accumulate high TAG even under conditions of rapid growth (Wijffels and Barbosa, 2010).

As is the case with crop plants, microalgae may be bred or engineered to be resistant to the antibiotics and pesticides used to control unwanted species. With this approach, a low-cost efficacious antibiotic or pesticide would also need to be identified for treatment of cultures (Georgianna and Mayfield, 2012). Alternatively, genes encoding for the expression of anti-microbial peptides could be introduced into microalgae to convey microbial protection; similarly genes expressing allelochemicals could be introduced to defend against predators (Qin *et al.*, 2012). Crop protection could be improved by breeding or genetic engineering to increase the cell size of particular microalgae strains, making them less ingestible by grazers. Increasing the cell size (and/or modifying other physical cell characteristics) may also improve the harvesting efficiency of the biomass.

10.5 HARVESTING OF MICROALGAL BIOMASS

After cultivation of microalgae, the biomass is separated from the bulk water for further downstream processing. The harvesting process is considered a major challenge and stumbling block in the microalgal production process. The combination of the small cell size of most microalgae and their dilution in culture means the majority of a microalgal culture is water. Hence, large volumes of culture suspensions need to be treated. Also, because of the complexity (reagents, energy and labor demands) the recovery process of microalgal biomass is costly, and may to constitute between 20 and 30% of the cost of microalgal biomass production (Molina Grima *et al.*, 2003).

The recovery of microalgal biomass can be achieved using one or more of several physical, chemical or biological methods: flotation, flocculation, sedimentation, filtration, centrifugation and ultrasound. The selection of the process depends on the size and properties of the microalgal strain and the culture system used, e.g. flocculation and flotation techniques often are applied to open pond systems, whereas filtration and centrifugation may apply to PBR systems. Typically, the harvest process is conducted as a two-stage process:

1) Bulk harvesting – biomass is separated from the harvested culture suspension, to reach a concentration between 2 and 7% of dry matter. Processes used include flocculation, flotation or gravity sedimentation.
2) Thickening – the slurry is further concentrated e.g. up to 15–20% of dry matter through techniques such as centrifugation and filtration. This step requires more energy that bulk harvesting.

10.5.1   *Flocculation*

Flocculation is a process of aggregating microalgal cells to facilitate their separation from the bulk culture media. It is generally promoted by the addition of a chemical (flocculant) into the medium, which disturbs the stability of microalgal particles causing them to aggregate (Suali and Sarbatly, 2012). The flocculants used with microalgae are either inorganic or polyelectrolyte. With respect to the former, lime, alum and ferric salts (chloride and sulfate) have been used, with efficacy related to dosage and pH, and the concentrations of magnesium and calcium. Recoveries of >85% have been reported for freshwater e.g. *Chlorella zofingiensis* (Wyatt *et al.*, 2012) and marine microalgae (*Attheya septentrionalis, Thalassiosira pseudonana*; Knuckey *et al.* (2006)) using ferric chloride and pH change for flocculation. Inorganic flocculants may have some disadvantages such as a poor efficiency to coagulate small particles, and low cost-efficiency when large amounts are required. Concentrations required to flocculate marine microalgae are five to ten times those required for freshwater species (Uduman *et al.*, 2010). However, their use in small amounts can improve the performance of polymeric flocculants in seawater (Suali and Sarbatly, 2012).

Polyelectrolyte flocculants are polymeric, and include ionic and non-ionic species, natural and synthetic polymers (Uduman *et al.*, 2010). Lower concentrations of polymer flocculants are needed for particle flocculation compared to inorganic flocculants. The natural polymer chitosan, produces over 98% flocculation efficiency in *Chaetoceros calcitrans* over the concentration range 10 to 100 mg/L in seawater (Harith *et al.*, 2009), whereas a concentration of 15 mg/L produced similar efficiencies in freshwater microalgae (Divakaran and Pillai, 2002). However, chitosan is expensive and its efficiency is typically reduced in saltwater (Suali and Sarbatly, 2012). Synthetic organic polyelectrolytes have been used in wastewater clarification and also proven to be good flocculants for microalgae. Of these MagnaFloc LT-25 at a concentration of 0.5 mg/L and at pH 10 to 10.6 flocculated marine microalgae including *Tetraselmis suecica, Skeletonema* sp., *C. calcitrans* and *Nitzschia closterium* with efficiencies of 80 to 95% (Knuckey *et al.*, 2006).

Autoflocculation can occur for certain microalgae where there is a natural elevation of pH during culture development e.g. to pH >10.5 for *Phaeodactylum tricornutum;* (Spilling *et al.*, 2011). Autoflocculation of *Skeletonema* has also been applied in a bioflocculation process to form flocs of *Nannochloropsis* (Schenk *et al.*, 2008). Another example of bioflocculation is the addition of microbes (e.g. 1 g/L) into cultures of *Pleurochrysis carterae* culture, whereby extracellular polymers released by the microbes during nutrient limitation induced microalga flocculation with efficiencies of 90% (Lee *et al.*, 2009). These latter variations represent interesting, potentially low-cost alternatives to the traditional flocculation methods, but can be considered as still under development.

10.5.2   *Gravity sedimentation*

Gravity sedimentation relies on the natural settling of particulate matter, including microalgae, from suspension. Settling rates are determined by the density and radius of microalgal cells and sedimentation velocity (Brennan and Owende, 2010). Gravity sedimentation is often used as for harvesting microalgae biomass in wastewater treatment because of the large volumes processed and the low value of biomass. However it is not widely used by other microalgae industries for biomass separation, though for some species and applications it is a reliable, cheap process to concentrate suspensions to 1.5% total suspended solids (Mohn, 1980). The method has proven effective for some of the larger microalgae (ca. 70 m) such as *Arthrospira* (*Spirulina*).

10.5.3 *Flotation*

Flotation is a form of gravity separation, in which air or gas is bubbled through a microalgal suspension and the gaseous molecules attach to cells, which are then carried to the liquid surface as a float for harvesting. There are three main flotation methods using bubble generation: dispersed air flotation, dissolved air flotation and electrolytic flotation (Uduman *et al.*, 2010). Dissolved air flotation is commonly used in industrial effluent treatment and is considered more efficient than sedimentation for microalgal separation. This process is often combined with other processes such as chemical flocculation, e.g. microalgal suspensions containing *Chlorella*, *Oscillatoria* and *Scenedesmus* were efficiently harvested as slurries containing up to 6% solids (Bare *et al.*, 1975). However, Brennan and Owende (2010) suggest there is limited evidence of technical or economic viability for the application of flotation in harvesting microalgae for biofuels.

10.5.4 *Centrifugation*

At the laboratory and small-scale e.g. up to several hundred liters, centrifugation provides rapid separation within minutes and microalgal biomass recoveries in excess of 90% are easily obtained. Culture biomass may be concentrated by up 150 times, reaching 15% of weight per volume (Mohn, 1980). The disadvantages of centrifugation are the high capital and operating costs; hence its main application has been for harvesting microalgae for high-value chemicals, or as concentrates used as aquaculture feeds (McCausland *et al.*, 1999). The efficiencies of different centrifuges (disc centrifuge, nozzle centrifuge, screw centrifuge, and hydrocyclones) for harvesting different microalgae were tested by Mohn (1980) with all displaying certain advantages and disadvantages. For harvesting 125 m^2 outdoor ponds containing mixed microalgae, a disc centrifuge (Laval models BRP207 SG7/P) was capable of harvesting 2000 L/h with high efficiency (Sim *et al.*, 1988). However, the same study found a continuous filtration process was more cost- and energy-efficient. For larger-scale separation, hydrocyclone systems may be more economical due to simplicity, but further development of current systems is necessary.

10.5.5 *Filtration*

Filtration involves passing the microalgal suspension through screens or filters, on which the microalgae will be retained, whereas the liquid media will pass through. The process operates continuously until the filter contains a thick microalgal paste (Harun *et al.*, 2010b). Filtration can provide high levels of water removal with the final concentrate between 10 and 27% of total weight per volume (Danquah *et al.*, 2009; Grima *et al.*, 2003).

Vacuum drum filters and chamber press filters are most often used for harvesting large micro-algae. They can provide relatively cheap, rapid and efficient separation of large microalgae e.g. >70 μm such as *Arthrospira* and *Coelastrum* (Brennan and Owende, 2010). However, certain types of filtration are ineffective for smaller species e.g. <30 μm for *Dunaliella*, *Chlorella* and *Scenedesmus*, because the filters rapidly clog (Mohn, 1980).

For the harvesting of smaller microalgae, membrane microfiltration and ultrafiltration (including tangential flow filtration; TFF) are technically feasible alternatives. An advantage of TFF over conventional filtration, flocculation, sedimentation and centrifugation is that superior recovery rates can be achieved (Uduman *et al.*, 2010). However, large scale recovery of microalgal biomass may be limited because of fouling, and the need to clean and replace membranes. TFF can be more cost effective than centrifugation for processing volumes up 2000 L/day, whereas the reverse may be true for volumes larger than 20,000 L/day due to the costs associated with membrane replacement (Mackay and Salusbury, 1988). A comparison of TFF and flocculation for harvesting *Tetraselmis suecica* concluded that both processes could be suitable for dewatering on an industrial scale, but TFF was more cost effective (Danquah *et al.*, 2009).

10.5.6   *Other separation techniques*

Ultrasound techniques are under development for harvesting microalgae (Bosma *et al.*, 2003). The separation is based on gentle acoustically induced aggregation followed by enhanced sedimentation. Suali and Sarbatly (2012) have described the concept of a positively charged surface material to attract and aggregate microalgal cells, as a harvest method warranting further investigation and development.

10.6   CONVERSION OF BIOMASS TO BIOFUELS

The harvesting of microalgae typically results in a concentrated biomass containing between 5 and 20% solids. Subsequent steps may involve drying to remove residual water, then extraction to recover the oil. There are various options for converting microalgae biomass to fuel. These include (modified from Huesemann and Benemann (2009)):

- Physical or chemical separation of lipids leading to production of biodiesel;
- Direct combustion of a dried biomass for power generation or thermochemical conversion to produce syngas, oils etc.;
- Fermentations to produce ethanol, methane or hydrogen; and
- Photobiological process to produce hydrogen.

These alternative processes are shown in Figure 10.1, and have been discussed in several reviews (Ghasemi *et al.*, 2012; Ghirardi *et al.*, 2000; Suali and Sarbatly, 2012). Some general comments with selected examples are given in the next sections.

10.6.1   *Drying of microalgae biomass*

After harvesting, microalgal biomass should be processed rapidly as chemical fractions may be subject to chemical degradation induced by the process itself, and by metabolic activity. Drying is an important step to extend the shelf life of the biomass and to maintain the quality of the constituent oil. Residual water can also reduce the efficiency of downstream processes such as lipid extraction or transesterification (Lam and Lee, 2012).

Methods to dry microalgae include sun drying, spray drying, drum drying, freeze-drying and low-pressure shelf drying (Richmond, 2004). As noted by Huesemann and Benemann (2009) the drying of harvested microalgae biomass may consume more energy to evaporate the water than is present in the biomass. Hence, when microalgae are being processed for biofuel alone, it is not cost-effective to completely dry the biomass using a thermal process utilizing fossil or biomass fuel. Sun-drying, where biomass is dried by laying out on drying pads, is regarded as the only low cost and practical process. However, the process may require a long-drying time and the risk of material loss. Also, sun-drying may not be a feasible method in temperate regions at certain times of the year, whereby the other more energy-consuming methods may be required to supplement the drying (Lam and Lee, 2012). These alternative drying methods remain applicable where the primary focus is extraction of high-value products, or such products are recovered as by-products from the process to partially offset the production cost of biofuel.

Strategies where biomass is processed for biofuels after partial drying, or without drying, are also feasible, as discussed in subsequent sections.

10.6.2   *Extraction of oil*

Various chemical and physical methods have been assessed for extracting oil from microalgae. Pretreatment methods, such as drying or cell disruption, are energy intensive and should only be considered if they improve extraction efficiency. Cell disruption steps (e.g. bead mill, homogenization, sonication, microwave) improve the efficiency of lipid extraction with chemical solvents, and should be performed before the drying step, if that is undertaken (Halim *et al.*, 2012). Addition of acid/alkali and enzymes can also be effective for some strains (Mercer and Armenta, 2011). The performance of the cell disruption methods in improving lipid extraction differs according to species and extraction method, but the microwave method appears the simplest and most effective, and is scalable (Amaro *et al.*, 2011).

Standard mechanical disruption techniques for oil extraction include pressing, bead milling and homogenization (Mercer and Armenta, 2011). Pressing is commonly used for seeds or nuts, and has also been applied to microalgae. Usually it is accompanied by a dewatering step before extraction, and has an advantage that no chemical solvents are required, with oil recoveries in the range from 70 to 75% (Suali and Sarbatly, 2012).

Supercritical-fluid extraction, using CO_2 as solvent, is rapid, efficient, organic-solvent free and safe for thermally-sensitive components. It has been applied for extraction of high-value produce from microalgae (Amaro *et al.*, 2011). Disadvantages are that samples need to be moisture-free for efficient extraction, the technique is time consuming, and therefore less efficient for large-scale application.

The Origin Oil Company has developed a novel process that does not require prior dewatering and can extract oil from microalgae at a rate of 22 L/min, i.e. with ≈95% efficiency (OriginOil, 2010). Cell walls are disrupted by applying an electromagnetic field combined with a pH adjustment, followed by a clarification to separate the oil, water and residue. This technique is the most efficient and promising commercial technique to date.

Standard extraction methods using organic solvents offer some advantages in that they can be applied to wet concentrated feedstock and have a low reactivity, but they are slow and can use large volumes of expensive, toxic solvents (Halim *et al.*, 2012). Solvents commonly used for chemical extraction, either singly or in combination, include hexane, toluene, chloroform and various alcohols (butanol, propanol, ethanol and methanol). Hexane is often preferred, because of its efficiency, reusability and low cost (Harun *et al.*, 2010b). The effect of residual water in the microalgae biomass on extraction efficiency is equivocal, and appears to depend on both the microalgae species and solvent combination (Halim *et al.*, 2012). For example, Halim *et al.* (2011) extracted dried and paste (70% moisture) preparations of *Chlorococcum* sp. and found the paste preparation were more efficiently extracted than dried product using a hexane: propanol mixture (3:2 v/v), whereas the reverse was true when hexane alone was used as solvent.

Solvent extraction can be combined with ultrasound or microwave to improve oil recovery (Halim *et al.*, 2012). Accelerated organic solvent extraction, whereby lipid is extracted at high temperature and pressure, has had success at the laboratory scale. However, neither this method nor microwave-assisted extraction has been assessed at an industrial scale because of their high energy requirements (Halim *et al.*, 2012).

For a few unique species e.g. *Botryococcus braunii*, which accumulate high concentration of hydrocarbons associated with the cell surface, it may be feasible to recover the oil from cultures without damaging the cell walls by washing with a non-toxic, biocompatible solvent such as hexane or decane (Banerjee *et al.*, 2002).

10.6.3   *Processes and biofuel products from microalgae*

10.6.3.1   *Biodiesel production*
Biodiesel is the most investigated and widely discussed of the biofuels obtainable from microalgae. Conversion of microalgal biomass to biodiesel occurs through transesterification. Biodiesel derived from microalgae (e.g. *Chlorella*) has been reported as having a heating value of 41 MJ/kg (Xu *et al.*, 2006) and complies with the US biodiesel standard ASTM6571 (Li *et al.*, 2007).

Transesterification is conducted within a reactor where the blended alcohol and catalyst are reacted with constituent fatty acids (e.g. bound as TAG, or free acids) from the microalgal oil. Methanol is most commonly used because of its low cost and physical and chemical advantages (polar and shortest-chain alcohol), but ethanol is also frequently used (Ghasemi *et al.*, 2012). Catalysts used are alkalis, acids or enzymes (lipases). After completion of the reaction, the products are a mixture of alkyl esters, glycerol, catalyst, alcohol, and mono-, di- and triacylglycerols. The biodiesel fraction (i.e. alkyl esters) is recovered by washing with water to remove alcohol and glycerol (Ghasemi *et al.*, 2012).

Biodiesel may also be produced from microalgal biomass through a direct extraction and transesterification. This route has been demonstrated on a laboratory scale to yield a high recovery of fatty acids esters from dried microalgae, but for moist microalgae results were variable (Belarbi *et al.*, 2000).

10.6.3.2   *Bio-oil production by hydrothermal liquefaction*
Hydrothermal liquefaction is a reduction process that has been applied to convert microalgal biomass to bio-oil (or "green crude"; Fig. 10.1) and other chemicals using low temperature and high pressure (e.g. 200–500°C and 5–20 MPa) (Suali and Sarbatly (2012), and reference therein). Reactors are complex and expensive, but importantly, liquefaction can be used to convert wet microalgal biomass ($\geq$60% moisture), therefore avoiding the energy-drying step. By the use of hydrogen in the presence of catalyst, the process utilizes the high water content of microalgae to transform biomass components to shorter, less complex molecules with a higher energy density (Brennan and Owende, 2010). Processing of *Dunaliella tertiolecta* biomass containing 78% moisture produced an oil of 42% DW, and a gross energy of 35 MJ/kg, and a positive energy balance of 2.94:1 (Minowa *et al.*, 1995). Bio-oil can be further treated to yield fuels such as green diesel or jet fuel, and the residue can be combusted directly or converted (e.g. by fermentation) in animal feed or fertilizer (Ghasemi *et al.*, 2012).

10.6.3.3   *Gasification for syngas*
In gasification, biomass is partially oxidized at high temperatures (Bulushev and Ross, 2011). In the normal process, the main product is syngas, a mixture of H_2, CO, CO_2, methane, water and tar vapors (long-chain aliphatics). Together with ash, this can account for 70 to 80% of the energy originally in the biomass. Syngas can be used for power or heat generation. It may also be further converted to liquid alkanes (e.g. biodiesel), or to dimethyl ether or gasoline via methanol formation using other processes (Bulushev and Ross, 2011). Conventional gasification is applied to decompose dry biomass ($\leq$20% moisture) at temperatures of 800 to 1000°C or higher) in the absence of oxygen. Alternatively, supercritical water gasification utilizes supercritical water to produce smaller molecules at lower temperature, i.e. 350 to 700°C (Ghasemi *et al.*, 2012).

Hirano *et al.* (1998) converted *Arthrospira (Spirulina)* biomass (containing 20% moisture) at 1000°C to a theoretical yield 0.64 g of methanol per gram of biomass. The energy balance was 1.1, indicating that this process was feasible as an energy-producing process. Dried samples of *Chlorella vulgaris* and *Arthrospira (Spirulina) platensis* processed by supercritical water gasification at 500°C yielded 22 g H_2/kg (in addition to significant amounts of methane and C_2-C_4 alkanes) with calorific values up to 36 MJ/m^3. Although the gasification process was regarded as efficient, the authors concluded that these microalgae may be better suited for processing for bio-oil production.

10.6.3.4   *Pyrolysis for bio-oil, biochar and syngas*
Pyrolysis, or thermal cracking, is a thermochemical processing for transforming dried biomass to bio-oil, syngas and biochar at temperatures from 200 to 750°C, in the absence of air or oxygen (Suali and Sarbatly, 2012). The quantity and quality of pyrolysis products depend on parameters such as feedstock, temperature, pressure, catalysts, heating rate and reaction time. Flash pyrolysis, e.g. 500°C, short hot-vapor residence time of about 1 s, can yield bio-oil in yields of up to 75% of DW of feedstock (Bridgwater, 2007). Because of this high conversion efficiency, flash pyrolysis

is considered to be a viable process for the future substitution of fossil fuels with biomass derived liquid fuels (Brennan and Owende, 2010). Slow pyrolysis (e.g. 400°C, very long residence times) typically produces similar proportions of bio-oil, biochar and gas (Bridgwater, 2007).

Pyrolysis of microalgae biomass has achieved promising results that could lead to commercial exploitation (Brennan and Owende, 2010). Bio-oils from microalgae are also considered more stable than bio-oils derived from lignocellulosic feedstock (Mohan *et al.*, 2006). Fast pyrolysis of heterotrophically-grown *Chlorella protothecoides* biomass resulted in a bio-oil yield of 58% of final product, and biochar (Miao and Wu, 2004). The same study showed yields of 18 and 24% of bio-oil from fast pyrolysis of autotrophically-grown *C. protothecoides* and *Microcystis aeruginosa* biomass, respectively.

10.6.3.5 *Direct combustion*
The simplest method to produce energy from biomass is a direct combustion, whereby biomass is burnt in the presence of air at high temperatures (e.g. > 800°C) to produce hot gases which provide power and heat. However, a direct combustion of microalgae biomass containing ≥80% moisture, requires that it first be dried, a very energy-demanding process. Apart from the energetic return on investment, combustion of dried microalgal biomass may present environmental problems due to its high nitrogen content (typically, 10% of ash-free dry weight). Direct combustion can cause the oxidation of nitrogen, leading to NO_x emissions, a major pollutant (Huesemann and Benemann, 2009).

Overall, there is scant literature evidence for a technically feasible application of the direct combustion of microalgal biomass, though co-firing microalgal biomass with coal may offer environmental benefits such as a lower emissions of GHGs and air pollution, comparing to coal-firing alone (Kadam, 2002).

10.6.3.6 *Fermentation processes to produce ethanol*
There are two major mechanisms by which ethanol can be produced from microalgae biomass. The first is from yeast fermentation of storage carbohydrate within microalgal biomass. The second is by a "self fermentation" of accumulated carbohydrate within microalgae under anaerobic conditions using utilizing endogenous enzymes.

Figure 10.6 shows a process pathway for the first of these methods. An initial step is pretreatment to make the carbohydrate component of the biomass more available for subsequent processing. Methods often used include ultrasonic disintegration or dilute acid treatment (Suali and Sarbatly, 2012). Thereafter, the polysaccharide components are hydrolyzed using either

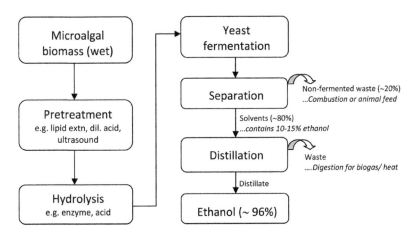

Figure 10.6.   Process steps leading to ethanol production from microalgae (modified from Suali and Sarbatly, 2012).

one or more enzymes (e.g. amylase, cellulose), alkali or acid (Ghasemi *et al.*, 2012). The pre-treatment and hydrolysis may occur as separate processes, or combined in one process (Suali and Sarbatly, 2012). Pretreatment can increase the efficiency of the subsequent fermentation process by more than 33% and ethanol production by more than 60%, though it may also increase the energy consumption for fermentation by up to 30% (Suali and Sarbatly (2012), and references therein). Fermentation of the treated biomass and metabolic conversion to ethanol usually takes place within an immobilized fermentor unit using the yeast *Saccharomyces cerevisiae*. Subsequent steps include separation of fermentation mixture, distillation of the liquid component to yield ethanol (approx. 96%) and processing of the waste through digestion, to yield biogas and heat as by-products. Advantages that microalgal biomass have over woody biomass as a potential feedstock for ethanol include the absence of lignin, and the requirement for less energy-intensive pretreatment. Also, many strains accumulate carbohydrate under specific culture conditions (e.g. to 30 to 60% of DW; Singh and Gu, 2010), and several have been experimentally assessed as feedstock. *Chlamydomonas reinhardtii* grown mixotrophically (60% carbohydrate) yielded 0.24 g ethanol/g DW of microalgal biomass (Choi *et al.*, 2010). Starch, which comprised 37% of the DW of phototrophically-grown *Chlorella vulgaris*, was extracted and converted to ethanol with 65% efficiency, using similar processes (Hirano *et al.*, 1997). Simultaneous biodiesel and ethanol production is also feasible and may prove more cost-effective, whereby lipid is extracted before fermentation. Using this approach, lipid-extracted *Chlorococcum* sp. was fermented with a yield of 0.38 g ethanol per g DW biomass (Harun *et al.*, 2010a). Moreover, the lipid-extraction pretreatment resulted in a 60% higher ethanol conversion, compared to the fermentation of non lipid-extracted biomass.

The second route to ethanol is via a direct production by the microalgae biomass. For example, cultures of the chlorophyte *Chlamydomonas reinhardtii* underwent self fermentation of carbohydrate storage product in the dark under anaerobic conditions, to produce ethanol concentration in solution of up to 1% (Hirano *et al.*, 1997). Several cyanobacteria grown phototrophically were also shown to produce ethanol under similar conditions (Ueda *et al.*, 1996). Recently, Joule Unlimited (Bedford, MA, USA) have developed a photosynthetic platform microorganism (most likely, an engineered cyanobacteria; (Rosgaard *et al.*, 2012)) to continuously produce and secrete ethanol (Berry *et al.*, 2011). The process converts >90% of the CO_2 consumed into ethanol at commercially viable costs (www.jouleunlimited.com/news/2011). Algenol (Bonita Springs, FL, USA) are using a similar approach using engineered cyanobacteria grown in closed PBRs to produce ethanol, and are currently assessing the commercial feasibility of the technology (Lane, 2012). Dilute solutions of ethanol in freshwater collect on the upper surface of the PBR (by evaporation from the culture and then condensation), which are then continuously removed and purified to fuel-grade ethanol.

10.6.3.7 *Hydrogen through fermentation or biophotolysis*

Hydrogen can be produced by microalgae through fermentation processes (which may either be dark-fermentation or photo-fermentation) and biophotolysis (Huesemann and Benemann (2009), and references cited therein). Dark fermentation can occur through self-fermentation whereby microalgae (e.g. *Chlamydomonas* sp.) can convert starch or glycogen into hydrogen, solvents and mixed acids. Organic acids produced during self-fermentation can be further converted to hydrogen using photosynthetic bacteria. This step-wise approach was demonstrated in a small outdoor system, albeit with low solar energy conversion efficiencies (Akano *et al.*, 1996).

Biophotolysis is the process whereby microalgae convert solar energy and water into hydrogen and fuel, with oxygen as a by-product. In direct biophotolysis, water is split into hydrogen and oxygen without the intermediate CO_2 fixation. However, this reaction which is catalyzed by the hydrogenase enzyme is inhibited by the simultaneous production of oxygen. With indirect biophotolysis, the conversion takes place in two stages, i.e. photosynthetic CO_2 fixation generates oxygen and carbohydrate, and the latter is then utilized for light-driven hydrogen production.

While the production of renewable hydrogen from photosynthesis is an attractive proposition, neither biophotolysis nor the fermentation pathways currently provide a mechanism for practical application (Huesemann and Benemann, 2009). With indirect biophotolysis, it is estimated that the yield of hydrogen production would need to be increased ten-fold or more to be viable commercially (Melis *et al.*, 2000). Combined approaches using classical breeding, recombinant DNA and gene technology may provide feasible pathways in the future by increasing the hydrogen productivity and/or the oxygen tolerance of the hydrogenase enzyme in green microalgae (Ghirardi *et al.*, 2000).

10.6.3.8 *Anaerobic digestion for methane production*
Anaerobic digestion is the conversion of biomass into a biogas, which consists mainly of methane and CO_2, with traces of other gases, e.g. H_2S and H_2. The process occurs in sequential stages of hydrolysis, bacterial fermentation and methanogenesis (Brennan and Owende, 2010). A wide range of biomass feedstock has been used to generate methane by this process including macroalgae (Huesemann and Benemann, 2009). For microalgae, this process was demonstrated by Oswald and Golueke (1960) who proposed a process linking pond production of microalgae utilizing sewage water and the energetic recovery of the biomass using anaerobic digestion. There have been few studies since, though interest has redeveloped recently given the demonstration of anaerobic digestion to convert a variety of organic wastes into renewable energy (Sialve *et al.*, 2009). Nevertheless, anaerobic digestion remains a less favored process for microalgal biomass conversion, because of several major bottlenecks (from Ghasemi *et al.* (2012)):

- In general, a low biodegradability of microalgae biomass,
- High protein concentrations which can result in ammonia release which can lead to toxicity and affect the efficiency of the digestor,
- The presence of sodium with marine species can also affect digestor performance.

10.7 TOWARDS COMMERCIAL PRODUCTION

10.7.1 *Current industry state*

Global production of biofuels had increased from 0.31 M to 1.9 M barrels per day between 2000 and 2011 (US Energy Information Administration; www.eia.gov). The increased demand for vegetable oils for producing biodiesel has given rise to significant pressure on the vegetable oil market. Against this background, there are tremendous opportunities for liquid biofuels generated from microalgal systems, albeit as a niche contributor within the broader biofuel market, based on their potential productivity, sustainability and environmental benefits. Over the same time, there has been a rapid expansion of companies directly investigating and investing in fuel production from microalgae (see also Table 10.1). Funding to support the pathway to commercialization has come from both the investment community and the petrochemical industry; based on 2008 data it was estimated more than US$ 1 billion (B) was committed by the private sector (Pienkos and Darzins, 2009). In 2009, the US Department of Energy (DOE) also funded three microalgae-based integrated biorefinery projects, i.e. by Sapphire Energy Inc. (US$ 50 million (M)); Algenol Biofuels, Inc. (US$ 25 M) and Solazyme Inc. (US$ 21.8 M). Within Europe US$ 27 M has been committed by the EU7th Framework Program (Algal Cluster) between 2007 and 2013. It has been projected that the public and private investment in microalgal biofuels in 2015 will give rise to a US$ 1.6 B market. Despite the investments, and the technological advancements that have been made, the current status is that there is no commercial-scale production of biofuel from microalgae, though several companies could be considered as producing at pre-pilot or pilot scale. The US companies listed above, i.e. Sapphire Energy, Algenol and Solazyme, are well advanced and provide interesting and instructive case studies because of their differences in production platforms:

- Solazyme is perhaps the closest to commercial biofuel production. Solazyme technology is based on heterotrophic production of lipid-rich microalgae in 0.5 megaliter (ML) fermentors. The company produces a high-value cosmetic range as a by-product of its oil production process. In 2010, the company delivered 1500 gallons of microalgal-derived jet fuel and 20,000 gallons of shipboard fuel to the US Navy for testing (OILGAE, 2011). It recently built its first commercial-scale microalgal fuel factory in Brazil, in conjunction with joint venture partner Bunge assisted by a US$ 120 M loan from the Brazilian Development Bank, and is projecting to produce 340 ML (2.2 M barrels) of fuel annually by 2016 (Fehrenbacher, 2013).

- Sapphire Energy employs open pond raceways. In 2008, it refined a high-octane fuel from microalgae chemically similar to crude oil. In 2009, Sapphire participated in a test flight using microalgae-based jet fuel in a Boeing 737–800 twin-engine aircraft. In 2012, the company announced that the first phase of its "Green Crude Farm", the world's first commercial demonstration microalgae-to-energy facility, was operational in Columbus. The 890 ha site incorporates a refinery on site, and reportedly has started producing low volumes of oil, with targets of 5.7 ML of microalgal crude per year by 2014, and 580 ML (3.7 M barrels) a year by 2018 (Fehrenbacher, 2013).
- Algenol are using engineered cyanobacteria grown in horizontal, polythene-bag PBRs to produce ethanol. They have reported a production of 0.065 ML of ethanol/ha/year, based at their 1.6 ha development unit. They are currently developing a 36 ha pilot-scale facility to assess the commercial feasibility of the technology (Lane, 2012). The plant aims to produce 0.38 ML (100,000 gallons) of fuel grade ethanol per year.

10.7.2   *Economics of biofuel production*

Despite the significant benefits arising from the potential use of microalgae as a biofuel feedstock, it has not reached a commercial scale to date, principally because of the high cost of producing microalgae biomass and oil compared to market alternatives. Broadly speaking, major contributing factors are the land costs, expensive bioreactors, and the cost of growing and harvesting microalgae biomass. For biodiesel, one report cites the production price may range from 1.60 to 4.80 US$/L, which is between 3 and 9 times as much as the production cost of alternatives such as fossil gasoline/diesel (US$ 0.53) and non-algal biofuels (US$0.60–0.73) (OILGAE, 2011).

There are many other estimates for the production of biodiesel derived from microalgae, and these are similarly wide-ranging (e.g. Table 10.7). These estimations are generally based on biomass productivity and lipid yields obtained at an experimental-scale, being replicated at a commercial level. Several of the estimates are based on techno-economic models where various production scenarios were tested, e.g. combinations of low to high biomass productivity and lipid content, or other system or site-specific factors influencing the model predictions (Benemann and Oswald, 1996; Davis *et al.*, 2011; Pienkos and Darzins, 2009; Richardson *et al.*, 2010; Williams and Laurens, 2010). A similar compilation of cost estimates – also highlighting significant variation – was reported by Carriquiry *et al.* (2010), with a mean value of 4.3 US$/L for microalgal triacylglycerol. The uncertainties with cost estimations are large, with the uncertainty of capital cost being more substantial that operating cost (Carriquiry *et al.*, 2010). Although several studies have estimates of near or below 2 US$/L, most of these are based on best-case scenarios (e.g. biomass productivities of $\geq$30 g DW/m^2/day and/or lipid content of $\geq$40% DW) which are yet to be demonstrated over sustained periods in large-scale systems, or reflect target costs based on efficiency or yield gains arising from future technological improvements (Benemann and Oswald, 1996; Huntley and Redalje, 2007; Pienkos and Darzins, 2009; Richardson *et al.*, 2010). Raceway pond data compiled from industry *Spirulina* producers has shown sustained productivities of 10 g DW/m^2/day are more typical (Benemann and Oswald, 1996), and other surveys have shown that while productivities in the range 30 to 40 g DW/m^2/day are possible for short periods, long-term average data rarely exceeds 20 g DW/m^2/day (Norsker *et al.*, 2011).

The most economic culture system for producing microalgal biomass and oil, i.e. open pond or PBR, will depend on a range of variable, often uncertain, factors. A consensus is that open pond production will usually be less expensive, e.g. as shown in Table 10.7, and from other data for microalgal biodiesel production, i.e. from 2.4 to 6.6 US$/L for ponds versus 4.0 to 10.6 US$/L for closed PBRs (Amaro *et al.*, 2011). However, under some scenarios closed PBRs may be able

Table 10.7. Estimated or potential costings for algal biomass and biodiesel production, based of various system and production scenarios. Modified from a table by Williams and Laurens (2010). Assumed specific gravity of biodiesel is 0.88. Adjusted to 2011 prices. Where data for crude algal oil is reported, the assumed conversion cost to biodiesel is 0.14 US$/L (Chisti, 2007). Assumed $US 1 = 0.75 Euro.

Reference	Assumed or estimated....		Biomass costs [US$/kg]			Biodiesel or FAME costs [US$/L]		
	Productivity [g DW/m²/day]	Lipid content [% DW]	PBR	Raceway	Hybrid	PBR	Raceway	Hybrid
Benemann and Oswald (1996)	30–60	50		0.15–0.25			0.27–0.45	
Molina Grima et al. (2003)	[1.25][1]	10	35			375		
Chisti (2007)	72	30	0.50			3.05		
Chisti (2007)	35	30		0.63			3.9	
Huntley and Redalje (2007)	19–60	40			0.11–0.36			0.25–0.81
Pienkos and Darzins (2009)	10–50	15–50					0.66–6.60	
Davis et al. (2011)	25	25					2.6	
Davis et al. (2011)	[1.25][1]	25				5.4		
Richardson et al. (2010)	20–30	20–40					1.8–7.2	
Richardson et al. (2010)	18–25	40–60					0.38–1.0	
Williams and Laurens (2010)	18–37	15–50			0.37–0.68		0.82–3.2	
Amer et al. (2011)	40	40				22		
Amer et al. (2011)	24	40					3.5	
Lundquist et al. (2010)	22	25					1.9–2.5	
Slade and Bauen (2013)	10–20	?	5–13					
Slade and Bauen (2013)	20–40	?		0.4–2.4				
Norsker et al. (2011)	6.4			6.6				
Norsker et al. (2011)	12.4–19.4		5.5–7.9					

[1]Volumetric productivity; unit is g/L/day

to produce biomass and oil at a cheaper final price (Chisti, 2007). Of note, based on modeling scenarios reported in Norsker *et al.* (2011), predicted microalgal biomass production costs were lower in raceway ponds (6.60 US$/kg) than in flat-panel PBRs (7.95 US$/kg), but least expensive in tubular PBRs (5.55 US$/kg). Few data are available for predicting production costs via a heterotrophic culture pathway, though a recent study modeled on cultures of *Chlorella protothecoides* grown in 11,000 L fermentors, estimates a cost of 1.2 US$/L for biodiesel (Tabernero *et al.*, 2012).

Techno-economic analyses have been applied to identify critical path elements that offer the best opportunities for cost reduction. Invariably, the two most significant factors identified are lipid content and biomass (Darzins *et al.*, 2010; Davis *et al.*, 2011), i.e. biological parameters that to a large degree are specific to the microalgae strain. Based on the study of Davis *et al.* (2011), lipid content was the most significant factor for open pond production, e.g. if the lipid content (e.g. 25% DW) was halved or doubled, the oil production cost either increases by ≈ 2 US$/L, or reduces by ≈ 1 US$/L, respectively. For similar adjustments in the assumed base biomass productivity (25 g/m²/day), the net cost impact was half as large. The PBR case from the same study also showed a greater sensitivity to lipid content than to the biomass production, though the cost savings were less pronounced. Additionally, the cost of PBR tubing exhibited very strong cost sensitivity, due to its large contribution to overall capital expenses (e.g. up to 70%).

10.7.3   *The concept of an integrated biorefinery*

Most of the economic analyses of biofuels from microalgae concur that a "fuel-only" option is unlikely to be economically feasible, and other revenue sources are needed to make production profitable (Chisti, 2007; Singh and Gu, 2010; Williams and Laurens, 2010). This introduces the concept of a microalgal-based biorefinery where all components of the biomass are used, effectively reducing the cost of any individual product. In the broad sense, a biorefinery is a facility that integrates biomass conversion processes and equipment to produce fuels, power, and value-added chemicals from biomass and it is analogous to today's petroleum refinery, which produce multiple fuels and products from petroleum (definition: www.oilgae.com). This approach has been adopted by operations in Canada, USA and Europe for processing crops such as beet, soybean and corn, as feedstock (Chisti, 2007). As one example, British Sugar (Wissington, UK) converts sugar beet to a product range including sugar, bioethanol and betaine (Wellisch *et al.*, 2010). Examples of microalgae-based companies using this approach have been mentioned elsewhere in this chapter (e.g. Algenol; also Table 10.1). Multiple models and configurations of biorefineries are possible (Wellisch *et al.,* 2010).

A schematic of a hypothetical microalgal-based biorefinery is shown in Figure 10.7. In the case of the lipid-extracted microalgae biomass, options could include the utilization of the protein fraction for animal feeds, and the conversion of the carbohydrate fraction for ethanol production. Nutrients (e.g. P, N, Si) and water could be recycled. Another possibility is that some of the biomass residue may be anaerobically digested to produce methane, to offset some of the electrical power requirements to run the production facility. The extraction of high-value co-products (e.g. pigments, cosmetics, toxins, nutraceuticals, etc.), depending on the particular microalgal species, may also contribute to offsetting production costs (Spolaore *et al.*, 2006).

The integration of wastewater treatment with biofuel production is an alternative model for a microalgae biorefinery. Based on a modeling study where wastewater treatment was the primary goal, production costs of microalgal oil could be <1 US$/L (Batten *et al.*, 2013). This modeling was based on a possible system integration at the Western Treatment Plant (Melbourne Water, Australia), a plant of 10,500 ha, processing about 400 ML/day. Integral factors contributing to the low production cost projected in this study were the availability of land at no cost, a generous supply of recyclable water at low cost, and sustainable and recyclable sources of CO_2, N and P at little cost. Potential products that were identified within four modeling scenarios included biocrude or microalgal oil, struvite and electricity (from either gasification or anaerobic digestion).

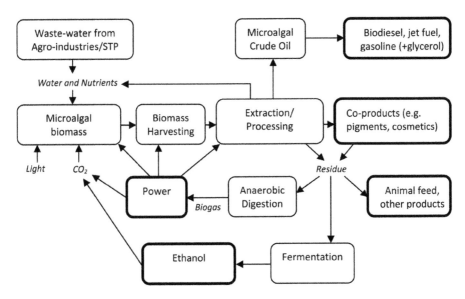

Figure 10.7. An integrated biorefinery concept, generating various energy products and co-products.
STP = sewage treatment plant.

Due to commercial sensitivities there is little validated industry data available for public examination, though a few production costs have been provided based on projections from pilot-scale
investigation. In 2009, it was reported Solix Biofuels Inc. were producing biofuel for 8.70
US$/L, though it was stated prices could be reduced to one-sixth by better energy utilization and
co-product development (Kanellos, 2009b). Algenol have targets to produce ethanol from 0.40–
0.45 US$/L based on their current production at their outdoor development unit (Lane, 2012),
whereas Sapphire Energy are aiming at commercial production of microalgal "green crude" oil
at 0.47–0.53 US$/L. Given the mismatch between these industry aspirational targets and cur-
rent best production-cost estimates based on modeling approaches using realistic data, it is clear
major advances are still required for commercial production of microalgal biofuel to be cost
competitive with market alternatives. A perspective from Hall and Benemann (2011), is that low
costs for microalgae biofuel could be achieved by increasing scales of production for pond production ten-fold (e.g. ≥400 ha per facility), developing strains that can sustain productivities of
30 g DW/m^2/day with high TAG content, in combination with significant reductions in harvesting
and oil extraction costs, through technological advances. Most, if not all, market analysts view
that the successful commercial implementation of microalgal biofuel will also rely on the parallel
development of high-value co-products through the biorefinery approach (Singh and Gu, 2010).

10.7.4 *Environmental sustainability and life cycle analysis (LCA)*

Some of the key drivers for the interest and investigation in microalgae as a feedstock for biofuel
production relate to their environmental benefits compared to other feedstock alternatives. Two
advantages of microalgae include their ability to grow utilizing wastewater and therefore aid in
nutrient removal/recycling, and their capacity to grow in non-arable regions utilizing less land
space than other biofuel feedstock, and thus avoiding competition with food production. An often
touted advantage is the ability of microalgae to utilize CO_2 from industrial flue gas. However,
CO_2 consumption by algal cultures in this circumstance is not CO_2 sequestration, nor even a
GHG abatement process. If an alga's oil is extracted to make biodiesel, and the residual biomass
is fed to a digester to make biogas for use in power generation, in both cases the alga's eventual
combustion releases CO_2 into the atmosphere. Thus one cannot claim all the CO_2 absorbed by

the alga as a carbon mitigation measure. However, it is valid to claim the CO_2 savings (as carbon offsets) from the fossil fuel that has been replaced by the biodiesel or bioelectricity. In the case of biodiesel, this is about 2.94 kg of CO_2 per kg of diesel fuel.

One of the common modeling tools used to assess the environmental sustainability of microalgae as a biofuel feedstock has been life cycle analysis (LCA). LCA quantifies the environmental impacts of all the processes contributing to the generation of goods or services, in keeping with the ISO14040 standard. The two factors of particular importance to microalgae-based biofuels (and biofuels generally), are (1) the energy depletion potential, i.e. how much fossil fuel energy is utilized over the life cycle to generate the biomass or biofuel, and (2) the global warming potential, i.e. the total amount of GHG liberated, measured as an equivalent mass of CO_2 (Scott et al., 2010).

Seven LCA studies that examined the microalgae to biofuel pathway were compared on the basis of net energy ratio (*NER*) of biomass production (Slade and Bauen, 2013). *NER* was calculated from the sum of energy used for cultivation, harvesting and drying, divided by the energy content of the dried biomass. Of eight raceway concepts compared, six had *NER* <1, i.e. the process would produce more energy than it consumes, though in some cases the benefit was marginal. In most cases the majority of energy (approximately two-thirds) consumed was in the biomass harvesting and dewatering, with the cultivation and harvesting phase accounting for most of the residual. All three PBRs examined consumed more energy than they produced, with the flat-plat system rating better (lower *NER*) than tubular systems. Also in PBRs, more energy was used in drying and dewatering, compared to cultivation and harvesting. This was attributed to the higher biomass densities achieved in PBR systems, and the higher energy consumption of PBRs at the cultivation stage. The study found that the levels of CO_2 emissions from biomass production in raceways were comparable to those associated with the cultivation and production stages of other biodiesel (rape methyl ester), whereas production in PBRs demonstrated greater emissions than conventional fossil diesel (Slade and Bauen, 2013). Key conclusions were (i) the energy balance for biomass production showed that the energy inputs could be high, thus limiting the value of microalgae biomass production as a source for energy and indicating production may be most attractive where energy is not the main product, and (ii) the carbon balance could be improved by using co-products to generate electricity to feed back into the system.

Another LCA analysis, assessing ethanol production from cyanobacteria (based on the Algenol process) indicated that this pathway potentially has more favorable (lower) CO_2 emissions than the microalgae to biodiesel pathway, by virtue of lower energy requirements because harvesting of the microalgal biomass is not required (Luo et al., 2010). Other environmental advantages this non-harvest approach offered compared to other microalgal biofuel processes were lower fertilizer requirements and water usage.

The integration of wastewater treatment with microalgae production has already been mentioned as a sustainable option to efficiently recover and recycle water and nutrients, and at the same time providing a microalgae feedstock for biofuel production. High-rate algal ponds, when supplemented with clarified domestic wastewater can recover nutrients with efficiencies similar or better than mechanical treatment methods. For example, laboratory studies demonstrated $\geq$96% removal of P and ammonium using microalgae cultured with CO_2-supplemented primary waste effluent (Woertz et al., 2009). Also, in the growing of freshwater microalgae, the recycling of water has added environmental benefits through a reduction in the requirement of valuable freshwater resources.

Even with recycling, water from production systems will have to be disposed of eventually, because of salt build-up (e.g. in saline systems, with evaporation) and other waste compounds such as autoinhibitors, that may reduce productivity (Borowitzka and Moheimani, 2013). As the waste water will still contain some nutrients, its disposal into waterways may have an environmental impact and be subject to regulatory control.

Biological characteristics of specific microalgae types can potentially give rise to other environmental concerns. The use of genetically modified (GM) microalgae is often advocated as a pathway to superior strain characteristics. However, before GM microalgae can be moved from the laboratory to outdoor production systems, the impacts that they may have on the environment would need to be established. Some species, at certain lifecycle-stages, can produce toxins whose effects may be chronic or acute (e.g. paralytic shellfish poisoning). Therefore use of such strains should be avoided or used under strictly controlled conditions to eliminate the risk of their release into the environment.

10.7.5 *Political and social factors*

The global production and consumption of biofuels has risen dramatically over the last few years. Key drivers for biofuel growth have been increasing fossil fuel prices coupled with their increasing price volatility, as well as energy security. Moreover, with the link between increasing atmospheric CO_2 and climate change, policy makers have been under increasing pressure to develop and deploy renewable, clean-energy technologies, including biofuels, to reduce greenhouse emissions. This has led to an increased public support for biofuels as a renewable liquid fuel source (e.g. mandated consumption, subsidies etc.) by many countries (Carriquiry *et al.*, 2010).

However the growth of the industry has become controversial, with claims that transforming cropland to produce the current biofuel feedstock incurs a carbon debt, i.e. releases more carbon than is saved by the biofuels (Dauvergne and Neville, 2010). As an alternative biofuel feedstock, microalgae could offer many potential advantages from the point of view of carbon impact. Microalgae production doesn't compete with food crops, has high productivity potential, and could utilize CO_2 within exhaust emissions from industry. At least with respect to the EU, current policies do not distinguish or preferentially support biofuels with better environmental credentials, though there is increasing momentum within government and non-government organizations which could change that in the near future (Kovacevic and Wesseler, 2010).

There have been tremendous investment, particularly in the US and Europe by government, industry and venture capitalists to develop a microalgae to liquid biofuel industry. Significant advances have been made through R&D and early industry activity, though there are still major bottlenecks to be overcome. Many experts believe commercialization is still a decade or more away, but it will be important to achieve tangible progress along the way to keep the momentum and maintain political and public support, research effort and investment that will be needed to achieve this goal.

Many of the social benefits or impacts of a microalgae-based biofuel industry are intrinsically linked to economic or environmental factors, many of which have been covered in previous discussion. Key points to reiterate are a sustainable microalgal production would place less demand on natural resources than alternative first-generation biofuel feedstock, would improve energy security, and has the potential to reduce GHG emissions. For social acceptance of a developing microalgae biofuel industry, it will be important for government and industry to be pro-active in providing the public with reliable information on issues including health and safety, production transparency and environmental sustainability. Some of the concerns of the public could include contamination of local waters with toxins, herbicides, or GM microalgae, and visual changes to the local landscape. The development of a new microalgae fuel industry may boost employment opportunities in regional areas particularly in the facility construction phase, but the net effect on employment may be marginal as microalgae may simply be replacing other feedstock in the fuel sector. Also, with the large-scale systems required for cost-efficiency, labor costs will need to be kept to a minimum with systems being highly automated. There may be additional employment opportunities from the development of co-products as part of the biorefinery approach.

10.8 CONCLUSION

The increases in atmospheric CO_2 and the depletion of fossil fuels have accelerated the need to develop alternative fuels that are carbon-neutral and renewable. While lignocellulosic biomass is presently the frontrunner for biofuel production, microalgae are considered technically viable and attractive alternatives. Microalgae are seen as having the potential to satisfy the need for higher energy density biofuels to displace our petroleum diesel and jet fuel usage. Microalgae biofuels are also likely to have less impacts on the environment and food supply than traditional biofuel-producing crops.

Consequently there has been an expanding R&D effort and investment during the last decade to develop technology to produce biofuels from microalgae. Advances have been made in the understanding of microalgal biology and technologies to grow, harvest and process microalgae to biofuels. Nevertheless, the economic feasibility of biofuel production from microalgae has yet to be demonstrated at a commercial level, although several companies are operating at pre-pilot or pilot scale, and are making positive projections of commercial-scale production within the next 5 years.

Though likely increases in crude oil production costs will improve the economics of liquid biofuel from microalgae production in the medium term, development of an industry in the shorter term, e.g. 5 to 10 years will require major reductions in production costs. Critically important will be further technology advances from R&D, including the development of high-lipid producing strains that are optimized to regional climate conditions, with high, sustainable growth rates in mass culture. Metabolic engineering and breeding techniques provide options for strain improvement to meet this need. Opinion is divided amongst experts as to which culture technology will be the most feasible and cost-effective for commercial scale-up, i.e. phototrophic production with ponds or enclosed PBRs, or heterotrophic production with fermentors. Each could prove viable under certain conditions or input scenarios, and each has specific advantages and disadvantages. Research efforts should also be directed at reducing the energy consumption and costs associated with the further downstream processing of microalgal biomass, particularly more efficient methods for harvesting and de-watering. The adoption of a bio-refinery approach, i.e. where all component of the microalgal biomass are processed to produce usable products, will also be important to further lower the cost of production.

Because of the multiple inputs and processes within the microalgae to biofuel pathway, and their interdependency, the hurdles that need to be overcome will require interdisciplinary research and collaboration between biologists, chemists, engineers and economists. Finally, the development of a microalgal biofuel industry will require a novel form of large-scale agriculture, likely to be established in regions that have not previously been developed for agricultural or industrial purposes. It is therefore crucial to also consider the environmental and ecological impacts of the industry, in addition to regulatory issues, societal effects and public support, especially while the technology is in the early stages of development.

ACKNOWLEDGMENTS

The authors acknowledge the constructive comments provided by Dr David Batten of CSIRO Energy Transformed Flagship, and other anonymous reviewers of this paper.

REFERENCES

Akano, T., Miura, Y., Fukatsu, K., Miyasaka, H., Ikuta, Y., Matsumoto, H., Hamasaki, A., Shioji, N., Mizoguchi, T., Yagi, K. & Maeda, I.: Hydrogen production by photosynthetic microorganisms. *Appl. Biochem. Biotechnol.* 57:8 (1996), pp. 677–688.

Amaro, H.M., Catarina Guedes, A. & Xavier Malcata, F.: Advances and perspectives in using microalgae to produce biodiesel. *Appl. Energy* 88:10 (2011), pp. 3402–3410.

Amer, L., Adhikari, B. & Pellegrino, J.: Technoeconomic analysis of five microalgae-to-biofuels processes of varying complexity. *Bioresour. Technol.* 102:20 (2011), pp. 9350–9359.

Andersen, R.A. & Kawachi, M.: Traditional microalgae isolation techniques. In: R.A. Andersen (ed): *Algal culturing techniques.* Academic Press, New York, USA, 2005, pp. 83–100.

Banerjee, A., Sharma, R., Chisti, Y. & Banerjee, U.C.: *Botryococcus braunii*: A renewable source of hydrocarbons and other chemicals. *Crit. Rev. Biotechnol.* 22:3 (2002), pp. 245–279.

Bare, W.F.R., Jones, N.B. & Middlebrooks, E.J.: Algae removal using dissolved air flotation. *Res. J. Water Pollut. Control Fed.* 47:1 (1975), pp. 153–169.

Batten, D.F.: Biofuels need flexi-fuel vehicles to achieve sustainable mobility. In: A. Subic, M. Leary & J. Wellnitz (eds): *Meeting the challenges to sustainable mobility*. RMIT University Press, Melbourne, Australia, 2008, pp. 219–230.

Batten, D.F. and O'Connell, D.: Biofuels in Australia – some economic and policy considerations. Report no. 07/177 for the RIRDC, 2007.

Batten, D.F., Campbell, P.K. & Threlfall, G.: Resouce potential of algae for sustainable biodiesel production in the APEC economies. Biofuels Task Force, Publ. No. APEC#211-RE-01.9, 2011.

Batten, D., Beer, T., Freischmidt, G., Grant, T., Liffman, K., Paterson, D., Priestley, T., Rye, L. & Threlfall, G.: Using wastewater and high-rate algal ponds for nutrient removal and the production of bioenergy and biofuels. *Water Sci. Technol.* 67:4 (2013), pp. 915–924.

Beattie, A., Percival, E. & Hirst, E.L.: Studies on metabolism of Chrysophyceae – Comparative structural investigations on leucosin (chrysolaminarin) separated from diatoms and laminarin from brown algae. *Biochem. J.* 79:3 (1961), pp. 531–537.

Bechet, Q., Munoz, R., Shilton, A. & Guieysse, B.: Outdoor cultivation of temperature-tolerant *Chlorella sorokiniana* in a column photobioreactor under low power-input. *Biotechnol. Bioeng.* 110:1 (2013), pp. 118–126.

Belarbi, E.H., Molina, E. & Chisti, Y.: A process for high yield and scaleable recovery of high purity eicosapentaenoic acid esters from microalgae and fish oil. *Enzyme Microb. Technol.* 26:7 (2000), pp. 516–529.

Benemann, J.: *Microalgae biofuels: A brief introduction*. 2009, www.adelaide.edu.au/biogas/renewable/biofuels_introduction.pdf (accessed March 2013).

Benemann, J.R. & Oswald, W.J.: Systems and economic analysis of microalgae ponds for conversion of CO_2 to biomass. Final Report, Pittsburgh Energy Technology Centre, Pittsburgh, PA, 1996.

Berry, D.A., Robertson, D.E., Skraly, F.A., Green, B.D., Ridley, C.P., Kosuri, S., Reppas, N.B., Sholl, M. & Afeyan, N.B.: Engineered CO_2 fixing microorganisms producing carbon-based products of interest. US Patent 7,981,647. Joule Unlimited, Inc., Cambridge, MA, 2011.

Bigogno, C., Khozin-Goldberg, I., Boussiba, S., Vonshak, A. & Cohen, Z.: Lipid and fatty acid composition of the green oleaginous alga *Parietochloris incisa*, the richest plant source of arachidonic acid. *Phytochemistry* 60:5 (2002), pp. 497–503.

Bondioli, P., Della Bella, L., Rivolta, G., Zittelli, G. C., Bassi, N., Rodolfi, L., Casini, D., Prussi, M., Chiaramonti, D. & Tredici, M.R.: Oil production by the marine microalgae *Nannochloropsis* sp. F&M-M24 and *Tetraselmis suecica* F&M-M33. *Bioresour. Technol.* 114 (2012), pp. 567–572.

Borowitzka, M.A.: Commercial production of microalgae: ponds, tanks, tubes and fermenters. *J. Biotechnol.* 70:1–3 (1999), pp. 313–321.

Borowitzka, M.A. & Borowitzka, L.J.: *Dunaliella*. In: M.A. Borowitzka & L.J. Borowitzka (eds): *Microalgal biotechnology*. Cambridge University Press, Cambridge, UK, 1988, pp. 27–58.

Borowitzka, M.A. & Moheimani, N.R.: Sustainable biofuels from algae. *Mitigat. Adapt. Strat. Global Change* 18:1 (2013), pp. 13–25.

Bosma, R., van Spronsen, W., Tramper, J. & Wijffels, R.: Ultrasound, a new separation technique to harvest microalgae. *J. Appl. Phycol.* 15:2–3 (2003), pp. 143–153.

Brennan, L. & Owende, P.: Biofuels from microalgae-A review of technologies for production, processing, and extractions of biofuels and co-products. *Renew. Sustain. Energy Rev.* 14:2 (2010), pp. 557–577.

Bridgwater, T.: Biomass pyrolysis. *Biomass Bioenergy* 31:4 (2007), pp. 7–18.

Brown, M.: The amino-acid and sugar composition of 16 species of microalgae used in mariculture. *J. Exp. Mar. Biol. Ecol.* 145:1 (1991), pp. 79–99.

Brown, M.R., Dunstan, G.A., Norwood, S.J. & Miller, K.A.: Effects of harvest stage and light on the biochemical composition of the diatom *Thalassiosira pseudonana*. *J. Phycol.* 32:1 (1996), pp. 64–73.

Bulushev, D.A. & Ross, J.R.H.: Catalysis for conversion of biomass to fuels via pyrolysis and gasification: A review. *Catal. Today* 171:1 (2011), pp. 1–13.

Carriquiry, M.A., Du, X. & Timilsina, G.R.: Second-generation biofuels: Economics and policies. The World Bank, Development Research Group, Environment and Energy Team, Policy Research Working Paper 5406, Washington DC, 2010.

Cheng-Wu, Z., Zmora, O., Kopel, R. & Richmond, A.: An industrial-size flat plate glass reactor for mass production of *Nannochloropsis* sp. (Eustigmatophyceae). *Aquaculture* 195:1/2 (2001), pp. 35–49.

Chisti, Y.: Biodiesel from microalgae. *Biotechnol. Adv.* 25:3 (2007), pp. 294–306.

Choi, S.P., Nguyen, M.T. & Sim, S.J.: Enzymatic pretreatment of *Chlamydomonas reinhardtii* biomass for ethanol production. *Bioresour. Technol.* 101:14 (2010), pp. 5330–5336.

Converti, A., Casazza, A.A., Ortiz, E.Y., Perego, P. & Del Borghi, M.: Effect of temperature and nitrogen concentration on the growth and lipid content of *Nannochloropsis oculata* and *Chlorella vulgaris* for biodiesel production. *Chem. Eng. Process.* 48:6 (2009), pp. 1146–1151.

Danquah, M.K., Ang, L., Uduman, N., Moheimani, N. & Fordea, G.M.: Dewatering of microalgal culture for biodiesel production: exploring polymer flocculation and tangential flow filtration. *J. Chem. Technol. Biotechnol.* 84:7 (2009), pp. 1078–1083.

Darzins, A., Pienkos, P. & Edye, L.: Current status and potential for algal biofuels production. Report T39-T2, 6 August 2010. A Report to IEA Bioenegy Task 39, Commercializing 1st and 2nd Generation Biofuels from Biomass 2010.

Dauvergne, P. & Neville, K.J.: Forests, food, and fuel in the tropics: The uneven social and ecological consequences of the emerging political economy of biofuels. *J. Peasant Stud.* 37:4 (2010), pp. 631–660.

Davis, R., Aden, A. & Pienkos, P.T.: Techno-economic analysis of autotrophic microalgae for fuel production. *Appl. Energy* 88:10 (2011), pp. 3524–3531.

De Oliveira, M., Monteiro, M.P.C., Robbs, P.G. & Leite, S.G.F.: Growth and chemical composition of *Spirulina maxima* and *Spirulina platensis* biomass at different temperatures. *Aquacult. Int.* 7:4 (1999), pp. 261–275.

Divakaran, R. & Pillai, V.N.S.: Flocculation of algae using chitosan. *J. Appl. Phycol.* 14:5 (2002), pp. 419–422.

Doucha, J., Straka, F. & Livansky, K.: Utilization of flue gas for cultivation of microalgae (*Chlorella* sp.) in an outdoor open thin-layer photobioreactor. *J. Appl. Phycol.* 17:5 (2005), pp. 403–412.

Dunstan, G.A., Volkman, J.K., Barrett, S.M. & Garland, C.D.: Changes in the lipid composition and maximization of the polyunsaturated fatty acid content of three microalgae grown in mass culture. *J. Appl. Phycol.* 5:1 (1993), pp. 71–83.

Eriksen, N.: The technology of microalgal culturing. *Biotechnol. Lett.* 30 (2008), pp. 1525–1536.

Fehrenbacher, K.: 2013 could be a make or break year for algae fuel – Tech news and analysis. 2013, http://gigaom.com/2013/01/17/2013-could-be-a-make-or-break-year-for-algae-fuel/ (accessed March 2013).

Gao, C., Zhai, Y., Ding, Y. & Wu, Q.: Application of sweet sorghum for biodiesel production by heterotrophic microalga *Chlorella prototothecoides. Appl. Energy* 87 (2010), pp. 756–761.

Georgianna, D.R. & Mayfield, S.P.: Exploiting diversity and synthetic biology for the production of algal biofuels. *Nature* 488:7411 (2012), pp. 329–335.

Ghasemi, Y., Rasoul-Amini, S., Naseri, A.T., Montazeri-Najafabady, N., Mobasher, M.A. & Dabbagh, F.: Microalgae biofuel potentials (Review). *Appl. Biochem. Microbiol.* 48:2 (2012), pp. 126–144.

Ghirardi, M.L., Zhang, J.P., Lee, J.W., Flynn, T., Seibert, M., Greenbaum, E. & Melis, A.: Microalgae: A green source of renewable H_2. *Trends Biotechnol.* 18:12 (2000), pp. 506–511.

Gonzalez-Rodriguez, E. & Maestrini, S.Y.: The use of some agricultural fertilizers for the mass production of marine algae. *Aquaculture* 36:3 (1984), pp. 245–256.

Gouveia, L. & Oliveira, A.C.: Microalgae as a raw material for biofuels production. *J. Ind. Microbiol. Biotechnol.* 36:2 (2009), pp. 269–274.

Grima, E.M., Belarbi, E.H., Fernandez, F.G.A., Medina, A.R. & Chisti, Y.: Recovery of microalgal biomass and metabolites: Process options and economics. *Biotechnol. Adv.* 20:7–8 (2003), pp. 491–515.

Grobbelaar, J.U.: Microalgal biomass production: challenges and realities. *Photosynthesis Res.* 106:1–2 (2010), pp. 135–144.

Guillard, R.R. & Ryther, J.H.: Studies of marine planktonic diatoms.1. *Cyclotella nana* Hustedt, and *Detonula confervacea* (Cleve) Gran. *Can. J. Microbiol.* 8:2 (1962), pp. 229–239.

Halim, R., Gladman, B., Danquah, M.K. & Webley, P.A.: Oil extraction from microalgae for biodiesel production. *Bioresour. Technol.* 102:1 (2011), pp. 178–185.

Halim, R., Danquah, M.K. & Webley, P.A.: Extraction of oil from microalgae for biodiesel production: A review. *Biotechnol. Adv.* 30:3 (2012), pp. 709–732.

Hall, C.A.S. & Benemann, J.R.: Oil from Algae? *Bioscience* 61:10 (2011), pp. 741–742.

Harith, Z.T., Yusoff, F.M., Mohamed, M.S., Din, M.S.M. & Ariff, A.B.: Effect of different flocculants on the flocculation performance of microalgae, *Chaetoceros calcitrans*, cells. *Afr. J. Biotechnol.* 8:21 (2009), pp. 5971–5978.

Harrison, P.J. & Berges, J.A.: Marine culture media. In: R.A. Andersen (ed): *Algal culturing techniques.* Academic Press, New York, USA, 2005, pp. 21–34.

Harun, R., Danquah, M.K. & Forde, G.M.: Microalgal biomass as a fermentation feedstock for bioethanol production. *J. Chem. Technol. Biotechnol.* 85:2 (2010a), pp. 199–203.

Harun, R., Singh, M., Forde, G.M. & Danquah, M.K.: Bioprocess engineering of microalgae to produce a variety of consumer products. *Renew. Sustain. Energy Rev.* 14:3 (2010b), pp. 1037–1047.

Hejazi, M.A. & Wijffels, R.H.: Milking of microalgae. *Trends Biotechnol.* 22:4 (2004), pp. 189–194.

Hemschemeier, A., Fouchard, S., Cournac, L., Peltier, G. & Happe, T.: Hydrogen production by *Chlamydomonas reinhardtii*: An elaborate interplay of electron sources and sinks. *Planta* 227:2 (2008), pp. 397–407.

Hirano, A., Ueda, R., Hirayama, S. & Ogushi, Y.: CO_2 fixation and ethanol production with microalgal photosynthesis and intracellular anaerobic fermentation. *Energy* 22:2–3 (1997), pp. 137–142.

Hirano, A., Hon-Nami, K., Kunito, S., Hada, M. & Ogushi, Y.: Temperature effect on continuous gasification of microalgal biomass: Theoretical yield of methanol production and its energy balance. *Catal. Today* 45:1–4 (1998), pp. 399–404.

Hu, Q.: Environmental effects on cell composition. In: A. Richmond (ed): *Handbook of microalgal culture: Biotechnology and applied phycology.* Blackwell Scientific, Oxford, UK, 2004, pp. 83–93.

Hu, Q., Kurano, N., Kawachi, M., Iwasaki, I. & Miyachi, S.: Ultrahigh-cell-density culture of a marine green alga *Chlorococcum littorale* in a flat-plate photobioreactor. *Appl. Microbiol. Biotechnol.* 49:6 (1998), pp. 655–662.

Hu, Q., Sommerfeld, M., Jarvis, E., Ghirardi, M., Posewitz, M., Seibert, M. & Darzins, A.: Microalgal triacylglycerols as feedstocks for biofuel production: Perspectives and advances. *Plant J.* 54:4 (2008), pp. 621–639.

Huerlimann, R., de Nys, R. & Heimann, K.: Growth, lipid content, productivity, and fatty acid composition of tropical microalgae for scale-up production. *Biotechnol. Bioeng.* 107:2 (2010), pp. 245–257.

Huesemann, M.H.: Can advances in science and technology prevent global warming? A critical review of limitations and challenges. *Mitigat. Adapt. Strat. Global Change* 11 (2006), pp. 539–577.

Huesemann, M.H. & Benemann, J.R.: Biofuels from microalgae: Review of products, processes and potential, with special focus on *Dunaliella* sp. In: A. Ben-Amotz, J. Polle & D. Subba Rao (eds): *The alga Dunaliella: Biodiversity, physiology, genomics and biotechnology.* Science Publishers Enfield, NH, USA, 2009, pp. 445–474.

Huntley, M.E. & Redalje, D.G.: CO_2 mitigation and renewable oil from photosynthetic microbes: A new appraisal. *Mitigat. Adapt. Strat. Global Change* 12:4 (2007), pp. 573–608.

Jeffrey, S.W., Leroi, J.M. & Brown, M.R.: Characteristics of microalgal species for Australian mariculture. G.L. Allan & W. Dall (eds): *Proceedings of the National Aquaculture Workshops*, 1991, Port Stephens, NSW, Australia, NSW Fisheries, 1992, pp. 164–173.

Jiang, Y. & Chen, F.: Effects of temperature and temperature shift on docosahexaenoic acid production by the marine microalga *Crypthecodinium cohnii*. *J. Am. Oil Chem. Soc.* 77:6 (2000), pp. 613–617.

Jones, C.S. & Mayfield, S.P.: Algae biofuels: Versatility for the future of bioenergy. *Curr. Opin. Biotechnol.* 23:3 (2012), pp. 346–351.

Kadam, K.L.: Environmental implications of power generation via coal-microalgae cofiring. *Energy* 27:10 (2002), pp. 905–922.

Kanellos, M.: Green fuel technologies closing down. 2009a, http://www.greentechmedia.com/articles/read/greenfuel-technologies-closing-down-4670 (accessed August 2013).

Kanellos, M.: *Algae biodiesel: It's $33 a gallon.* 2009b, www.www.greentechmedia.com/articles/read/algae-biodiesel-its-33-a-gallon-5652 (accessed March 2013).

Kazamia, E., Aldridge, D.C. & Smith, A.G.: Synthetic ecology – a way forward for sustainable algal biofuel production? *J. Biotechnol.* 162:1 (2012), pp. 163–169.

Kessler, E.: Upper limits of temperature for growth in *Chlorella* (Chlorophyceae). *Plant Syst. Evol.* 151:1–2 (1985), pp. 67–71.

Khan, S.A., Rashmi, Hussain, M.Z., Prasad, S. & Banerjee, U.C.: Prospects of biodiesel production from microalgae in India. *Renew.e Sustain. Energy Rev.* 13:9 (2009), pp. 2361–2372.

Knothe, G.: A technical evaluation of biodiesel from vegetable oils vs algae. Will algae-derived biodiesel perform? *Green Chem.* 13:11 (2011), pp. 3048–3065.

Knuckey, R.M., Brown, M.R., Barrett, S.M. & Hallegraeff, G.M.: Isolation of new nanoplanktonic diatom strains and their evaluation as diets for juvenile Pacific oysters (*Crassostrea gigas*). *Aquaculture* 211:1–4 (2002), pp. 253–274.

Knuckey, R.M., Brown, M.R., Robert, R. & Frampton, D.M.F.: Production of microalgal concentrates by flocculation and their assessment as aquaculture feeds. *Aquacult. Eng.* 35:3 (2006), pp. 300–313.

Konur, O.: The scientometric evaluation of the research on the algae and bio-energy. *Appl. Energy* 88:10 (2011), pp. 3532–3540.

Kovacevic, V. & Wesseler, J.: Cost-effectiveness analysis of algae energy production in the EU. *Energy Policy* 38:10 (2010), pp. 5749–5757.

Lam, M. & Lee, K.: Microalgae biofuels: A critical review of issues, problems and the way forward. *Biotechnol. Adv.* 30:3 (2012), pp. 673–690.

Lane, J.: Take it to the limit: Algenol and rising yields in advanced biofuels,http://www.biofuelsdigest.com/bdigest/2012/09/25/take-it-to-the-limit-algenol-and-rising-yields-in-advanced-biofuels/ (accessed March 2013).

Laurens, L.M.L. & Wolfrum, E.J.: Feasibility of spectroscopic characterization of algal lipids: Chemometric correlation of NIR and FTIR spectra with exogenous lipids in algal biomass. *Bioenergy Res.* 4:1 (2011), pp. 22–35.

Laws, E.A., Taguchi, S., Hirata, J. & Pang, L.: High algal production rates achieved in a shallow outdoor flume. *Biotechnol. Bioeng.* 28:2 (1986), pp. 191–197.

Lee, A.K., Lewis, D.M. & Ashman, P.J.: Microbial flocculation, a potentially low-cost harvesting technique for marine microalgae for the production of biodiesel. *J. Appl. Phycol.* 21:5 (2009), pp. 559–567.

Li, X.F., Xu, H. & Wu, Q.Y.: Large-scale biodiesel production from microalga *Chlorella protothecoides* through heterotropic cultivation in bioreactors. *Biotechnol. Bioeng.* 98:4 (2007), pp. 764–771.

Li, X., Hu, H.-Y. & Zhang, Y.-P.: Growth and lipid accumulation properties of a freshwater microalga *Scenedesmus* sp. under different cultivation temperature. *Bioresour. Technol.* 102:3 (2011), pp. 3098–3102.

Li, Y., Horsman, M., Wang, B., Wu, N. & Lan, C.Q.: Effects of nitrogen sources on cell growth and lipid accumulation of green alga *Neochloris oleoabundans*. *Appl. Microbiol. Biotechnol.* 81:4 (2008), pp. 629–636.

Li, Y., Zhou, W., Hu, B., Min, M., Chen, P. & Ruan, R.R.: Integration of algae cultivation as biodiesel production feedstock with municipal wastewater treatment: Strains screening and significance evaluation of environmental factors. *Bioresour. Technol.* 102:23 (2011), pp. 10,861–10,867.

Li, Y.T., Han, D.X., Hu, G.R., Sommerfeld, M. & Hu, Q.A.: Inhibition of starch synthesis results in overproduction of lipids in *Chlamydomonas reinhardtii*. *Biotechnol. Bioeng.* 107:2 (2010), pp. 258–268.

Liang, Y., Sarkany, N. & Cui, Y.: Biomass and lipid productivities of *Chlorella vulgaris* under autotrophic, heterotrophic and mixotrophic growth conditions. *Biotechnol. Lett.* 31:7 (2009), pp. 1043–1049.

Lundquist, T.J., Woertz, I.C., Quinn, N.W.T. & Benemann, J.R.: Realistic technology and engineering assessment of algae biofuel production. Report. October 2010. Energy Biosciences Institute, University of California Berkeley, CA, 2010.

Luo, D., Hu, Z., Choi, D.G., Thomas, V.M., Realff, M.J. & Chance, R.R.: Life cycle energy and greenhouse gas emissions for an ethanol production process based on blue-green algae. *Environ. Sci. Technol.* 44:22 (2010), pp. 8670–8677.

Mackay, D. & Salusbury, T.: Choosing between centrifugation and cross-flow microfiltration. *Chem. Eng.* (London) 447 (1988), pp. 45–50.

McCausland, M., Brown, M., Barrett, S., Diemar, J. & Heasman, M.: Evaluation of live microalgae and microalgal pastes as supplementary food for juvenile Pacific oysters (*Crassostrea gigas*). *Aquaculture* 174:3–4 (1999), pp. 323–342.

Melis, A., Zhang, L., Forestier, M., Ghirardi, M. & Seibert, M.: Sustained photobiological hydrogen gas production upon reversible inactivation of oxygen evolution in the green alga *Chlamydomonas reinhardtii*. *Plant Physiol.* 122:1 (2000), pp. 127–135.

Mercer, P. & Armenta, R.E.: Developments in oil extraction from microalgae. *Eur. J. Lipid Sci. Technol.* 113:5 (2011), pp. 539–547.

Miao, X.L. & Wu, Q.Y.: High yield bio-oil production from fast pyrolysis by metabolic controlling of *Chlorella protothecoides*. *J. Biotechnol.* 110:1 (2004), pp. 85–93.

Minowa, T., Yokoyama, S., Kishimoto, M. & Okakura, T.: Oil production from algal cells of *Dunaliella tertiolecta* by direct thermochemical liquefaction. *Fuel* 74:12 (1995), pp. 1735–1738.

Mišurcová, L., Kráčmar, S., Klejdus, B. & Vacek, J.: Nitrogen content, dietary fiber, and digestibility in algal food products. *Czech J. Food Sci.* 28:1 (2010), pp. 27–35.

Mohan, D., Pittman, C.U., Jr. & Steele, P.H.: Pyrolysis of wood/biomass for bio-oil: A critical review. *Energy Fuels* 20:3 (2006), pp. 848–889.

Moheimani, N.R. & Borowitzka, M.A.: The long-term culture of the coccolithophore *Pleurochrysis carterae* (Haptophyta) in outdoor raceway ponds. *J. Appl. Phycol.* 18:6 (2006), pp. 703–712.

Moheimani, N.R. & Borowitzka, M.A.: Limits to productivity of the alga *Pleurochrysis carterae* (Haptophyta) grown in outdoor raceway ponds. *Biotechnol. Bioeng.* 96:1 (2007), pp. 27–36.

Mohn, F.H.: Experience and strategies in the recovery of biomass from mass cultures of microalgae. In: G. Schelef & C. J. Soeder (eds): *Algae biomass*. Elsevier, Amsterdam, The Netherlands, 1980, pp. 547–571.

Molina-Grima, E., Sanchez-Perez, J.A., Garcia-Camacho, F., Garcia-Sanchez, J.L., Acien-Fernandez, F.G. & Lopez-Alonso, D.: Outdoor culture of *Isochrysis galbana* ALII-4 in a closed tubular photobioreactor. *J. Biotechnol.* 37:2 (1994), pp. 159–166.

Molina Grima, E., Belarbi, E.H., Acien Fernandez, F.G., Robles Medina, A. & Chisti, Y.: Recovery of microalgal biomass and metabolites: process options and economics. *Biotechnol. Adv.* 20:7–8 (2003), pp. 491–515.

Moreno, J., Vargas, M.A., Rodriguez, H., Rivas, J. & Guerrero, M.G.: Outdoor cultivation of a nitrogen-fixing marine cyanobacterium, *Anabaena* sp. ATCC 33047. *Biomol. Eng.* 20:4–6 (2003), pp. 191–197.

Mujtaba, G., Choi, W., Lee, C.-G. & Lee, K.: Lipid production by *Chlorella vulgaris* after a shift from nutrient-rich to nitrogen starvation conditions. *Bioresour. Technol.* 123 (2012), pp. 279–283.

Norsker, N.H., Barbosa, M.J., Vermue, M.H. & Wijffels, R.H.: Microalgal production – a close look at the economics. *Biotechnol. Adv.* 29:1 (2011), pp. 24–27.

OILGAE: Algae status report: How far are we from commercialization? Published as a prelude to *3rd Algae World Europe 2011*, 2011, 4–5 May 2011, Madrid, CMT – Centre for Management Technology, pp. 1–23.

OriginOil: Algae harvesting, dewatering and extraction: A breakthrough technology to transform algae into oil. *World biofuel markets*. Amsterdam, The Netherlands, 2010.

Oswald, W.J. & Golueke, C.G.: Biological transformation of solar energy. *Adv. Appl. Microbiol.* 2 (1960), pp. 223–262.

Ötleş, S. & Pire, R.: Fatty acid composition of *Chlorella* and *Spirulina* microalgae species. *J. AOAC Int.* 84:6 (2001), pp. 1708–1714.

Park, Y., Je, K.W., Lee, K., Jung, S.E. & Choi, T.J.: Growth promotion of *Chlorella ellipsoidea* by co-inoculation with *Brevundimonas* sp. isolated from the microalga. *Hydrobiologia* 598:1 (2007), pp. 219–228.

Pate, R., Klise, G. & Wu, B.: Resource demand implications for US algae biofuels production scale-up. *Appl. Energy* 88:10 (2011), pp. 3377–3388.

Perez-Garcia, O., Escalante, F.M.E., de-Bashan, L.E. & Bashan, Y.: Heterotrophic cultures of microalgae: Metabolism and potential products. *Water Res.* (Oxford) 45:1 (2011), pp. 11–36.

Petkov, G., Ivanova, A., Iliev, I. & Vaseva, I.: A critical look at the microalgae biodiesel. *Eur. J. Lipid Sci. Technol.* 114:2 (2012), pp. 103–111.

Pienkos, P.T. & Darzins, A.: The promise and challenges of microalgal-derived biofuels. *Biofuel. Bioprod. Bior.* 3:4 (2009), pp. 431–440.

Pulz, O.: Photobioreactors: production systems for phototrophic microorganisms. *Appl. Microbiol. Biotechnol.* 57:3 (2001), pp. 287–293.

Pulz, O.: Performance summary report: Evaluation of GreenFuel's 3D matrix algae growth engineering scale unit. APS Red Hawk Power Plant, AZ. 2007, http://go-greener.com/snaffled_pdfs/greenfuels%20the%20best%20doc%20ever.pdf.

Pulz, O. & Scheibenbogen, K.: Photobioreactors: Design and performance with respect to light energy input. *Adv. Biochem. Eng.* 59 (1998), pp. 123–152.

Qin, S., Lin, H., Jiang, P., Qin, S., Lin, H.Z. & Jiang, P.: Advances in genetic engineering of marine algae. *Biotechnol. Adv.* 30:6 (2012), pp. 1602–1613.

Radakovits, R., Jinkerson, R.E., Darzins, A. & Posewitz, M.C.: Genetic engineering of algae for enhanced biofuel production. *Eukaryot. Cell* 9:4 (2010), pp. 486–501.

Ratha, S.K. & Prasanna, R.: Bioprospecting microalgae as potential sources of "green energy" – Challenges and perspectives (Review). *Appl. Biochem. Microbiol.* 48:2 (2012), pp. 109–125.

Ratledge, C.: Fatty acid biosynthesis in microorganisms being used for single cell oil production. *Biochimie* 86:11 (2004), pp. 807–815.

Richardson, J.W., Outlaw, J.L. & Allison, M.: The economics of microalgae oil. *AgBioForum* 13:2 (2010), pp. 119–130.

Richardson, K., Beardall, J. & Raven, J.A.: Adaptation of unicellular algae to irradiance – an analysis of strategies. *New Phytol.* 93:2 (1983), pp. 157–191.

Richmond, A.: The challenge confronting industrial microagriculture: High photosynthetic efficiency in large-scale reactors. *Hydrobiologia* 151:1 (1987), pp. 117–121.

Richmond, A.: Open systems for the mass-production of photoautotrophic microalgae outdoors – Physiological principles. *J. Appl. Phycol.* 4:3 (1992), pp. 281–286.

Richmond, A.: Microalgal biotechnology at the turn of the millennium: A personal view. *J. Appl. Phycol.* 12:3–5 (2000), pp. 441–451.

Richmond, A.: *Handbook of microalgal culture: Biotechnology and applied phycology.* Blackwell Scientific Ltd ,Oxford, UK, 2004.

Rodolfi, L., Zittelli, G.C., Bassi, N., Padovani, G., Biondi, N., Bonini, G. & Tredici, M.R.: Microalgae for oil: Strain selection, induction of lipid synthesis and outdoor mass cultivation in a low-cost photobioreactor. *Biotechnol. Bioeng.* 102:1 (2009), pp. 100–112.

Rosgaard, L., de Porcellinis, A.J., Jacobsen, J.H., Frigaard, N.U., Sakuragi, Y. & de Porcellinis, A.J.: Bioengineering of carbon fixation, biofuels, and biochemicals in cyanobacteria and plants. *J. Biotechnol.* 162:1 (2012), pp. 134–147.

Ruxton, C.H.S., Reed, S.C., Simpson, M.J.A. & Millington, K.J.: The health benefits of omega-3 polyunsaturated fatty acids: A review of the evidence. *J. Hum. Nutr. Diet.* 20:3 (2007), pp. 275–285.

Salih, F.M.: Microalgae tolerance to high concentrations of carbon dioxide: A review. *J. Environ. Prot.* 2:5 (2011), pp. 648–654.

Sánchez, J.F., Fernández-Sevilla, J.M., Acién, F.G., Cerón, M.C., Pérez-Parra, J. & Molina-Grima, E.: Biomass and lutein productivity of *Scenedesmus almeriensis*: influence of irradiance, dilution rate and temperature. *Appl. Microbiol. Biotechnol.* 79:5 (2008), pp. 719–729.

Sarin, R., Sharma, M., Sinharay, S. & Malhotra, R.K.: *Jatropha*-palm biodiesel blends: An optimum mix for Asia. *Fuel* 86:10–11 (2007), pp. 1365–1371.

Schenk, P.M., Thomas-Hall, S.R., Stephens, E., Marx, U.C., Mussgnug, J.H., Posten, C., Kruse, O. & Hankamer, B.: Second generation biofuels: High-efficiency microalgae for biodiesel production. *Bioenergy Res.* 1:1 (2008), pp. 20–43.

Scott, S.A., Davey, M.P., Dennis, J.S., Horst, I., Howe, C.J., Lea-Smith, D.J. & Smith, A.G.: Biodiesel from algae: Challenges and prospects. *Curr. Opin. Biotechnol.* 21:3 (2010), pp. 277–286.

Sharma, K.K., Schuhmann, H. & Schenk, P.M.: High lipid induction in microalgae for biodiesel production. *Energies* 5:5 (2012), pp. 1532–1553.

Sharma, N.K. & Rai, A.K.: Biodiversity and biogeography of microalgae: Progress and pitfalls. *Environ. Rev.* 19 (2011), pp. 1–15.

Sheehan, J., Dunahay, T., Benemann, J. & Roessler, P.: A look back at the U.S. Department of Energy's Aquatic Species Program: Biodiesel from algae. National Renewable Energy Laboratory Golden, CO, USA, 1998.

Shi, X., Liu, H., Zhang, X. & Chen, F.: Production of biomass and lutein by *Chlorella protothecoides* at various glucose concentrations in heterotrophic cultures. *Process Biochem.* 34:4 (1999), pp. 341–347.

Sialve, B., Bernet, N. & Bernard, O.: Anaerobic digestion of microalgae as a necessary step to make microalgal biodiesel sustainable. *Biotechnol. Adv.* 27:4 (2009), pp. 409–416.

Sim, T.S., Goh, A. & Becker, E.W.: Comparison of centrifugation, dissolved air flotation and drum filtration techniques for harvesting sewage-grown algae. *Biomass* 16:1 (1988), pp. 51–62.

Singh, J. & Gu, S.: Commercialization potential of microalgae for biofuels production. *Renew. Sustain. Energy Rev.* 14:9 (2010), pp. 2596–2610.

Slade, R. & Bauen, A.: Micro-algae cultivation for biofuels: Cost, energy balance, environmental impacts and future prospects. *Biomass Bioenergy* (2013), pp. 29–38.

Smith, V.H., Sturm, B.S.M., DeNoyelles, F.J. & Billings, S.A.: The ecology of algal biodiesel production. *Trends Ecol. Evol.* 25:5 (2010), pp. 301–309.

Spilling, K., Seppala, J. & Tamminen, T.: Inducing autoflocculation in the diatom *Phaeodactylum tricornutum* through CO_2 regulation. *J. Appl. Phycol.* 23:6 (2011), pp. 959–966.

Spolaore, P., Joannis-Cassan, C., Duran, E. & Isambert, A.: Commercial applications of microalgae. *J. Biosci. Bioeng.* 101:2 (2006), pp. 87–96.

Stephens, E., Ross, I.L., Mussgnug, J.H., Wagner, L.D., Borowitzka, M.A., Posten, C., Kruse, O. & Hankamer, B.: Future prospects of microalgal biofuel production systems. *Trends Plant Sci.* 15:10 (2010), pp. 554–564.

Suali, E. & Sarbatly, R.: Conversion of microalgae to biofuel. *Renew. Sustain. Energy Rev.* 16:6 (2012), pp. 4316–4342.

Tabernero, A., Martin del Valle, E.M. & Galan, M.A.: Evaluating the industrial potential of biodiesel from a microalgae heterotrophic culture: Scale-up and economics. *Biochem. Eng. J.* 63 (2012), pp. 104–115.

Thi Thai Yen, D., Sivaloganathan, B. & Obbard, J.P.: Screening of marine microalgae for biodiesel feedstock. *Biomass Bioenergy* 35:7 (2011), pp. 2534–2544.

Thomas, W.H., Seibert, D.L.R., Alden, M., Neori, A. & Eldridge, P.: Yields, photosynthetic efficiencies and proximate composition of dense marine microalgal cultures. 3. *Isochrysis* sp. and *Monallantus salina* experiments and comparative conclusions. *Biomass* 5:4 (1984a), pp. 299–316.

Thomas, W.H., Seibert, D.L.R., Alden, M., Neori, A. & Eldridge, P.: Yields, photosynthetic efficiencies and proximate composition of dense marine microalgal cultures. 1. Introduction and *Phaeodactylum tricornutum* experiments. *Biomass* 5:3 (1984b), pp. 181–209.

Thomas, W.H., Seibert, D.L.R., Alden, M., Neori, A. & Eldridge, P.: Yields, photosynthetic efficiencies and proximate composition of dense marine microalgal cultures. 2. *Dunaliella primolecta* and *Tetraselmis suecica* experiments. *Biomass* 5:3 (1984c), pp. 211–225.

Thompson, P.A., Guo, M.X. & Harrison, P.J.: Effects of variation in temperature.1. On the biochemical composition of 8 species of marine phytoplankton. *J. Phycol.* 28:4 (1992), pp. 481–488.

Tornabene, T.G., Holzer, G., Lien, S. & Burris, N.: Lipid composition of the nitrogen starved greenalga *Neochloris oleoabundans*. *Enzyme Microb. Technol.* 5:6 (1983), pp. 435–440.

Tredici, M.R.: Mass production of microalgae: Photobioreactors. In: A. Richmond (ed): *Handbook of microalgal culture: Biotechnology and applied phycology*. Blackwell Science, Oxford, UK, 2004, pp. 178–214.

Uduman, N., Qi, Y., Danquah, M.K., Forde, G.M. & Hoadley, A.: Dewatering of microalgal cultures: A major bottleneck to algae-based fuels. *J. Renew. Sustain. Energy* 2:1 (2010), pp. 012701-01–012701-15.

Ueda, R., Hirayama, S., Sugata, K. & Nakajama, H.: Process for the production of ethanol from microalgae, UP Patent 5578,472, USA 1996.

van Beilen, J.B.: Why microalgal biofuels won't save the internal combustion machine. *Biofuel. Bioprod. Bior.* 4:1 (2010), pp. 41–52.

van Den Hoek, C., Mann, D.G. & Jahns, H.M. (eds): *Algae. An introduction to phycology*. Cambridge University Press, Cambridge, UK, 1995.

Vesk, M., Jeffrey, S.W. & Hallegraeff, G.M.: Golden-brown algae: Prymnesiophyta and Chrysophyta. In: M.N. Clayton & R.J. King (eds): *Biology of marine plants*. Longman Cheshire, Melbourne, Australia, 1990, pp. 96–114.

Volkman, J.K. & Brown, M.R.: Nutritional value of microalgae and applications. In: D.V. Subba Rao (ed): *Algal cultures, analogues of blooms and applications,* Volume 1. Science Publishers, Plymouth, UK, 2006, pp. 407–457.

Volkman, J.K., Jeffrey, S.W., Nichols, P.D., Rogers, G.I. & Garland, C.D.: Fatty acid and lipid composition of 10 species of microalgae used in mariculture. *J. Exp. Mar. Biol. Ecol.* 128:3 (1989), pp. 219–240.

Wang, B., Li, Y., Wu, N. & Lan, C.Q.: CO_2 bio-mitigation using microalgae. *Appl. Microbiol. Biotechnol.* 79:5 (2008), pp. 707–718.

Watanabe, M.M.: Freshwater culture media. In: R. A. Andersen (ed): *Algal culturing techniques*. Academic Press, New York, USA, 2005, pp. 13–20.

Weissman, J.C., Goebel, R.P. & Benemann, J.R.: Photobioreactor design: Mixing, carbon utilization, and oxygen accumulation. *Biotechnol. Bioeng.* 31:4 (1988), pp. 336–344.

Wellisch, M., Jungmeier, G., Karbowski, A., Patel, M.K. & Rogulska, M.: Biorefinery systems – potential contributors to sustainable innovation. *Biofuel. Bioprod. Bior.* 4:3 (2010), pp. 275–286.

Wijffels, R.H. & Barbosa, M.J.: An outlook on microalgal biofuels. *Science* 329:5993 (2010), pp. 796–799.

Williams, P.J.L. & Laurens, L.M.L.: Microalgae as biodiesel & biomass feedstocks: Review & analysis of the biochemistry, energetics & economics. *Energy Environ. Sci.* 3:5 (2010), pp. 554–590.

Woertz, I.C., Fulton, L. & Lundquist, T.J.: Nutrient removal and greenhouse gas abatement with CO_2-supplemented algal high rate ponds. Paper written for the *WEFTEC annual conference*, Water Environment Federation, 2009, pp. 7924–7936.

Wu, H., Li, Y., Chen, L. & Zong, M.: Production of microbial oil with high oleic acid content by *Trichosporon capitatum. Appl. Energy* 88 (2011), pp. 138–142.

Wyatt, N.B., Gloe, L.M., Brady, P.V., Hewson, J. C., Grillet, A.M., Hankins, M.G. & Pohl, P.I.: Critical conditions for ferric chloride-induced flocculation of freshwater algae. *Biotechnol. Bioeng.* 109:2 (2012), pp. 493–501.

Xiong, W., Li, X., Xiang, J. & Wu, Q.: High-density fermentation of microalga *Chlorella protothecoides* in bioreactor for microbio-diesel production. *Appl. Microbiol. Biotechnol.* 78:1 (2008), pp. 29–36.

Xu, H., Miao, X.L. & Wu, Q.Y.: High quality biodiesel production from a microalga *Chlorella protothecoides* by heterotrophic growth in fermenters. *J. Biotechnol.* 126:4 (2006), pp. 499–507.

Zhang, C.W. & Richmond, A.: Sustainable, high-yielding outdoor mass cultures of *Chaetoceros muelleri* var. *subsalsum* and *Isochrysis galbana* in vertical plate reactors. *Mar. Biotechnol.* 5:3 (2003), pp. 302–310.

Zheng, Y., Chi, Z., Lucker, B. & Chen, S.: Two-stage heterotrophic and phototrophic culture strategy for algal biomass and lipid production. *Bioresour. Technol.* 103:1 (2012), pp. 484–488.

Zittelli, G.C., Pastorelli, R. & Tredici, M.R.: A Modular Flat Panel Photobioreactor (MFPP) for indoor mass cultivation of *Nannochloropsis* sp under artificial illumination. *J. Appl. Phycol.* 12:3–5 (2000), pp. 521–526.

Zmora, O. & Richmond, A.: Microalgae for aquaculture: Microalgae production for aquaculture. In: A. Richmond (ed): *Handbook of microalgal culture: Biotechnology and applied phycology*. Blackwell Science, Oxford, UK, 2004, pp. 365–379.

CHAPTER 11

Biodiesel emissions and performance

Syed Ameer Basha

11.1 INTRODUCTION

Biodiesel is a clean burning alternative fuel produced from renewable resources. The fuel is a mixture of fatty acid alkyl esters made from vegetable oils, animal fats or recycled greases. Biodiesel can be used in compression-ignition (diesel) engines in its pure form with little or no modification. Biodiesel is simple to use, biodegradable, nontoxic, and essentially free of sulfur and aromatics. It is usually used as a petroleum diesel additive to reduce levels of particulates, carbon monoxide, hydrocarbons and air toxics from diesel-powered vehicles. When used as an additive, the resulting diesel fuel may be called B5, B10 or B20, representing the percentage of the biodiesel that is blended with petroleum diesel. In the United States, most biodiesel is made from soybean oil or recycled cooking oils. Animal fats, other vegetable oils, and other recycled oils can also be used to produce biodiesel, depending on their cost and availability. In the future, blends of all kinds of fats and oils may be used to produce biodiesel. Biodiesel is made through a chemical process called transesterification whereby the glycerin is separated from the fat or vegetable oil. The process leaves behind two products – methyl esters (the chemical name for biodiesel) and glycerin (a valuable byproduct usually sold to be used in soaps and other products). Biodiesel is the only alternative fuel to have fully completed the health effects testing requirements of the 1990 Clean Air Act Amendments (US Department of Energy, n.d.).

11.1.1 *Need of biodiesel*

Agriculture is basically an energy conversion process, as it is a producer as well as consumer of energy. The production process involves the conversion of solar energy into biomass by the way of photosynthesis. Agricultural technology depends upon direct sources of energy such as petroleum and electricity, whereas indirect energy sources are chemical fertilizers and farm machinery. Land preparation and harvesting account for the bulk of diesel consumption whereas electricity is largely used for irrigation purposes. Diesel engines are widely used for irrigation from canals, pumping water from wells, and for agricultural farm machinery such as tractors, and thrashers. Diesel engines are also used for transportation of the harvested crops from the fields to the markets. They play a vital role in our modernized agricultural sector. These engines are used in tractors and for running small motor pumps. It is impossible to do away with internal combustion engines at this juncture and alternative fuels must be sought to ensure safe survival of the existing engines. Identification of appropriate alternatives to the conventional petroleum-based fuels has been subjected to a variety of technical, political, geographical and economic considerations. During the World Wars I and II and the oil crisis periods (1970s and 1990s) various alternative fuels were tried and partial success was achieved. A lot of research and development is being conducted in several parts of the world. More than 350 oil-bearing crops are identified, some of which are considered as potential alternative fuels for diesel engines. The average annual real oil prices over the five years 2007–11 were 220% above the average for 1997–2001; for coal the increase was 141% and for gas 95%. These long run price movements inevitably lead to demand and supply responses. World primary energy consumption is projected to grow by 1.6% p.a. from 2011 to 2030, adding 36% to global consumption by 2030 (BP, 2013).

Table 11.1.   Oil consumption of the world (BP, 2012).

Thousand barrels per day	2001	2002	2003	2004	2005	2006	2007	2008	2009	2010	2011
USA	19649	19761	20033	20732	20802	20687	20680	19498	18771	19180	18835
Canada	2008	2051	2115	2231	2229	2246	2323	2288	2179	2298	2293
Mexico	1939	1864	1909	1983	2030	2019	2067	2054	1995	2014	2027
Total South & Central America	4945	4930	4778	4966	5111	5233	5582	5788	5763	6079	6241
Total Europe & Eurasia	19593	19571	19776	19935	20095	20342	19984	20002	18123	19039	18924
Total Middle East	5260	5467	5707	6100	6365	6615	6895	7270	7510	7890	8076
Total Africa	2510	2560	2629	2247	2864	2855	3006	3160	3243	3377	3336
Total Asia & Pacific	21343	21983	22738	24053	24429	24875	25783	25720	26047	27563	28301
Total World	77245	78187	79686	82746	83925	84873	86321	85768	84631	87439	88034

Table 11.2.   Production of biofuels (BP, 2012).

Thousand tons oil equivalent	2001	2002	2003	2004	2005	2006	2007	2008	2009	2010	2011
USA	3288	3987	5226	6357	7478	9746	13456	19096	21670	25467	28251
Canada	111	113	113	113	133	160	461	500	721	746	961
Total South & Central America	5638	6280	7227	7291	8093	9405	12303	15788	15942	17863	16129
Total Europe & Eurasia	889	1206	1619	2031	3157	5052	6820	8196	10243	10811	9837
Total Middle East	–	–	–	–	–	–	–	–	–	–	–
Total Africa	6	6	6	6	6	6	6	10	14	29	29
Total Asia & Pacific	89	238	491	603	834	1280	1563	2468	3207	3528	3649
Total World	10021	11830	14682	16401	19701	25648	34613	46063	51802	58457	58868

The oil consumption and bio-fuels production are shown in Tables 11.1 and 11.2, respectively.
As the fossil fuels are depleting day by day, there is a need to find an alternative fuel to meet global energy demands. In addition to the energy crisis, environmental pollution is another problem faced by the world. Noxious pollutant concentration in the atmosphere is mainly responsible for global warming and its concentration is increasing as the consumption of fossil fuels is increasing at an alarming rate year by year. These problems forced us to look for alternative fuels, which can satisfy ever-increasing demands for energy as well as protect the environment by repressing the levels of noxious pollutants.

11.1.2 *Biofuel*

A biofuel is a type of fuel whose energy is derived from biological carbon fixation. Biofuels include fuels derived from biomass conversion, as well as solid biomass, liquid fuels and various biogases. Biofuels are gaining increased public and scientific attention, driven by factors such as oil price hikes and the need for increased energy security. Biofuels do not fully address global warming concerns.

Types of biofuels:

1) First-generation biofuels: 'First-generation' or conventional biofuels are made from sugar, starch, or vegetable oil, which can be easily extracted using conventional technology. They comprise:
   - Bioalcohols
   - Biodiesel
   - Green diesel
   - Vegetable oil
   - Bioethers
   - Biogas
   - Syngas
   - Solid biofuels
2) Second-generation (advanced) biofuels: Second generation biofuels, also known as advanced biofuels, are fuels that can be manufactured from various types of biomass. Biomass is a wide-ranging term meaning any source of organic carbon that is renewed rapidly as part of the carbon cycle. Biomass is derived from plant materials but can also include animal materials.

Second generation biofuels are made from lignocellulosic biomass or woody crops, agricultural residues or waste, which makes it harder to extract the required fuel.

Biodiesel is an oxygenated fuel which contains 10 to 15% oxygen by weight. It is renewable and can be said to be sulfur free. These facts lead biodiesel to complete combustion and emit fewer exhaust fumes than diesel engines. If the fuel properties of biodiesel and diesel fuel are compared, the biodiesel has higher viscosity, density, pour point, flash point and cetane number than diesel fuel. Also the energy content or net calorific value of biodiesel is about 12% less than that of diesel fuel on a mass basis. Using biodiesel can help reduce the world's dependence on fossil fuels and it also has significant environmental benefits. The reasons for these environmental benefits are: using biodiesel instead of the conventional diesel fuel reduces the exhaust emissions such as the overall life cycle of carbon dioxide (CO_2), particulate matter (PM), carbon monoxide (CO), sulfur oxides (SO_x), volatile organic compounds (VOCs), and unburned hydrocarbons (HC) significantly. Depending on the abundant availability of feedstock in local region, the different feed stocks are focused for the biodiesel production. In the United States, the primary source for biodiesel production is soybean oil, while European Union nations prefer to utilize rapeseed oil, and in South East Asia regions, palm oil, coconut oil and jatropha oil are utilized mainly for biodiesel production.

Advantages of biodiesel:

- Biodiesel can be used in pure form (B100) or may be blended with petroleum diesel at any concentration in most injection pump diesel engines.
- Biodiesel has better ignition and combustion characteristics (due to a higher cetane index), which allows the engine to run more smoothly with less of the "knocking" sounds typical of diesel engines.
- Biodiesel substantially reduces exhaust emissions (unburned hydrocarbons, carbon monoxide and particulate matter). It contains naturally occurring oxygen, which enables the fuel to burn more completely and all but eliminates the black smoke normally associated with diesel engines.

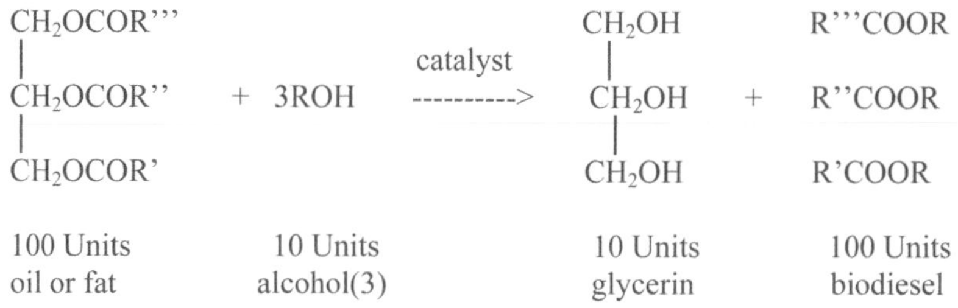

$$
\begin{array}{c}
\mathrm{CH_2OCOR'''} \\
| \\
\mathrm{CH_2OCOR''} \\
| \\
\mathrm{CH_2OCOR'}
\end{array}
\quad + \quad 3\mathrm{ROH}
\quad \xrightarrow[]{\text{catalyst}} \quad
\begin{array}{c}
\mathrm{CH_2OH} \\
| \\
\mathrm{CH_2OH} \\
| \\
\mathrm{CH_2OH}
\end{array}
\quad + \quad
\begin{array}{c}
\mathrm{R'''COOR} \\
\mathrm{R''COOR} \\
\mathrm{R'COOR}
\end{array}
$$

| 100 Units | 10 Units | 10 Units | 100 Units |
| oil or fat | alcohol(3) | glycerin | biodiesel |

Figure 11.1.   Chemical reaction of base catalyzed biodiesel production.

- Biodiesel has very good intrinsic lubrication properties. Even blends as low as B1 (1% in ULSD) can improve the lubricity in highly de-sulfurized mineral diesel. In engines approved for operation with B100 biodiesel, the engine wear is significantly reduced.
- Using biodiesel as a B20 or B30 blend (20/30% mix with normal diesel) has no reported effect on fuel consumption.
- Biodiesel has a higher flashpoint than mineral diesel.
- It is also readily biodegradable and non-toxic, which makes it a safer and more environmentally friendly fuel to handle, particularly in sensitive areas.
- Biodiesel is naturally free of sulfur and so produces no sulfur dioxide, considered to be one of the main precursors to acid rain.
- Biodiesel is made from renewable resources, which means it reduces the contribution of carbon dioxide (one of the main greenhouse gases) to the atmosphere.

11.1.3   Production of biodiesel

There are three basic routes to biodiesel production from oils and fats:

- Base catalyzed transesterification of the oil.
- Direct acid catalyzed transesterification of the oil.
- Conversion of the oil to its fatty acids and then to biodiesel.

Most of the biodiesel produced today uses the base catalyzed reaction for several reasons:

- It yields high conversion (98%) with minimal side reactions and reaction time.
- It is a direct conversion to biodiesel with no intermediate compounds.
- No exotic materials of construction are needed.

The chemical reaction for base catalyzed biodiesel production is shown in Figure 11.1. One hundred units of fat or oil (such as soybean oil) are reacted with 10 units of a short chain alcohol in the presence of a catalyst to produce 10 units of glycerin and 100 units of biodiesel. The short chain alcohol, signified by ROH (usually methanol, but sometimes ethanol) is charged in excess to assist in quick conversion. The catalyst is usually sodium or potassium hydroxide that has already been mixed with the methanol. R', R'', and R''' indicate the fatty acid chains associated with the oil or fat which are largely palmitic, stearic, oleic, and linoleic acids for naturally occurring oils and fats.

Production of oil from various crops has given the Table 11.3. Oil palm gives the highest amount of oil per acre and *Zea mays* gives the lowest.

11.2   BIODIESEL EMISSIONS

Biodiesel is the first and only alternative fuel to have a complete evaluation of emission results. Biodiesel reduces particulate matter (PM), hydrocarbon (HC), and carbon monoxide (CO)

Table 11.3. Oil per acre production for various crops in liters (Kurki, 2010).

Plant	Latin name	Oil/acre [Liter]	Plant	Latin name	Oil/acre [Liter]
Oil palm	*Elaeis guineensis*	2309	Rice	*Oriza sativa* L.	322
Macauba palm	*Acrocomia aculeate*	1745	Buffalo gourd	*Cucurbitafoetidissima*	307
Pequi	*Caryocar brasiliense*	1450	Safflower	*Carthamus tinctorius*	303
Buriti palm	*Mauritia flexuosa*	1268	Crambe	*Crambe abyssinica*	273
Oiticia	*Licania rigida*	1162	Sesame	*Sesamum indicum*	269
Coconut	*Cocos nucifera*	1045	Camelina	*Camelina sativa*	227
Avocado	*Persea americana*	1022	Mustard	*Brassica alba*	223
Brazil nut	*Bertholletia excelsa*	927	Coriander	*Coriandrum sativum*	208
Macadamia nut	*Macadamia terniflora*	870	Pumpkin seed	*Cucurbita pepo*	208
Jatropa	*Jatropha curcas*	734	Euphorbia	*Euphorbia lagascae*	204
Babassu palm	*Orbignya martiana*	711	Hazelnut	*Corylus avellana*	185
Jojoba	*Simmondsia chinensis*	704	Linseed	*Linum usitatissimum*	185
Pecan	*Carya illinoensis*	693	Coffee	*Coffea arabica*	178
Bacuri	*Platonia insignis*	553	Soybean	*Glycine max*	174
Castor bean	*Ricinus communis*	548	Hemp	*Cannabis sativa*	140
Gopher plant	*Euphorbia lathyris*	519	Cotton	*Gossypium hirsutum*	125
Piassava	*Attalea funifera*	515	Calendula	*Calendula officinalis*	117
Olive tree	*Olea europaea*	469	Kenaf	*Hibiscus cannabinus* L.	106
Rapeseed	*Brassica napus*	462	Rubber seed	*Hevea brasiliensis*	98
Opium poppy	*Papaver somniferum*	119	Lupine	*Lupinus albus*	24
Peanut	*Ariachis hypogaea*	109	Palm	*Erythea salvadorensis*	23
Cocoa	*Theobroma cacao*	105	Oat	*Avena sativa*	22
Sunflower	*Helianthus annuus*	98	Cashew nut	*Anacardium occidentale*	18
Tung oil tree	*Aleurites fordii*	96	Corn	*Zea mays*	18

Table 11.4. Biodiesel emissions compared to diesel fuel emissions (US EPA, 2002).

	B100	B20	B5	B2
CO	−45%	−11%	−3.2%	−1.3%
CO_2	−48%	−12%	−3.9%	−1.5%
PM	−47%	−12%	−3.1%	−1.2%
NO_x	+10%	−2% to +2%	−1% to +1%	0%/negligible
PAH (polycyclic aromatic hydrocarbons)	−80%	−13%	Not reported	Not reported
nPAH (nitrated PAHs)	−90%	−50%	Not reported	Not reported
Ozone potential of speciated hydrocarbons	−50%	−10%	Not reported	Not reported

emissions from most modern four-stroke CI engines. These benefits occur because the fuel (B100) contains 12% oxygen by weight. The presence of fuel oxygen allows the fuel to burn more completely, so fewer unburned fuel emissions result. This same phenomenon reduces air toxics, because the air toxics are associated with the unburned or partially burned HC and PM emissions. Testing has shown that PM, HC, and CO reductions are independent of the feedstock used to make biodiesel. Table 11.4 shows the decrease in the emissions with the increase of biodiesel. Pure biodiesel (B100) emits less emissions in comparison with the diesel.

Different factors affecting the biodiesel emissions (i.e. NO_x, CO_x, HC and particulate matter) and performance characteristics (brake specific fuel consumption and efficiency) are:

- Biodiesel content.
- Biodiesel properties and its feed stock.

- Type of engine and its operation conditions.
- Catalyst/additive.

11.2.1 NO_x

Nitrogen makes up almost 80% of the atmosphere. It is normally inert and not directly involved in the combustion process itself. But flame temperatures above 1371°C cause nitrogen and oxygen to combine. This combination forms various compounds called "oxides of nitrogen" or NO_x. This typically occurs when the engine is under load and combustion temperatures soar. Most of the NO_x that comes out of the tailpipe is in the form of nitric oxide (NO), a colorless poisonous gas. It then combines with oxygen in the atmosphere to form nitrogen dioxide (NO_2), which creates a brownish haze in badly polluted areas. To minimize the formation of NO_x in the engine, exhaust gas recirculation (EGR) is used. Recirculating a small amount of exhaust gas back into the intake manifold to dilute the air/fuel mixture has a "cooling" effect on combustion. This process keeps temperatures below the NO_x formation threshold. In many engines produced after 1980, particularly those with computerized controls, a special three-way catalytic converter is also used to further reduce NO_x in the exhaust. The first chamber of the converter contains a special "reduction" catalyst that breaks NO_x down into Oxygen and Nitrogen. The second chamber contains the "oxidation" catalyst that reburns CO and HC.

The well-known Zeldovich mechanism (1946) clearly describes the NO formation by the following reactions:

$$O + N_2 \rightarrow NO + N$$
$$N + O_2 \rightarrow NO + O$$

The NO formed during combustion can be further converted to NO_2 by reaction. Afterwards it can also be converted back to the form of NO by reaction.

$$NO + HO_2 \rightarrow NO_2 + OH$$
$$NO_2 + O \rightarrow NO + O_2$$

The use of pure biodiesel causes the increase in NO_x emissions (Hazar, 2009; Ozsezen *et al.*, 2009). For example, a maximum of 15% increase in NO_x emissions for B100 was observed at high load condition as the result of 12% oxygen content of the B100 and higher gas temperature in the combustion chamber (Nabi *et al.*, 2009).

NO_x emissions will increase when using biodiesel. This increase is mainly due to higher oxygen content for biodiesel. Moreover, the cetane number and different injection characteristics also have an impact on NO_x emissions for biodiesel. The content of unsaturated compounds in biodiesel could have a greater impact on NO_x emissions. The larger the content of unsaturated compounds, the more NO_x emissions will be reduced, which is a matter of concern. The larger the engine load, the higher will be the level of NO_x emissions for biodiesel, which is in line with the mechanism of NO_x formation. A further study is needed to determine the effect of injection timing and injection pressure on NO_x emissions of biodiesel. The use of EGR will reduce NO_x emissions of biodiesel, but due to the change of combustion characteristics for biodiesel.

Biodiesel content: It could be observed that the increasing content of biodiesel in the blends resulted in the increased NO_x emissions.

Biodiesel properties and its feed stock: Properties of biodiesel such as cetane number, advance in injection and combustion, especially higher oxygen content, and feedstock of biodiesel have an important effect on NO_x emissions for biodiesel. The higher cetane number of biodiesel shortens ignition delay and thus advances combustion. Also, the higher oxygen content in biodiesel enhances the formation of NO_x.

Type of engine and its operation conditions: NO_x emissions increased in the Low heat rejection engine compared with the original engine due to a higher combustion temperature (Banapurmath and Tewari, 2008). NO_x concentration varies linearly with load (Godiganur *et al.*, 2009; 2010). As load increased, the overall fuel-air ratio increased which resulted in an increase in the average gas temperature in the combustion chamber and hence NO_x formation which is sensitive to temperature increases.

11.2.2 CO_x

Carbon monoxide is formed when the fuel mixture is rich and there is insufficient oxygen to burn all the fuel completely. The richer the fuel mixture the greater the quantity of CO produced. That makes CO a good diagnostic indicator of incomplete combustion, carburetor maladjustments, clogged air filter, sticking choke, defective heated air intake system, and leaky fuel injectors.

The CO emissions reduce when using biodiesel due to higher oxygen content and lower carbon to hydrogen ratio in biodiesel compared to diesel. With content of pure biodiesel increasing in blended fuel, CO emissions of blends reduce. CO emissions for biodiesel are affected by its feedstock and other properties of biodiesel such as cetane number and increase in combustion.

Content of biodiesel: With content of pure biodiesel increasing in blends fuel, CO emissions of blends reduce due to increasing in oxygen content.

Feedstock and properties of biodiesel: Feedstock of biodiesel affects CO emissions (Wu *et al.*, 2009) because of the different oxygen content and cetane number between them.

Engine type and its operating conditions: Different engines affect CO emissions. Karabektas (2009) tested biodiesel and diesel fuel on a direct injection diesel engine in naturally aspirated and TU conditions. CO emissions in the NA conditions for both biodiesel and diesel were all higher than those in the TU conditions, which increases air to the diesel engine and enables mixing of fuel and air more easily in the combustion chamber. CO emissions reduced when this low heat loss engine was used. CO emissions increased with engine load increasing (Utsa, 2005a; 2005b). The main reason for this increase is that the air/fuel ratio decreases with an increase in load, which is typical of all internal combustion engines. CO emissions for biodiesel decrease with an increase in engine speed, as the result of the better air/fuel mixing process and/or the increased fuel/air equivalence ratio with the increased engine speed (Lin *et al.*, 2006; Qi *et al.*, 2009).

11.2.3 *HC emissions of biodiesel*

Hydrocarbon emissions result when fuel molecules in the engine do not burn or burn only partially. Hydrocarbons react in the presence of nitrogen oxides and sunlight to form ground-level ozone, a major component of smog. Ozone can irritate the eyes, damage lungs, and aggravate respiratory problems. It is our most widespread urban air pollution problem. Some kinds of exhaust hydrocarbons are also toxic, with the potential to cause cancer.

HC emissions reduce when pure biodiesel is used instead of diesel (Wu *et al.*, 2009). The feedstock of biodiesel and its properties have an effect on HC emissions, especially for the different chain length or saturation level of biodiesels. The advance in injection and combustion of biodiesel favors the lower HC emissions.

Content of biodiesel: HC emissions decreased with increasing biodiesel percentage in the blend (Ghobadian *et al.*, 2009; Gumus and Tsolakisa, 2007). Godiganur *et al.* (2010) found that the reduction in HC was linear with the addition of biodiesel for the blends.

Feedstock and properties of biodiesel: HC emissions decreased by 45–67% on average due to the combined effect of oxygen content and cetane number (Wu *et al.*, 2009). An increase in chain length or saturation level of several biodiesels led to a higher reduction in HC emissions (Graboski *et al.*, 2003).

Engine operating conditions: HC emissions increased with load increase, due to high fuel consumption in high load.

11.2.4   *Particulate matter (PM) emissions*

"Particulate matter", also known as particle pollution or PM, is a complex mixture of extremely small particles and liquid droplets. Particle pollution is made up of a number of components, including acids (such as nitrates and sulfates), organic chemicals, metals, and soil or dust particles.

The size of particles is directly linked to their potential for causing health problems. EPA is concerned about particles that are 10 μm in diameter or smaller because those are the particles that generally pass through the throat and nose and enter the lungs. Once inhaled, these particles can affect the heart and lungs and cause serious health problems. EPA groups particle pollution into two categories:

- "Inhalable coarse particles," such as those found near roadways and dusty industries, are larger than 2.5 μm and smaller than 10 μm in diameter.
- "Fine particles," such as those found in smoke and haze, are 2.5 μm in diameter and smaller. These particles can be directly emitted from sources such as forest fires, or they can form when gases emitted from power plants, industries and automobiles react in the air.

The use of biodiesel instead of diesel causes the reduction in PM emissions by 53–69% (Wu et al., 2009). This reduction is proportional to the amount of biodiesel in the blended fuel. PM emissions are reduced because of the lower aromatic and sulfur compounds and the higher cetane number for the biodiesel, but the more important factor is the higher oxygen content. It should be noted that the advantage of no sulfur characteristics for biodiesel will disappear as the sulfur content in diesel is becoming less and less. The larger the engine load, the greater will be the PM emissions of biodiesel. Furthermore, the higher the engine speed, the lower the PM emissions will be. The value of an injection mechanism in biodiesel engines is uncertain at this stage. It is necessary to study further the matching characteristics of biodiesel and/or its blends with the engine. The use of EGR might decrease PM emissions of biodiesel, although the PM emissions level is still very low relative to diesel. But PM emissions of biodiesel compared to diesel will increase abnormally in the case of low temperature conditions. Oxygenates can increase PM emissions of biodiesel, but it would not be useful for power recovery. The metal-based additives may be effective to reduce PM emissions of biodiesel due to their catalytic effect.

Biodiesel content: PM emissions decrease remarkably with an increase in biodiesel content in blends (Sahoo et al., 2009).

Biodiesel properties and its feed stock: The higher the oxygen content in biodiesel, the greater the reduction of PM emissions because of complete combustion, and this further increases the oxidation of soot. The reduction in cetane number resulted in the decrease of particulates at high load (Yoshiyuki, 2000). PM emissions decreased with shorter fatty acid carbon-chain lengths (Lin et al., 2009). The smoke emission from ethyl ester is more than that of methyl ester due to the presence of more oxygen for methyl ester.

Engine type and its operating conditions: PM emissions increase as load increases (Buyukkaya, 2010; Deshmukh and Bhuyar, 2009). It is mainly due to the decreased air–fuel ratio at higher loads when larger quantities of fuel are injected into the combustion chamber, much of which goes unburnt into the exhaust. The higher the engine speed, the lower are the PM emissions (Kaplan et al., 2006; Song and Zhang, 2008; Ulusoy et al., 2009). The improved combustion efficiency should be attributed to an increase in turbulence with an increase in engine speed, which enhances the extent of complete combustion.

11.3   BIODIESEL PERFORMANCE

11.3.1   *Brake specific fuel consumption*

Fuel consumption increases when using biodiesel, but this trend will be weakened as the proportion of biodiesel reduces in the blend of fuel with diesel. The increase in biodiesel fuel consumption is mainly due to its low heating value, as well as its high density and high viscosity. The different feedstock of biodiesel with different heating value and carbon chain length, or different production processes and quality, also have an impact on engine economy. The use of a turbocharged engine or a low heat release engine, will improve biodiesel engine economy. Engine operating conditions, such as load, speed, injection timing and injection pressure, etc., are also influential to biodiesel engine economy, and although these influences are not essential, the further study on these conditions should be carried out to improve the engine and its control systems in order to obtain the optimal match.

Biodiesel content: With an increase in the content of biodiesel, engine fuel consumption will increase.

Biodiesel properties and its feedstock: The lower heating value, higher density and higher viscosity play a primary role in the fuel consumption of biodiesel. Consumption increased for biodiesel compared to diesel, which contributed to the loss in heating value of biodiesel.

Engine type and its operating conditions: Biodiesel engine economy is affected by the engine type and its operating conditions, such as load, speed, and injection timing and injection pressure.

11.3.2 *Efficiency*

The use of biodiesel will lead to reduced engine power. The main reason for power loss is the reduced heating value of biodiesel compared to diesel. The high viscosity and high lubricity of biodiesel also have certain effects on engine power. Feedstock of biodiesel is not an important factor which affects engine power.

Content of biodiesel: Engine power will decrease with the increase of content of biodiesel (Aydin and Bayindir, 2010; Raheman and Phadatare, 2004; Reyes and Sepúlveda, 2006).

Properties of biodiesel and its feedstock: Properties of biodiesel, especially in heating value, viscosity and lubricity, have an important effect on engine power. The heating value of fuels is an important measure of the amount of energy it releases to do work. So, the lower heating value of biodiesel is attributed to the decrease in engine power. The higher viscosity of biodiesel, which enhances fuel spray penetration, and thus improves air–fuel mixing, is used to explain the recovery in torque and power for biodiesel related to diesel (Monyem *et al.*, 2001). The high lubricity of biodiesel might result in the reduced friction loss and thus improve the brake effective power.

Engine type and its operating conditions: The engine type and its operating conditions, such as engine load, engine speed, injection timing and injection pressure, have an effect on biodiesel engine power. The difference of BTE between biodiesel and pure diesel tended to increase with the increase of fuel injection pressure (Sharma *et al.*, 2009). Power and torque were increased up to almost pure diesel levels by reducing injection advance because it was possible to optimize combustion, and by improving performances especially at low and medium speeds, with respect to nominal injection advance operation (Carraretto *et al.*, 2004).

11.4 EFFECT OF A CATALYST OR ADDITIVE

An additive is usually a chemical introduced into a solution with fuel. Additives must be replenished at each refueling. A catalyst affects the fuel but does not become part of it. The Fitch fuel catalyst is a once-only treatment and will last up to 400,000 km or virtually the life of the engine.

11.4.1 *Effect of a catalyst on biodiesel emissions*

Metal-based additives (Gürü *et al.*, 2010; Ryu, 2010), alcohol (methanol and ethanol) (Kwanchareon *et al.*, 2007), cetane number improver (Kim and Choi, 2010) and emulsifiers (Lin and Lin, 2007), can be added into biodiesel to improve NO_x emissions. The NO_x emissions were most effectively reduced by burning the O/W/O three-phase biodiesel emulsion that contained aqueous ammonia, particularly at lower engine speeds. Metallic additives, oxide additives, emulsifier, etc. seem to be useful to improve NO_x emissions of biodiesel. CO emissions of biodiesel reduce with metal based additives. The oxidation converter might play an important role in CO emission for biodiesel. Lujan *et al.* (2009) found that the oxidative catalytic converter reduced CO emissions greater than that without the converter, but the conversion efficiency of the converter declined slightly. Although an oxidative catalytic converter has a positive impact on HC emissions for biodiesel, its function seems to be weakened. Metal based additives have less efficiency to improve HC emissions for biodiesel than the others emissions. And a small proportion of ethanol and methanol added into biodiesel and its blends with diesel may be advantageous to HC emissions. The addition of oxygenates such as ethanol, methanol and alcohol into biodiesel causes a decrease in PM emissions due to the enrichment of oxygen content in the fuel (Bhale *et al.*, 2009; Cheung *et al.*, 2009; Lapuerta *et al.*, 2008). The biodiesel blends with Mg and Mo had a better effect on PM emissions due to their catalytic effect (Keskin *et al.*, 2008).

11.4.2   *Effect of catalysts and additives on biodiesel performance*

Additives used to improve the properties of biodiesel may further improve the combustion performance of a biodiesel engine, thus it will promote economy, and this will also improve engine power.

11.4.2.1   *Brake specific fuel consumption*
The BSFC of biodiesel fuel with antioxidants decreased more than that without antioxidants, although no specific trends were detected according to the type or amount of antioxidants.

11.4.2.2   *Efficiency*
The use of B20X with 1% 4-nonyl phenoxy acetic acid (NPAA) additive produced higher brake power over the entire speed range in comparison to B20 and B0 (diesel) (Kalam and Masjuki, 2008).

11.5   CONCLUSIONS

Even though 350 oil bearing crops are identified, only few have the potential to make biodiesel. These are sunflower, rapeseed, palm, jatropha, etc. Overall, biodiesel, especially for the blends with small portion of biodiesel, is technically feasible as an alternative fuel in compression ignition engines with no or minor modification to the engine.

REFERENCES

Aydin, H. & Bayindir, H.: Performance and emission analysis of cottonseed oil methyl ester in a diesel engine. *Renew. Energy* 35 (2010), pp. 588–592.
Banapurmath, N.R. & Tewari, P.G.: Performance of a low heat rejection engine fuelled with low volatile Honge oil and its methyl ester (HOME). Proceedings of the Institution of Mechanical Engineers, Part A: *J. Power Energy* 222 (2008), pp. 323–230.
Bhale, P.V., Deshpande, N.V. & Thombre, S.B.: Improving the low temperature properties of biodiesel fuel. *Renew. Energy* 34 (2009), pp. 794–800.
BP: BP statistical review of world energy. June 2012, www.bp.com/statisticalreview (accessed August 2013).
BP: BP World Energy Outlook 2030. 2013, www.bp.com/content/.../BP_World_Energy_Outlook_booklet_ 2013.pdf (accessed August 2013).
Buyukkaya, E.: Effects of biodiesel on a DI diesel engine performance, emission and combustion characteristics. *Fuel* 89 (2010), pp. 3099–3105.
Carraretto, C., Macor, A., Mirandola, A., Stoppato, A. & Tonon, S.: Biodiesel as alternative fuel: experimental analysis and energetic evaluations. *Energy* 29 (2004), pp. 2195–2211.
Cheung, C.S., Zhu, L. & Huang, Z.: Regulated and unregulated emissions from a diesel engine fueled with biodiesel and biodiesel blended with methanol. *Atmos. Environ.* 43 (2009), pp. 4865–4872.

Deshmukh, S.J. & Bhuyar, L.B.: Transesterified Hingan (Balanites) oil as a fuel for compression ignition engines. *Biomass Bioenergy* 33 (2009), pp. 108–112.

Ghobadian, B., Rahimi, H., Nikbakht, A.M., Najafi, G. & Yusaf, T.F.: Diesel engine performance and exhaust emission analysis using waste cooking biodiesel fuel with an artificial neural network. *Renew. Energy* 34 (2009), pp. 976–982.

Godiganur, S., Murthy, C.H.S. & Reddy, R.P.: 6BTA 5.9 G2-1 Cummins engine performance and emission tests using methyl ester mahua (*Madhuca indica*) oil/diesel blends. *Renew. Energy* 34 (2009), pp. 2172–2177.

Godiganur, S., Murthy, C.H.S. & Reddy, R.P.: Performance and emission characteristics of a Kirloskar HA394 diesel engine operated on fish oil methyl esters. *Renew. Energy* 35 (2010), pp. 355–359.

Graboski, M.S., McCormick, R.L., Alleman, T.L. & Herring, A.M.: The effect of biodiesel composition on engine emissions from a DDC series 60 diesel engine. NREL/SR-510-31461, National Renewable Energy Laboratory (NREL), 2003.

Gumus, M. & Kasifoglu, S.: Performance and emission evaluation of a compression ignition engine using a biodiesel (apricot seed kernel oil methyl ester) and its blends with diesel fuel. *Biomass Bioenergy* 34 (2010), pp. 134–139.

Gürü, M., Koca, A., Can Ö, Çınar, C. & Şahin, F.: Biodiesel production from waste chicken fat based sources and evaluation with Mg based additive in a diesel engine. *Renew. Energy* 35 (2010), pp. 637–643.

Hazar, H.: Effects of biodiesel on a low heat loss diesel engine. Renew. *Energy* 34 (2009), pp. 1533–1537.

Kalam, M.A. & Masjuki, H.H.: Testing palm biodiesel and NPAA additives to control NO_x and CO while improving efficiency in diesel engines. *Biomass Bioenergy* 32 (2008), pp. 1116–1122.

Kaplan, C., Arslan, R. & Sürmen, A.: Performance characteristics of sunflower methyl esters as biodiesel. *Energy Source* A 28 (2006), pp. 751–755.

Karabektas, M.: The effects of turbocharger on the performance and exhaust emissions of a diesel engine fuelled with biodiesel. *Renew. Energy* 34 (2009), pp. 989–993.

Keskin, A., Gürü, M. & Altıparmak, D.: Influence of tall oil biodiesel with Mg and Mo based fuel additives on diesel engine performance and emission. *Bioresource Technol.* 99 (2008), pp. 6434–6438.

Kim, H. & Choi, B.: The effect of biodiesel and bioethanol blended diesel fuel on nanoparticles and exhaust emissions from CRDI diesel engine. *Renew. Energy* 35 (2010), pp. 157–163.

Kurki, Al, Hill, A., Morris, M. & Lowe, A.: Biodiesel: The sustainabilty dimensions. ATTRA Publication, National Centre for Appropriate Technology, Butte, MT, 2010, pp. 4–5, http://www.attra.org/attra-pub/biodiesel_sustainable.html, (accessed August 2013).

Kwanchareon, P., Luengnaruemitchai, A. & Jai-In, S.: Solubility of a diesel–biodiesel–ethanol blend, its fuel properties, and its emission characteristics from diesel engine. *Fuel* 86 (2007), pp. 1053–1061.

Lapuerta, M., Herreros, J.M., Lyons, L.L., García-Contreras, R. & Brice, Y.: Effect of the alcohol type used in the production of waste cooking oil biodiesel on diesel performance and emissions. *Fuel* 87 (2008), pp. 3161–3169.

Lin, B.-F., Huang, J.-H. & Huang, D.-Y.: Experimental study of the effects of vegetable oil methyl ester on DI diesel engine performance characteristics and pollutant emissions. *Fuel* 88 (2009), pp. 1779–1785.

Lin, C.-Y. & Lin, H.-A.: Diesel engine performance and emission characteristics of biodiesel produced by the peroxidation process. *Fuel* 85 (2006), pp. 298–305.

Lin, C.-Y. & Lin, H.-A.: Engine performance and emission characteristics of a three phase emulsion of biodiesel produced by peroxidation. *Fuel Process. Technol.* 88 (2007), pp. 35–41.

Luján, J.M., Bermúdez, V., Tormos, B. & Pla, B.: Comparative analysis of a DI diesel engine fuelled with biodiesel blends during the European MVEG-A cycle: Performance and emissions (II). *Biomass Bioenergy* 33 (2009), pp. 948–956.

Monyem, A., Van Gerpen, J.H. & Canakci, M.: The effect of timing and oxidation on emissions from biodiesel-fueled engines. *Trans. ASAE* 44 (2001), pp. 35–42.

Nabi, M.N., Najmul Hoque, S.M. & Akhter, M.S.: Karanja (Pongamia Pinnata) biodiesel production in Bangladesh, characterization of karanja biodiesel and its effect on diesel emissions. *Fuel Process. Technol.* 90 (2009), pp. 1080–1086.

Ozsezen, A.N., Canakci, M., Turkcan, A. & Sayin, C.: Performance and combustion characteristics of a DI diesel engine fueled with waste palm oil and canola oil methyl esters. *Fuel* 88 (2009), pp. 629–636.

Qi, D.H., Geng, L.M., Chen, H., Bian, Y.Z.H., Liu, J. & Ren, X.C.H.: Combustion and performance evaluation of a diesel engine fueled with biodiesel produced from soybean crude oil. *Renew. Energy* 34 (2009), pp. 2706–2713.

Raheman, H. & Phadatare, A.G.: Diesel engine emissions and performance from blends of karanja methyl ester and diesel. *Biomass Bioenergy* 27 (2004), pp. 393–397.

Reyes, J.F. & Sepúlveda, M.A.: PM-10 emissions and power of a diesel engine fueled with crude and refined biodiesel from salmon oil. *Fuel* 85 (2006), pp. 1714–1719.

Ryu, K.: The characteristics of performance and exhaust emissions of a diesel engine using a biodiesel with antioxidants. *Bioresource Technol.* 101, pp. S78–82.

Sahoo, P.K., Das, L.M., Babu, M.K.G., Arora. P., Singh, V.P., Kumar, N.R. & Varyani, T.S.: Comparative evaluation of performance and emission characteristics of jatropha, karanja and polanga based biodiesel as fuel in a tractor engine. *Fuel* 88 (2009), pp. 1698–1707.

Sharma, D., Soni, S.L. & Mathur, J.: Emission reduction in a direct injection diesel engine fueled by neem-diesel blend. *Energy Sources* A 31 (2009), pp. 500–508.

Song, J.-T. & Zhang, C.-H.: An experimental study on the performance and exhaust emissions of a diesel engine fuelled with soybean oil methyl ester. Proceedings of the Institution of Mechanical Engineers, *J. Automob. Eng.* 222 (2008), pp. 2487–2496.

Tsolakisa, A., Megaritis, A., Wyszynski, M.L. & Theinnoi, K.: Engine performance and emissions of a diesel engine operating on diesel-RME (rapeseed methyl ester) blends with EGR (exhaust gas recirculation). *Energy* 32 (2007), pp. 2072–2080.

Ulusoy, Y., Arslan, R. & Kaplan, C.: Emission characteristics of sunflower oil methyl ester. *Energy Sources* A 31 (2009), pp. 906–910.

US Department of Energy: Biofuels – Biomass energy data book, http://www.cta.ornl.gov/bedb/biofuels.shtml (accessed August 2013).

US EPA: A comprehensive analysis of biodiesel impacts on exhaust emissions. Draft technical report, United States Environmental Protection Agency, EPA 420-P-02-001, October 2002.

Usta, N.: An experimental study on performance and exhaust emissions of a diesel engine fuelled with tobacco seed oil methyl ester. *Energy Convers. Manage.* 46 (2005a), pp. 2373–2386.

Usta, N.: Use of tobacco seed oil methyl ester in a turbocharged indirect injection diesel engine. *Biomass Bioenergy* 28 (2005b), pp. 77–86.

Usta, N., Öztürk, E., Can, Ö., Conkur, E.S., Nas, S., Çon, A.H., Can, A.Ç. & Topcu, M.: Combustion of biodiesel fuel produced from hazelnut soapstock/waste sunflower oil mixture in a diesel engine. *Energy Convers. Manage.* 46 (2005), pp. 741–755.

Wu, F., Wang, J., Chen, W. & Shuai, S.: A study on emission performance of a diesel engine fueled with five typical methyl ester biodiesels. *Atmos. Environ.* 43 (2009), pp. 1481–1485.

Yoshiyuki, K.: Effects of fuel cetane number and aromatics on combustion process and emissions of a direct injection diesel engine. *JSAE Rev.* 21 (2000), pp. 469–475.

CHAPTER 12

Biogas

Paul Harris & Hans Oechsner

12.1 INTRODUCTION

Anaerobic digestion (AD) is a natural, biological waste conversion process that humans have harnessed, initially to treat human and animal wastes and more recently to generate renewable energy from residues and fuel crops. AD microbes use the readily degradable part of the waste to provide their nutrients and energy but less degradable parts of the waste stream will be largely unaffected – AD is not a waste "disposal" process. In fact there will be little reduction in the volume of material but the treated output will be less smelly, contain fewer pathogens and be better as a plant fertilizer than the input. There is also have the benefit of renewable energy, in the form of methane, released during the digestion process. Because of the development from a simple, natural system (that actually also occurs in stomachs) to a modern industrialized process AD can be implemented in a range of configurations, from the quite simple systems used at household level in developing countries to large, complex industrial systems favored in the western world.

As a result anaerobic digestion may be regarded as part of a waste treatment system or an energy production process, assisting in the "reduce, reuse, recycle" approach to lessening the pollution load currently placed on our planet. An anaerobic digester is really a "primary" treatment, taking material and making it easier to handle in later stages of secondary and tertiary treatment (which may both yield other products) prior to release back into the biosphere.

12.2 WHAT IS BIOGAS?

Biogas, also known as marsh gas or natural gas, is a mixture of methane (CH_4) and carbon dioxide (CO_2) formed biologically from biological sources. The formation of biogas occurs when microbes degrade easily rotted, moist biological material in the absence of oxygen, a process known as anaerobic digestion, so it is a renewable fuel produced from either waste or other organic material by a natural process. El Bassam (2010) and Scheckel (2013) provide information on possible biogas yields for digesting vegetable matter.

"Producer gas" or "syngas", which is produced from organic material by a controlled combustion (thermal/chemical) process with restricted oxygen resulting in the formation of carbon monoxide (CO – quite poisonous), hydrogen (H_2) and possibly some methane (CH_4) is quite different to biogas. Dry materials that are not so readily degradable, like wood, are better suited to thermal processes while moist materials that do degrade quickly (almost anything other than wood) provide better biogas substrates.

Anaerobic digestion is basically a simple process carried out in a number of steps by many different bacteria that can use almost any readily degradable organic material as a substrate – it occurs in digestive systems, marshes, oceans, rubbish dumps, septic tanks and the Arctic tundra. Humans tend to make the anaerobic digestion process as complicated as possible by trying to improve on nature in complex machines, but a simple approach is still possible.

As methane is very hard to compress biogas is best used as a stationary fuel (used in ovens, furnaces, boilers or stationary engines for pumps, mills and generators), rather than a mobile/transport fuel. Methane will not liquefy at ambient temperature, so has to be stored at high pressure, like oxygen and nitrogen (or refrigerated to get liquefied natural gas, LNG) to get the high energy density needed for transport applications. It takes a lot of energy to compress the gas (this energy is usually just wasted), plus there is the hazard of high pressure. Variable volume storage (flexible bag or floating drum are the two main variants) is much easier and cheaper to arrange than high-pressure cylinders, regulators and compressors, but does take up more room.

12.3   BRIEF HISTORY

To give an accurate history of biogas development is difficult, as there are conflicting dates in some of the following, some of which are gathered from e-mail correspondence.

Marco Polo mentions the use of covered sewage tanks. It probably goes back 2000–3000 years ago in ancient Chinese literature (Kurt Roos, *person. commun.*, 2013).

Evidence indicates that biogas was used for heating bath water in Assyria during the 10th century BC and in Persia during the 16th century AD (Sabaonnadiere, 2009) while Klass (1998) reports natural gas wells in Asia as early as 615 AD and the Chinese using bamboo pipes to transport natural gas for lighting in 900 AD.

In 1630 Van Helmont observed flammable gases emitted from decaying vegetable matter, the direct correlation between matter decayed and gas produced was determined by Volta in 1776, the chemistry of methane was first studied in 1786 by Berthollet and methane was observed in the gas from decomposing cattle manure by Davey in 1808 (Klass, 1998; Sabaonnadiere, 2009; Tietjen, 1975).

A digester was built in the 1840s in the city of Otago, New Zealand. This may have been for rendering or "digesting" whale blubber rather than anaerobic digestion (Christian Couturier, *person. commun.*, 2013).

The "first" digestion plant was built at a leper colony in Bombay, India in 1859 (Sabaonnadiere, 2009). According to Klass (1998) this plant was powering a gas engine by 1897.

According to Abbasi *et al.* (2012) a Frenchman, Mouras, applied anaerobic digestion for the first time to treat wastewater in his invention of a crude version of a septic tank in 1881, that he called an "automatic scavanger" (*sic*).

AD reached England in 1896 when biogas was recovered from a "carefully designed" sewage treatment facility and used to fuel street lamps in Exeter (Klass, 1998; Sabaonnadiere, 2009).

By 1925 biogas was being supplied to municipal users by pipeline in Essen (Klass, 1998) and also put into cylinders for vehicle fuel in German cities by the 1930s (Sabaonnadiere, 2009). The development of microbiology as a science led to research by Buswell and others in the 1930s to identify anaerobic bacteria and the conditions that promote methane production (Philip D. Lusk, *person. commun.*, 2013).

Interest in biogas/anaerobic digestion continued in the following years, with people like Harold Bates (an English inventor remembered for his "chicken-powered car") reported in the Press and Fry (1974) and Spargo (1979) publishing books for the general public audience, prompted at least in part by the energy crises of the 1970s. Academic work was also continuing with D.T. Hill, Y.R. Chen and many others publishing papers in a range of journals including waste treatment, environmental and agricultural engineering fields. A number of books were also published over the years, for example Stafford, Wheatly and Hughes (1980), Meynell (1982) and Hobson and Wheatley (1993). More recently books by House (2010) and Scheckel (2013) have become available, aimed at the general public rather than technical audiences.

It was not until the 1960s that high rate digesters were being developed. McCarty and co-workers at Stanford University developed the anaerobic filter and then the upflow anaerobic sludge blanket (UASB) was developed at the Agricultural University in Wageningen in 1970 (Fang, 2010). A number of other innovations such as packed bed and suspended growth digesters have occurred more recently.

Abbasi *et al.* (2012) and Fulford (1988) go on to look at development of biogas in different developing countries and in the West.

Modern Chinese biogas began in the 1930s through a private company set up by Lo Guorui but progress was slow until Chairman Mao's Great Leap Forward in the 1950s. After a slowing of interest new designs appeared in Sichuan Province in the 1970s but the Cultural Revolution slowed adoption until later in the 1970s and by 1979 designs had been standardized (Fulford, 1988). The "Chinese Biogas Manual" is now available online at http://www.fastonline.org/CD3WD_40/JF/432/24-572.pdf. More recently plastics have been used to replace the masonry or concrete dome and factory produced plastic digesters have been developed (Abbasi *et al.*, 2012).

In India, the other "traditional" home of biogas, interest in applications outside the first leper colony came to fruition in 1937 as a result of work by S.V. Desai, a microbiologist (Abbasi *et al.*, 2012). By 1951 the work had taken up by J.J. Patel at the Khadi and Village Industries Commission (KVIC), developing the steel floating drum digester. In the 1960s the government of Uttah Pradesh established a "Gobar Gas" program, led by Ram Bux Singh. A ferro-cement gas holder has been developed by the Structural Engineering Research Centre, Roorkee, to replace the steel gas drum, reducing initial cost and extending drum life (Abbasi *et al.*, 2012; Fulford, 1988).

12.4   ANAEROBIC DIGESTION

Anaerobic digestion breaks down readily degradable organic matter in a series of steps facilitated by a consortium of microbes, where a product of one step becomes the substrate for the next step. The initial step is usually considered to be "hydrolysis" – where extra cellular enzymes break complex organic molecules like fats and starches into simpler molecules like fatty acids and glucose. These simpler molecules are then utilized by a group of "acetogenic" bacteria to produce acetic acid, with carbon dioxide as another product of the breakdown. "Methanogens", which are archaea, are then able to use the acetic acid and produce methane. There are also other groups of "methanogens" that convert carbon dioxide and methanol to methane, but the acetate path is the major route (Abbasi *et al.*, 2012; Sabaonnadiere, 2009):

- acetotrophic pathway ($CH_3COOH \rightarrow CH_4 + CO_2$)
- hydrogenotrophic pathway ($CO_2 + 4H_2 \rightarrow CH_4 + 2H_2O$)
- methylotrophic pathway ($CH_3OH + H_2 \rightarrow CH_4 + H_2O$).

As a result of these steps "biogas" is mainly methane (typically 60%, but less if the digester is not operating properly and sometimes up to about 80%) and carbon dioxide with traces of hydrogen sulfide, ammonia, water vapor, other organic volatiles and possibly some nitrogen gas. In single stage processes, gas quality is mainly dependent on input composition for the digester. Carbohydrates are degraded to a gas with around 50% methane-content, protein around 60% and fat up to 70% (Czepuck *et al.*, 2006).

There are three temperature ranges where anaerobic digestion appears optimized, although some digestion may occur over the entire temperature range from arctic regions to inside volcanoes. The common temperature ranges are approximately (Sabaonnadiere, 2009):

- psychrophyilic – ambient temperatures up to 20°C
- mesophylic – body temperatures of 25–35°C
- thermophylic – temperatures above 45°C (upper limit about 60°C)

For mesophylic operation the commonly accepted optimum is 35°C and the optimum is 55°C in the thermophylic range (Abbasi *et al.*, 2012). It is possible to run a digester successfully at all temperatures between 35 and 55°C, provided a constant temperature is maintained over a long period (and any change is made slowly), so that the microbiological population has a chance to adapt to the operating temperature. At temperatures lower than 35°C, the rate of the digestion process is not very good and it needs a very long retention time to get process stability and an acceptable conversion.

Some digesters are two stage, where the acidogenesis and CO_2 production occur in the first stage and methanogenesis occurs in the second tank to give higher methane percentages. This means that each stage can be operated with conditions to maximize the performance of the microbes involved without having to compromise so the other stage can also exist. The disadvantages are that the capital cost of two tanks is higher than one larger tank and the advantage of symbiosis is lost, so better control is needed. Dividing the process into two stages is tested in laboratory scale but is not state of the art for practice (Oechsner and Lemmer, 2009; Zielonka *et al.*, 2010).

Many text books will emphasize close control of carbon:nitrogen (C:N) ratio, pH, temperature, alkalinity and possibly other parameters to gain optimum performance, but microbes are tough and will still operate with less than optimum conditions. In defining "optimum" the effort spent maintaining optimum conditions must be balanced by a corresponding increase in gas production or effluent quality that more than offsets the effort. This applies to any aspect of anaerobic digestion – whatever component is added must significantly increase gas production (or effluent quality) and/or reduce operational/maintenance problems and costs to justify its inclusion and capital cost.

Being a microbiological process the presence of any toxic or inhibitory substances will reduce, or possibly end, biogas production. Antibiotics are a potential problem, as are some cleaning chemicals and heavy metals (Abbasi *et al.*, 2012). Keep in mind that ammonia and Volatile Fatty Acids (VFAs) that are produced in the anaerobic digestion process are also inhibitory at higher than normal levels, but microbes can acclimate to substances like ammonia. High concentrations of VFAs are mostly caused by overloading of the digester or by lack of trace elements (Lemmer, *et al.*, 2010, pp. 45–77).

One of the advantages of AD is that the process destroys a large proportion of pathogens that may be present in the waste (Klass, 1998), so reduces the risk of disease transmission, and weed seeds are also left unviable (Hoferer, 2006; Schrade *et al.*, 2003). On the other hand the anaerobic breakdown of organic matter is a much slower progress than aerobic digestion (composting) but aerobic processes often require additional energy and produce a larger volume of sludge.

12.5   USES OF BIOGAS

Small scale biogas is best used directly for cooking/heating, light or even absorption refrigeration, rather than the complication and energy waste of using it as a transport fuel or trying to make electricity from biogas. Pumps and other equipment can be run off a gas powered engine rather than using electricity. The biogas equivalents of various common fuels are given in Table 12.1. For larger plants an important consideration is the utilization of biogas produced. As methane is 23 times worse than carbon dioxide as a greenhouse gas any gas not used must be flared off and leaks should be minimized. In Germany are more than 7500 big biogas plants in work, where combined heat and power units (CHP) are used to produce electricity and heat. Their permanent running electric average power is about $450\,kW_{el}$. About 4% of the total electricity consumption in Germany is produced in these plants (Oechsner *et al.*, 2012).

Both spark ignition and diesel engines can use biogas. A spark ignition engine only needs a suitable carburettor to handle gas rather than liquid, but keep in mind that biogas will probably be supplied at a lower pressure than liquified petroleum gas (LPG) and definitely compressed natural gas (CNG). It is important that the gas flow will be stopped if the engine stalls, to avoid a gas build up and possible explosion on restarting. As well as needing a gas carburettor a diesel engine requires at least 10% diesel to initiate combustion and to cool the injectors (so they will still work on liquid fuel if necessary), but manufacturers often specify higher diesel levels to maintain warranty.

Table 12.1.   Biogas equivalents (Palmer, 1988).

Imperial (US)		biogas (65% CH₄) is equivalent to	Metric	
1000	cubic feet		28.3	cubic meters
600	ft^3	natural gas	17	m^3
6.6	US gal	propane	25	liters
5.9	US gal	butane	22	liters
4.7	US gal	gasoline	18	liters
4.3	US gal	#2 fuel	16	liters
44	lb	bituminous coal	20	kg
100	lb	medium dry wood	45	kg

12.6   USES FOR LIQUID/SLUDGE

As anaerobic digestion is not a method of "waste disposal" but a "primary" treatment method the amount of material out of the digester is almost equal to the amount in, but the properties do change. In fact only carbon, oxygen and hydrogen leave via the gas stream in any quantity and each cubic meter of gas removes less than two kilograms of mass (from 20 kg of pig manure or 40 kg of influent, for example), so all the other nutrients remain in the liquid/sludge.

Soluble elements like potassium and nitrogen are found mainly in the liquid phase, along with fine suspended particles, but larger particles and less soluble elements such as phosphorus remain largely in the sludge. The anaerobic treatment process tends to break down a lot of the finer material, so the treated solids are relatively free draining. It also makes a lot of the elements more plant available and less likely to cause damage when applied to vegetation.

This means that the output of a digester is a good source of nutrients for plants and there are papers showing better performance for AD fertilized plants compared to synthetic fertilizers or raw manure (Müller and Müller, 2012). Digester output can also be used in aquatic systems as well as land based systems and is safer for those involved in field work.

12.7   MODELING DIGESTER PERFORMANCE

A number of anaerobic digestion models have been proposed over the years (Lyberatos and Skiadas, 1999). Some are "black box" empirical models, which are usually simple to use, while others try to model the many processes involved so require a lot of calibration data and are more suited to research applications. One of the simpler models is a "steady state" model proposed by Chen (1983) which looks at how much methane would be produced from a given waste under a given set of Volatile Solids, Temperature and Retention Time conditions. Batstone *et al.* (2002), as part of the IWA Anaerobic Digestion Modelling Task Group, have developed ADM1, which models a number of processes within a digester and has been modified for different situations by other people.

12.8   DIGESTER PERFORMANCE

When looking at an anaerobic digester as a biogas production unit the performance depends on the type of waste used (as shown in Fig. 12.1), the amount of volatile solids (VS) put in per day, the operating temperature (T) and the retention time (RT). Since VS reduction is directly related to both biogas production and chemical/biological oxygen demand (COD/BOD) reduction (the pollution load) the same variables apply to the digester as a pollution treatment unit anyway.

A number of publications recommend 35°C as the optimum temperature and most designs aim for maximum gas per unit volume of digester at a retention time of about 6 days, when the "easy" material has been digested (high rate designs to minimize digester size/cost), but digesters operate quite well at lower temperatures and longer retention times, which also give better gas production per volume of influent as less digestible material is degraded. The two diagrams of Figure 12.2 and Figure 12.3, based on a model proposed by Chen (1983) illustrate this.

	oDM-Content %	Methane Yield		Methane yield [l/kg oDM]
		l/kg oDM	l/kg fresh	
Basic substrate: farm manure				
Cattle	7.7	130 - 300	10 - 23	
Pig	5.6	210 - 320	12 - 18	
Poultry	11.5	250 - 400	29 - 46	
Agricultural products				
Straw	75 - 80	71 - 240	55 - 185	
Greencut forage	38	230 - 410	87 - 156	
Maize silage	32	320 - 400	102 - 128	
Agroindustrial wastes				
Potato pulp	5 - 15	250 - 400	25 - 40	
Vegetable waste	4 - 18	400	40 - 50	
Beer draff (brewer grains)	20	370 - 390	74 - 78	
Municipal wastes				
Biowastes	30 - 70	200 - 600	100 - 300	
Rumen content (slaughterhouse)	9 - 17	160 - 400	21 - 52	
flotate fat	4 - 23	600 - 800	78 - 104	
kitchen waste	9 - 22	350 - 600	60 - 100	

0   200   400   600   800

Figure 12.1.   Methane yields of different substrates for the biogas process (Oechsner, 2013).

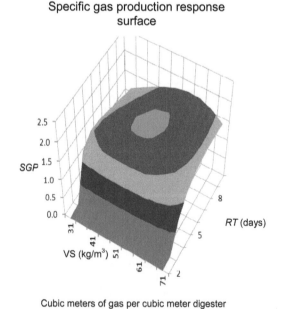

Specific gas production response surface

Cubic meters of gas per cubic meter digester

■ 0.0-0.5   ■ 0.5-1.0   ▨ 1.0-1.5   ■ 1.5-2.0   ▨ 2.0-2.5

Figure 12.2.   Gas production per unit digester volume at 35°C (piggery waste).

12.9 TYPES OF DIGESTERS

Digesters can come in all types, shapes and sizes. The common versions for small-scale rural use are the Indian "floating drum" type (Fig. 12.4) and the Chinese "fixed dome" type (Fig. 12.5), both of which have many thousands of installations. The "poly plug flow" design (Fig. 12.6) is a relative newcomer but is widely used in several countries, due largely to the work of Dr Reg Preston of the University of Tropical Agriculture, Columbia.

The next level up is a "continuous flow stirred tank" (CFST) digester (Fig. 12.7) and then there are anaerobic filters and packed bed digesters as well as "upflow anaerobic sludge blanket" (UASB) digesters, (Fig. 12.8) but these are more suitable for high rate industrial applications and usually have heating and possibly stirring devices. Another development is the "suspended bed digester" (Fig. 12.9) where solid substates of various forms are placed in the digester so biofilms can develop, holding the microbes in the tank at higher waste throughput rates (similar to a packed bed digester).

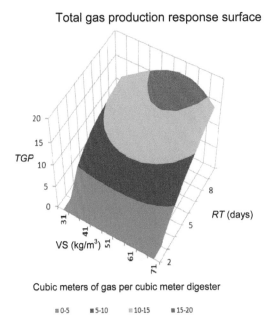

Figure 12.3.   Gas production per unit influent added at 35°C (piggery waste).

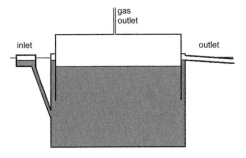

Figure 12.4.   Indian type digester.

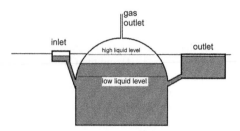

Figure 12.5.   Chinese type digester.

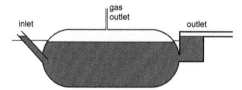

Figure 12.6.   Plug flow digester.

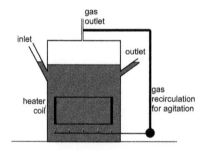

Figure 12.7.   CFST type digester.

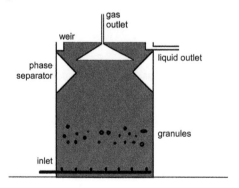

Figure 12.8.   UASB digester.

12.10   GAS STORAGE

The digester itself provides the simplest gas storage. In the Chinese fixed dome digester (Fig. 12.10) the supply pressure will fall as gas is used and liquid from the outlet flows back to the digester space. The gas storage equals the liquid displaced.

The floating drum storage (Fig. 12.11), whether as a digester or separate storage, maintains a steady pressure during use as the drum falls to displace gas. The drum rises to provide storage as

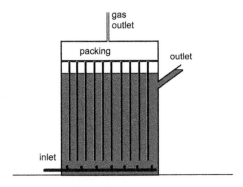

Figure 12.9.   Suspended bed digester.

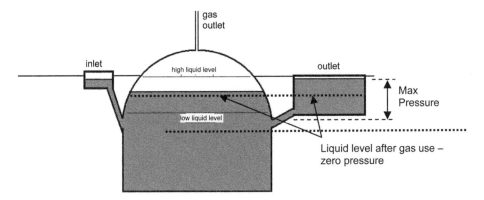

Figure 12.10.   Gas storage in a Chinese fixed dome digester.

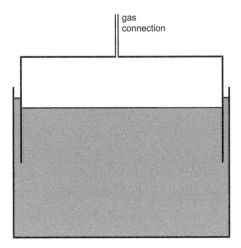

Figure 12.11.   Floating drum gas storage.

biogas is generated. Floating drum storage may need extra weight to develop enough pressure for the burner.

Flexible bag storage (Fig. 12.12) may need some weights on the bag to develop more pressure for the burner, but be careful of puncturing the storage.

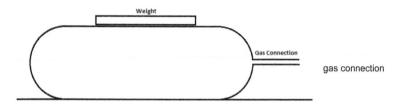

Figure 12.12.   Flexible bag storage.

If the use point is a fair way from the digester better burner performance may be obtained by having a separate Gas storage near the use point. This can be either a flexible, gas tight bag or floating drum type storage. With the second storage it is a good idea to allow low pressure collection of the gas (to minimize the possibility of gas leaks) and to use weight to increase pressure during burner operation, but a "back flow preventer" is required to make best use of this.

The storage should be sized to hold one day's use of gas if possible, which should be as closely matched as possible to digester gas production. If production does exceed use then any excess gas needs to be flared, as the methane in biogas is about 23 times worse as a greenhouse gas than the carbon dioxide produced during combustion.

12.11 SAFETY

There are a number of safety issues to be considered when working with a biogas system.

12.11.1 *Fire/explosion*

Methane (CH_4), which is makes up from 0 to 80% of biogas, forms explosive mixtures in air, the lower explosive limit being 5% methane and the upper limit 15% methane in air. Biogas mixtures containing more than 50% methane are combustible, while lower percentages may support, or fuel, combustion as long as there is an ignition source present. With this in mind no naked flames should be used in the vicinity of a digester and electrical equipment must be of suitable quality, normally "explosion proof". Other sources of sparks are any iron or steel tools or other items, power tools (particularly those with commutators and brushes), normal electrical switches, mobile phones and static electricity.

If conducting a flammability test take a small sample well away from the main digester, or incorporate a flame trap (see Figs. 12.13 and 12.14) in the supply line, which must be of suitable length (minimum 20 m).

Burn back can also be prevented by having a section of pipe filled with steel wool, which will also absorb hydrogen sulfide (H_2S) so needs monitoring but will not prevent reverse gas flow. Commercial units of sintered material are also available and must be used on larger units.

12.11.2 *Disease*

As Anaerobic Digestion relies on a mixed population of bacteria of largely unknown origin, but often originating from animal/human wastes, to carry out the waste treatment process care should be taken to avoid contact with the digester contents and to wash thoroughly after working around the digester (and particularly before eating or drinking). This also helps to minimize the spread of odors that may accompany the digestion process. The digestion process does reduce the number of pathogenic (disease causing) bacteria, particularly at higher operating temperatures, but the biological nature of the process needs to be kept in mind.

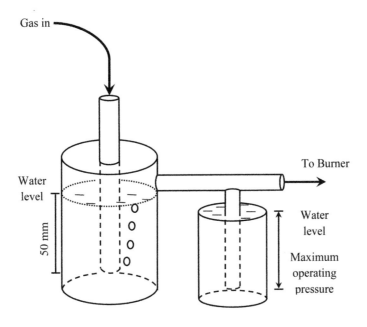

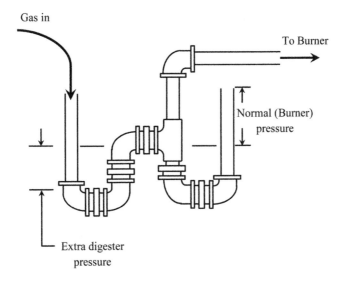

Figure 12.13.   Two designs for a flame trap/backflow preventer.

12.11.3 *Asphyxiation*

As biogas displaces air it reduces the oxygen level, restricting respiration, so any digester area needs to be well ventilated to minimize the risks of fire/explosion and asphyxiation.

Biogas consists mainly of CH_4 and CO_2, with low levels of H_2S, ammonia (NH_3) and other gases. Each of these components has its own problems, as well as displacing oxygen:

- CH_4 – lighter than air (will collect in roof spaces, etc.), explosive (see above).
- CO_2 – heavier than air (will collect in sumps, etc.), slightly elevated levels affect respiration rate, higher levels displace oxygen as well.

Figure 12.14.   Commercial metallic flame trap (Flammer GmbH, http://www.flammergmbh.de/index.htm).

- NH_3 – caustic and hazardous, affects eyes, heart, stomach, airways and skin.
- H_2S – (rotten egg gas) destroys olfactory (smelling) tissues and lungs, becomes odorless as the level increases to the dangerous range and fatalities have occurred. The odor threshold is a few parts per billion in air and it smells of bad eggs. However, at higher concentrations 1–3000 mg/L H_2S numbs the olfactory nerves and effectively at these concentrations is odorless. If the H_2S is above 3 mg/L it may cause health problems. At 10 mg/L the law limits the exposure to 8 hours per day – and this level of exposure will cause considerable discomfort. At 15 mg/L the law limits exposure to 15 minutes per day. At a few hundred mg/l – H_2S in air unconciousness may occur in less than an hour At a few thousands of mg/l – H_2S in air contact with the gas may cause death as the nerves around the heart and lungs are paralyzed. H_2S when it enters the human body may split up to give a free HS^- ion. The HS^- ion has the same destructive reaction in the body as CN (cyanide). Just think of H_2S as a gas which may have the same consequences as cyanide to get the general picture. H_2S detectors are available on most sewage works sites so borrow one (Les. Gornall, Director, Practically Green Environmental Services, *pers. commun.*, 2013).

12.11.4 *Summary*

Adequate ventilation, suitable precautions and adequate protective equipment will minimize the dangers associated with biogas, making it a good servant rather than a bad master.

Like water, electricity, automobiles and most of life biogas is not completely safe, but by being aware of the dangers involved a safe and happy digestion experience is possible. http://www.adelaide.edu.au/biogas/safety/

12.12 ADVANCED DIGESTION

As interest in renewable energy and waste treatment/utilization has increased, over the last 45 years in particular, a number of advances have been made in anaerobic digestion technology. High rate digesters operating in either the mesophylic or thermophylic range have become more common, two stage digestion has been utilized in some instances and the realization that microbes attach to solids has led to the development of upflow anaerobic sludge blanket (UASB) and packed bed digesters. More recently digestion using salt water has been investigated and companies are looking at developing packaged systems for smaller applications.

12.12.1   *High rate digesters*

High rate CFST digesters are usually heated to around 37°C (body temperature, because a lot if anaerobic microbes live in the digestive tracts of mammals) and stirred to give good contact between microbes and their substrate. There is some debate as to whether the agitation should be continuous or intermittent. There may also be equipment to control pH to about 7.5 (optimum for microbes) (Sabaonnadiere, 2009) and high rate digesters may operate at Retention Times as low as one to six days.

Heating may be carried out by different approaches, but it is important that the heating surface is kept below 60°C to avoid caking of the heating surface:

- *Internal heating coil* – a heat exchanger is placed inside the digester and hot water is circulated to provide heating. This provides reasonably uniform heating but if a problem develops the digester may have to be emptied.
- *External heating jacket* – the heat exchanger is placed around the wall digester, so it can be accessed from outside the digester. Heat transfer and surface area may be adversely affected by this arrangement and convection/mixing is needed to carry heat through the digester.
- *External heat exchanger* – digestate is pumped from the digester through a heat exchanger and returned to the digester. A reasonably large heat exchanger will be needed to provide adequate heating and minimize the likelihood of blocking.

Several methods of agitation are used when necessary, as some digester types are essentially self-agitating:

- *Gas recirculation* – biogas is taken from the gas line by a compressor and released at the bottom of the digester through a distribution system. The rising gas bubbles provide agitation and may help release micro bubbles of biogas. The agitation efficiency may be improved by using "draft tubes" in the digester.
- *Mechanical stirring* – A large propeller or several smaller units are used to push digestate around in the digester. Mechanical mixers are the most effective system for distribution of substrates in the digester. They are usable up to dry matter (DM) content of 10% in the digester. In Europe, most CSTR biogas digesters are using this system (Naegele *et al.*, 2013).
- *Digestate recirculation* – digestate is removed from the digester (possibly to be passed through a heat exchanger, as above) and returned to provide mixing. It is hard to provide proper mixing in larger digesters and "dead space" effectively reduces the hydraulic retention time (HRT).

To make use of the higher growth rates possible at thermophylic temperatures some digesters are operated at up to 60°C, which brings the retention time down to 3.5 days or less. Under these conditions extra energy is required for heating the digester, but this may come from waste heat from generator engines (known as "combined heat and power" or CHP units), and digester operation may be less stable, requiring more supervision of the digester.

12.12.2   *Two stage digesters*

Since different stages of the anaerobic digestion process prefer different conditions and microbes grow at different rates the idea of separating the acidogenesis (acid forming) stage and methanogenesis (methane forming) is sometimes adopted. This approach also has the benefit of reducing the CO_2 content in biogas since CO_2 is generated in the earlier phases of digestion, but as one of the pathways to methane is from CO_2 there may be some loss of methane and the processes need to be carefully managed. A practical disadvantage of two stage digestion is that because of their greater surface area two smaller tanks will cost more than a single larger one and the two tanks will also lose more heat. The two stage systems are more robust than single stage processes, but they are still only relevant for laboratory research and not yet practiced on a commercial scale (Oechsner and Lemmer, 2009).

12.12.3   *Anaerobic filters*

The first advance in digestion technology, in the 1960s, was the development of the anaerobic filter by McCarty and co-workers at Stanford University (Fang, 2010). The anaerobic filter contains a porous media which will support the growth of micro-organisms so they are retained in the digester instead of being washed out with the substrate. The support media can be anything that provides a large surface area for microbes to attach to and that has passages for the waste to flow through. Bio balls, wooden slats, stone, plastic, ceramic, and fired clay have been used as support media. Because of restricted passages through the media only soluble wastes with low particulate loads can be treated.

12.12.4   *Upflow anaerobic sludge blanket (UASB) digesters*

When treating wastes with low suspended solids the concept of having microbes form granules that can be retained in the digester was developed by Lettinga at Wageningen University in the 1970s, following on from the anaerobic filter work (Fang, 2010). The digester tank is provided with baffles near the top liquid level to retain the sludge granules and waste is pumped into the base of the digester through a diffuser at a rate to keep the granules suspended in the tank, rather than having them settle. Sometimes it is difficult to get granules to form, so seed granules may need to be bought in to inoculate the digester initially (Abassi *et al.*, 2012)

12.12.5   *Suspended growth digesters*

Another approach to reducing retention time, allowing a smaller digester to be used to treat a given waste stream but reducing the clogging of anaerobic filters, is to provide a support media that microbes can attach to, forming a biofilm. Because the microbes are held in the digester by the support media pumping rates can be increased and it is possible to get retention times of just one day. Plastic grids, ropes and other materials have all been used as suspension media. One of the problems with suspended growth digesters is that the biofilm can grow so thick that it restricts the flow of waste and one way of dealing with this is to shut off the waste feed so the biofilm starts to eat itself.

12.12.6   *Salt water digesters*

In some areas there may be limited fresh water but plentiful supplies of salt water, so it makes sense to use the fresh water for stock and domestic supplies and use salt water for waste disposal. Recent work has shown that some methanogens and their supporting microbial ecosystem can live happily in water twice as saline as sea water, so anaerobic digestion of wastes in sea water is quite possible. There is a critical disadvantage of using salt water; the digested residue, which normally includes a lot of nutrients, cannot be used as fertilizer and the nutrients will get lost in the system. As with all digester types it is critical to look at the whole system to get an integrated solution rather than creating a series of problems.

12.12.7   *Solid digestion*

Most people think of anaerobic digestion as a liquid process, as discussed above, but many wastes, such as chicken and feedlot manure, municipal solid waste (MSW) and market waste, are collected as moist solids rather than as liquids and slurries. It has been realized for some time that biogas can be "mined" from MSW dumps and more recently a number of facilities for processing solid waste have been developed and are in commercial operation. MSW poses some practical challenges as it usually includes plastics and metals. This can be dealt with by "source separation" (where the householder uses different bins for different wastes), pre-separation (where waste goes through a sorting process prior to digestion) or post digestion separation (if the digestion process is not hindered by the contaminants).

The digesters used may be containers loaded by conveyers or other machinery, then sealed and inoculated with leachate from previous digestions. Another approach is to use "garages" loaded by truck and/or front end loader then sealed and inoculated by leachate. In either case the liquid from the digestion process is collected in drains for use as inoculum, or may be further digested in a conventional liquid digester. Once the solid digestion process is complete the container or garage is emptied ready for the next batch, so this process is often carried out in a number of smaller units that are at different stages of the process. In Germany some garage digesters (around 30) are in practical work with waste and/or energy crops (Kusch and Oechsner, 2008; Kusch *et al.*, 2004; 2005). The DiCom process, developed in Western Australia, uses aerobic, anaerobic and aerobic stages over a 21 day cycle, so three units are utilized (Walker, 2011).

12.13   PACKAGED UNITS

A number of companies are selling or developing small, self-contained units for domestic or small industrial applications. ARTI and GreenBox in India are two examples, there is Puxin Biogas in China and there are BioBowser and Portagester in Australia and England respectively.

12.14   STARTUP

Starting an anaerobic digester is relatively easy if a few points are kept in mind:

- You need to have the appropriate bacteria present. In cattle and pig waste there are usually the right bacteria but poultry waste and vegetable waste may need inoculation, so startup may have to begin with cattle or pig manure and gradually introduce the other waste form.
- Startup will be slow. The actual process is fairly slow and proceeds in a series of steps, just like AD itself. First the oxygen in headspace air has to be used up to get anaerobic conditions, so the first gas collected will be carbon dioxide (which will extinguish a flame). Acetogenesis then generates more CO_2 and acetic acid, so if this stage gets too advanced methanogenesis can be inhibited. Once there is some acetic acid methanogenesis can start, but of course some initial bacteria must be present.
- Any sudden changes can upset the process. If a change is required to the amount of waste added per day, the concentration or the type of waste it is better to make the change gradual so the bacterial population has time to adjust. While the change is underway monitor gas quality (see Monitoring, Section 12.15 below), as a drop in quality (more CO_2 and/or less CH_4) is often the first indication of possible digester failure. If the gas quality does deteriorate it is best to stop feeding and give the digester a chance to recover.

The preferred way to start a digester is to part fill with water (to check operation and make any alterations necessary without having to deal with effluent) and then add half the working volume of dairy or piggery effluent (or active digester sludge if available). Wait until the digester is producing flammable gas, which may take a few weeks if temperatures are cool, before starting feeding at half the design rate. Once operation has settled down increase to full rate, still using the initial waste and wait for operation to settle down again. If changing to a different waste type use a 50/50 mix for at least one retention time before changing over completely, again monitor the digester performance and wait until satisfied that the operation has adjusted before making any further change.

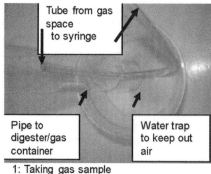

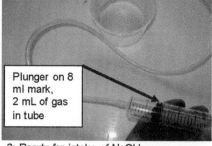

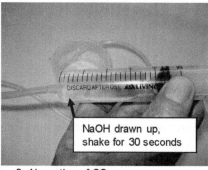

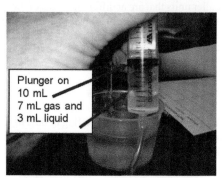

Figure 12.15.   Syringe protocol.

12.15   MONITORING DIGESTER OPERATION

One of the best indicators of proper digester operation is that the gas volume produced per day is consistent with the waste input and that the gas will support combustion (which indicates at least 50% methane).

12.15.1   *Indication of CO_2 percentage*

A syringe body fitted with some flexible tube and some dilute sodium hydroxide solution can be used to estimate carbon dioxide percentage, as NaOH absorbs CO_2 but not methane. The Protocol is available at http://bit.ly/SyringeProtocol and the procedure is shown in Figure 12.15. Draw up 20–30 mL of biogas and put the end of the tube into the NaOH solution, then push out excess gas to get a 10 mL gas sample (you have to allow for the gas in the tube, which may be 4–5 mL). Now draw up approximately 20 mL of solution and keep the end of the tube submerged while shaking the syringe for 30 seconds. Point the syringe downwards and push out excess liquid, so the syringe plunger reaches 10 mL. Now read the volume of liquid, which should be 3–4 mL indicating about 30–40% of gas absorbed so the balance is mainly methane. If there is over 50% methane (a reading of less than 5 mL of liquid) and the flame will still not burn properly there must be nitrogen or some other gas present.

As the digester is a well buffered system a simple pH measurement will really only indicate that the digester is already in trouble, usually well after the gas quality/volume has dropped.

Another very good way to monitor/control process stability is to measure the content of volatile fatty acids (VFA) in the digester. The total acid concentration (HAc) should not increase to values higher than 3000 mg/L. The content of acetic acid should always have the double value of propionic

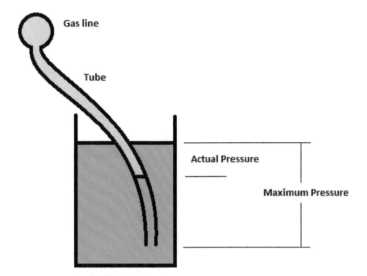

Figure 12.16. Measuring gas pressure.

acid. Increasing acid concentration, especially the increasing of propionic acid, is an alarm that the process is becoming unstable (Stockl *et al.*, 2012).

12.15.2 *Measuring gas pressure*

To measure pressure in a digester simply put a clear plastic tube into a container of water (a glass container is easiest) and connect the tube to the digester gas line (Fig. 12.16). The digester pressure will push down the water surface in the tube and the difference between inside and outside levels is known as "water gauge". If the pressure is too great gas will bubble out the bottom of the tube, which acts as a safety valve to protect the digester from over pressure.

If installed as shown (with the tube from the bottom of the gas line) at the lowest point of the gas line (which must be laid on a steady grade to allow condensation to drain) this arrangement will also act as a condensate trap.

Putting the tube into a container of water is a better pressure relief than the manometer below as overpressure may blow all the water out of the manometer so it cannot reseal and also sediment may block the manometer, restricting or blocking gas flow.

Another method of measuring pressure is make a "manometer" by bending a plastic tube into a U and putting water in – when one end is connected to the gas line and the other is open to atmosphere (a restriction in the tube will help the levels settle quickly) measure the difference in water heights (Fig. 12.17).

12.16 BURNERS

Satisfactory burner operation is obtained when the flame is large enough to heat the item fairly quickly but is stable – in other words the flame "attaches" to the burner.

Flame size is set by the combination of gas jet size and operating pressure, which set the gas flow rate. At low operating pressures a larger jet is required to give reasonable gas flow. For a given jet size, as pressure is increased the gas flow rate increases and so does the flame vigor – until the flame "blows out". Blow out is caused by gas velocity being greater than the flame velocity of about 0.25 m/s. If the operating pressure is causing "blow out" the actual burner opening needs to be bigger for the same jet size, as this will reduce the gas exit velocity so the flame will stay

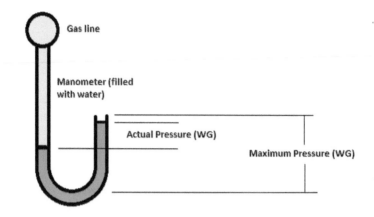

Figure 12.17.   Manometer arrangement.

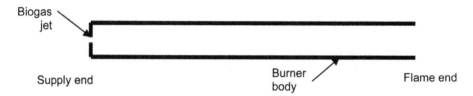

Figure 12.18.   Simple diffusion burner.

attached. Of course using a larger jet will allow more gas flow, but a larger burner opening may be required as well and the mixer length must be enough to allow the velocity to drop off.

A simple "diffusion" type burner (Fig. 12.18) will work satisfactorily provided the gas velocity is not too high. This is achieved by restricting the biogas flow rate through a small opening and letting the gas exit through a larger diameter metal (or ceramic) tube. A 1 mm jet and a 12 mm body work quite well.

Commercial burners are "premixed" type and will work very well with biogas provided the gas jet is sized for the lower supply pressure.

Combustion of biogas is basically the combination of a methane molecule (CH_4 which has one carbon and four hydrogen atoms) with two oxygen molecules (O_2 each with two oxygen atoms) to form a carbon dioxide molecule and two water molecules (ignoring all the trace gases). This can be written as:

$$CH_4 + 2O_2 \rightarrow CO_2 + 2H_2O$$

12.17   FAULT FINDING

Possible faults will be minimized by careful construction initially and regular checking as part of a routine – a digester is a living system that does need some care and attention, just like plants and animals.

As with all fault finding try the easy things first and try to work by elimination:

- If there is a loss of pressure check that there is water in any water traps and that all hoses are connected and outlet taps are off. Then look for damage to the digester and gas holder (if separate). Use soapy water to check for possible gas leaks. If there are no obvious leaks isolate the digester (if there is no backflow preventer) – if the digester inflates over the next few days there is a leak in the gasholder section, if not look at the digester system.

- If no leaks are present think about how much gas has been used and what has been put into the digester – possibly do a check on gas quality.
- Poor burner operation may be caused by low supply pressure or poor gas quality.
- Low pressure may result from insufficient weight on the gas storage or damaged/blocked supply lines, particularly condensed water in low pipe sections. If the supply line has not been installed properly on a grade with a drain at the lowest point water may have collected and be restricting gas flow.

12.18   CONSTRUCTION TIPS

There are some ideas on the Biogas Wiki at http://biogas.wikispaces.com/Online+Course to make a small plug flow digester, where the Excel model is also available. As well as the obvious things to look out for, such as avoiding punctures/leaks and having the inlet higher than the outlet, there are a few other things to watch. One of the most important is to have gas lines laid on a fall with a drain point in the low spot, to allow any condensation to get out of the way of gas flow. When building a plug flow poly digester make sure the digester body at both the inlet and outlet projects far enough below liquid operating level to hold in the gas pressure (but not too much further, so they provide extra over pressure protection) and that the gas outlet will remain above liquid level, even when there is no gas pressure.

The "flametrap" design given in the Safety Section (12.11) also acts as a pressure relief, drain point (more so if installed at a low point) and backflow preventer, which is useful for stopping the burner supply pressure pushing liquid out of the digester while using the biogas, if a separate gas storage with variable weight is used.

12.19   CONCLUSIONS

Biogas is a natural process that produces energy and organic fertilizer from readily degradable, moist organic matter. Both liquid and solid processes are possible and the concept is scalable from household level up to regional level if necessary, but localized systems minimize waste transport and material distribution costs. A number of systems have been developed, some more suited to particular substrate types, so make a proper investigation before starting a project – a small digester is a good start.

REFERENCES

Abbasi T., Tauseef, S.M. & Abbasi, S.A.: Biogas energy. *SpringerBriefs in Environmental Science* Vol. 2, 2012.

Batstone D.J., Keller, J., Angelidaki, I., Kalyuzhnyi, S.V., Pavlostathis, S.G., Rozzi, A., Sanders, W.T.M., Siegrist, H. & Vavili, V.A.: The IWA Anaerobic Digestion Model No 1 (ADM1). *Water Sci. Technol.* 45:10 (2002), pp. 63–73, IWA Publishing.

Chen, Y.R.: Kinetic analysis of anaerobic digestion of pig manure and its design implications. *Agr. Wastes* 8 (1983), pp. 65–81.

Czepuck, K., Oechsner, H., Schumacher, B. & Lemmer, A.: Hohenheim biogas yield test. *Landtechnik* 61:2 (2006), pp. 82–83.

El Bassam, N.: *Handbook of bioenergy crops : a complete reference to species, development and applications.* Earthscan, 2010.

Fang, H.H.P.: *Environmental anaerobic technology: applications and new developments.* Imperial College Press, London, Hackensack, NJ, distributed by World Scientific Pub., 2010.

Fry, L.J.: *Practical building of methane power plants for rural energy independence.* Andover (Staddlestones, Penton Mewsey, Andover, Hants. SP11 0RQ): D.A. Knox, 1974.

Fulford, D.: *Running a biogas programme: a handbook*. Intermediate Technology Publications, London, UK, 1988.

Hobson, P.N. & Wheatley, A.D.: *Anaerobic digestion: modern theory and practice*. Elsevier Applied Science, London; New York, 1993.

Hoferer, M.: H*ygienic studies on the inactivation of selected bacteria and viruses in the mesophilic and thermophilic anaerobic digestion of organic and kitchen waste and other residual and waste materials of animal origin*. [Seuchenhygienische Untersuchungen zur Inaktivierung ausgewählter Bakterien und Viren bei der mesophilen und thermophilen anaeroben Faulung von Bio- und Küchenabfällen sowie anderen Rest- und Abfallstoffen tierischer Herkunft]. PhD Thesis, University of Hohenheim, Shaker-Verlag, Berlin, Germany, 2001.

House, D.: *The complete biogas handbook*. Alternative House Information, 2010.

Klass, D.L.: *Biomass for renewable energy, fuels, and chemicals*. Academic Press, San Diego, 1989.

Kusch, S. & Oechsner, H.: Biogas production out of solid substrates. [Biogas gewinnen aus stapelfähigen Feststoffen]. *Ökologie&Landbau* 132:4 (2004), pp. 23–25.

Kusch, S., Oechsner, H. & Jungbluth, T.: Biogas production in discontinuously operated solid-phase digestion systems. *Proceedings of 7th FAO/SREN-Workshop The Future of Biogas for Sustainable Energy Production in Europe*, 30 Nov.–2 Dec. 2005, Uppsala, Sweden, 2005.

Kusch, S., Oechsner, H., & Jungbluth, T.: Biogas production with horse dung in solid-phase digestion systems. *Bioresour. Technol.* 99 (2008), pp. 1280–1292.

Lyberatos, G. & Skiadas, I.V.: Modelling of anaerobic digestion – a review. *Global Nest: Int. J.* 1:2 (1999), pp. 63–76.

Meynell, P.J.: *Methane: planning a digester*. Prism Press, Dorchester, UK, 1982.

Müller, K. & Müller, T.: Effects of anaerobic digestion on digestate nutrient availability and crop growth: a review. *Eng. Life Sci.* 12:3 (2012), pp. 242–257.

Naegele, H.-J., Lemmer, A. Oechsner, H. & Jungbluth, T.: Electric energy consumption of the full scale research biogas plant "Unterer Lindenhof": results of longterm and full detail measurements. *Energies* 5 (2013), pp. 5198–5214.

Oechsner, H.: Experiences with biogas in Germany. Seminar, CAU, Beijing, 26.04.2013.

Oechsner, H. & Lemmer, A.: What brings the hydrolysis in the biogas fermentation? [Was kann die Hydrolyse bei der Biogasvergärung leisten?] *VDI-Berichte* Nr 2057, VDI-Verlag GmbH, Düsseldorf, Germany, 2009.

Oechsner, H., Jungbluth, T., Kranert, M. & Kusch, S.: Editorial: Progress in biogas – State of the art and future perspectives. *Eng. Life Sci.* 12:3 (2012), pp. 239–240.

Palmer, D.G.: Biogas energy from animal waste. Solar Energy Research Institute, 1981 in Handbook of Biogas Utilization, 1988.

Sabonnadiere, J.-C.: *Renewable energy technologies*. ISTA/Wiley, London, UK, 2009.

Scheckel, P.: *The homeowners energy handbook*. Storey Publishing, North Adams, MA, 2013.

Schrade, S., Oechsner, H., Pekrun, C. & Claupein, W.: Influence of the biogas process on the germinability of seeds. *Landtechnik* 58:2 (2003), pp. 90–91.

Spargo, R.F.: Methane CH_4, the replaceable energy. Australian Methane Gas Research, Tomerong, NSW 1979.

Stafford, D.A., Wheatley, B.I. & Hughes, D.E.: *Anaerobic digestion*. Elsevier Science & Technology.

Stockl, A. & Oechsner, H.: Near-infrared spectroscopic online monitoring of process stability in biogas plants. *Eng. Life Sci.* 12:3 (2012), pp. 295–305.

Tietjen, C.: From biodung to biogas – a historical review of European experience. In: W.J. Jewell (ed): *Energy, agriculture, and waste management, Proceedings of the 1975 Cornell Agricultural Waste Management Conference*. Ann Arbor Science Publishers, Ann Arbor, MI, 1975, pp. 207–260.

Walker, L.R., Cord-Ruwisch, R. & Sciberras, S.: Performance of a commercial-scale DiCom™ demonstration facility treating mixed municipal solid waste. *Proceedings of the International Conference on Solid Waste 2011 – Moving Towards Sustainable Resource Management*, Hong Kong SAR, P.R. China, 2–6 May 2011, pp. 417–421.

Zielonka, S., Lemmer, A., Oechsner, H. & Jungbluth, T.: Energy balance of a two-phase anaerobic digestion process for energy crops. *Eng. Life Sci.* 10:6 (2010), pp. 515–519.

CHAPTER 13

Thermal gasification of waste biomass from agriculture production for energy purposes

Janusz Piechocki, Dariusz Wiśniewski & Andrzej Białowiec

13.1 INTRODUCTION

Gasification of biomass/waste transformation into gaseous fuels to produce biogas is an alternative technology to the anaerobic fermentation process employing methane bacteria. Gasification should be understood as denoting a whole range of thermodynamic processes, heat and mass exchange as well as multi-directional exo- and endothermic chemical reactions, which run at high temperatures and result in solid fuel conversion to the gaseous form. Apart from the material being gasified, a gasification agent (steam, air, oxygen or carbon dioxide) also takes part in the process.

The amount and composition of synthesis gas produced in the biomass gasification process depends mainly on the type of biomass as well as on the gasification agent, temperature, pressure and the method of gasification as it will be discussed in detail in this chapter.

Biomass-to-energy conversion is the fastest growing segment of the energy business in many countries of the European Union. The progressive depletion of fossil fuels and their negative environmental impacts have spurred the search for renewable energy sources and the most effective methods of processing alternative fuels. The prices of electricity, heat and fuels continue to increase globally. Higher prices of fossil energy carriers and policies that call for a systematic increase in the share of renewable energy in total generation increase the demand for renewable energy, even though it is generally perceived to be more expensive than energy from conventional sources.

This chapter overviews the current trends in waste biomass thermal gasification. The main gasification technologies, the associated processes and the use of gaseous fuels as energy sources will be discussed, including the sources, parameters and possible applications of waste biomass, elemental composition and organic content of biomass and their effect on gasification. Biomass contaminants and their impact on biomass processing equipment and emission levels will be analyzed.

The successive chapter focuses on the main thermal processes in biomass upgrading, including drying, torrefaction, pyrolysis and gasification. The main technologies and equipment for thermal gasification of biomass will be presented including new solutions for processing the organic fraction of municipal solid waste through biological gasification.

13.2 BIOMASS WASTE

Biomass is the organic matter in agricultural crops which is produced by photosynthesis. In this process, solar energy is accumulated by plant organisms (phyto-mass), and the energy transferred between organisms in the food chain is accumulated in the zoo-mass. Another type of biomass is microbial biomass produced by, for example, plankton and algae. The main energy parameters of biomass are heating value, moisture content and ash content. Biomass is characterized by low heating value, low carbonization, low ash content per unit volume and high content of

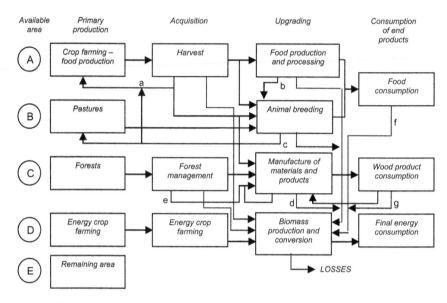

Figure 13.1.   Block diagram of the conversion of primary biomass to end-use products: (a) residues from grain production and other byproducts from the food industry, (b) residues from food production and processing, (c) residues from animal production, (d) residues from the manufacture of wood products, (e) residues from forest management, (f) post-consumption residues, (g) residues from discarded wood products and materials (Domanski *et al.*, 2007).

volatile organic compounds. The quality of biomass-derived energy varies considerably, and it is determined by moisture content, content of volatile substances and ash, heating value, grade and density.

Biomass materials can be divided into the following groups based on their origin:

- Herbaceous and woody plants,
- Waste from plant and food production,
- Waste from the wood processing industry,
- Waste from biofuel and biomaterial production,
- Waste from animal production,
- Waste from the processing of municipal solid waste.

The biomass conversion diagram (Fig. 13.1) proposes an approach for estimating biomass resources from production to end-use. Biomass waste from various technological processes can be converted to energy. The area available for the production of non-energy biomass can be used to estimate the area needed for growing energy crops.

When the above energy sources are taken into account, Poland's potential energy output is estimated at 755 PJ/year (Bilitewski *et al.*, 2006). The main sources of biomass in Poland are:

- agricultural produce: crops, food production, animal breeding,
- agricultural residues: manure, poultry litter, spent substrate from oyster mushroom and button mushroom farms,
- municipal solid waste,
- residues from the wood processing industry,
- residues from forest management,
- energy crop farms: grass, willow, poplar, eucalyptus, straw,
- other biofuels: paper, cardboard, municipal solid waste, refuse-derived fuel, sewage sludge.

Table 13.1. Energy densities of various biofuels (Chmielniak *et al.*, 2008).

Biofuel	Energy density [kWh/m^3]
Wood pellets	2756
Wood chips	785–1094
Bark	727
Sawdust	518
Straw, straw bales	482

13.2.1 *Properties of biomass*

Biomass has low energy density (Table 13.1), which makes energy generation from biomass unprofitable in sites with conventional energy distribution systems due to high transport and processing costs.

The chemical energy of biomass can be converted in various processes to heat and electricity. The choice of the conversion method is determined by biomass parameters which provide the starting point for developing an appropriate processing technique. The list of technological parameters of biomass, usually considered for choosing between methods of biomass conversion into energy, is given in Table 13.2.

Technical and elemental analyses provide the most important information about the energy properties of biomass. A technical analysis supplies data about volatile and flammable components, ballast in the form of mineral substances and moisture, heat of combustion and heating value. Ash content, bulk density and hardness are important considerations during the conversion of biomass to other types of fuel. The organic composition of biomass plays an important role in biochemical and microbiological conversion. Biomass composition is determined by biochemical analysis which evaluates the share of cellulose, hemicellulose, lignin, pectin, starch and polysaccharides.

13.2.2 *Biomass for energy production*

Biomass for energy production may be used in the form of solid fuels (in primary form) such as wood, straw, chips, mechanically processed pellets and briquettes, mechanically and thermally (torrefaction, pyrolysis) processed biocoal. Biogas, syngas produced by pyrolysis and gasification are gaseous fuels from biomass treatment. Biomass can be chemically (e.g. oil, alcohol) and thermally (e.g. pyrolysis oil) processed to produce liquid fuels. The main chemical elements found in biomass are carbon, hydrogen and oxygen (Fig. 13.2).

The main components which determine the energy content of biomass are cellulose and hemicellulose, the basic building blocks of plant tissue. The typical composition of plant biomass is as follows:

- cellulose: 40–50%,
- hemicellulose: 20–30%,
- lignin: 20–25%,
- ash: 1–5%.

The above components differ in their heating value which amounts to 16.2 MJ/kg for hemicellulose, 17.3 MJ/kg for cellulose, 28.8 MJ/kg for lignin and 36.0 MJ/kg for resins. The average heating value of biomass depends on share of mentioned constituents, and differs between specific types of biomass (Table 13.3).

Table 13.2.   Biomass parameters (Domanski *et al.*, 2007).

Parameter	Symbol	Unit	Remarks
Moisture content	W	%	
Dimensions		mm	Sawdust, chips, slices, shavings, logs, stumps, etc.
Shape			Briquettes, pellets, fibers, sticks, etc.
Heating value	W_d	GJ/kg	
Ash content	A		
Ash composition			Oxides: Fe_2O_3, CaO, SiO_2, Al_2O_3, etc.
Bulk density	ρ	kg/m^3	
Real density	ρ	kg/m^3	
Angle of repose		degrees	
Volatile matter	V		
Hardness and cohesiveness of feedstock			Mainly for determining feedstock quality
Dust explosion			Ignition energy of dust released during drying, grinding, etc.
Landfill fermentation			Degradation of energy value and increased risk of spontaneous combustion
Composition:			
Nitrogen	N	%	
Chlorine	Cl	%	
Fluorine	F	%	
Sulfur	S	%	
Carbon	C	%	
Hydrogen	H	%	
Oxygen	O	%	
Major contaminants		mg/kg	Al, Si, K, Na, Ca, Mg, Fe, P, Ti and other
Minor contaminants		mg/kg	As, Ba, Cd, Be and other
Heavy metals		mg/kg	Pb, Ni, Sb, Cr, Cu, Mn, Se, Te, Sn and other
Metal vapor		mg/kg	Cd, Hg (regardless of solid fraction content)
Energy density		kJ/m^3	
Characteristic temperatures	T	°C	Deformation temperature, ash fusion temperature, flow temperature, oxidation and reduction temperature
Potential fly ash utilization			
Fixed carbon	FC	%	
C/N ratio	C/N	–	

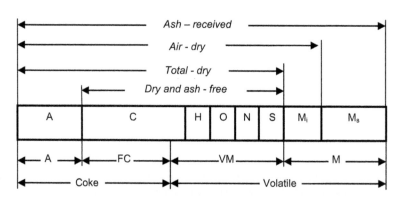

Figure 13.2.   Share of biomass components: A – ash, C – carbon, FC – fixed carbon, H – hydrogen, O – oxygen, N – nitrogen, S – sulfur, M_i – inherent moisture, M_s – surface moisture.

Table 13.3. Heating value of biomass materials (Domanski *et al.*, 2007).

Biomass material	Moisture content (raw material) [%]	Heating value of dry material [MJ/kg]	Heating value of wet material [MJ/kg]	Bulk density [kg/m]
Branch chips	50–60	18.5–20	6–9	250–400
Whole-tree chips	45–55	18.5–20	6–9	250–350
Trunk chips	30–55	18.5–20	6–9	200–350
Stump chips	50–65	18.5–20	6–9	250–350
Chips from timber residues	10–50	18.5–20	6–9	150–300
Sawing residues	45–60	18.5–20	6–10	250–350
Sawdust	45–60	19–19.2	6–10	250–350
Chips from timber cuts	5–15	19–19.2	13–16	80–120
Timber milling dust	5–15	19–19.2	15–17	100–150
Plywood residues	5–15	19–19.2	15–17	200–300
Timber	15–30	18–19	12–15	150–250
Yellow straw (chopped)	15	18.2	14.4	50–80
Gray straw (chopped)	15	18.2	15	50–80

Biomass is a difficult fuel for power generation systems because of its uneven structure and high moisture content. The chemical composition and physical attributes of biomass are highly significant for thermal conversion processes such as pyrolysis, gasification, torrefaction and combustion. In most cases, biomass has to be upgraded before it is converted to energy. Preliminary treatment involves drying, homogenization and processes that increase bulk density. Biomass is rarely suited for direct use, and it requires conditioning. Thermal processing takes place in dryers. Dried material is homogenized mechanically by grinding and milling. Ground and dried biomass is combusted in fluidized bed boilers, pulverized fuel burners, fast pyrolysis systems and fluidized-bed gasifiers. In fixed-bed gasifiers, bulk density of biomass has to be increased through granulation.

The suitability of biomass for energy generation is determined by the following parameters:

- *moisture content:* generally high,
- *content of volatile matter:* generally high, around 70–85% in dry matter,
- *ash content:* generally low, around 25% in sludge and fermented residues,
- *mineral content:* including calcium, sodium, potassium, phosphorus and chlorine,
- *ash fusion temperature which is lower than in coal:* highly significant parameter for gasification processes,
- *formation of tarry substances:* higher-order hydrocarbons, alkali metal vapors, ammonia, sulfur and chlorine compounds during thermal processing,
- *low bulk density,*
- *logistic problems with continuous biomass supplies,*
- *storage and transport problems:* biomass decay and decomposition, loss of energy attributes.

The above parameters are very significant for thermal gasification. Pyrolysis plays a crucial role in thermal processing owing to high concentrations of volatile substances in biomass. The efficiency of gasification is also influenced by lower temperatures of physicochemical reactions in the reactor. Biomass contaminants contribute to corrosion, which is why reactors are made of materials which are highly resistant to corrosion and erosion. Undesirable components such as tarry substances, sulfur components, ammonia and alkaline substances pose a threat for turbine engines, piston engines and boilers, and they have to be removed from generator gas. Ash fusion temperature is lower than in coal, which poses an additional problem during the gasification process. Low ash fusion temperatures lead to the formation of slag which restricts biomass

Table 13.4.   Biomass contamination (Domanski *et al.*, 2007).

Pollutant	Woody biomass	Willow, straw	Sewage sludge, municipal solid waste
Ash [%]	1–5	4–11	10–45
Sulfur [%]	<0.1	0.1–0.3	0.1–1
Nitrogen [%]	0.4–0.7	0.5–3	0.5–6
Chlorine [%]	<0.1	0.1–0.25	0.1–1

Table 13.5.   Contamination of biomass waste from the paper industry (Domanski *et al.*, 2007).

Pollutant	Hydrolyzed lignin from the paper industry	Waste from an agricultural biogas plant
Ash [%]	26.77	26.62
Sulfur [%]	0.37	0.87
Nitrogen [%]	0.47	3.76
Chlorine [%]	0.01	0.43

movement in the bed, blocks the flow of gas and the gasification agent. The above problems are encountered in fixed-bed reactors and boilers.

Ash content varies considerably across different biomass types. The lowest ash content is observed in woody biomass, and the highest – in biomass waste such as fermentation residues, sewage sludge and organic fraction of municipal solid waste containing the highest level of pollutants (Table 13.4).

In selected types of biomass, pollution levels significantly exceed the values given in Table 13.4. Strongly contaminated organic matter is supplied by industrial plants, sewage treatment plants and plants that process municipal solid waste. For example, the level of contamination of biomass waste from an agricultural biogas plant is higher than in case of biomass waste – hydrolyzed lignin from the paper industry (Table 13.5).

Biomass waste from the processing of municipal solid waste has high energy potential. This type of biomass is obtained from advanced mechanical and biological treatment systems. The organic fraction of municipal solid waste is mechanically separated and biodried. After treatment, it can be used as fuel for thermal or biological gasification. The use of the above biofuel (Solid Recovered Fuel) is more costly in thermal gasification (grinding, drying) than in biological gasification where the input material has to be ground only. Thermal processing requires highly corrosion-resistant equipment, and treatment temperature has to be precisely controlled to prevent the production of dioxins and furans which appear at high chlorine concentrations. Sewage sludge is a source of organic waste from homes and industrial plants. Subject to treatment phase, sewage sludge can pose a significant risk of bacteriological contamination. Sludge contains pathogens responsible for typhoid fever, dysentery, tetanus and tuberculosis, which is why it has to be stabilized and purified prior to use. Sewage sludge is also rich in heavy metals. High-temperature corrosion resulting from elevated chlorine concentrations in fuel poses the greatest danger. The physicochemical properties of municipal solid waste and sewage have been given in Table 13.6.

The intensity of thermal processes is determined by moisture content and volatile matter content with consideration:

- water evaporation at the temperature of 170°C,
- release of CO and CO_2 at the temperature of 170 to 270°C; this process also involves the release of tarry substances containing higher-order hydrocarbons, including C_6H_6 benzene, which are characterized by rapid condensation at temperatures below 170°C,

Table 13.6.   Physicochemical properties of sewage sludge and municipal solid waste
(Domanski *et al.*, 2007).

Parameter	Municipal solid waste	Sewage sludge
Moisture content [%]	0.5	6.8
Ash content [%]	–	16.2
Content of volatile matter [%]	–	57.2
C content [%]	54.95	30.33
H content [%]	6.56	3.87
N content [%]	0.32	4.92
S content [%]	0.12	1.16
Cl content [%]	0.231	0.06
H/C ratio [%]	1.42	1.52
Heating value of biomass waste [kJ/kg]	20682	11836

- inhibited release of CO_2 and CO and intensified release of methanol, acetic acid and hydrocarbons at the temperature of 270 to 280°C,
- release of hydrogen and large amounts of hydrocarbons at the temperature of 280 to 400°C,
- hydrogen enrichment of gas and intensified release of oxygen above 500°C.

13.3   THERMAL GASIFICATION

Thermal gasification involves the conversion of organic solids into gas under the influence of temperature. The required substrates for the process are solids with high carbon content and a gasification agent, usually oxygen, air, steam, hydrogen or carbon dioxide. The product of that process is synthesis gas (syngas). Thermal gasification is highly energy-consuming, and it requires temperatures of around 800°C. When exposed to high temperature, the gasification agent reacts with the carbonization product to release flammable process gas, but the residues are only mineral substances. Gasification is a difficult and costly process, but the wide range of syngas applications contribute to its growing popularity. The practical applications for gas from biomass gasification are listed in Figure 13.3.

The gasification process takes place in reactors referred to as gas generators. The reactor relies on the gasification agent to convert the chemical energy of biomass to chemical energy of gas. A series of chemical reactions take place in the generator to produce gas. In principle, gasification is a process of chemical endothermic reactions under sub-stoichiometric (oxygen deficient) conditions which involve carbon, carbon dioxide, carbon monoxide, hydrogen, steam and methane. In addition to flammable gas containing flammable compounds such as:

- carbon monoxide CO,
- hydrogen H_2,
- methane CH_4,

the process also produces liquid substances, tarry and solid substances such as coke breeze and slag. Gasification involves thermal processing of biomass which leads to the production of generator gas. Gasification has the following advantages in comparison with other methods of biomass conversion:

- wide range of uses for the generated gas (heat, electricity, raw material for methanol production),
- lower carbon dioxide emissions,
- significantly higher electric generation efficiency,
- option of using unprocessed biomass, including biomass with high moisture content (in fluidized-bed reactors).

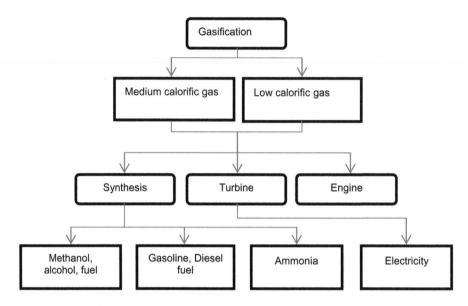

Figure 13.3.   Syngas applications (Kordylewski, 2008).

Biomass gasification in reactors involves several stages. The first stage is biomass drying. Unlike coal, biomass is characterized by high moisture content which increases hydrogen and methane concentrations at successive stages of the process. The following stage is pyrolysis during which volatile matter is removed from biomass at high temperatures and under anaerobic conditions. Pyrolysis is carried out at the temperature of 200 to 600°C, and it produces a volatile mixture of CO, CO_2, H_2, CH_4, steam and aromatics. Gas also contains high-order hydrocarbons from tarry substances. The solid products of the pyrolysis process are char, coke and mineral ballast. A series of chemical exothermic and endothermic reactions take place under sub-stoichiometric (oxygen deficient) conditions (coefficient $0 < \lambda < 1$) in the last phase of gasification. Gasification also involves heat and mass exchange processes which produce flammable gases: CO, H_2 and CH_4. The conversion of pyrolysis products to syngas requires temperatures higher than 750°C. The gasification process itself can be autothermal or allothermal:

- in autothermal gasification, exothermic reactions cater to the needs of endothermic reactions,
- in allothermal gasification, external energy caters to the needs of endothermic reactions.

During autothermal gasification of biomass (with oxygen or air), the energy for drying, degasifying and gasifying the carbonization product is provided by incomplete combustion of the carbonization product from the pyrolytic zone (Medcalf *et al.*, 1988).

13.3.1   *Pyrolysis as the basic process of biomass gasification*

During gasification, carbonized solid and liquid residues from pyrolysis are gasified to produce flammable gases. Pyrolysis is the main process in gasification because biomass contains high levels of volatile matter. Pyrolysis itself is often referred to as dry distillation, cracking or carbonization, and it takes place at relatively low temperatures from 300 to 650°C. As the result of pyrolysis, the cellulose found in biomass is decomposed to methane, carbon dioxide, carbon monoxide, steam and such organic compounds like phenol, acetic acid, benzene.

During pyrolysis, the direction of heat transfer is opposite to the flow of pyrolytic products, i.e. gasses and tarry substances. Heat enters fuel particles through conduction and radiation. Volatile matter is degasified, producing a solid carbonization product which is characterized by higher

Table 13.7. Process parameters for pyrolysis variants.

Process parameters	Conventional pyrolysis	Fast pyrolysis	Rapid pyrolysis
Temperature [°C]	300–700	600–1000	800–1000
Heating rate [°C/s]	0.1–1	10–200	≥1000
Residence time at final temperature [s]	600–6000	0.5–5	<0.5
Particle size [mm]	5–50	<1	dust

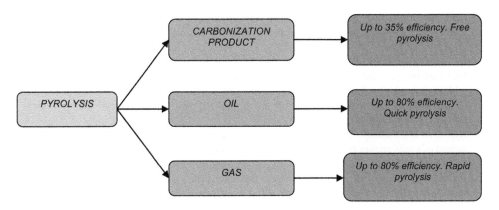

Figure 13.4.   The yield of end products, subject to pyrolytic conditions.

carbon content. Pyrolysis is one of the many processes of gasification, but it is very often deployed separately. One of the best known examples of industrial pyrolysis is char production where wood is decomposed in externally heated retorts under anaerobic conditions at the temperature of around 500°C. Several variants of pyrolysis are used in practice. They differ mainly with regard to process temperature and particle heating rate (Table 13.7).

Process parameters can be adjusted to maximize the yield of end products such as gas, oil and solid fraction. The yield of end products depends on the pyrolytic conditions. Slow pyrolysis is traditionally used in char production. Quick pyrolysis is applied industrially to manufacture fuels, solvents, chemical agents and other biomass end-products. Quick pyrolysis carried out at moderate temperature supports the production of high-quality liquids (Fig. 13.4).

13.3.2 *Biomass torrefaction*

Torrefaction is a thermo-chemical treatment, in a narrow temperature range from 200 to 300°C, where mostly hemicellulose components of biomass depolymerise (Dhungana, 2011). This treatment is carried out under atmospheric conditions in a non-oxidizing environment at low heating rates (<50°C/min) and for a relatively long reactor residence time (typically 1 hour) (Bergman *et al.*, 2005). During the process, the biomass partly decomposes giving off various condensable and non-condensable gases. The final product is a carbon rich solid, which is referred to as torrefied biomass, biochar, biocarbon (Lehmann *et al.*, 2011). In the literature, the torrefaction process is also referred as roasting, slow and mild pyrolysis, wood-cooking and high-temperature drying (Bergman *et al.*, 2005). Torrefaction science is easier to understand by relating it to the pyrolysis process, as torrefaction reactions are essentially the first step of decomposition reactions of pyrolysis; although the process conditions are similar, the heating rate is much lower in torrefaction (Dhungana, 2011). In the torrefaction process, biomass loses mass due to the escape of moisture, light hydrocarbon volatiles and some gases. Torrefaction of 1 kg of biomass produces about 0.3 kg of volatiles and 0.7 kg of torrefied solids (Gaunt and Lehmann, 2008).

The principal characteristics of torrefied products are as follows:

- *High energy density:* Torrefied biomass contains 70–80% of the original weight while retaining 80–90% of the original energy of the biomass. In effect, there can be an increase of around 30% in its energy density.
- *Hydrophobicity:* Torrefied biomass becomes hydrophobic, i.e., it does not absorb moisture or its equilibrium moisture percentage drops. The equilibrium moisture content of torrefied biomass is very low (from 1 to 3%) (Lipinsky *et al.*, 2002).
- *Increased fixed carbon:* The fixed carbon content of torrefied biomass is high. For example, depending on the treatment temperature and duration, it is between 25 and 40%, while the ash content is low. This property makes the torrefied material a very attractive reducing agent (Bergman, 2005).
- *Reduced oxygen:* Torrefaction reduces the O/C ratio through reduction in oxygen. This makes torrefied biomass better suited for gasification due to its lower O/C ratio (Prins, 2008). In addition to its higher heating value, torrefied biomass also produces less smoke when it burns. This is because the smoke-causing volatiles have already been driven off during the torrefaction process and the biomass is also dry.
- *Improved grindability:* Torrefied biomass grindability is superior to that of raw biomass. The output of a pulverizing mill can increase by 3–10 times (Bridgemann *et al.*, 2010; Phanphanich and Mani, 2011).
- *Combustion properties:* Torrefied biomass takes less time for ignition due to less moisture and it burns longer due to larger percentage of fixed carbon compared to raw biomass (Bridgemann *et al.*, 2010).

Torrefaction has following steps:

- *Initial heating:* biomass is heated until drying starts.
- *Pre-drying:* process undergo with constant temperature of 100°C, when unbound water molecules evaporate.
- *Post-drying and heating:* biomass temperature increases to 200°C. Physically bounded water is released, light organic fractions evaporate.
- *Torrefaction:* when temperature rises above 200°C. During that phase most mass lost occurs.
- *Cooling:* temperature of the product decreases (Hardy *et al.*, 2004; van der Stelt, 2011).

Torrefaction is a relatively new area in scientific research. The torrefaction of biomass may be explained by the degradation of its polymeric constituent (hemicellulose, cellulose, lignin, xylan and dextran). Such degradation is influenced by time, temperature, constituents and many other factors. While some work has already been done, the field is relatively unexplored, especially in terms of the design and feedstock. Torrefaction science is based on the behavior of the major lignocellulose material of the biomass with hemicellulose undergoing major thermochemical degradation. Feedstocks currently used on a commercial-scale or in research facilities include wood chip and wood pellets, tree bark, crop residues (straw, nut shells and rice hulls), switch grass, organic wastes including distillers' grain, bagasse from the sugarcane industry, olive mill waste, chicken litter, dairy manure, sewage sludge and paper sludge (Soshi *et al.*, 2009).

Various forms of woody and herbaceous biomass constitute the substrate for torrefaction. Substrate biomass contains three main polymers: cellulose, hemicellulose and lignin which contribute to biomass density and structural strength. Such biomass is difficult to use because thermal processing requires substrates with the smallest particles. Hemicellulose is most reactive, and its structure is modified at temperatures higher than 250°C, whereas cellulose is the most stable polymer, and its structure remains fairy unchanged. The reactivity of lignin is average, and it ranges between that of hemicellulose and cellulose. Biomass loses its structural strength due to damage to the hemicellulose matrix which binds cellulose fibers. Cellulose depolymerization

Table 13.8. Parameters of wood pellets and torrefied wood pellets (Janowicz, 2006).

Parameter	Wood pellets	Torrefied wood pellets
Bulk density [kg/m]	650	750
Energy density [MJ/m^3]	11.3	14.25
Heating value [MJ/kg]	17.3	19.3
Moisture content [%]	10	3

reduces fiber length and lowers structural strength. A reduction in the particle size of torrefied biomass has been found to save 80–90% of energy in comparison with unprocessed biomass. This is a highly significant consideration because pelletization is a very energy-consuming process (Kalina and Skorek, 2007).

Subject to torrefaction conditions, the heating value of torrefied biomass ranges from 20 to 24 MJ/kg^{-1}, which is higher than the heating value of unprocessed wood (17–19 MJ/kg). Torrefied biomass is nearly completely dry, and it has moisture content of 1–6%. Due to low moisture content and hydrophobicity, torrefied biomass is easier and less costly to store and transport. The parameters of pellets obtained from woody biomass and torrefied woody biomass are presented in Table 13.8.

Furthermore, not all biomasses are lignocellulosic in nature. Some waste biomasses, like municipal solid waste, sewage waste and agricultural animal waste are made up of fats, proteins and other organic matter, with very little lignocellulose content. Due to the wide-scale industrialization of agriculture, production of such wastes has substantially increased, and the torrefaction process may help utilize this large volume of non-lignocellulose biomass.

13.3.3 Gasification – basic reactions

Gasification involves the use of a gasification agent. The composition and heating value of the resulting gas is largely determined by the type of gasification agent. The major gasification reactions are:

- incomplete combustion in air or oxygen,
- reactions with steam,
- reactions with hydrogen,
- reactions with carbon dioxide.

When air is used as the gasification agent, flammable gaseous substances in the produced gas account for around 40% of its molar composition (the remaining 60% are inflammable substances, mainly carbon dioxide and nitrogen). The heating value of the resulting gas is generally relatively low, and it ranges from 4 to 12 MJ/m^3 for biomass. For this reason, gasification gases are classified as low caloric gases. The heating value of syngas can be increased (to around 17 MJ/m^3) through the use of special gasification technologies (only the three main elements are gasified: carbon, hydrogen and oxygen). Various reactions take place in a reactor, both endothermic and exothermic. Every chemical reaction leads to changes in enthalpy (Nadziakiewicz *et al.*, 2007). The key gasification reactions take place in the reaction zone which can be further divided into the combustion (oxidation) zone and the reduction zone. The following oxidation reactions are observed in the combustion zone:

$$C + O_2 \rightarrow CO_2 \qquad (13.1)$$

$$C + 1/2O_2 \rightarrow CO \qquad (13.2)$$

$$H_2 + 1/2O_2 \rightarrow H_2O \qquad (13.3)$$

The resulting oxygen-free gas flows to the reduction zone where a series of chemical endothermic and exothermic reactions take place. The basic gasification reactions are:

$$CO_2 + C \rightarrow 2CO \tag{13.4}$$

$$H_2O + C \rightarrow H_2 + CO \tag{13.5}$$

$$CO_2 + H_2 \rightarrow CO + H_2O \tag{13.6}$$

$$C + 2H_2 \rightarrow CH_4 \tag{13.7}$$

$$2CO + 2H_2 \rightarrow CH_4 + CO_2 \tag{13.8}$$

$$CO_2 + 4H_2 \rightarrow CH_4 + 2H_2 \tag{13.9}$$

$$CH_4 + H_2O \rightarrow CO + 3H_2 \tag{13.10}$$

Equilibrium composition and reaction rate are determined by the temperature and pressure inside the reactor. The equilibrium state of each reaction significantly influences the composition of the resulting gas. To illustrate, a pressure increase in an endothermic reaction (13.4) shifts the reaction equilibrium to obtain CO_2, whereas a pressure drop will lead to the production of CO. A temperature rise increases reaction efficiency (13.4) and leads to the production of CO. A cooling effect is observed in heating and degasification zones outside the reaction zone.

Higher methane CH_4 concentrations in gas increase its heating value. The efficiency of methane formation in reaction (13.7) increases with a rise in pressure and a drop in temperature. The above conditions also contribute to the generation of CO_2 and H_2O, but those compounds can be relatively easily removed from gas. Pressure gasification produces gas with relatively high heating value.

It is assumed that when gases are cooled below a certain temperature threshold, chemical equilibrium is not achieved because the reaction may slow down drastically or stop entirely. This is an important consideration in calculations of syngas quantity. The equilibrium composition of syngas is largely dependent on the following process parameters:

- gasification temperature,
- gasification pressure,
- initial reagent composition (gasified fraction and gasification agent).

At higher reaction temperatures, the equilibrium composition of endothermic reactions moves to the right towards a higher share of flammable components: CO, H_2 and CH_4. For this reason, the process should be carried out at the highest possible temperature (the achievable temperature is limited by structural design and ash fusion). High temperature also lowers the dioxin content of gas. Higher gasification pressure contributes to the efficiency of methane production, thus increasing the heating value and methane number of gas. The discussed technologies are highly recommended when gas is used for combustion in internal combustion engines and gas turbines.

The gasification process is applied to process substances with specific parameters, such as chemical composition, moisture content, heating value and ash content. The key parameter is granulation, which makes various gasification technologies difficult to compare. Minerals are the least suitable substances for gasification. The gasification process takes place on the surface of grains, and chemical reactions are reversible, unlike in pyrolysis which takes place within the entire substrate mass. Grain surface is the phase boundary between solids and gases (Rybak, 2006). In reality, gasification is a much more complex phenomenon which involves many more chemical reactions. Fuels containing sulfur compounds are gasified at the temperature of 600–900°C. The following reactions are observed:

$$2SO_2 + 2C \rightarrow S_2 + 2CO_2 \tag{13.11}$$

$$S_2 + 2C \rightarrow CS_2 \tag{13.12}$$

$$2SO_2 + 4CO \rightarrow 4CO_2 + S_2 \tag{13.13}$$

$$2SO_2 + 4H_2 \rightarrow S_2 + 4H_2O \tag{13.14}$$

$$2SO_2 + CH_4 \rightarrow S_2 + 4H_2O + CO_2 \tag{13.15}$$

The aim of gasification is to remove dangerous sulfur compounds. This process produces only hydrogen sulfide, carbon dioxide, carbon monoxide, hydrogen and methane (Sciazko and Zielinski, 2003).

13.3.4 *Biomass gasification methods*

The development of gas generators, gas scrubbing systems and methods for managing process residues contributes to progress in biomass gasification technology. At the current level of technological advancement, the majority of gasification processes are analyzed experimentally. There are no universal methods for mathematical modeling of gasification processes which would support the analytical design of a reactor with the required output and parameters. A variety of biomass gasification technologies have been proposed, but most of them are not applied on a commercial scale. The above results from numerous structural and operational problems as well as relatively low profitability of the proposed solutions. Further research and development is needed to improve reactors, gas scrubbing systems and process gas receivers. Biomass gasification technologies are being developed separately from coal gasification methods. Various structural solutions have been designed for reactors and systems, but most methods for the gasification of biomass, in particular biomass waste, have only entered the development stage, and some have reached the level of demonstration installations. Purely commercial technologies are still rarely encountered.

The majority of commercially available biomass gasification technologies rely on various forms of wood. Gas generators can be divided into the following groups depending on their structure:

- fixed-bed reactors,
- fluidized-bed reactors,
- entrained-flow bed reactors,
- countercurrent-flow reactors.

Fixed-bed reactors are the most popular type of gasifiers owing to their simple structure. The following variants are available:

- downdraft cocurrent-flow fixed-bed reactors,
- updraft countercurrent-flow fixed-bed reactors,
- downdraft open-core fixed-bed reactors.

Fixed-bed reactors share the following features:

- they are atmospheric reactors: insignificant excess pressure is forced by a fan,
- minor drop in gas pressure in the bed,
- residence time from several hours to several days,
- can be operated at 20–120% power rating,
- gasified feedstock is slowly decomposed,
- power scaling is difficult: almost impossible,
- feedstock with high ash content can be used,

- uneven temperature distribution in the reactor,
- separate process zones in the reactor,
- low unit yield: relative to reactor volume,
- high coal conversion factor,
- low quantity of process residues: ash, slag,
- long cold start-up,
- simple structure, reliable performance when regularly serviced,
- possible presence of channels for gas phase flow which lowers heat and mass exchange,
- char is generally required for first start-up.

In a fixed-bed reactor, various chemical and physical reactions take place in reaction zones, including:

- feedstock heating,
- release of moisture: feedstock drying,
- release of volatile substances: degasification (pyrolysis),
- homogeneous oxidation reactions: volatile substances with oxygen,
- heterogeneous oxidation reactions: solid phase (semicoke) with oxygen,
- heterogeneous reactions without oxygen in the char bed,
- homogenous reactions without oxygen in the gas phase.

Subject to reactor type, those processes take place in different parts of the reactor and at different stages of conversion. The composition and properties of the resulting gas are determined by the applied gasification technology. The most popular fixed-bed reactors are downdraft (cocurrent-flow) and updraft (countercurrent-flow) gasifiers. Countercurrent-flow reactors and downdraft open-core reactors are rarely used, and commercial applications that rely on the above solutions are practically non-existent:

- downdraft cocurrent-flow reactor:
  o solid phase and gas phase move down the reactor,
  o lowest output in comparison with other gasification methods,
  o suitable only for selected types of feedstock,
  o feedstock has to be evenly distributed in the reactor,
  o slagging tendency,
  o long gasification process,
  o biomass with low moisture content is required,
  o produces relatively pure gas (low tar content),
  o high exit gas temperature,
- updraft countercurrent-flow reactor:
  o solid phase moves down the reactor, gas phase moves upwards,
  o reactor output ranges between that of downdraft and fluidized-bed reactors,
  o coarse wood can be used,
  o feedstock is dried inside the reactor, feedstock with high moisture content can be used,
  o polluted gas with high tar content is produced,
  o relatively low exit gas temperature,
  o high efficiency,
- downdraft open-core reactor:
  o a variant of a downdraft reactor,
  o feedstock and air are fed together at the top of the reactor,
  o fixed internal diameter along the entire reactor (without a constricting throat),
  o feedstock with low bulk density is required,
  o exit gas temperature ranges between that of downdraft and updraft reactors.

Downdraft reactors are the most commonly applied gasifiers in electricity generation systems due to low pollution of gas. The energy input of most generators is around 3 MW (around 1 MW

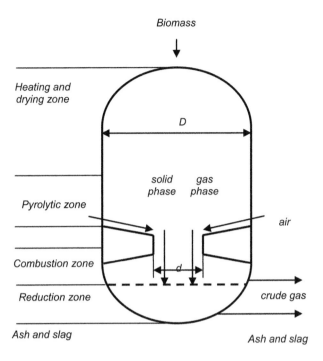

Figure 13.5.   Downdraft reactor.

of the system's electric power), which approximates the lower threshold of commercially available technologies. A simplified diagram of a reactor is presented in Figure 13.5.

In a downdraft reactor, biomass and gas move in the same direction. The most important part of a reactor is the constricting throat in the center with a d/D diameter ratio. Owing to the presence of a throat, the produced gas flows through the combustion zone where hot-spot temperature can exceed 1000°C. Tarry substances flowing through the system are thermally decomposed (cracked) in the reactor. The optimal value of the d/D ratio is determined by the type and size of biomass. In comparison with other types of gasifiers, downdraft reactors are characterized by the lowest tar content of crude gas. For this reason, they are most commonly used to power internal combustion engines and turbines, in particular low-power devices. The energy input of most generators is around 3 MW (around 1 MW of the system's electric power). Downdraft reactors may be equipped with a double jacket which directs hot gas from the lower part of the reactor upwards to the groove between jackets, and the feedstock is heated in drying and pyrolytic zones. Reactor diameter is a vital parameter of downdraft gasifiers.

Downdraft reactors, in particular high power devices, are not highly popular on account of numerous operational problems. Typical annual operating time is around 6000 hours, which lowers economic efficiency. Downdraft reactors work optimally with selected types of biomass only. The main weaknesses of downdraft reactors are:

• biomass of selected type and size is required,
• the size of feedstock particles has to be carefully chosen to ensure even flow through the reactor,
• due to a limited internal drying zone, biomass with low moisture content is required, the optimal moisture content is 20–25%,
• relatively high dust content of gas – before exiting the reactor, gas flows through the reduction zone filled with coke breeze (char),

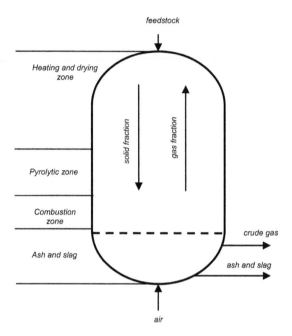

Figure 13.6.   Updraft reactor.

- high loss of hot exit gas lowers the reactor's overall efficiency,
- slagging tendency in choked flow,
- regular service is required due to wear in the reactor's hot-spot,
- highly sensitive to changes in load.

Downdraft gasifiers require relatively dry biomass with low content of ash and small particles (<1 cm). The maximum moisture content is 20% (30% according to some manufacturers). In various solutions, biomass first passes through a drying zone heated with flue gas from the gas engine before it is introduced to the reactor. Subject to the size of biomass particles and the requirements of the gas generator, feedstock may have to be additionally ground or briquetted. Most downdraft gasifiers are atmospheric reactors where air is used as the gasification agent to lower process costs. An open-core reactor is a variant of a downdraft gasifier. Open-core reactors are used for biomass with low bulk density (high porosity). This structural variant does not feature a constricting throat which limits feedstock flow. An open-core reactor has a fixed internal diameter along its entire length. Feedstock and the gasification agent (air) are fed together at the top of the reactor, and gas is collected in the lower part of the reactor (typical cocurrent flow). Exit gas has the temperature of 250–500°C. Pyrolysis plays a more important role in the gasification process because biomass has significantly higher volatile matter content (70–86% of dry matter) than coal (30% of dry matter). Fuels with a high content of alkaline compounds, such as grass, straw and other agricultural residues, can contribute to corrosion, erosion and the formation of deposits on the surface of channels and heat exchangers. For this reason, alkaline compounds have to be removed from gas in the scrubbing process. The ash fusion temperature of biomass is lower than that of other fuels, which causes additional problems.

The gasification process is different in an updraft and a downdraft reactor. The gasifying agent (air) is characterized by countercurrent flow, and it is supplied under the furnace grate, whereas biomass is fed at the top of the reactor. The gasification process in a countercurrent-flow reactor is presented in Figure 13.6.

Table 13.9.   Composition of gas from biomass degasification in a fixed-bed reactor and the produced quantity of solid residues (Domanski *et al.*, 2007).

Component	Countercurrent-flow (updraft) reactor	Cocurrent-flow (downdraft) reactor
CO [%]	15–20	10–22
H_2 [%]	10–14	15–21
CO_2 [%]	8–10	11–13
CH_4 [%]	2–3	1–5
H_2O [%]	10–20	10–20
N_2 [%]	residual	residual
W_d [MJ/m_n^3]	3.7–5.1	4.0–5.6
Solid particles [mg/m_n^3]	100–3000	20–8000
Tarry substances [mg/m_n^3]	10000–150000	10–6000

Table 13.10.   Composition of gas from waste gasification in a pressure reactor (Domanski *et al.*, 2007).

Component	H_2 [%]	O_2 [%]	N_2 [%]	CH_4 [%]	CO [%]	CO_2 [%]	C_nH_m [%]	H_2S [mg/m_n^3]
Crude gas	45.90	0.01	1.8	6.89	10.33	34.4	0.62	1520
Scrubbed gas	64.09	0.06	1.40	8.38	19.63	6.32	0.12	<0.05

Updraft reactors do not feature a constricting throat, which eliminates problems with the gravitational movement of feedstock. Biomass in various size categories may be used. Hot gas exiting at the top of the reactor speeds up biomass drying, which is an additional advantage. Owing to this solution, the maximum moisture content of biomass can reach 40%. The main weakness of updraft reactors are high concentrations of dust and tarry substances in the produced gas. The composition of gas from biomass degasification in updraft and downdraft fixed-bed reactors is presented in Table 13.9. The composition of gas from waste gasification in a pressure reactor is shown in Table 13.10.

In all types of gasifiers, gasification should take place at the highest enthalpies for chemical substances in the produced gas to maximize process efficiency. Efficiency can be described by an equation with a heating value parameter (Wandrasz and Wandrasz, 2006):

$$\eta = \frac{W_{\text{dry gas}} \cdot W_{d\,\text{dry gas}}}{(mW_d)_{\text{dry feedstock}} + (VW_d)_{\text{gasification agent}}} \tag{13.16}$$

where:
$V_{\text{dry gas}}$ – volumetric flow rate of dry gas [m^3/s],
$W_{d\,\text{dry gas}}$ – heating value of dry gas [J/kg],
$m_{\text{dry feedstock}}$ – mass flow rate of dry feedstock [kg/s],
$W_{d\,\text{dry feedstock}}$ – heating value of dry feedstock [J/kg],
$V_{\text{gasification agent}}$ – volumetric flow rate of gasification agent [m^3/s],
$W_{d\,\text{gasification agent}}$ – heating value of gasification agent [J/kg].

Gasifiers also differ in their feedstock conversion (reaction) rates. The conversion rate is determined based on the content of flammable substances entering and exiting the reactor, and it is expressed by the following formula:

$$\text{conv.rate} = \frac{m_{p.p} - m_{p.k}}{m_{p.p}} = 1 - \frac{m_{p.k}}{m_{p.p}} \tag{13.17}$$

where:
$m_{p.k}$ – percentage content of flammable substances entering the reactor,
$m_{p.p}$ – percentage content of flammable substances exiting the reactor.

Advanced biomass gasification systems are currently provided by several dozen suppliers in Europe, USA and Canada, of which:

- 75% supply downdraft fixed-bed reactors,
- 20% supply fluidized-bed reactors,
- 2.5% supply updraft fixed-bed reactors,
- 2.5% supply other types of reactors.

Most suppliers offer only a single type of reactors, and some specialize in individual devices. Very few manufacturers supply entire lines of gasifiers. Most of them have developed pilot systems only, and no commercial applications have been put to use. Relatively few manufacturers offer wood gasification methods that have been tested in practice. The recent progress in gasification technologies indicates that manufacturers focus mainly on fluidized-bed reactors (stationary or circulating), and the above trend is particularly visible on the market of high power gasifiers. Due to cost concerns, the majority of low power reactors of up to several hundred kW are fixed-bed devices. Tarry substances pose the greatest challenge to the development of efficient gasification systems. Tar contains mostly higher-order hydrocarbons (higher than benzene) which are condensed at lower temperatures (below 150°C). Two tar management techniques have been proposed:

- thermal decomposition (cracking),
- tar removal.

Thermal decomposition of tarry substances is a costly process which is rarely deployed in low power reactors. Tar is separated in the gas scrubbing system and evacuated for further processing (usually through combustion). Syngas produced by countercurrent-flow fixed-bed reactors is generally characterized by higher tar levels. Other pollutants are also removed from gas, including:

- solid particles
- alkaline compounds: mainly potassium and sodium,
- chlorine and fluorine compounds: halogenated compounds,
- ammonia and other nitrogen compounds,
- sulfur compounds.

Syngas is generally used in combined heat and power plants with piston engines or gas turbines. At the current level of technological advancement, a combined heat and power plant with piston engines or gas turbines cannot be integrated with a biomass gasifier without a gas scrubbing system. In many installations, gas exiting the reactor is fired directly in the boiler. This solution eliminates the need for high-precision purification of gas, and gas flows only through cyclone separators. The nominal electric power of combined heat and power plants integrated with a gasifier ranges from 150 to 1500 kW, and heat power – from 300 to 3000 kW. Gas turbines are installed in high power systems with several to several dozen MW, which are generally coupled with steam turbines. Net energy efficiency reaches 24–30% in an integrated system and up to 40% in a combined cycle facility. Subject to the size of biomass particles and the requirements of the gas generator, feedstock may have to be additionally ground or briquetted. In some systems,

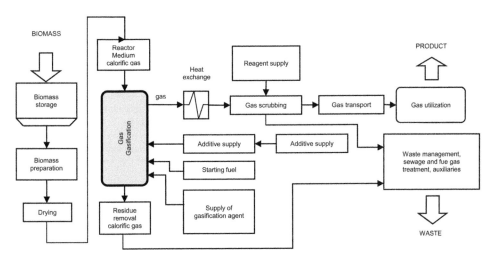

Figure 13.7.   Power generation system integrated with a biomass gasifier.

biomass first passes through a drying zone heated with flue gas from the gas engine before it is introduced to the reactor. Gasification residues are ash and tarry substances which have to be evacuated and neutralized (or landfilled). Ash from the gas generator accounts for 1–2% of dry feedstock. A typical power generation system integrated with a biomass gasifier comprises the following elements:

- biomass feed system (feedstock supply),
- gasification and combustion devices (reactor),
- gas scrubbing system: cyclones, gas coolers (heat recovery), gas filters, gas scrubbers and washers,
- ash evacuation system,
- gas compressor (blower),
- device for converting gas to energy (turbine, gas engine, boiler, etc.),
- system for heat recovery from an engine or a turbine,
- chimney stack,
- auxiliary systems.

A simplified diagram of a power generation system integrated with a biomass gasifier is presented in Figure 13.7.

Gasification efficiency reaches 20% in the simplest systems and 90% in complex installations (Skorek and Kalina, 2005). The concentrations of the oxidizing agent are lowered and a complete combustion cycle is introduced to increase the efficiency of biofuel gasification. Gasifiers are characterized by much higher efficiency than conventional combustion systems, which is comparable to that of high-efficiency installations fired with bituminous coal. Higher efficiency lowers SO_x, NO_x, and dust emissions to ambient. Unlike combustion which requires the supply of air, gasification takes place under anaerobic conditions which minimizes oxide production and reduce the migration of heavy metals to flue gas. Syngas has to be cooled and purified before it is used to power a piston engine. Gas exiting the generator is strongly polluted with solid particles, tarry substances, hydrogen sulfide, trace compounds containing chlorine, fluorine, magnesium, silicon, calcium, sodium and other elements. Syngas is usually scrubbed. The required gas purity levels are determined by the utilization method. Gas is purified in wet or dry scrubbing systems with the use of ceramic filters. The chemical degradation of scrubbing residues can pose a serious problem. The gasification process may also produce ash.

Table 13.11.   Comparison of various gasifier types (Skorek and Kalina, 2005).

	Content		Variation in gas quality	Typical heat input [MW]	
Gasifier type	Tarry substances	Ash		min.	max.
Countercurrent-flow fixed-bed reactor	very high	low	very high	0.05	1.5
Cocurrent-flow fixed-bed reactor	very low	low	high	0.5	10
Bubbling fluidized-bed reactor	average	high	very low	0.5	30
Circulating fluidized-bed reactor	low	very high	very low	1	100

Higher-order hydrocarbons, including tarry substances and small amounts of solid particles, are difficult to remove from crude gas, and they pose one of the greatest challenges in the development of thermal gasification technologies (Vaisanen *et al.*, 2001). The heating value of syngas ranges from 4 to 20 MJ/m^3, and it is highly dependent on the applied gasification agent (Van der Stelt *et al.*, 2011). In most systems, air is used as the gasification agent to reduce operating costs. Syngas generated in systems where air is applied as the gasification agent has the lowest heating value in the range of 4 to 7 MJ/m^3. The above results from the fact that the supplied air contains 70% nitrogen which is an inflammable gas. Pure oxygen or air-oxygen mixtures are used to increase the heating value of syngas. The application of pure oxygen, oxygen-enriched air or steam removes nitrogen from the gasification process and increases the calorific value of syngas. Higher steam content increases hydrogen and methane concentrations in the generated gas. Process efficiency (reaction yield) is determined by the contact area between the gas phase and the solid phase in the reduction zone and the residence time of gas. In oxygen gasification systems, the heating value of syngas ranges from 10 to 13 MJ/m^3, and when steam is used as the gasification agent, heating value is determined in the range of 15 to 20 MJ/m^3. Oxygen production uses 20% of the generated energy (Skorek and Kalina, 2005). The existing methods of oxygen generation rely on absorbers. There are two types of adsorption systems: vacuum swing adsorption (VSA) and pressure swing adsorption (PSA) which relies on a zeolite adsorber. Adsorption uses around 0.4 kWh of energy to generate 1 kg of oxygen. Membrane oxygen generators consume far less energy, and this technology has recently entered the commercialization phase. It is expected to become the leading oxygen generation method for the gasification of biomass waste in the future.

The heating value of syngas can also be increased by expanding the surface area of the reduction zone. This goal can be achieved in a fluidized-bed gasifier. Gas is additionally recirculated to extend reaction time. The above treatments increase the heating value of syngas and maximize process efficiency, but they imply higher costs and higher auxiliary energy consumption (generation of pure oxygen, fluidized-bed technology, higher process pressure).

A comparison of biomass combustion and gasification technologies indicates that alkaline compounds can be easily removed from gasification systems before syngas combustion. Straw has a high content of alkaline compounds, and conventional boilers fired with straw are highly susceptible to corrosion, erosion and sediment deposition on the surface of heat exchangers (Tomeczek, 1991). The content of tar and dust is largely determined by the applied gasification technology. The heat input and gas contamination of various gasifier types are compared in Table 13.11.

The required gas quality in various energy conversion systems is presented in Table 13.12.

Syngas has high temperature, and it cannot be directly used in a piston engine due to its low density. At lower temperatures, gas has higher density, and it contains more flammable elements per unit of volume. The optimal temperature of syngas is around 20°C. Every increase in temperature by 6 degrees above 20°C decreases engine output by around 1%. Cooled gas has to be purified to the level indicated in Table 13.9 to comply with the requirements set for

Table 13.12. The required gas quality in various energy conversion systems (Skorek and Kalina, 2005).

Parameter	Combustion chamber		Internal combustion engine	Gas turbine
	Boiler	Cofiring		
Heating value [MJ/m³]	>4	no requirements	>4	>4
Dust [mg/m³]	no requirements	no requirements	<5–50	<5–7
Tar [g/m³]	no requirements	no requirements	<0.5	<0.1–5
Alkali metals [mg/kg]	no requirements	no requirements	<1–2	<0.2–1

Table 13.13. Heat sources in systems integrated with a piston engine (Verhoeff *et al.*, 2011).

Heat source	Source temperature, [°C]	Ratio of heat recovery to chemical energy of fuel [%]
Cooling of water jacket, oil pan and air at the exit of the turbocharger compressor	95	30
Flue gas	500	25
Flue gas (condensers)	120	8

conversion systems that rely on a piston engine. There are two types of methods for scrubbing crude gas:

- wet scrubbing,
- dry scrubbing.

In a wet scrubber, tar particles suspended in gas are removed in water or other organic liquid. In a dry scrubber, tarry substances are not removed from gas, but are converted to gaseous products such as hydrogen, steam, methane, carbon monoxide or carbon dioxide.

Cooled and scrubbed gas is most often used in combined systems. Power/heat coupling in a CHP plant has several advantages, including:

- high overall efficiency of power generation,
- very low emission levels,
- significant reduction of transmission losses: the power source can be developed at the place of consumption or in close vicinity,
- combined systems can be fired with low caloric gas,
- combined systems are cost-effective with a short payback period (even under 3 years),
- energy incentive programs support the development of cogeneration systems.

Cogeneration systems are very well suited for small applications as well as sites which have a high demand for electricity and heat throughout the year. Piston engines are characterized by high efficiency and low investment spending per unit. The heat produced during gas combustion can be used for biomass conditioning and for heating. Internal combustion engines have several heat sources with different temperatures. The heat sources of systems integrated with an internal combustion engine, the temperature and efficiency of the source are presented in Table 13.13.

The data presented in Table 13.13 indicates that flue gas is well suited for convective drying of biomass, including in drum dryers, whereas low-temperature heat from the cooling system can be used in low-temperature dryers. The cooling system and the flue gas/water heat exchanger are characterized by similar efficiency. When exhaust gas is directly applied in the drying process, heat recovery efficiency will increase due to an absence of heat loss in the exchanger, and it may exceed 30%. Based on their structure and fuel source, engines integrated with gasification systems can be divided into three main groups:

- *Diesel engines:* highest thermal efficiency,
- *dual-fuel engines fired with gaseous fuel and small amounts of liquid fuel to ignite the mixture:* medium efficiency,
- *spark-ignition gas engines:* low efficiency.

Non-standard gaseous fuels, such as syngas, have to meet various requirements. Their heating value can fluctuate in the range of 5 to 10%, and the maximum rate of changes in heating value is 0.5%/30 s (Verhoeff *et al.*, 2011).

13.3.5   *Byproducts of biomass gasification and elimination methods*

In addition to the reactions discussed in Section 13.3.3, organic matter gasification involves many side reactions that produce contaminants. Most side reactions take place in the pyrolytic zone, including:

$$C + 2S \rightarrow CS_2 \tag{13.18}$$

$$CO + S \rightarrow COS \tag{13.19}$$

$$H_2 + S \rightarrow H_2S \tag{13.20}$$

$$2CO_2 + S \rightarrow SO_2 + 2CO \tag{13.21}$$

$$3H_2 + N_2 \rightarrow 2NH_3 \tag{13.22}$$

$$C + NH_3 \rightarrow HCN + H_2 \tag{13.23}$$

$$O_2 + N_2 \rightarrow 2NO \tag{13.24}$$

Pyrolysis contributes most to the production of contaminants such as tarry substances, ash, ammonia, hydrogen cyanide, hydrogen chloride and acid gases (H_2S, COS). Tar is removed by aerosol separators, and ash is evacuated by dust removal systems (cyclones, filters, spray towers). Spray towers also eliminate ammonia and acid gas. In power generation systems, gas has to be purified at high temperatures to improve generation efficiency. High-temperature filters for removing solid pollutants are being developed. The share of acid gases is minimized by adding limestone or dolomite to feedstock or applying metal oxides as sorbents (ZnO, NiO) (Kordylewski, 2008):

$$CaCO_3 \rightarrow CaO + CO_2 \tag{13.25}$$

$$CaO + H_2S \rightarrow CaS + H_2O \tag{13.26}$$

$$MO + H_2S \rightarrow MS + H_2O \tag{13.27}$$

where: M = Zn, Ni.

The tar content of gas from biomass combustion is determined by biomass type, the applied gasification technology and its parameters. The products of biomass gasification, including tarry substances, are listed in Table 13.14.

Tarry substances have been directly removed from hot gas because they have a tendency to condense and settle on other mechanical parts, such as turbine blades or engine cylinders.

The tar content of syngas fed to a piston engine may not exceed 100 mg/m^3. The ash limit is set at 50 mg m^{-3}, and the optimal ash content is 5 mg/m^3. Different methods of removing tarry substances from syngas are presented in Table 13.15.

The data shown in the table indicates that complete purification of gas is difficult to achieve, if not impossible, with the sole use of filters. Filter cleaning poses an additional problem. High-temperature methods based on cracking offer a much more effective alternative. Those techniques

Table 13.14. Products of biomass gasification (Domanski *et al.*, 2007).

Volatile compounds		Tarry substances	
Symbol	Share [%]	Component	Share [%]
CO	16.4	Benzene	37.9
CO_2	12.6	Toluene	14.3
CH_4	4.1	Single-ring hydrocarbons	13.9
H_2	5.7	Naphthalene	9.6
H_2O	15.3	Double-ring hydrocarbons	7.8
N_2	45.9	Phenolic compounds	4.6
		Heterocyclic compounds	6.5

Table 13.15. Filters for removing tarry substances from syngas (Bergman and Kiel, 2005).

Filter	Temperature [°C]	Reduction [%]
Bag filter	~200	max. 25
Sand bed filter	10–20	60–95
Scrubber with rotating tower	50–60	10–25
Venturi scrubber	~60	50–90
Wet electrostatic precipitator	40–50	<60

rely on fixed and fluidized beds filled with catalyst materials such as limestone, dolomite and nickel compounds. High-temperature methods deliver gas purification efficiency in excess of 99.5%, but they are very costly to implement (Bergman and Kiel, 2005).

In addition to mechanical methods, syngas is also treated catalytically. Catalytic methods can be applied both inside and outside the reactor, and they can also be used to modify the chemical composition of process gas. The most widely used catalysts are zinc and nickel. Zinc enhances the generation of hydrogen and methane which increases the heating value of syngas. Nickel is a catalyst for the treatment of hot gas. It decomposes hydrocarbons, and it is used in methane reforming. Nickel converts NH_3 (even up to 95%) and reduces the content of light hydrocarbons: C_2H_6, C_6H_6, and C_7H_6. Nickel may be applied with other catalysts, including aluminum, when higher-order hydrocarbons (even higher than propane) have to be decomposed. Another combination are tri-metallic La-Ni-Fe catalysts which can reduce the tar content of syngas by even 90% (Rakowski, 2002).

13.3.6 *Design parameters of gasification reactors*

Design parameters of gasification reactors will be discussed on the example of downdraft gasifiers. There are two types of downdraft reactors: gasifiers with a constricting throat and open-core gasifiers. The former type is used for firing biomass with low ash content and homogenous size, whereas the latter has a wider range of tolerance for different types of fuel. Open-core gasifiers do not feature a constricting throat, therefore, they can process biomass characterized by higher moisture and ash content and less uniform size. In gasifiers with a throat, gas flow rates are higher in oxidation and reduction zones, which lower tar levels but increases the dust content of synthetized gas. Reactors with larger oxidation and reduction zones have smaller hot zones,

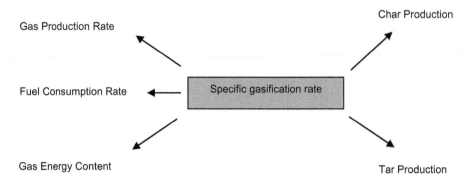

Figure 13.8.   The effect of SGR on the operating parameters of a downdraft reactor (Sivakumar *et al.*, 2006).

which increases tar content. Fuels with high ash content (rice husks: 21.3%) cause congestion and slagging in the combustion zone.

The dimensions of structural elements play a key role in the design of downdraft gasifiers. Dimensions are calculated based on empirical relationships. The most important design parameters are specific gasification rate (*SGR*), gas residence time (*GRT*) and surface area of air supply nozzles. Derivative parameters include dimensions of the furnace and the constricting throat, total length of combustion and reduction zones, air flow rate, dimensions and number of nozzles. The key design parameters are:

Equivalence ratio (ER): ratio of the quantity of oxygen supplied per 1 kg of biomass to stoichiometric value. A theoretically optimal value of *ER* is 0.3. When the value of *ER* approximates 1, the main reaction is combustion, and when it moves closer to 0, pyrolysis is the main process.

Specific gasification rate (SGR): volumetric flow rate of gas passing through a surface unit of a constricting throat where gas volume is measured under normal conditions. The recommended values of *SGR* range from 0.3 to 1.0. This indicator is of fundamental importance in the design of downdraft gasifiers because it influences other reactor parameters, as shown in Figure 13.8.

Specific solid flow rate (SSR): fuel mass passing through a constricting throat. This parameter is derived from *SGR* because 1 kg of fuel produces approximately 2.4 m^3 of gas, therefore $SSR = SGR/2.4$.

Gas residence time (GRT): Mean time of gas residence in the reaction zone. If '*V*' is total reactor volume, '*ε*' is porosity defined as the ratio of empty reactor volume to total reactor volume, and '*G*' is gas flow intensity, then *GRT* equals:

$$GRT = \frac{V \cdot s}{G} \cdot \frac{273}{T} \cdot 360 \qquad (13.28)$$

where:
V – total reactor volume,
ε – porosity (degree of filling),
T – average temperature in a reactor,
G – intensity of gas flow through a reactor.
The recommended value of *GRT* is 0.5 [s].

Air blast velocity (Vb): linear velocity of air under normal conditions. The recommended value of *Vb* is 5 to 30 m/s. At higher speeds, air better penetrates the combustion zone which prevents the formation of hot spots (Reed *et al.*, 1999).

The materials for the construction of gasifiers have to be resistant to high temperature. When exposed to high temperature, organic matter, even substances with low moisture content, produce organic acids which contribute to corrosion. The sites where water and organic acids from biomass drying are condensed in the reactor, such as the biomass hopper, have to be made of acid-resistant

Table 13.16.   Symptoms of carbon monoxide poisoning at various gas concentrations.

CO concentrations in air [% by volume]	Poisoning symptoms
0.01–0.02	Mild headache after 2–3 hours of exposure;
0.04	Severe headache beings approximately after 1 hour of exposure;
0.08	Dizziness, vomiting and convulsions after 45 minutes of exposure; permanent coma after 2 hours of exposure;
0.16	Severe headache, vomiting and convulsions after 20 minutes of exposure; death after 2 hours of exposure;
0.32	Severe headache and vomiting after 5–10 minutes of exposure; death after 30 minutes;
0.64	Headache and vomiting after 1–2 minutes of exposure, death in less than 20 minutes;
1.28	Loss of consciousness after 2–3 breaths, death after 3 minutes.

steel (moisture from fuel is condensed). The elements which are most susceptible to corrosion under the influence of high temperature include the main body of the reactor with a constricting throat, rotating grate and nozzles. Those parts have to be made of stainless steel or inocel alloys for high temperature applications. Regular steel can be used in sites where temperature does not exceed 480°C and where condensation of water and organic acids does not occur. Connections where temperature does not exceed 120°C can be made of plastic pipes which are resistant to corrosion.

Gasification systems have to be air-tight to ensure optimal operation and safety. An air leak into the space under the grate could damage the entire gasifier as the result of gas combustion and uncontrolled rise in temperature. System leaks could lead to the release carbon dioxide, the main flammable gas produced in a gasifier. Carbon monoxide is an odorless and highly toxic gas which poses a serious environmental threat. Carbon monoxide first affects organs that are most susceptible to hypoxia – the circulatory system and the central nervous system. Acute poisoning impairs carbohydrate metabolism, causes internal bleeding and extensive tissue necrosis. If leakage takes place outside the gasifier, e.g. in filters or connecting pipes, gas will become diluted, and it will prevent engine start-up. For this reason, gasification systems have to be completely air-tight. The symptoms of carbon monoxide poisoning at various gas concentrations are presented in Table 13.16.

13.4   SUMMARY

Gasification is a technology of biomass/waste transformation into gaseous fuels and finally into energy. This process has been widely investigated for many years. Works were involved in optimization of the technological parameters: temperature, pressure, searching the new solution of reactor types and configuration, types of gasification agents, as well new type of biomass processing: sewage sludge, municipal solid waste, RDF. All these activities of scientists and engineers have contributed to an improvement of construction and materials used for gasification reactors, improvement of economical aspects of biomass conversion into gaseous fuels, and energy both electricity and heat, with consideration of environmental impact assessment of gasification process. Still some problems have to be overcome, like heterogeneity of biomass composition and properties, high content of ash in some types of biomass causing problems with slag management, high content of pollutants in some type of biomass: sewage sludge, municipal solid waste, generation of hazardous waste when polluted biomass is gasified. Therefore intensive work will need to be carried on biomass upgrading, including drying, torrefaction, and pyrolysis. Interesting technology seems to be the torrefaction of biomass, which allows the production of high quality, relatively uniform second generation fuel – biochar, which may be then gasified.

REFERENCES

Bergman P. & Kiel J.H.A.: Torrefaction for biomass upgrading. *14th European Biomass Conference & Exhibition*, ECN Report, ECN-RX—05-180, 2005.

Bergman, P.C.A.: Combined torrefaction and pelletisation – the TOP process. ECN Report, ECN-C—05-073, 2005.

Bergman, P.C.A., Boersma, A.R., Zwart, R.W.H. & Kiel, J.H.A.: Development of torrefaction for biomass co-firing in existing coal-fired power stations "biocoal". ECN Report, ECN-C-05-013, 2005.

Bilitewski, B., Hardtle, G., & Marek, K.: *Podrecznik gospodarki odpadami. Teoria i praktyka (Handbook of waste management – the theory and practice)*. Wydawnictwo „Seidel-Przywecki" Sp. z o.o., Warszawa, Poland, 2006.

Bridgemann, T.G., Jones, J.M., Shield, I. & Williams, P.T.: Torrefaction of reed canary grass, wheat straw and willow to enhance solid fuel qualities and combustion properties. *Fuel* 87 (2008), pp. 844–856.

Bridgemann, T.G., Jones, J.M., Williams, A. & Waldron, D.J.: An investigation of the grindability of two torrefied energy crops. *Fuel* 89:12 (2010), pp. 3911–3918.

Chmielniak, T., Skorek, J., Kalina, J. & Lepszy, S.: *Układy energetyczne zintegrowane ze zgazowaniem biomasy* (An integrated energetical systems with biomass gasification). Wydawnictwo Politechniki Slaskiej, Gliwice, Poland, 2008.

Dhungana, A.: *Torrefaction of biomass*. MSc Thesis, Dalhousie University, Halifax, Nova Scotia, 2011.

Domanski, J., Dzurenda, L., Jablonski, M. & Osipiuk, J.: *Drewno jako materiał energetyczny* (Wood as and energetical material). Wydawnictwo SGGW, Warszawa, Poland, 2007.

Gaunt, J. & Lehmann, J.: Energy balance and emissions associated with biochar sequestration and pyrolysis bioenergy production. *Environ. Sci Technol.* 42 (2008), pp. 4152–4158.

Hardy, T., Kordylewski, W. & Stojanowska, G.: Katalityczne Zgazowanie Biomasy (Catalitic gasification of biomass). *Gospodarka Paliwami i Energia* 1 (2004), pp. 23–27.

Janowicz, L.: Biomasa w Polsce (Biomass in Poland). *Energetyka* 8 (2006), pp. 601–604.

Kalina, J. & Skorek, J.: Uwarunkowania technologiczne budowy układow energetycznych zintegrowanych z termicznym zgazowywaniem biomasy (Technological conditions of constrtuction of energetic blocks integrated with thermal gasification of biomass). *Energetyka* 7 (2007), pp. 537–545.

Kordylewski, W. (ed): *Spalanie i paliwa* (Incineration and fuels). Oficyna Wydawnicza Politechniki Wroclawskiej, Wroclaw, Poland, 2008.

Lehmann, J., Rillig, M.C., Thies, J., Masiello, C.A., Hockaday, W.C. & Crowley, D.: Biochar effects on soil biota – A review. *Soil Biol. Biochem.* 43:9 (2011), pp. 1812–1836.

Lipinsky, E.S., Arcate, J.R. & Reed, T.B.: Enhanced wood fuels via torrefaction. *Fuel Chemistry Division Preprints* 47:1 (2002), pp. 408–410.

Medcalf, B.D., Manahan, S.E. & Larsen, D.W.: Gasification as an alternative method for the destruction of sulphur contaning waste. *Waste Manage.* 18 (1988), pp. 197–201.

Nadziakiewicz, J., Waclawiak, K. & Stelmach, S.: *Procesy termiczne utylizacji odpadow* (Processes of thermal treatment of waste). Wydawnictwo Politechniki Slaskiej, Gliwice, Poland, 2007.

Phanphanich, M. & Mani, S.: Impact of torrefaction on the grindability and fuel characteristics of forest biomass. *Bioresource Technol.* 102:2 (2011), pp. 1246–1253.

Prins, M.J.: *Thermodynamic analysis of biomass gasification and torrefaction*. PhD Thesis, Technische Universiteit Eindhoven, Netherlands, 2005.

Rakowski, J.: Możliwości zgazowania biomasy dla potrzeb energetycznych (Possibilities of biomass gasification for energetical purposes). *II Konferencja Naukowo techniczna: Energetyka Gazowa*. Prace IMiUE i ITC Politechniki Slaskiej, 2002, p. 293.

Reed, T.B., Walt, R., Ellis, S., Das, A. & Deutch, S.: Superficial velocity – the key to downdraft gasification. *4th Biomass Conference of the Americas*, Oakland, CA, 1999.

Rybak, W.: *Spalanie i wspolspalanie biopaliw stałych* (Incineration, and co-incineration of solid fuels). Oficyna Wydawnicza Politechniki Wroclawskiej, Wroclaw, Poland, 2006.

Sciazko, M. & Zielinski, H.: *Termochemiczne przetworstwo wegla i* biomasy (Termochemical coal and biomas treatment). Wydawnictwo Instytutu Chemicznej Przerobki Wegla, Zabrze, Poland, 2003.

Sivakumar, S., Pitchandi, K. & Natarajan, E.: Design and analysis of downdraft biomass gasifier using computional fluid dynamics. *International Congress on Computational Mechanics and Simulation (ICCMS 06)*, IIT Guwahati, India, 2006.

Skorek, J. & Kalina, J.: *Gazowe układy kogeneracyjne* (Gaseous cogenerative systems). Wydawnictwa Naukowo-Techniczne, Warszawa, Poland, 2005.

Sohi, S., Lopez-Capel, E., Krull, E. & Bol, R. :Biochar, climate change and soil: a review to guide future research. In '*CSIRO Land and Water Science Report*', series ISSN: 1834-6618, 2009.

Tomeczek, J.: *Zgazowanie wegla* (Coal gasification). Wydawnictwo Politechniki Slaskiej, Gliwice, Poland, 1991.

Vaisanen, P., Hietanen, L. & Syska, G.: New strategy in the energy use of waste. *Paliwa z odpadow* III, 2001.

Van der Stelt, M.J.C., Gerhauser, H., Kiel, J.H.A. & Ptasinski, K.J.: Biomass upgrading by torrefaction for the production of biofuels. *Biomass Bioenerg.* 35 (2011), pp. 3748–3762.

Verhoeff, F., Adell, A., Arnuelos, Boersma A.R., Pels, J.R., Lensselink, J., Kiel, J.H.A. & Schukken, H.: Torrefaction technology for the production of solid bioenergy carriers from biomass and waste. ECN Report, ECN-E–11-039, 2011.

Wandrasz, W. & Wandrasz, J.: *Paliwa formowalne* (Refused derived fuels). Wydawnictwo „Seidel-Przywecki" Sp. z o.o., Warszawa, Poland, 2006.

CHAPTER 14

An innovative perspective: Transition towards a bio-based economy

Nicole van Beeck, Albert Moerkerken, Kees Kwant & Bert Stuij

14.1 INTRODUCTION: WHY WE NEED A BIO-BASED ECONOMY

14.1.1 *Towards a sustainable future*

This chapter will illustrate that, in the transition towards a sustainable future, the development of energy systems on the one hand, and the development of the agricultural sectors on the other are mutually dependent. The main underlying factor in this mutual dependency is the fact that we live in a carbon based society. Humanity consumes vast amounts of carbon each year in the form of food, feed, materials or energy. Most of this carbon is fossil carbon, and most of this fossil carbon is used for energy. Obviously, a considerable amount of bio-based carbon is associated with agricultural production.

Evidence is accumulating to indicate that, with current production patterns, we are exceeding the planetary boundaries in such a way that humanity is threatened (see Section 14.1.3). To secure a sustainable future for humanity, many challenges need to be faced. One of the major challenges is to replace fossil carbon with sustainable bio-based carbon, implying a transition towards a bio-based economy. The agricultural sector (including forestry) is the key provider of such biomass, and thus plays a vital role in this transition process (see Section 14.2). The (potential) availability of sustainable biomass is the subject of Section 14.3, while Section 14.4 deals with the cascading approach in biomass deployment. Cascading is a precondition in the transition towards a bio-based economy, as part of the concept of the 'Trias Biologica', which is a logical concept to feed both people and livestock, to produce enough materials and energy to meet demand, *and* at the same time stay well within the planetary boundaries. Section 14.5 illustrates, supported by Dutch case studies, the challenges and opportunities related to the transition process. Section 14.6 discusses prospects and impacts of a bio-based economy, followed by conclusions.

14.1.2 *Relationship between agriculture and energy*

Energy and agriculture are interdependent. Energy is required for any economic activity, including those in the agricultural sector. The agricultural sector consumes considerable amounts of energy to produce food and materials for people and feed for livestock. Today, the world food production not only accounts for 30% of our land-use and 70% of our freshwater production, but also 20% of our energy demand (Aiking, 2006; FAOSTAT, 2013). With an increasing population, food production and associated energy consumption are likely to increase to meet growing world demand. For instance, the Food and Agriculture Organization of the United Nations (Bruinsma, 2009) estimates that food production will need to increase by 70% to meet demand in 2050.

The agricultural sector, while on the one hand a major energy consumer, also plays an important role as supplier of energy. Biomass from agriculture has long been used for energy production, albeit that early use was mainly restricted to small scale applications for cooking and heating. And as early as 1925, the New York Times quoted Henry Ford pronouncing bio-based transport fuels (i.e., ethyl alcohol) as 'the fuel of the future' (Kovarik, 1998). In recent years, policy makers in many countries (such as The Netherlands) have declared biomass as a viable alternative to replace fossil fuels, in order to counteract depletion of fossil fuel reserves as well as mitigate environmental problems.

However, even today, the claim on biomass for energy production is already interfering with biomass demand for food and feed (and materials). The competing claims on biomass are invoking a great deal of controversy. Meanwhile, world demand for both energy and food continue to increase. For instance, the International Energy Agency, in the World Energy Outlook 2012, states that global primary energy demand is expected to rise by over one-third in the period to 2035. And the United Nations (2011) estimate that world population will grow to 9.3 billion in 2050, and to over 10 billion people before the turn of the century, resulting in a substantial increase in mouths to feed. This implies a fundamental change is needed to meet both food demand and energy demand in the future. In addition to the food/energy controversy of biomass consumption, humanity faces several other challenges that often emphasize the relationship between energy and agriculture. These challenges are the topic of the next section. The question whether agriculture can produce enough biomass to serve all of human needs is addressed in Section 14.3.

14.1.3   *What are the challenges?*

Throughout history, our planet has shown periods of significant environmental change. Nonetheless, for over 10,000 years, the earth has offered an agreeable and stable environment for humanity in which to flourish. Since the Industrial Revolution however, human actions appear to be the main driver for environmental change. Galloway *et al.* (2009) have shown that human activities now convert more nitrogen from the atmosphere into reactive forms than all of the earth's terrestrial processes combined. The main cause for the human induced imbalance in the nitrogen cycle is the large-scale use of fertilizers. Smil (2001) states that without fertilizers, the world population would be capped at about 3.3 billion people. While today, mankind is able to feed almost 6 billion people (see also Section 14.2.2). However, fertilizers also cause severe environmental impacts that threaten humanity. In 2009, Nature published an article of Rockström *et al.* (2009a) containing an overview of the most relevant environmental threats to humanity today. Rockström *et al.* identified and quantified nine parameters for planetary boundaries within which humanity can continue to develop in a sustainable manner. They also listed the current status as well as the pre-industrial values (see Table 14.1). As Table 14.1 shows, three parameters already exceed the boundaries of a safe operating space for humanity, with the risk of generating abrupt or irreversible environmental changes. One of these parameters is the amount of nitrogen removed from the atmosphere for human use, which is closely related to the manner in which fertilizers are used today. Another limit exceeded is the carbon dioxide concentration in the atmosphere, caused most notably by the combustion of fossil fuels for energy production.

Although not all parameter values in Table 14.1 are based on ample scientific evidence, and there is debate on the position of some boundaries, the work of Rockström *et al.* (2009a; 2009b) does provide insight and frames the current environmental dilemma. A complicating factor is the interdependence of the planetary boundaries; transgressing one boundary can strongly affect other boundaries. For instance, Townsend and Howard (2010) consider the human interference in the nitrogen cycle to be one of the main threats to global biodiversity.

The example of The Netherlands as a relatively small country with a rather high population density of 494/km^2 (CBS, 2012), shows how the boundaries of the environmental processes are closely related. The large-scale use of fertilizers enabled The Netherlands to become the world's second-largest exporter of agricultural products, after the United States (Agentschap NL, 2012; PBL, 2012). The human interference in the nitrogen cycle is, however, at the same time affecting overall resilience of ecosystems via acidification of terrestrial ecosystems and eutrophication of coastal and freshwater systems (Rockström *et al.*, 2009b). In addition, the associated emissions of nitrous oxide and methane have a strong greenhouse gas effect.

Looking more closely at the planetary boundaries in Table 14.1, it appears that agriculture and energy supply play a significant role in the three boundaries that are currently being exceeded. To return to a safe operating space for humanity, a fundamental change in our management of

Table 14.1. Environmental boundaries of planet earth. Adapted from Rockström *et al.* (2009a), and Rockström *et al.* (2009b).

Earth-system process	Parameters	Proposed boundary	Current status*	Pre-industrial value
Climate change	• Atmospheric CO_2 concentration [part per million]	350	*387*	280
	• Energy imbalance at Earth's surface, [W per m^2]	1	*1.5*	0
Rate of biodiversity loss	• Extinction rate [number of species per million species per year]	10	*>100*	0.1–1
Nitrogen cycle (part of a boundary with the phosphorus cycle)	• Amount of N_2 removed from atmosphere for human use [Mton per yr]	35	*121*	0
Phosphorus cycle (part of a boundary with the nitrogen cycle)	• Quantity of inflow of phosphorus to ocean (increase compared with natural background weathering) [Mton per yr]	11	8.5–9.5	–1
Stratospheric ozone depletion	• Stratospheric O_3 concentration [Dobson unit]	276	283	290
Ocean acidification	• Average global saturation state of aragonite in surface ocean water [Ωarag]	2.75	2.90	3.44
Global freshwater use	• Consumption of fresh water by humans [km^3 per yr]	4000	2600	415
Land-system change	• Percentage of global land cover converted to cropland	15	11.7	low

*Values in italics indicate that the boundaries have already been crossed.

natural resources is required, especially in light of a growing world population. A smart approach is needed redesigning both energy supply and food (and non-food) production in order to meet future needs. In short, we need a bio-based economy.

14.1.4 *The smart approach: a bio-based economy*

A bio-based economy is not an objective in itself. It is a smart approach to meeting a number of environmental challenges, while simultaneously providing green economic growth. WWF Denmark in 2009 defined a bio-based economy as an economy based on production paradigms that rely on biological processes and that, as with natural ecosystems, uses natural inputs, expends minimum amounts of energy and does not produce waste as all materials discarded by one process are inputs for another process or reused in the ecosystem. Basically, the bio-based economy (BBE) uses biomass for non-food applications such as components, chemicals, materials, transport fuels, electricity, and heat. Bio-energy from biogas, biofuels or solid biomass is thus a subset of the bio-based economy. In turn, the bio-based economy is a subset of a bio-economy, with the latter also including biomass for food applications (in addition to the non-food applications of the bio-based economy). This relationship between bio-energy, bio-based economy and bio-economy is illustrated in Figure 14.1, which also shows that the bio-economy together with the fossil carbon constitute the carbon economy[1].

---

[1]Note that all carbon in principle has a biological origin. However, in this chapter, we refer to bio-based carbon as the short cycle carbon originating from recently grown biomass as opposed to the fossil carbon originating from coal, oil and gas.

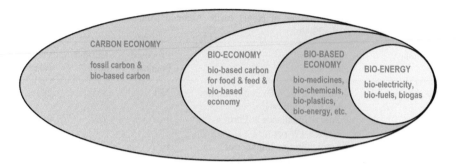

Figure 14.1.   The relationship between bio-energy as a subset of the bio-based economy for non-food applications of biomass, while the bio-based economy is in turn a subset of the bio-economy, which
also includes the food & feed applications of biomass. The bio-economy encompasses the part
of the carbon economy that uses bio-based carbon, as opposed to fossil based carbon.

According to the European Commission (EC, 2012) the bio-economy includes the sectors of
agriculture, forestry, fisheries, food and pulp and paper production, as well as parts of chemical,
biotechnological and energy industries. Bio-based products are defined as products that are wholly
or partly derived from materials of biological origin, excluding materials embedded in geological
formations and/or fossilized. The EC believes that a bio-economy holds great potential: it can
ensure food security, manage natural resources sustainably, reduce dependence on non-renewable
resources, help mitigate and adapt to climate change, and create jobs and competitiveness. According to the EC, the EU bio-economy already has a turnover of nearly 2 trillion euros and employs
more than 22 million people, accounting for 9% of total employment in the EU. These figures
include the production of sustainable biological resources and their conversion, as well as the
conversion of waste streams into bio-based products, biofuels and bio-energy.

The transition of current economies to bio-based ones implies the substitution of fossil-based
resources with sustainable biomass (in solid, gaseous, or liquid form). The aim of this chapter
is to investigate the conditions under which this transition towards a bio-based economy can
develop sustainably without interfering with the requirements of the bio-economy or the planetary
boundaries. Figure 14.1 indicates that it might be wise to reduce the carbon footprint of humanity
before trying to substitute fossil-based carbon with bio-based ones. This logical order is part
of the concept of the 'Trias Biologica', as discussed in Section 14.4. Our investigation into the
bio-based economy is closely related to food supply issues and agricultural policy, and includes
issues related to waste reduction, circular process cycles, bio-refinery, introduction of new protein
products, and innovations in land use.

14.2   AGRICULTURE: THE FOUNDATION OF A BIO-BASED ECONOMY

14.2.1   *Agriculture and food*

At present, the primary goal of the agricultural sector is the supply of food for the human population, either directly or through fodder for animals that provide food products. The FAOSTAT (2013)
database shows world food supply steadily increasing from 390.5 Mton in 1961 to 976.7 Mton in
2009 (see Fig. 14.2). The annual average per capita supply peaked in 1989 at 152.9 kg/capita/year
(coming from 128 kg/capita/year in 1961). Since 1989, however, the per capita consumption has
dropped to 146.7 kg/capita/year in 2009. Matching food demand and supply is currently not so
much a matter of increasing production as it is a matter of distribution and efficiency: Over
800 million of people are malnourished (Alexandratos, 2009) and 1.6 billion people are overweight (WHO, 2006). In addition, one third of all transport in the world is food-related (Smil,
2002b).

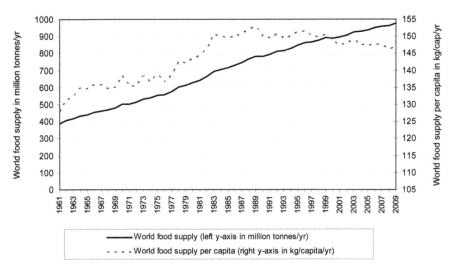

Figure 14.2.   Total world food supply and food supply per capita between 1961 and 2009. Source: FAOSTAT (2013).

Table 14.2.   Meat production worldwide 1950–2050. Based on Bruinsma (2002a; 2002b).

Year	World population [billion]	Global meat production [Mton]	Average per capita meat production [kg/capita]
1950	2.7	45	16.7
2000	6.0	233	38.8
2050	9.0	450	50.0

Food demand is not only a function of the world population, it also depends on national diets. The metabolic rate of humans differs with age, sex, weight and daily activity, but an average of 11.7 MJ per capita a day (2800 kcal/capita/day) can be used as a world average (FAOSTAT, 2013). Note that the per capita food supply in most developed countries is much higher, for instance in The Netherlands it amounts to 13.6 MJ/capita/day and in the USA even 15.5 MJ/capita/day. An average diet should consist of approximately 30% protein, 20% fat and 50% carbohydrates. Protein is vital in our diet, among others because it contains nitrogen, which is an essential component to build up DNA and RNA. The form in which proteins, but also fat and carbohydrates, are consumed depend among others on personal taste, genetic variation, availability of products and economic aspects. For instance, meat as a source of protein is typically consumed more in western populations than eastern or southern ones. Generally, meat consumption goes up with increasing welfare, as does feed consumption: as a result of animal metabolism, on average 4–8 kg of plant protein is required to yield 1 kg of meat protein (depending on the type of meat). To illustrate, Table 14.2 shows the meat production for 1950, 2000, and (estimated) 2050, taken from Bruinsma (2002a; 2002b). Clearly, the average per capita meat consumption has increased over the years, implying a considerable increase in associated feed production.

So agriculture today is primarily focused on food supply. However, in a bio-based economy, agriculture has to provide sufficient bio-carbon for non-food applications as well. The question is whether this can be done without impeding food supply. The transition towards a bio-based economy implies a substantial increase in agricultural production. The following sections address the current issues that hamper increased agricultural production and challenge the transition towards a bio-based economy.

14.2.2   Soil fertility

The fertility of soil in the natural environment is generally low, and probably ever since mankind turned to farming, humans have tried to raise soil fertility. Basically, soil fertility is the capacity of the soil to supply nutrients to plants. Although many factor influence soil fertility, Troeh and Thompson (2005) distinguish the following dominating factors:

- Chemical composition: sufficient amounts of organic matter in the soil allow for metabolic processes and energy conversion. Complex organisms also require macro nutrients such as nitrogen, phosphorous, and potassium, as well as micro nutrients such as zinc, selenium, and manganese.
- Biological activity: plants live in a symbiosis with bacteria and fungi. Bacteria and fungi feed themselves with excrements of the roots, while protecting these roots from diseases. Diseases negatively affect quality and quantity of crops.
- Physical parameters: good soil structure allows for sufficient water supply and draining of excess water.

Already early in history, farmers learned that livestock dung and guano from birds considerably improve soil fertility. In the 19th century, scientists discovered that (reactive) nitrogen is one of the key components in manure that enhances crop growth. In 1909 Fritz Haber and Carl Bosch of BASF developed an industrial way of producing ammonia, the bioactive form of nitrogen. The Haber-Bosch process initiated the large scale production of fertilizer, and is still used as the standard conversion process for fertilizer today. Galloway (2008) estimates that the current annual anthropogenic nitrogen production is about 170–187 Mton. The earth's total annual nitrogen production is estimated at 300 Mton, implying that human activities now convert more nitrogen from the atmosphere into reactive nitrogen forms than all of the earth's terrestrial processes combined.

Unmistakably, humanity has benefited from the large-scale use of fertilizers. Smil (2001) calculated that global population would have been capped at 3.3 billion people if large-scale application of fertilizers had not been introduced, because of nitrogen limitations. Instead, today we are able to feed about 7 billion people. Figure 14.3 shows the vital role of nitrogen and fertilizers in human population growth.

However, the large scale use of fertilizers also has severe environmental effects. The production of fertilizers through the Haber-Bosch process is associated with the use of considerable amounts of natural gas, a fossil fuel, to transform inactive gaseous nitrogen into bioactive ammonia that crops can absorb. The process is associated with large-scale emissions of greenhouse gases (although modern catalysts are able to reduce these emissions with more than 90%). Also, on average only 50% of the nitrogen in the soil is absorbed by plants (Crews and Peoples, 2004). The remaining nitrogen in the soil eventually leads to overall resilience of ecosystems via acidification of terrestrial ecosystems and eutrophication of coastal and freshwater systems (Rockström et al., 2009b).

In a bio-based economy, the use of fertilizers is vital to ensure enough bio-carbon production to meet demand for food *and* produce the necessary raw materials for bio-based non-food applications. The question is how to handle the large-scale use of fertilizers to minimize the detrimental environmental effects. The answer lies in the cascading approach, which will be addressed in Section 14.4.

14.2.3   Land use

Bruinsma (2009), in a FAO expert meeting on how to feed the world in 2050, estimated that meeting world food demand in 2050 requires a 70% increase in global food production. More food requires more soil. However, arable land is disappearing rapidly in developed countries. In the EU, each year an additional 0.1 million hectares of agricultural land is 'lost' to housing, industry and infrastructure. Bruinsma states that the increase in arable land between 2005 and 2050 is estimated

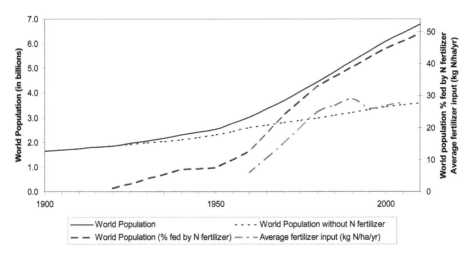

Figure 14.3. Developments in world population and fertilizer use (taken from Smil, 2001). The sharp increase in world population since 1950 is enabled by the large-scale use of N fertilizer. Without the use of N fertilizer, world population would have shown a much more gradual increase.

to be just 5%, as the net balance resulting from an expansion of 120 million hectares in developing countries and a decrease of 50 million hectares in developed countries such as Western Europe. This implies that the efficiency of food production per hectare has to increase considerably to feed the future population. Intensification is generally seen as the answer. Life cycle analyses shows that intensive agriculture requires less energy and raw materials per unit production, although these benefits are partly counterbalanced by the larger transportation distances often associated with large-scale high-intensity agriculture (FAO, 2011). Intensive agriculture is also subject to intensive social debate concerning health issues, animal welfare, prevention of diseases, use of antibiotics, and ecological issues. For instance, in the densely populated Netherlands, 55% of the total surface of 41,543 km^2 is now used as arable land. During the past decade, the arable land has decreased with 7%. However, the average standard yield in agriculture (in euros) has increased with 41% between 2000 and 2011 (CBS, 2013), with the increase in standard yield in cattle farming reaching over 80%. The livestock, however, does produce 3.5 Mton of carbon from animal manure each year, causing serious environmental problems, mainly due to the excess of minerals such as nitrogen and phosphorous.

Meat versus crops

Aiking *et al.* (2006) used the following calculation to illustrate the efficiency of meat production. Approximately 400 million hectares of present global land use (about the land surface of the EU28) are used for feed crops. Estimating the protein content of the feed crops to be 144 million tons, and assuming 5 kg plant protein for 1 kg of meat protein, the maximum yield of meat protein is 29 Mton. Aiking states that the same amount of plant protein (29 Mton) would require not 400 million hectares, but only about 25 million hectares (which is about the land surface of the UK). Smil (2000) estimates at least a billion more people and part of the current livestock could be fed if the area now devoted to feed crops would be used to grow a mixture of food crops, with only their milling residues used for feed. So more efficient use of land in agriculture is possible if it goes hand in hand with efficient production, using residues of crops as feed. Feed and non-food applications of biomass should be derived from residues of food production as much as possible to leave more land available for food crops.

Source: Based on Aiking et al. (2006), Smil (2000)

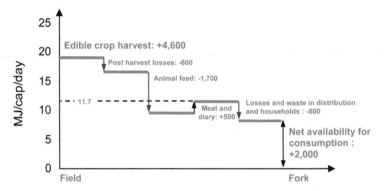

Figure 14.4.   Estimate of worldwide average losses and conversions in the food supply chain (Lundqvist, 2008). As a global average, Smil (2000) states that farmers produce the equivalent of 19.3 MJ (4600 kcal) per capita per day, i.e., before conversion of food to feed. Losses, conversions and wastage at the various stages leave roughly 11.7 MJ/capita/day for supply (mixture of animal and vegetal foods). At the end of the chain, on average 8.4 MJ/capita/day are available for consumption.

14.2.4   *Wastes in the food chain*

More efficiency in food supply requires attention to losses over the entire food chain. Indeed, considerable losses in the food chain occur. Lundqvist (2008) estimates that less than half of the agricultural yield is available for net consumption, as shown in Figure 14.4. On a global scale, the conversion of feed to meat protein accounts for a net average loss per capita per day of almost half of the total loss. Note that meat consumption is relatively low in developing countries and countries such as India, where the national diet is less meat oriented. So the global average presented here obscures the fact that the losses as a result of meat consumption account for a substantially higher share in the meat eating regions, as we have seen in Section 14.2.1. So reducing meat consumption seems an obvious starting point for increasing efficiency in the food chain.

Using biomass more efficiently allows claims for food, feed and industrial use to coexist without competing with each other. For instance, the cellulose in biomass (often more than 50%) cannot be digested by humans, while the lignin in biomass cannot be digested by humans or animals. These non-digestible parts can be used for many purposes, such as ethanol production (by separating the hydrocarbon), production of textile and building materials from fibers, soil fertility improvement with nitrogen and phosphorous, and finally energy production. This issue is further addressed in Section 14.4 (and Section 14.5) about the cascading approach.

14.2.5   *Agrification policy at the origin of non-food industrial applications of biomass*

Sustainable biomass is the main resource for a bio-based economy, and agriculture (including forestry) is the sector that has to produce this resource in sufficient amounts to meet demand for food and non-food applications. The development of non-food applications of biomass in Europe has been largely driven by the effects of agricultural policies in the 1980s. In response to the EU Common Agricultural Policy, the agro-industry in Europe has created new markets for the surplus of agricultural products.

At the origin of the European Union was the desire of the six founding Member Sates (Belgium, France, Luxembourg, Italy, The Netherlands, and West-Germany) to develop a common market not only for coal and steel, but also for agricultural products. In 1962 the Common Agricultural Policy (CAP) came into force for the Member Sates, promoting self-sufficiency and supply

Table 14.3. Overview of bio-based products as a result of agricultural policy between 1980 and 2000 (taken from Bos *et al.*, 2008).

Product	Application	Component	Crops
Glue	Plastic, paper, hot melts	Gluten, starch, inulin	Wheat, potato, chicory
Chemical additive	Crosslinker for coating, softener	Oil, fatty acid, gluten, starch	Flax oil, calendula, potato, wheat
Coating	Paper, textile, plastics, food	Starch, flour, gluten, fatty acid	Potato, wheat, oilseed
Composite	Motorcars, building material, wind turbine, molds	Fibers	Flax, hemp, straw
Fiber	Paper, textile, building material, insulation	Fibers	Flax, hemp, miscanthus
Film	Packaging, agricultural film	Starch, gluten	Potato, wheat, chicken feathers
Fine chemicals	Emulsions, lubricants, detergents, germ inhibitors	Oil, fatty acids, inulin, limonene	Oilseed, chicory, caraway
Plastics and rubber	Packaging, molds, foams	Starch, oil, lactic acid	Potato, flax oil, maize

security for its community. The CAP guaranteed prices for surplus of production inside the union, and import tariffs and quotas for some products outside the union. Today in the EU28, the CAP is still very much in place, accounting for about one third of total EU budget.

Bos *et al.* (2008) describe that in the early 1980s, the surplus of production of sugar, milk, and wheat (among others) increased to a level that made the CAP extremely expensive, instigating the search for alternative markets for agricultural products. This also spurred Dutch research in non-food applications of biomass, supported by policy that was termed 'agrification' policy. The high oil prices at the time further incited the search for non-food applications of biomass. The conversion of biomass to energy was viewed as an interesting alternative to energy production from fossil fuels. This view was further supported by the adoption of the Kyoto Protocol in 1997, which committed nations to reduce carbon dioxide (CO_2) emissions. In addition, problems with contaminated feed causing BSE ('mad cow disease') and high dioxin levels in milk led to a ban on using certain byproducts of the agro-food industry in feed. This created another driver for the agro-food industry to investigate non-food applications of bio-based byproducts, all the more since European livestock was decreasing as a result of EU legislation on fertilizers.

In the 1990s, Dutch companies in the protein- and carbohydrate processing industry, such as Avebe, Cargill, and Nedalco, actively investigated non-food applications. Also, DSM, Akzo-Nobel, Shell, Dow Chemicals and Uniqema invested in research on vegetable oil and chemicals, while the paper industry investigated fibers from biomass. Table 14.3 gives an overview of products developed in the period 1980–2000.

As we have seen in Section 14.2.1, with a growing world population, meeting future food demand is a challenge in itself for the agricultural sector. However, in the transition towards a sustainable future, the agricultural sector also has to provide the building blocks of a bio-based economy. In addition to increased food demand, the increased use of bio-based feedstock in industrial sectors will put a further strain on agricultural production. And with the current government policies promoting the substitution of fossil energy with bio-energy, the claim on agricultural production rises to a maximum level. The following section addresses the role of biomass in a sustainable energy supply, while Section 14.2.4 deals with the sustainable use of biomass itself. Section 14.4 explores the necessary changes in current agricultural practice by means of the cascading approach, to allow the different claims on biomass to coexist without competing with each other.

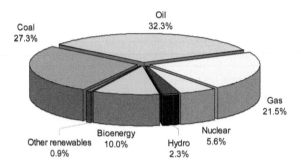

Figure 14.5.   World primary energy demand by source in 2010 (excluding electricity trade). Total amount is 533 EJ. Source: IEA (2012).

14.3   BIOMASS AT THE BASIS OF SUSTAINABLE ENERGY SUPPLY

14.3.1   *Current energy demand*

The world today relies heavily on the use of fossil fuels to meet energy demand. The IEA (2012) state that world primary energy demand amounts to 533 EJ in 2010, of which 81% is supplied by fossil fuels: oil, natural gas and coal/peat (see Fig. 14.5). Most countries have to import these fossil fuels to meet their domestic demand, thereby increasing the risks associated with security of energy supply. Data from Eurostat (2012) show, for instance, that the EU's dependence on imports of fossil fuels from non-EU countries rose from 47.8% of primary energy consumption in 2000 to 55.5% in 2009. On top of import dependency, many countries face depletion of their national fossil resources within the coming decades, further increasing energy import dependency if they fail to switch to national (renewable) alternatives. The Netherlands for instance, expects to run out of its conventional natural gas reserves within the next 15 years, assuming that the net annual production (extraction) remains constant at its 2010 level (CBS, 2011a).

Interestingly, import dependency in energy supply shows similarities with the import dependency in food supply: about 70% of the EU-demand for proteins is imported (USDA 2011). EU import dependency of protein is largely a result of minimizing production costs in agriculture, implying that protein crops can best be grown where the circumstances are best, i.e. outside the EU, even if this means that EU self-sufficiency decreases.

The vast amounts of fossil fuels used today cause various environmental problems, of which climate change is one with global impact, also for agricultural production. Current anthropogenic emissions of CO_2 and equivalent greenhouse gases amount to 30.2 Gton in 2010. According to climate experts (Chum *et al.*, 2011), the long-term concentration of greenhouse gases in the atmosphere needs to be limited to around 450 parts per million of carbon-dioxide equivalent in order to limit the global increase in average temperature to 2°C in the long term, compared with pre-industrial levels. However, similar to food demand, energy demand is expected to increase in the future. The IEA in World Energy Outlook 2012 state that rising incomes and population growth are expected to push total primary energy demand up from 533 EJ in 2010 to 642 EJ in 2020 and further up to 782 EJ in 2035, assuming current policies stay in place. The associated CO_2 emissions in 2020 amount to 36.3 Gton worldwide and increase further to 44.1 Gton in 2035. According to the IEA, these levels are well beyond the 450 parts per million limit of 22.1 Gton in 2035. IEA estimates that only with radical new policies, there is a 50% chance of limiting the global increase in average temperature to 2°C in the long term (compared with pre-industrial levels). According to the IEA, with continuation of current policies, the share of fossil fuels in 2020 and 2035 remains high at about 80%.

As Figure 14.1 illustrates (and Fig. 14.5 demonstrates for the energy sector), current demand for fossil carbon is far greater than current supply of bio-based carbon. In order to secure the long term supply of food and energy, and at the same time respond to increasing environmental problems, we need to reduce the fossil carbon footprint of the economy, first by increasing efficiency in production and supply chains, and secondly by substituting the remaining fossil carbon supply with sustainable carbon as much as possible. De-carbonizing the current world economy implies rethinking conversion processes, for instance by using solar energy conversion techniques.

Converting solar energy into green vegetation is actually not a very efficient process: only about 0.5% of the solar irradiance on the earth's surface is fixed in biomass (Leopoldina, 2012). The solar irradiance is approximately 140 W/m^2. Krausmann (2008) states that at least two third of the terrestrial surface of the earth is used by humans to produce biomass, yielding a vegetated surface of about 10 billion hectares. This implies that annually approximately 2.2 thousand EJ of the total of 440 thousand EJ of energy is stored in green vegetation. The figure of 2.2 thousand EJ corresponds with data from Haberl (2007) stating that the net amount of biomass annually assimilated by vegetation (i.e., the net primary production) is 118.4 Pg (Petagrams) per year (with an upper heating value of 18.3 MJ/kg this yields 2.2 thousand EJ/year). Note that the conversion into fossil fuels is even less efficient, as fossil fuels appear only after millions of years of putting biomass under high pressure. To improve energy conversion efficiency of the entire global system, it would be better to use solar energy directly for heating or electricity, or use the kinetic energy of water and wind, or even use non-solar resources such as tides and geothermal. All these resources are considered renewable energy sources, but a proper treatment of their potential lies outside the scope of this chapter. However, biomass can under specific circumstances be considered a sustainable resource, as we will discuss in Section 14.3.4. The major advantage of biomass over the renewable resources mentioned above is the fact that biomass (like fossil fuels) is an energy resource that can be stored for later use. The other renewable resources all require backup systems to guarantee energy supply at all times, even when there is no wind or sunlight available.

The 2012 report of Leopoldina 'Bioenergy: Chances and Limits' recommends setting efficiency in energy supply as a first priority, and using biomass initially as a resource for food, feed and materials, before deploying it as an energy source. Viable options for bio-energy would typically include biogenic waste such as manure. This line of reasoning corresponds with the 'Trias Biologica' and the cascading approach of the bio-based economy, which will be discussed in Section 14.4. First, we turn to the food versus energy controversy, illustrating with a simple calculation what this controversy might lead to.

14.3.2   *Food for thought: energy demand versus food demand*

Basically, food provides energy for the human body. Based on FAO Statistics for 2006–2008, current per capita food consumption averages 4.26 GJ per capita per year (2790 kcal per day), but consumption differs greatly per country (and per person). For instance, the average for developed countries is 5.24 GJ/capita/year (where the USA tops the list with 5.73 GJ/capita/year), while the least developed countries only consume 3.24 GJ/capita/year. Note that the actual food consumption may be lower than the quantities mentioned above, depending on the amount that is wasted during storage, preparation, cooking, and as leftovers.

The average energy demand per capita, however, proves to be much higher than food consumption. With world energy demand in 2010 at 533 EJ (IEA, 2012), and a world population of 6.9 billion people, the average annual per capita energy consumption is 77.29 GJ. In addition, the difference in energy consumption between developed and a developing country is even larger than for food. This is clearly illustrated by Nonhebel (2012) when plotting total primary energy use against GDP per capita for several countries (see Fig. 14.6). Nonhebel (2012) concludes that developed countries use 200–400 GJ of primary energy per person per year, nearly 5 times more than developing countries. As mentioned in Section 14.3.1, currently 81% of the primary energy demand is accounted for by fossil fuels.

This back-of-the-envelope calculation shows that overall, the need for energy exceeds food demand by far, especially in developed countries. Note that particularly these developed countries

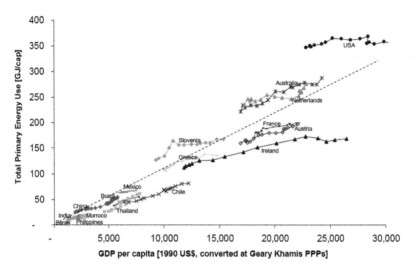

Figure 14.6.   Energy use per capita as a function of GDP per capita. Data are for the years 1990–2005. Taken from Nonhebel (2012).

have actively been promoting the substitution of fossil fuels for bio-energy in recent years, thereby creating a strong drive to use biomass for energy instead of food. This example also reinforces the assertion that de-carbonizing the economy (through efficiency improvements) should precede the substitution of fossil resources with sustainable biomass. Nonetheless, the question rises how much bio-carbon is (potentially) available for food and non-food applications. This is in fact the topic of the next section.

14.3.3   *The carbon balance: the theoretical potential for a bio-based economy*

This section explores the possibilities for substituting fossil carbon with sustainable carbon from biomass. As mentioned in Section 14.3.1, the world primary energy demand is currently 533 EJ (IEA, 2012a), with 81% coming from fossil fuels (coal, natural gas and oil). About half of the primary energy demand (275 EJ) is accounted for by the transformation of fossil fuel into transport fuels, chemicals and plastics and other materials, of which about 33 EJ is for non-energy use (IEA, 2012b). The other half of total primary energy demand is accounted for by the production of electricity and heat (see Fig. 14.7).

Krausmann *et al.* (2008) give a detailed description of global biomass flows in 2000, including the amounts of biomass harvested from cropland, grazing and forestry. They state that primary production of biomass from photosynthesis totals 118,434 Mton (or 2167 EJ, using a heating value of 18.3 MJ/kg dry matter). The total amount extracted and used is 12,139 Mton (222 EJ), while unused extraction equals 6558 Mton (120 EJ). So only two-third of harvested biomass is used after extraction, either directly for food, indirectly for animal feed, or for industrial- and energy purposes.

Krausmann *et al.* (2008) state that the share of total extracted primary biomass used for food and feed harvest (edible crop harvest and grazed biomass) amounts to 187 EJ/year. With an average annual per capita food consumption of 4.26 GJ, the global end-use of biomass for food is about 30 EJ. This means about 157 EJ is wasted in the food chain each year. The chain from primary use to net consumption of biomass is represented in Figure 14.8.

Combining the data of IEA and Krausmann *et al.*, we get an overview of the global carbon balance, with the demand for fossil and bio-based resources as presented in Table 14.4. Data on total biomass available in the global system are uncertain, so the figures in the table are estimates

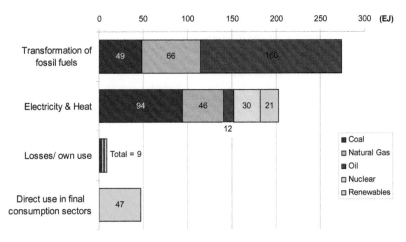

Figure 14.7. The supply side of the global energy system in 2010 (in EJ). Based on data from IEA (2012a). Transformation of fossil fuels is from primary energy into a form that can be used in the final consumption sectors.

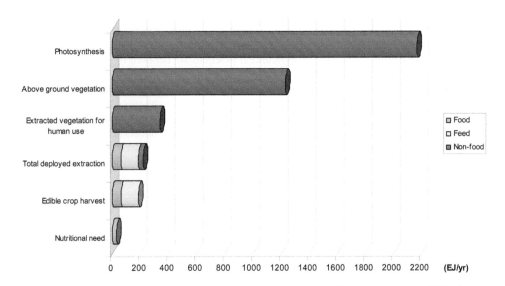

Figure 14.8. Annual energy content (in ExaJoules per year) of global biomass sources available through extraction for human utilization. Based on data from Krausmann *et al.* (2008) and Haberl (2007).

and should be used with caution. Nonetheless, their order of magnitude gives sufficient insight in the fossil and bio-based carbon flows in the world.

Indeed, Table 14.4 clearly shows that fossil fuels dominate the demand for carbon resources in our society. However, the biomass extraction seems in theory sufficient to cover the primary use of the bio-based carbon and a substantial part of the fossil based demand for carbon. Also, it is clear that a lot can be gained from improving efficiency in the chain between extraction and end use. This is indeed the philosophy underlying the cascading approach that is at the heart of the bio-based economy: bio-based harvest is deployed initially for high value components and food, while the residues are as much as possible used for the lower value applications such as chemicals

Table 14.4.   Estimates of annual world consumption of fossil and bio-based resources. Based on data from Krausmann *et al.* (2008), and IEA (2012).

World carbon based resource consumption	Fossil based primary use [EJ/year]	Fossil based end use [EJ/year]	Bio-based primary use [EJ/year]	Bio-based end use [EJ/year]
Total extracted	432		342	
Total deployed	426		222	
Edible crop harvest (in 2000)			187	30
Of which feed			*124*	
Electricity & heat	152	57	30	Na
Non-food application (fuels/chemicals/ plastics/materials)	275	235	6	Na
*Of which non-energy use**	*33*		*Na*	*Na*

EJ = ExaJoule = 10^{18} Joule. Fossil based use is based on IEA data for 2010 (IEA 2012a; 2012b). Edible crop harvest is based on data for 2000 taken from Krausmann *et al.* (2008).
*Non-energy use covers those fuels that are used as raw materials in the different sectors and are not consumed as a fuel or transformed into another fuel. Non-energy use also includes petrochemical feedstocks.

and energy. In addition, innovations in agricultural practices are believed to further improve the yield per hectare of existing agricultural land in the future.

The global average picture presented here is in line with studies of for instance the IPCC (2011), estimating a potential of 100–300 EJ for bio-energy. There seems to be enough biomass available to fulfill the world carbon demand, for now and in the future, provided that efficiency is improved and the economy is de-carbonized as much as possible. However, Nonhebel (2012) points out that a global average picture does not allow for the existing differences between various economies with respect to their needs and production possibilities, nor does it take into account the capacity of the global transport system (harbors, railways, roads). As Nonhebel puts it: 'the globe is able to supply the needs, but mankind is not able to distribute this production.' To make matters even more complex, the issue of sustainability of biomass production puts another constraint on the global average picture, and this issue is the topic of the next section.

14.3.4   *Sustainability of biomass*

Sustainable production and use of biomass is a prerequisite for the development of a bio-based economy. Clearly, the sustainability criteria should cover the non-food application of biomass in a way that does not interfere with food supply. However, as we have seen in the previous sections, this is not self-evident with current and projected developments in bio-energy. As Nonhebel (2012) mentions, the increased use of cereals as feedstock for biofuel was one of the reasons for world market food prices in 2008 to reach highest levels ever, causing among others food riots in 33 countries. Even though other causes were attributed to the rise of food prices (such as high oil prices, speculation, and bad weather), the food-energy controversy is an issue to be dealt with when dealing with sustainability of biomass. In 2005 the Gleneagles Plan of Action of the G8+5 initiated the launch of the Global Bioenergy Partnership to support wider, cost effective, biomass and biofuels deployment. One of the main aims of the Partnership is to favor the transformation of biomass use towards more efficient and sustainable practices, and to support national and regional bioenergy policy-making and market development (GBEP, 2013). The Global BioEnergy Partnership has developed a set of 24 sustainability criteria for bioenergy projects on social, environmental and economic themes. According to GBEP, bioenergy is sustainable only if its

Table 14.5. Table of the 24 sustainability criteria for bioenergy developed by the Global BioEnergy Partnership. Source: FAO (2011).

PILLARS
GBEP's work on sustainability indicators was developed under the following three pillars, noting interlinkages between them:

Environmental	Social	Economic

THEMES
GBEP considers the following themes relevant, and these guided the development of indicators under these pillars:

Environmental	Social	Economic
Greenhouse gas emissions, Productive capacity of the land and ecosystems, Air quality, Water availability, use efficiency and quality, Biological diversity, Land-use change, including indirect effects.	Price and supply of a national food basket, Access to land, water and other natural resources, Labor conditions, Rural and social development, Access to energy, Human health and safety.	Resource availability and use efficiencies in bioenergy production, conversion, distribution and end use, Economic development, Economic viability and competitiveness of bioenergy, Access to technology and technological capabilities, Energy security/Diversification of sources and supply, Energy security/Infrastructure and logistics for distribution and use.

INDICATORS:

1. Lifecycle GHG emissions 2. Soil quality 3. Harvest levels of wood resources 4. Emissions of non-GHG air pollutants, including air toxics 5. Water use and efficiency 6. Water quality 7. Biological diversity in the landscape 8. Land use and land-use change related to bioenergy feedstock production	9. Allocation and tenure of land for new bioenergy production 10. Price and supply of a national food basket 11. Change in income 12. Jobs in the bioenergy sector 13. Change in unpaid time spent by women and children collecting biomass 14. Bioenergy used to expand access to modern energy services 15. Change in mortality and burden of disease attributable to indoor smoke 16. Incidence of occupational injury, illness and fatalities	17. Productivity 18. Net energy balance 19. Gross value added 20. Change in consumption of fossil fuels and traditional use of biomass 21. Training and requalification of the workforce 22. Energy diversity 23. Infrastructure and logistics for distribution of bioenergy 24. Capacity and flexibility of use of bioenergy

entire production chain (feedstock production, refining and conversion) and end-use practices are sustainable. Table 14.5 summarizes the 24 criteria, allocated to their respective pillars and themes. Note that criteria no. 10 deals with the price and supply of a national food basket.

Based on the 24 sustainability indicators for bioenergy of GBEP, we can conclude that making the biomass supply chain sustainable includes at least the following elements:

• Improving yields in agriculture (with better crops, cultures, nursing, care)
• Nutrient recycling

- Optimal use of water (drip irrigation)
- Minimal pesticide (organic pest control)
- Minimal energy use in production chains.

These elements are actually very much in line with the Good Agricultural Practices (GAP) codes, standards and regulations promoted by the FAO (2013). The general principles for Good Agricultural Practices were first presented to the FAO Committee on Agriculture (COAG) in 2003. GAP was defined as 'practices that address environmental, economic and social sustainability for on-farm processes, and result in safe and quality food and non-food agricultural products'. The GAP principles are included in most private and public sector standards, although the FAO states that they vary in the scope that they cover. The objectives of GAP include:

- Ensuring safety and quality of produce in the food chain
- Capturing new market advantages by modifying supply chain governance
- Improving natural resources use, workers health and working conditions, and/or
- Creating new market opportunities for farmers and exporters in developing countries.

With the criteria and codes for sustainable use of biomass in place, the question arises how this can be put into practice to accomplish the transition to a bio-based economy. The cascading approach that is discussed in the next section provides a logical framework to do just that. Of course, the proof of the pudding is in the eating, so Section 14.5 deals exclusively with case studies on how the theory of cascading is currently implemented in practice.

14.4 A CASCADING APPROACH FOR SUSTAINABLE DEPLOYMENT OF BIOMASS AND THE TRIAS BIOLOGICA

Biomass is used in many different sectors for a variety of applications other than food. The traditional use of biomass, which is still widely used in developing countries, is to burn it in order to generate heat (used for cooking or space heating). Also, wood is used on a large scale as construction material in the building sector, and liquid biofuels are more and more used in the transport sector. The chemical sector uses biomass for making soap or fine chemicals such as cosmetics. And the paper sector is of course a bulk user of biomass. In a bio-based economy, the use of biomass has to be optimized and prioritized. The different biomass applications need to be ordered in an efficient way to allow for the competing claims on biomass to co-exist. The cascading approach as described by among others LNV (2007), IEA (2010a; 2010b), and CE (2012) provides practical ordering principles. Various definitions of cascading are used, but basically cascading is about making smart choices in applying a resource, *and* about thinking in supply chains. This is due to the fact that the choice for a specific application influences future possibilities. According to CE (2012), cascading can be applied with respect to time, value, or function. Optimal cascading makes use of all three types as much as possible:

- *Cascading in time* allows for a long(er) life span of biomass; preferably, the initial applications of biomass should leave open as many options as possible at the end-use stage. A typical example is paper recycling.
- *Cascading in value* can be used to optimize cascades in time. Choosing between alternatives based on their added values ensures the highest value added is generated over the entire life cycle. For instance, straw can be used for bio-energy production, but has higher added value when used for ethanol production (and subsequent production of plastics).
- *Cascading in function* is actually co-production i.e., the production of different functional streams from one biomass stream (e.g. protein, oil and an energy carrier), maximizing total functional use. Cascading in function can be achieved by using bio-refinery, such as the refinery of grass. For further optimization, cascading in function should be followed by cascading in value and/or time.

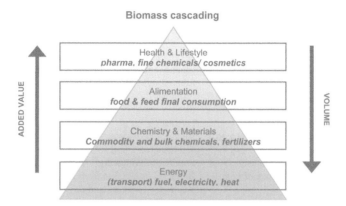

Figure 14.9. Value pyramid of biomass cascading from smaller supplies of high value applications to bulk supplies of low value products. Based on LNV (2007).

Using these optimizing principles, a logical order for using biomass appears. First, biomass should be applied for pharmaceuticals and fine chemicals such as cosmetics. Consequently, biomass is used for food and feed, in slightly more volumes and with slightly less added value. One step further down the cascade, biomass is applied in commodity and bulk chemicals and fertilizers, and at the bottom of the cascade, the low-value bulk biomass is used for biofuels and production of electricity and heat. This value pyramid of cascaded biomass is illustrated in Figure 14.9.

All applications use bio-based feedstock to produce a product. Basically, cascading implies the use of residues of one process as feedstock for the next process in the chain. In the cascade chain, we distinguish primary feedstock (harvest from agriculture, aquaculture or forestry), secondary feedstock (processing wastes) and tertiary feedstock (such as consumer wastes). Most of the dedicated feedstock is used at the top of the value pyramid. The different kinds of feedstock have different characteristics; they differ in concentrations of functional components such as nutrients, and the concentration of sugar, starch, oil or protein. In their study on cascading of biomass, CE (2012) presented 13 potential cases for a sustainable bio-based economy, which are listed in Table 14.6. The cascading alternatives mentioned in the table are the preferred applications of the biomass; the current system would benefit from a shift towards these cascading alternatives. According to CE, the selected cascading cases could result in greenhouse gas emissions reductions of 10–12% of the targeted emission reduction in the EU in 2030. However, they also state that current policies promoting bio-energy and emissions reduction actually hamper existing cascading approaches. For instance, recycling of paper for application in building materials is impeded by the alternative use of paper waste for energy production. The same holds for waste wood, which could be applied better in particleboard before using it to generate energy.

Cascading is not new: straw is for example a residue of the production of grains and oil seeds. Traditionally, farmers use straw for feed and beddings, and surplus of straw for soil improvement. And the conversion of sugar crops to produce sugar also creates a high-protein by-product which is typically used as feed. However, most of these applications stay in the food chain. Cascading in a bio-based economy requires cross-sector optimization, and thus cooperation, between the chains of different sectors such as the chemical industry, energy companies, agriculture, waste treatment industry and the food industry.

Basically, the transition towards a bio-based economy represents a development similar to the Trias Energetica. The Trias Energetica follows three logical steps: (i) reduce energy use with efficiency; (ii) use renewable energy as much as possible; (iii) use non-renewable resources as efficiently as possible). Expanding this philosophy to cross-sector chains in a bio-based economy

Table 14.6.   Selected biomass cascading options for a sustainable bio-based economy (CE, 2012).

Residues

1. Straw for ethanol production	*Current use*	Green manure *or* feed *or* bedding
	Cascade alternative	Ethanol *and* fertilizer-rich (N, P) lignin stream → Ethanol-derived products e.g. biofuel, ethylene
2. Bio-ethanol to chemistry	*Current feedstock*	Naphtha
	Cascade alternative	Straw
3. Manure for biogas production	*Current use*	Fertilizer
	Cascade alternative	Biogas *and* digestate/lignin-rich fraction as fertilizer/ soil enhancer
4. Waste for production of chemicals	*Current use*	Landfill/MSWI/energy
	Cascade alternative	Biodiesel production from fatty acids in residue streams (such as cooking oil and C1 slaughter-waste)
5. Grass refinery	*Current*	Feed
	Cascade alternative	Grass-refinery to produce protein, fiber, bio-energy, fertilizer
6. CO_2 as feedstock in greenhouses	*Currently*	Emission
	Cascade alternative	CO_2 capture at production plants (for e.g. biogas, bio-cokes and ethanol) combined to use in greenhouses or algae farms
7. CHP vs. small scale bio-energy production	*Currently*	Emission
	Cascade alternative	Use for district heating or industrial heating or fish farms

Woody biomass

8. Bio-cokes for chemistry	*Current*	Production with fossil cokes
	Cascade alternative	Production with bio-cokes made from waste-wood

Consumer waste

9. Electricity and heat from bio-waste	*Currently*	Landfill
	Cascade alternative	Combined heat and power production from MSWI
10. Recycling of bio-plastic	*Currently*	Composting by biological degradation
	Cascade alternative	Production of bio-products (e.g. bio-plastics) which can be recycled within the current infrastructure
11. Additional recycling of paper	*Currently*	Waste
	Cascade alternative	Recycling of waste paper

results in a 'Trias Biologica' for an integrated production system for food, energy and materials, which includes the following logical steps:

- De-carbonizing, reducing the carbon footprint
- Substituting fossil carbon with sustainable bio-carbon (produced in a sustainable way, i.e. respecting planetary boundaries)
- Cascading, making circular chains and applying bio-refinery.

How this way of thinking can be put into practice is shown with the help of case studies from The Netherlands, the topic of the next section.

14.5   CASE STUDIES OF CASCADING IN THE NETHERLANDS

14.5.1   *Facts and figures of The Netherlands*

According to Agentschap NL (2012), The Netherlands is the second-largest exporter of agri-food products in the world, surpassed only by the US (which has a surface area 296 times greater). More than 80% of exports are destined for Europe, primarily Germany and other neighboring countries. Also, production efficiency is high in The Netherlands: the added value per hectare is up to five times higher than the European average. This is partly the result of high standard research at the Dutch universities; two Dutch universities feature in the European top 10 with regard to the number of publications related to the agri-food sector. Table 14.7 contains more facts and figures of The Netherlands related to a bio-based economy. These figures are also used in the Dutch case studies in the following subsections.

14.5.2   *The Trias Biologica: the sugar case*

Sugar is a much used food product. In 2012, the world sugar production totaled 168 Mton, with a projected increase in demand of about 2% annually until 2020 (OECD/FAO, 2011). About 80% of the sugar is produced from sugar cane, while beet now represents 20% of the world's sugar production. To illustrate the transition towards a bio-based economy, the handling of sugar by Suiker Unie is an interesting case to illustrate the Trias Biologica. Suiker Unie is part of Royal Cosun, a cooperative of about 9000 Dutch sugar beet growers that deals with processing of agricultural raw materials into valuable products and components for other businesses. Suiker Unie is currently one of the five largest producers of beet sugar in Europe with a production of about 1 Mton of sugar in 2011 (Cosun, 2012a). Suiker Unie has found a way to allow for optimal plant utilization, improve efficiency along the production chain, and at the same time be competitive in an international market.

Traditionally, the production chain of beet sugar includes the sowing to harvest phase on agricultural land (March–September), which involves the use of fertilizers and herbicides. After the harvest, the yield of sugar beets is transported to the factory for processing, and the resulting products are subsequently transported to clients. These parts of the production chain all require raw material and fossil fuels, at least until now. During the past two decades, Suiker Unie has transformed the production chain in line with the Trias Biologica, in order to reduce the carbon footprint of their production chain, substitute fossil carbon with bio-based ones where possible, and use a cascading approach.

14.5.2.1   *De-carbonization*

The de-carbonization of the production chain of Suiker Unie between 1990 and 2010 largely took shape through an increase of sugar beet yield, a reduction in fertilizer use, and energy efficiency improvements in processing. The efforts in these fields have resulted in a relatively low carbon footprint of the production chain of Suiker Unie. Note that a high sugar content of the beet can also positively influence the carbon footprint. However, the sugar content largely depends on (uncontrollable) weather conditions. Although the sugar yield per hectare also depends on weather conditions, research efforts have improved yield per hectare considerably in The Netherlands, as Figure 14.10 shows for the years since 1950. Since the middle of the 1980s, the sugar yield per hectare has increased by 40–50%. In 2011, Suiker Unie (2012b) reported a sugar yield of 13.6 ton per hectare.

Especially since 2006, after a major reform of the EU sugar market, the research efforts on sugar yield took flight. Sugar production in the EU is subject to strong regulation with quota allowances, and the reform in 2006 led to a centralisation of production in The Netherlands, going from 28 small scale plants to presently 2 large-scale plants. Although the yield per hectare cannot increase infinitely, Suiker Unie expects to achieve a yield of on average 24 ton per hectare in The Netherlands (compared to an average yield of 32 ton of sugar per hectare in countries with considerably more sun-hours). The expected future yield results in a potential production capacity in the Neterhlands of 2–3 thousand ton of sugar per year.

Suiker Unie has also successfully managed to reduce the use of fertilizers with approximately 40%, as Figure 14.11 shows (Suiker Unie, 2012). This positively affects the carbon footprint, as the use of nitrous oxide soil as fertilizer accounts for a substantial share of the total carbon footprint.

Improvements in energy efficiency by Suiker Unie are supported by the voluntary agreements that industrial branch organizations have with the Dutch government since 1992 (Agentschap NL, 2013a). Suiker Unie reports in the 2011 sustainability report that since 1990, the energy

Table 14.7. Facts and figures of The Netherlands.

Geography

Population [mio, in 2012][a]	16.8
Total surface [km^2][a]	41543
Land area [mio ha][a]	3.4
Population density [inhab/ km^2][a]	494

Economy

Domestic product 2011 [in mio €][b]	555271
Share of agriculture in employment in 2010 [%][a]	10.2%
Share of agriculture (incl. wood, food) in added value in 2010[c]	10.0%
Average price of agricultural land in 2011 [€ per ha][c]	49000

Land use[d]	**Mio ha**	
Total surface	4.15	100%
Agriculture	2.30	55%
Grasslands	*1.00*	
Other agriculture	*1.30*	
Horticulture	*0.01*	
Forest and wildlife area	0.48	12%

Biomass Net Primary Production[e]	**Mton C/year**	**PJ**
Agriculture	14.7	539
Grasslands	*7.5*	*275*
Other agriculture	*7.2*	*264*
Horticulture	*na*	*na*
Forest and wildlife area	2.5	92
Total biomass	17.2	630

Agricultural production & consumption	**Mton/year**	
Imported soy in 2009	3554	
Manure production in 2011 [in Mton/year][f]	71	
Cattle		*75%*
Pigs		*17%*
Poultry		*2%*
Amount of manure exported (as share of transported amount)		27%
Material consumption biomass 2010 [in Mton][h]		50.6
Daily food consumption [MJ/capita/day]		13.6

Energy

Net domestic energy use in 2011 [PJ][h]	2.9
Energy intensity in 2011 [GJ/€][h]	6
Extraction natural gas in 2011 [in billion Sm^3][h]	79
Mineral reserves gas [in billion Sm^3][h]	1230
Energy consumption in agriculture 2011 [in PJ]	135.7

Waste & Emissions

CO_2 emissions 2011 [Mton $CO_{2\text{-eq}}$/year][g]	212
CO_2 emissions from livestock in 2009 [Mton $CO_{2\text{-eq}}$/year]	25
Excess P [kg P_2O_5/ha/year][f]	28
Excess N [kg/ha/year][f]	119

Sources:
[a] CBS Statline (2013).
[b] CBS (2011), gross domestic product based on gross, market prices, price level 2005.
[c] LEI (2012).
[d] Koppejan *et al.* (2009).
[e] NPP is the amount of biomass that grows in a year. It is generally given in units of grams (g) of carbon (C) per square meter (m^2) per year, with 1 gram C corresponding to about 2 gram biomass (dry matter). Based on production factors as averages for Europe, taken from Leopoldina (2012). The Higher Heating Value (*HHV*) of biomass (dry matter) is 18.3 MJ/kg.
[f] CBS (2012a). Note that 2.29 kg P_2O_5 is 1 kg P.
[g] CBS (2012d)
[h] CBS (2012e)
[i] Agentschap NL (2011).

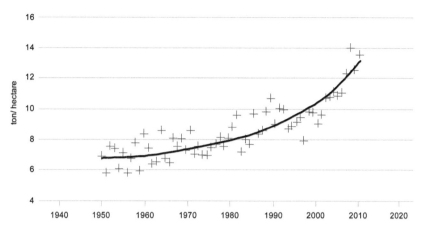

Figure 14.10. Development of sugar yield in The Netherlands (in ton/hectare). The variation over the years reflects the weather conditions during the growing season. Source: Cosun (2012b).

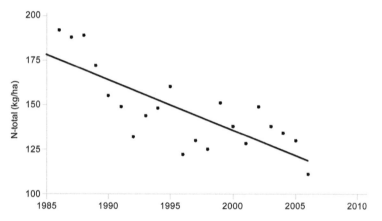

Figure 14.11. Developments in the of application of N-fertilizers for sugar beets by Suiker Unie between 1985 and 2010. Source: IRS Bietenstatistiek (2012).

efficiency has improved with about 50% (see Fig. 14.12), resulting in an energy consumption per ton of sugar of 937 kWh in 2010 (excluding thick juice processing).

The improvements in energy efficiency of the past decades contribute to a substantial decrease of the carbon footprint. The fact that Suiker Unie uses natural gas as an energy source is also favorable to a lower carbon footprint, as natural gas produces far less CO_2 emissions than oil or coal. The higher sugar yield is also beneficial to the carbon footprint, as does a reduction in fertilizer use. As a result of the de-carbonization efforts in the past decades, Suiker Unie (2012b) now reports a relatively small carbon footprint for their sugar production in The Netherlands: 480 kg CO_2 per ton of granulated sugar. The current carbon footprint is largely determined by the consumption of natural gas for energy production (43%) and use of nitrous oxide soil as fertilizer (23%), and to a lesser extent by transport fuels and the use of herbicides. The composition of the carbon footprint is shown in Figure 14.13.

14.5.2.2 *Substitution of fossil carbon with bio-based carbon*
After de-carbonization, the next step in the Trias Biologica stipulates the substitution of fossil carbon for bio-based carbon. Suiker Unie has in recent years taken several measures to do so,

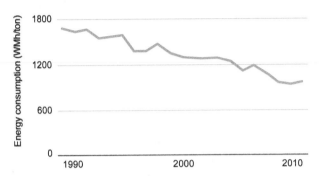

Figure 14.12.   Energy consumption (in MWh/ton) by Suiker Unie for processing sugar beets in The Netherlands between 1990 and 2010. The improvement in energy efficiency in 2010 relative to 1990
is about 50%. Source: Suiker Unie (2012a).

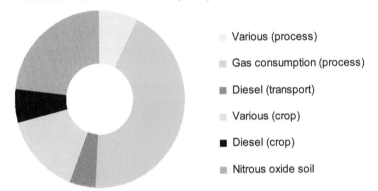

Figure 14.13.   Composition of the carbon footprint in $CO_{2\text{-eq}}$ per ton of granulated sugar produced by Suiker
Unie in 2010. The total carbon footprint equals 480 kg CO_2 per ton sugar. Source: Suiker
Unie (2012a).

in particular concerning the substitution of natural gas for biogas, as well as the substitution of
diesel for biofuels as a transport fuel.

The processing of beet results in residual flows such as beet tops and leaves. Since 2011,
Suiker Unie ferments the residual flows on site, with a biomass digester in Dinteloord and as
of 2012 also one in Vierverlaten. The fermentation process results in biogas, while preserving
all the valuable minerals in the residual material i.e., the digestate. The digestate can be sold to
farmers as a fertilizer, thus closing the mineral cycle and maintaining soil fertility. Most residues
originate from the company's own production processes, although suitable residual flows from
other businesses will also be used as a source. Suiker Unie has also identified opportunities to
ferment surplus beet pulp instead of drying it. Drying requires a great deal of energy whereas
fermentation produces energy.

Each biomass digester will process more than 100,000 tons of vegetable residues annually, and
produce approximately 10 million m³ of biogas. So the total production of biogas by the digesters
at full capacity will be more than 20 million m³ of biogas per year. The biogas is used for energy
production or applied in a liquefied form as a transportation fuel for Suiker Unie's vehicles that
run on biogas. These biogas fuelled vehicles emit 90% less CO_2 and NO_x than diesel vehicles,
and 95% less soot. Furthermore, the biogas fuelled vehicles reduce noise with 70%.

14.5.2.3   Cascading

Traditionally, the Dutch sugar industry only used the sugar as a commercial product, while the
beet pulp was used as animal feed. However, as a result of considerable research effort, today

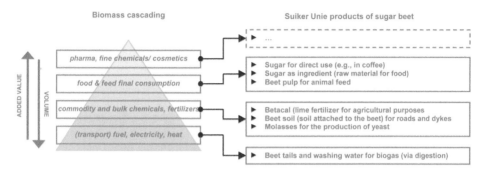

Figure 14.14.　Schematic overview of cascading principles in the sugar production of the Agro-Food-Cluster 'Nieuw Prinsenland' in Dinteloord. Source: Suiker Unie (2012c).

Suiker Unie uses as much parts of the beet as possible, using a cascading approach as illustrated in Figure 14.14.

About 8 years ago, Suiker Unie decided to go one step beyond. With partners, SuikerUnie started the development of Nieuw Prinsenland, a large-scale area dedicated to greenhouse horticulture combined with an industrial agro-food cluster in Dinteloord, The Netherlands. Nieuw Prinsenland can be regarded as a showcase in the transition towards a bio-based economy and the Trias Biologica. Although the site is still in a development phase, Suiker Unie has already published a first Sustainability Report 2011 of the site (Suiker Unie, 2012d). Underlying principle of the agro-food cluster is the shared use of materials and energy for all companies at the site. The development of Nieuw Prinsenland includes the following projects:

14.5.2.4　*De-carbonization*
- Combined heat and power (CHP) production by greenhouses. Many greenhouses use decentral combined heat and power (CHP) units. The CHP units are fired with natural gas or biogas to produce electricity, heat and CO_2. All three products are essential in greenhouse horticulture. The fact that both the heat and CO_2 from the CHP units are used, makes them more sustainable than conventional energy production. The excess electricity is fed to the national grid. *In operation.*
- A 150/20 kV transformer substation, to facilitate the supply of electricity to and from the site. The current electricity grid has not enough capacity to facilitate the local production and consumption of electricity by the greenhouses. Many greenhouses with CHP are net suppliers of electricity. The substation will increase the capacity of the local grid and make it more reliable, also enabling additional generation of electricity from wind turbines and PV solar systems. *Operational in 2013.*
- In total 7 wind turbines with a combined capacity of 21 MW (7 × 3 MW) will be placed on site to produce electricity that is fed to the national grid. *Under development.*

14.5.2.5　*Substitution*
- A biomass digester at the Dinteloord site is in operation since 2011, using beet tails and other residual waste (including washing water) to produce about 10 million m^3 of biogas annually. Part of the biogas is used as transportation fuel for Suiker Unie's vehicle fleet. The remainder of the biogas is fed into the national gas grid. *In operation.*
- A gas substation to transform the pressure in the gas distribution grid, which enables the biomass digester of Suiker Unie to supply biogas to the national gas grid. The construction of the substation uses bio-composite material. *In operation.*
- A CO_2 distribution grid to supply CO_2 to the greenhouse as fertilizer in the horticulture. The CO_2 comes from the biomass Suiker Unie's digester at the site (about 1500 ton of CO_2 per year) and other surrounding sources. *Under development.*

14.5.2.6   Cascading

- Beet soil is soil attached to the beet that enters Suiker Unie's processing plants. The soil is collected and re-used for infrastructure construction purposes on site. In total, Suiker Unie has used more than 300,000 m^3 of beet soil, thereby avoiding the energy and costs of transporting soil from elsewhere. *In operation.*
- Purified process water (effluent) of Suiker Unie is used as complementary water supply (approximately 300,000 m^3 of effluent per year, accounting for 15% of total water consumption at the site) for irrigation in greenhouses, to complement the insufficient rainwater supply. *In operation.*
- Residual heat of beet processing (53 MW$_{th}$) used for heating greenhouses. *Under development.*
- Separate sewage system for drain water of the greenhouses, with a purification system at the site that removes traces of herbicides from the drain water. *Under construction.*

The Nieuw Prinsenland project will provide Suiker Unie many additional opportunities for cascading, from bulk-chemistry (plastics) to fine chemicals and pharma-applications. An additional advantage is that Suiker Unie can use the distribution channels of Royal Cosun, which already supplies animal feed products and bio-based chemicals to the market.

14.5.3   Bio-refinery: the grass cascading case

Bio-refining is the sustainable processing of biomass into a range of marketable bio-based products and bio-energy (IEA, 2013). Conventional biomass conversion technologies applied in the production of for instance sugar, starch and paper already use certain bio-refinery steps. However, the dedicated bio-refineries make use of the full potential of biomass cascading by applying a process similar as the one used in oil refineries. This allows for improved integration and optimization of all the bio-refinery sub-systems. A bio-refinery can use various types of biomass, including wood and agricultural crops, forest residues, organic residues (both plant and animal derived), aquatic biomass (algae and sea weeds) and industrial wastes. The products derived from a bio-refinery can be intermediates or end-use products, and cover the entire range of the biomass cascade: from pharma and fine chemicals, food and feed, materials and bulk chemicals, to (transport) fuels, electricity and heat.

In The Netherlands, bio-refinery is seen as an essential step to realize the bio-based economy. In order to enhance this development, the Dutch government has supported various pilot projects (Agentschap NL, 2013b). These pilot projects differ in their approach to bio-refinery, including the enzymatic approach, the thermo-chemical approach, the treatment of waste streams (e.g., waste water), and the choice of feedstock, such as sugar beet or grass. A common product as grass is actually an interesting feedstock for bio-refinery. Traditionally, grass is used to feed cows in order to produce milk, meat (and dung). However, due to the milk quota in the EU, not all of the 1 million hectare of grassland in The Netherlands can be used to feed cows. This has spurred farmers to find other applications for grass. Compared to cows, bio-refineries can convert grass with much higher efficiencies into a much wider range of products, including: fibers, fertilizer, organic acids and biogas (see Fig. 14.15).

Fresh grass contains about 80–90% water. Especially in spring and wet summers, the water content can be very high, which makes some of the grass unsuited as feed. CE (2012) states that because of the high water content, about 7–15% of annual grass yield is lost. With 1 Mha fertilized grassland in The Netherlands, and an average of 7 metric tons of dry matter per hectare, this would result in a loss of about 0.5–1.0 Mton of grass each year. However, by pressing and milling fresh grass and consequently extracting the fibers, you get a juice not unlike potato juice. The grass juice contains proteins (15–20%), amino acids, organic acids (5%), sugars (27%), and minerals (10%) that can all be extracted from the grass through bio-refinery.

In The Netherlands, Grassa! has built a mobile unit as a pilot for grass refinery that is operational since September 2011 (Grassa, 2013). The unit has a capacity of processing 500 kg of grass per hour. It separates grass into fibers for the paper and pulp industry and proteins to replace (imported)

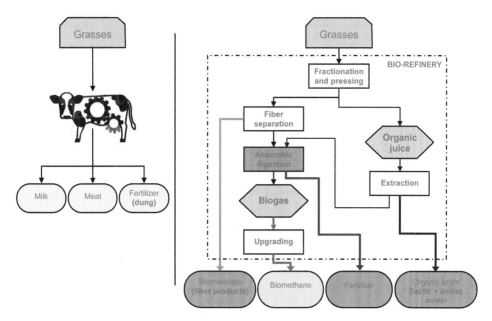

Figure 14.15.   Schematic overview of the conversion of grass by a cow (left) and a bio-refinery according to the IEA classification system (right, derived from Cherubini *et al.*, 2009).

soy as animal feed. Grassa! already has a viable business case on the basis of this fiber and protein extraction, in part because there is no need to transport grass to a central unit for processing. The profitability can be increased further by refining the grass juice (that otherwise would be returned to the land) in order to produce amino acids, organic acids and sugars. Also, application of the grass proteins in human food as a replacement of meat and dairy products is possible, similar to the use of soy proteins in soy drinks and rice proteins in rice 'milk'. In addition, the grass that is used as feedstock for the refinery can be supplemented with other bio-resources such as beet leaves, and carrot and tomato leaves in order to overcome the seasonality in availability of grass.

Grassa! is a cross-sector consortium of companies, including fodder companies, paper and pulp industry, and the construction and engineering industry. The consortium also cooperates with the chemical industry. This approach of cooperation between the chains of different sectors is indeed an essential first step to achieve optimized cascading, a prerequisite in the transition towards a bio-based economy.

Grass refinery potential

In order to determine the potential of grass refinery in Europe, CE (2012) presents the following case based on the additionality to the original function of grass as feed:

Assuming an availability of 30 Mton (dry matter) of surplus grass in the EU per year, grass refinery could potentially produce 6 Mton protein, 9 Mton fibre, 1 Mton fat, and 14 Mton sugars (Van Zijderveld, 2012). Assuming that the protein and part of the sugar replace feed; assuming the remaining sugars and fat are used to produce biogas; and assuming fibre is used as fuel in coal fired power plants, the avoided carbon emissions are estimated as follows:

- The high-protein concentrate substitutes soy as animal feed, part of the sugars are added to realize an appropriate VEM3-value. Based on protein content and VEM value, substitution of soy could avoid around 5.8 Mton $CO_{2\text{-eq}}$.

- The remaining sugars (7.8 Mton) and fat (1 Mton) are used to produce biogas, around 136 PJ/year. Substituting for natural gas, would result in avoided emission of 8.85 Mton $CO_{2\text{-eq}}$.
- Because market volume for cardboard filler which could be substituted for grass fibers is small, the fibers will more likely be used as fuel in coal fired power plants, with associated avoided emissions of 17 Mton CO_2.

These three applications together amount to a total CO_2 benefit of around 31.6 Mton per year. Benefits could be even larger if the grass protein is used to produce meat alternatives.

Source: CE (2012)

14.5.4   *Making circular chains: the manure case*

Animal manure is essential for soil fertility. Traditionally, agricultural practice was more or less circular: land was used to feed cattle, which produced manure that was used to fertilize the land. However, the livestock industry has developed into a highly intensive industry, with a large share of the livestock being kept indoors and a considerable amount of the animal feed and artificial fertilizer being imported. This has resulted in a more linear agricultural economy, as illustrated in Figure 14.16.

The vast amounts of manure that the livestock industry currently produces, have detrimental effects on the environment because of the excess amounts of nitrogen and phosphorous (and potassium) in soil and water. Data of CBS (2012a) show that, since the turn of the century, approximately 70 Mton of manure is produced each year in The Netherlands, with cattle accounting for more than 75% of the manure, pigs 17%, and poultry 2% (see Fig. 14.17). Note that these amounts are based on the wet manure streams and not on dry matter (the dry matter of cattle manure is significantly lower than the dry matter of chicken manure). The costs of handling the animal manure in The Netherlands are estimated at 300 million euro annually (LTO, 2011). About 60% of the handled manure is re-used in the Dutch agricultural sector, about 27% is exported and the remainder is processed (CBS, 2012b).

CBS (2012c) data for 2011 show that there is excess of nitrogen and phosphorous in The Netherlands of respectively 119 kg/ha and 28 kg P_2O_5/ha (2.29 kg P_2O_5 is 1 kg P). These surplus values amount to about one third of the annual supply of both nitrogen and phosphorous. Partly due to increasingly stringent regulation concerning mineral disposition (and the increasing scarcity of phosphorous), the livestock industry is looking for alternative ways to process manure, including:

- *Bio-refinery of manure* separates manures in different types of useful organic matter, fibers and minerals that can replace fertilizer. Also CO_2 can be used as fertilizing gas in horticulture. Expected to be profitable within 3–5 years.

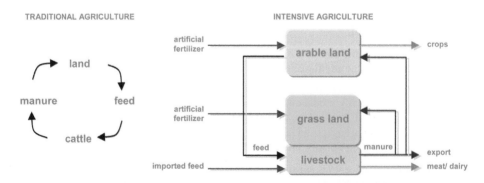

Figure 14.16.   Schematic overview of manure flows in traditional versus intensive agriculture.

- *Manure digesting* to produce biogas (a mixture of CO_2 and CH_4) and concentrate of minerals. The biogas can consequently be liquefied to LBG (liquefied biogas) to be used as transport fuel. Profitability is an issue.
- *(Jack screw-) press* to separate the fluid and thick fraction of manure. Usually most of the nitrogen and potassium is in the fluid fraction and most of the phosphorous is in the thick fraction.
- *Fertilizer replacement* with concentrates of minerals from animal manure to reduce fossil energy use (the production of (N-)fertilizers requires significant amounts of natural gas).
- *Algae and duckweed* production on manure-culture to produce proteins and to regain minerals. Not profitable yet, possible long-term solution.
- *Burning* or gasification. Only profitable for chicken manure ('stackable manure').
- *Reversed osmosis* to concentrate minerals and purifying water.
- *Drying and graining* of the thick fraction to produce phosphorous rich concentrates. Only viable when heat is locally available at no or low cost.
- *Evaporation* to concentrate minerals. Requires a lot of energy, only viable when heat is locally available at no or low cost.
- *Stripping* of ammonia to separate and recycle nitrogen. Very expensive.
- *Pasteurization or sterilization.* Requires a lot of energy and transportation.
- *Aerating* for nitrification and de-nitrification are common techniques in water purification. Associated with high emissions.

The viability of each alternative strongly depends on local circumstances, such as ownership of land, the presence of crop farmers, or the demand for fertilizing minerals. According to CE (2012), optimizing the cascade for manure in Europe would involve anaerobic digestion and subsequent separation of the residual digestate in a wet and a dry fraction. Digesting manure has the beneficial effect that the manure does not have to be stored, thereby avoiding methane emissions (and ammonia and N_2O emissions). In addition, digested manure can substitute fertilizers, even better than untreated manure, due to the fact that the nutrients in digested manure are better available to plants. The flow diagram of manure digestion is shown in Figure 14.18. Residual digestate separation allows for separately managing the nitrogen (liquid fraction) and phosphorus (solid fraction) present in the manure. During digestion decomposable organic components are converted into biogas, which can subsequently be used as feedstock (for e.g. methanol or ammonia production), natural gas substitution (by separating methane and injecting it in the natural gas grid), diesel substitution in automotive applications, and heat and/or power generation.

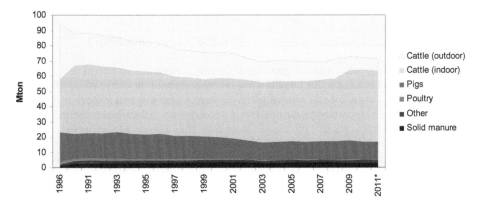

Figure 14.17.   Production of animal manure (in million ton) in The Netherlands between 1986 and 2011. Data of 2011 are preliminary. Based on CBS (2012a).

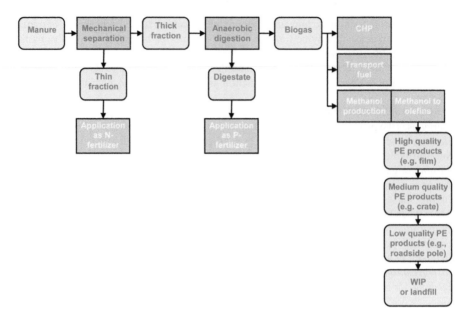

Figure 14.18.   Flow diagram of manure digestion.

A practical example of making the agricultural chain more circular is soon to be found in the village Vragender in The Netherlands (Agentschap NL, 2013b). Vragender suffers from a surplus of manure, which threatens the natural parks in the surroundings. Local dairy and pig farmers started an initiative in 2010 to install a digestion unit that converts manure into biogas and digestate. The biogas is upgraded to natural gas quality, while the digestate is further refined into concentrates with high levels of minerals such as phosphorous, nitrogen and potassium. These concentrates will be used as replacements of artificial fertilizer. This is expected to reduce energy demand for the production and transport of fertilizers by 10 TJ per year. The digestion unit is expected to produce 43.5 TJ of biogas annually, which is sufficient to meet local demand from households and small-scale industry. In addition, 13 TJ of biogas can be used as transport fuel for the farmers' trucks and tractors. In total, approximately 10 kiloton of CO_2 emissions are expected to be avoided each year.

14.6   DISCUSSION AND CONCLUSIONS ON IMPACT AND PROSPECTS

We have discussed the path towards a 'bio-based economy', an economic system that is far more than today based on resources that literally grow on the surface of the earth. We have identified several challenges, for instance the challenges on the supply side: we seem to touch sustainability boundaries already with current production levels of food, feed, materials and energy, while the population is expected to steadily increase to more than 10 billion people before the turn of the century. Can we produce even more? Or is the answer more subtle? There are also challenges on the demand side: the current demand for fossil carbon is far exceeding the total demand of carbon for food. Is it realistic to meet such a vast energy demand with bio-based resources? Or, again, is the answer more subtle? In addition, there are challenges of interaction and competition: there is a deep sense that the ability to feed mankind should not be compromised by whatever path is chosen towards a 'bio-based economy'. These challenges are real and must be faced. Yet, as we argue in this chapter, there are ways forward. In this section, we briefly recapitulate the line of reasoning that has emerged in this chapter, for a sound, responsible, feasible and innovative transition to a bio-based economy.

It is important to ascertain that our economy is already largely based on biological resources, most of which are however not renewable in any practical sense. The fossil stock that was formed in countless eons is now spent in a matter of decades. It is therefore perhaps more precise to say that the current economy is 'carbon based'. Carbon feeds and clothes us, and meet much of our needs in terms of material and energy. Although exact numbers are difficult to obtain, our data search indicates that currently about 70% of the carbon in the world economy are fossil based, i.e., they are derived from oil, natural gas or coal. About 90% of the carbon are directly or indirectly (i.e. via the intermediate step of refining) converted into energy – mainly into power, into heat, and via transport fuels into kinetic energy. This not only results in an irrecoverable loss of resources, it also results in a steady rise in atmospheric CO_2. This in turn will derail the earth's climate and acidify the oceans, to name but two long-term worldwide environmental effects. The amount of fossil resources used today (and expected in the future) makes it unrealistic to imply a straightforward substitution of fossil carbon with short cycle bio-based carbon, even if that would be technically feasible. Our data indicate that this would require the agricultural production to treble, while current agricultural production and consumption patterns already strain the natural boundaries of system earth. Issues such as bio-diversity, the nitrogen cycle, soil fertility, land use, and wastes in the food chain have been touched upon in this chapter, and stress the complexity of the problem at hand.

Against this background, the transition towards a bio-based economy contains three logical but essential components, presented in this chapter as the Trias Biologica. First of all, a *de-carbonization* of the economy is necessary. Secondly, fossil carbon need to be substituted with sustainably produced bio-based carbon. And thirdly, cascading of the biomass is required. De-carbonization as a first step is essential, to align society's carbon demand with the amount of bio-based carbon that can be harvested, while keeping well within the planetary boundaries. De-carbonizing is mainly targeted at the energy system, which has a high carbon intensity, while de-carbonizing food is impossible, and materials make up only a fraction of the total consumption. De-carbonizing the energy system starts with minimizing energy demand through efficiency improvements. Subsequently, fossil resources should be substituted with non-carbon renewable resources such as hydro, wind, solar, geothermal energy and tidal energy (and in some views including nuclear energy). Given the vast availability of non-carbon energy resources, theoretically there is no reason why this could not be done. Clearly, however, this will not happen overnight, as the challenges are great. Even in the long run fossil carbon will still be part of the energy system, for instance in transport. Once the demand for fossil carbon has been brought down well below the level of bio-based carbon that can be sustainably harvested, it becomes possible to substitute the entire amount of fossil carbon with the sustainable bio-based carbon without interfering in the food supply for a growing population. This *substitution* is the second component of the Trias Biologica. It is essential that this bio-based carbon is produced and consumed in a sustainable manner, in order to stay well within the planetary boundaries. Although an in-depth treatment of sustainable agriculture falls outside the scope of this chapter, it is obviously crucial for a sustainable future for mankind. The third component of the Trias Biologica is *cascading*, an essential element in sustainable agriculture, in order to optimize the use of bio-based resources with bio-refinery and circular process chains. Biomass cascading implies the application of high value bio-based carbon before deploying lower ones, and using only the remaining bulk residual to produce energy. This approach involves bio-refinery to extract minerals and other useful products in waste streams, making optimum use of every next step in the refinery process. Cascading also implies production chains being made more circular, by reusing minerals at the place of extraction as much as possible to avoid local depletion or eutrophication of the soil. The second and third component of the Trias Biologica are strongly related, and therefore should not be seen as sequential, because the options for substituting fossil carbon with bio-based ones largely depend on bio-cascades and the effective use of waste streams.

The Trias Biologica and the cascading approach are not mere theoretical concepts. In this chapter we have highlighted Dutch cases as examples pointing towards a sustainable future. The example of grass refinery, for instance, makes clear that very high value products can be extracted from a source as common as grass. These products include fibers and proteins (which already make a viable case), and can easily be expanded to amino acids, fertilizer minerals and –by digesting part of the intermediate products– also bio-methane and ethanol. The sugar case shows how the 'product portfolio' from sugar beets has dramatically expanded since the eighties. Initially producing only sugar and animal feed (beet pulp), the range of products today has expanded extensively beyond sugar and feed to very high value products such as pharma and cosmetics, as well as products further down the cascading pyramid such as biofuels and biogas. The Dutch case of manure digestion illustrates a practical way to make production chains more circular.

Innovations such as the ones presented in this chapter are, in our view, absolutely essential in the move towards a bio-based economy in which bio-based carbon provide enough food, materials and energy without exceeding the planetary boundaries, thereby providing a sustainable future.

REFERENCES

Agentschap NL: Energie- en klimaatmonitor Agrosectoren 2011. 2011, http://www.agentschapnl.nl/content/energie-en-klimaatmonitor-agrosectoren-2011 (accessed March 2013).

Agentschap NL: Made in Holland: Agri-Food. Bits & bites. 2012, http://www.hollandtrade.com/publications/made-in-holland/ (accessed March, 2013).

Agentschap NL: Meerjarenafspraken energie-efficiency. 2013a, http://www.agentschapnl.nl/nl/programmas-regelingen/meerjarenafspraken-energie-efficiency (accessed March 2013).

Agentschap NL: Energie-inovatiecatalogus (search phrase 'Bioraffinage'). 2013b, http://www.agentschapnl.nl/programmas-regelingen/energie-innovatiecatalogus-0 (accessed March, 2013).

Aiking, H., de Boer, J. & Vereijken, J.M.: Sustainable protein production and consumption: pigs or peas? *Environment & Policy*, Vol. 45, Springer, Dordrecht, The Netherlands, 2006.

Alexandratos, N.: World food and agriculture to 2030/2050: Highlights and views from mid 2009. Expert meeting on How to feed the world in 2050 (12–13 October 2009), 2009, http://www.fao.org/wsfs/forum2050/wsfs-background-documents/wsfs-expert-papers/en/ (accessed June 2013).

Alexandratos, N. & Bruinsma, J.: World agriculture towards 2030/2050: the 2012 revision. ESA Working paper No. 12-03, Food and Agriculture Organization of the United Nations, Rome, Italy, 2012.

Bos, H.L., Slingerland, M., Elbersen, W. & Rabbinge, R.: Beyond agrification: twenty five years of policy and innovation for non-food application of renewable resources in The Netherlands. *Biofuels Bioprod. Bioref.* 2:4 (2008), pp. 343–357.

Bruinsma, J.: World agriculture: Towards 2015/2030, An FAO perspective. Food and Agriculture Organization of the United Nations, Rome, Italy, 2002a, http://www.fao.org/fileadmin/user_upload/esag/docs/y4252e.pdf (accessed December 2012).

Bruinsma, J.: The resource outlook to 2050: by how much do land, water and crop yields need to increase by 2050? FAO expert meeting 'How to Feed the World in 2050', Food and Agriculture Organization of the United Nations, Rome, Italy, 2002b, http: www.fao.org/docrep/012/ak542e/ak542e00.htm (accessed December 2012).

CBS: Mestproductie door de veestapel, 1986–2011 (indicator 0104, version 13, 27 maart 2012). Compendium voor de Leefomgeving. 2012a, http://www.compendiumvoordeleefomgeving.nl/indicatoren/nl0104-Mestproductie-door-de-veestapel.html?i=3–17 (accessed March, 2013).

CBS: Transport en verwerking van mest, 2000–2010 (indicator 0403, version 13, 20 april 2012). Compendium voor de Leefomgeving. 2012b, http://www.compendiumvoordeleefomgeving.nl/indicatoren/nl0403-Transport-en-verwerking-van-mest.html?i=11–60 (accessed March, 2013).

CBS: Stikstof – en fosfaatbalans voor landbouwgrond, 1980–2011. Compendium voor de Leefomgeving. 2012c, http://www.compendiumvoordeleefomgeving.nl/indicatoren/nl0093-Stikstof–en-fosforbalans.html?i=11–60 (accessed March, 2013).

CBS: Environmental accounts of The Netherlands: Environmental accounts of The Netherlands Greenhouse gas emissions by Dutch economic activities 2012, The Hague /Heerlen, The Netherlands, 2012d, http://www.cbs.nl/NR/rdonlyres/BC7B85C6-1E78-4A22-9DC6-613E3E0A28EF/0/2012environmental-accountsgreenhousegasemissionspub.pdf (accessed March 2013).

CBS: Environmental accounts of The Netherlands 2011. The Hague /Heerlen, The Netherlands, 2012e, http://www.cbs.nl/NR/rdonlyres/3F5F2C12-CB59-4C59-AE1A-FD46AF6D4DAD/0/2011c174pub.pdf (accessed March 2013).

CBS: Land- en tuinbouwcijfers 2012. 2012f, http://www.cbs.nl/nl-NL/menu/themas/landbouw/publicaties/ publicaties/archief/2012/2012-land-en-tuinbouwcijfers-2012-pub.htm (accessed March 2013).

CBS: Landbouw in Vogelvlucht. 2013, http://www.cbs.nl/nl-NL/menu/themas/landbouw/publicaties/ landbouw-vogelvlucht/default.htm (accessed Feb 2013).

CE: Cascading of biomass: 13 solutions for a sustainable bio-based economy. Making better choices for use of biomass residues, by-products and wastes. 2012, http://www.ce.nl/publicatie/cascading_of_biomass% 3Cbr%3E13_solutions_for_a_sustainable_bio-based_economy7/1276 (accessed February 2013).

Cherubini, F., Jungmeier, G., Wellisch, M., Willke, T., Skiadas, I., Van Ree, R. & de Jong, E.: Toward a common classification approach for biorefinery systems. In: Biofuels, Bioproducts & Biorefining (Biofpr, 2009), 2009, http://www.iea-bioenergy.task42-biorefineries.com/publications/papers/ (accessed March, 2013).

Chum, H., Faaij, A., Moreira, J., Berndes, G., Dhamija, P., Dong, H., Gabrielle, B., Goss Eng, A., Lucht, W., Mapako, M., Masera Cerutti, O., McIntyre, T., Minowa, T. & Pingoud, K.: Bioenergy. In: O. Edenhofer, R. Pichs-Madruga, Y. Sokona, K. Seyboth, P. Matschoss, S. Kadner, T. Zwickel, P. Eickemeier, G. Hansen, S. Schlömer & C. von Stechow (eds): IPCC Special Report on Renewable Energy Sources and Climate Change Mitigation, Cambridge University Press, Cambridge, United Kingdom and New York, NY, USA, 2011, http://srren.ipcc-wg3.de/report (Chapter 2 – Bioenergy) (accessed July 2013).

Cosun: Annual Report 2011. 2012a, http://www.cosun-jaarverslag.nl/index.php?language_id=2 (accessed February 2013).

Cosun: Sustainability Report 2011. 2012b, http://www.cosun-jaarverslag.nl/index.php?language_ id=2 (accessed February 2013).

Crews, T.E. & Peoples, M.B.: Legume versus fertilizer sources of nitrogen: Ecological tradeoffs and human needs. *Agricul. Ecosyst. Environ. Change* 19 (2004), pp. 292–305.

EC: Innovating for sustainable growth: a bioeconomy for Europe. COM(2012) 60 final, 2012, http://ec.europa.eu/research/bioeconomy/pdf/201202_innovating_sustainable_growth.pdf (accessed January 2012).

Erisman, J.W., van Grinsven, H., Leip, A., Mosier, A. & Bleeker, A.: Nitrogen and biofuels; an overview of the current state of knowledge. 2010, http://dare2.ubvu.vu.nl/bitstream/handle/1871/34001/249919.pdf? sequence =1 (accessed December 2012).

Eurostat: Energy statistics, Imports (by country of origin). EU27 net imports of fossil fuels (natural gas, oil, and solid fuels) as a % of fuel-specific gross inland energy consumption. 2012, http://www.eea.europa.eu/data-and-maps/indicators/net-energy-import-dependency/net-energy-import-dependency-assessment-2 (accessed June 2013).

FAO: Energy and protein requirements. 1991, http://www.fao.org/docrep/003/AA040E/AA040E00.htm# TOC (accessed December 2013).

FAO: Food security and agricultural mitigation in developing countries: options for capturing syner-gies. Food and Agriculture Organization of the United Nations, Rome, Italy, 2009, http://www.fao. org/docrep/012/i1318e/i1318e00.pdf (accessed December 2013).

FAO: The global bioenergy partnership sustainability indicators for bioenergy – First edition. 2011, http://www.globalbioenergy.org/fileadmin/user_upload/gbep/docs/Indicators/The_GBEP_Sustainability_ Indicators_for_Bioenergy_FINAL.pdf (accessed December 2012).

FAO: Website Good Agricultural Practices. 2013, http://www.fao.org/prods/gap/ (accessed February 2013).

FAOSTAT: Statistical database of the Food and Agriculture Organization (FAO) of the United Nations, Rome, Italy, 2013, http://faostat3.fao.org/home/index.html#VISUALIZE_BY_DOMAIN (accessed March 2013).

Galloway, J.N., Townsend, A.R., Erisman, J.W., Bekunda, M., Cai, Z., Freney, J.R., Martinelli, L.A., Seitzinger, S.P. & Sutton, M.A.: Transformation of the nitrogen cycle: recent trends, questions and potential solutions. *Science* 320 (2008), pp. 889–892.

GBEP: Global BioEnergy Partnership Website: Purpose and Functions. 2013, http://www.globalbioenergy. org/aboutgbep/purpose0/en/ (accessed February 2013).

Grassa!: Grassa! website. 2013, http://www.grassanederland.nl/ (accessed March 2013).

Haberl, H., Erb, K.H., Krausmann, F., Gaube, V., Bondeau, A., Plutzar, C., Gingrich, S., Lucht, W. & Fischer-Kowalski, M.: Quantifying and mapping the human appropriation of net primary production in earth's terrestrial ecosystems. *Proceedings of the National Academy of Sciences (PNAS)* vol. 104 – no. 31

(July 31, 2007), 2007, pp. 12,942–12,947, http://www.pnas.org/content/104/31/12942 (accessed December 2012).

IEA: Better use of biomass for energy. Position Paper by IEA RETD and IEA Bioenergy, 2010a, http://www.ieabioenergy.com/LibItem.aspx?id=6476 (accessed February 2013).

IEA: BUBE: Better use of biomass for energy. Background Report to the Position Paper of IEA RETD and IEA Bioenergy, 2010b, http://www.ieabioenergy.com/LibItem.aspx?id=6476 (accessed February 2013).

IEA: World Energy Outlook 2012. 2012a, http://www.worldenergyoutlook.org (accessed January 2013).

IEA: Key World Energy Statistics 2012. 2012b, http://www.iea.org/publications/freepublications/publication/kwes.pdf (accessed May 2013).

IEA: IEA Bioenergy, Task 42: Biorefinery. 2013, http://www.iea-bioenergy.task42-biorefineries.com/ (accessed March 2013).

Koppejan, J., Elbersen, W., Meeusen, M. & Bindraban, P.: Beschikbaarheid van Nederlandse biomassa voor electriciteit en warmte in 2020. 2009, http://library.wur.nl/WebQuery/clc/1929592 (accessed March 2013).

Kovarik, B.: Henry Ford, Charles Kettering and the fuel of the future. *Automotive History Review* 32 (1998), pp. 7–27, http://www.environmentalhistory.org/billkovarik/research/henry-ford-charles-kettering-and-the-fuel-of-the-future/ (accessed June 2013).

Krausmann, F., Erb, K.-H., Gingrich, S., Lauk, C. & Haberl, H.: Global patterns of socioeconomic biomass flows in the year 2000: a comprehensive assessment of supply, consumption and constraints. *Ecol. Econ.* 65:3 (2008), pp. 471–487, http://www.uni-klu.ac.at/socec/downloads/ 2008_KrausmannErbGingrich2008_globalPatterns_EE_65_15.pdf (accesssed Jan 2012).

LEI: Lanbouw Economisch Bericht. 2012, http://edepot.wur.nl/213938 (accessed March 2013).

Leopoldina-Nationale Akademie der Wissenschaften: Bioenergy: chances and limits. Halle (Saale), Germany, 2012, http://www.leopoldina.org/uploads/tx_leopublication/201207_Stellungnahme_Bioenergie_LAY_en_final.pdf (accessed January 2013).

LNV: Overheidsvisie op de bio-based economy in de energietransitie: de keten sluiten. Ministerie van Landbouw, Natuur en Voedselkwaliteit (LNV), Den Haag, The Netherlands, 2007, http://www.groenegrondstoffen.nl/downloads/Overheidsvisie%20op%20de%20Bio-based%20Economy%20in%20de%20energietransitie.pdf (accessed February 2013).

LTO: Achtergrondrapport Integrale visie duurzame drijfmestverwaarding: Visie van LTO Nederland, 2011, http://www.lto.nl/media/default.aspx/emma/org/10760430/achtergrondrapport+integrale+visie+duurzame+drijfmestverwaarding+visie+van+lto+nederland+augustus+2011.pdf (accessed March 2013).

Lundqvist, J., de Fraiture, C. & Molden, D.: Saving water: from field to fork – Curbing losses and wastage in the food chain. SIWI Policy Brief, SIWI, 2008.

Nonhebel, S.: Global food supply and the impacts of increased use of biofuels. *Energy* 37 (2012), pp. 115–121.

PBL (Planbureau voor de Leefomgeving): Balans van de Leefomgeving 2012: Stikstofverlies per Nederlander bijna gehalveerd, 2012, http://themasites.pbl.nl/balansvandeleefomgeving/2012/integraal-stikstof/trend-stikstofintensiteit-nederlandse-economie%2c-eu-en-wereld (accessed May 2013).

Rockström, J., Steffen, W., Noone, K., Persson, Å., Chapin III F.S., Lambin, E., Lenton, T.M., Scheffer, M., Folke, C., Schellnhuber, H., Nykvist, B., De Wit, C.A., Hughes, T., van der Leeuw, S., Rodhe, H., Sörlin, S., Snyder, P.K., Costanza, R., Svedin, U., Falkenmark, M., Karlberg, L., Corell, R.W., Fabry, V.J., Hansen, J., Walker, B., Liverman, D., Richardson, K., Crutzen, P. & Foley, J.: A safe operating space for humanity. *Nature* 461 (2009a), pp. 472–475.

Rockström J., Steffen, W., Noone, K., Persson, Å., Chapin III, F.S., Lambin, E., Lenton, T.M., Scheffer, M., Folke, C., Schellnhuber, H., Nykvist, B., De Wit, C.A., Hughes, T., van der Leeuw, S., Rodhe, H., Sörlin, S., Snyder, P.K., Costanza, R., Svedin, U., Falkenmark, M., Karlberg, L., Corell, R.W., Fabry, V.J., Hansen, J., Walker, B., Liverman, D., Richardson, K., Crutzen, P. & Foley, J.: Planetary boundaries: exploring the safe operating space for humanity. *Ecol. Society* 14:2 (2009b), article 32 [online], http://www.ecologyandsociety.org/vol14/iss2/art32/ (accessed January 2013).

Smil, V.: *Feeding the world: a challenge for the twenty-first century.* MIT Press, Cambridge, MA, 2000.

Smil, V.: *Enriching the Earth: Fritz Haber, Carl Bosch and the transformation of world food production.* MIT Press, Cambridge, MA, 2001.

Smil, V.: Nitrogen and food production: proteins for human diets. *Ambio* 31 (2002a), pp. 126–131.

Smil, V.: *The Earth's biosphere: evolution, dynamics and change.* MIT Press, Cambridge, MA, 2002b.

Suiker Unie: Naturally Sustainable, Sustainability Report 2011. 2012a, http://www.suikerunie.nl/Sustainability-report.aspx (accessed March 2013).

Suiker Unie: Sustainability website. 2012b, http://www.suikerunie.nl/Sustainability.aspx (accessed March 2013).

Suiker Unie: About Suiker Unie > Products. 2012c, http://www.suikerunie.nl/Products.aspx (accessed March 2013).

Suiker Unie: Duurzaamheidverslag Nieuw Prinsenland 2011. Agro & Food cluster Nieuw Prinsenland: knooppunt voor de groene economie, 2012d, http://www.nieuwprinsenland.nl/static/files/documenten/120507_Duurzaamheidverslag_NP_2011_vastgesteld_def.pdf (accessed March 2013).

Townsend, A.R. & Howarth, R.W.: Fixing the global nitrogen problem. *Scientific American* 302 (2010), pp. 64–71.

Troeh, F.R. & Thompson, L.M.: *Soils and soil fertility*. Blackwell Publishing, Oxford, 2005.

UN: The World at six billion. 1999, http://www.un.org/esa/population/publications/sixbillion/sixbilpart1.pdf (accessed May 2013).

UN Internet database: World Population 1950–2100. Medium Variant, http://esa.un.org/wpp/unpp/panel_population.htm (accessed January 2013).
Source: Population Division of the Department of Economic and Social Affairs of the United Nations Secretariat, *World Population Prospects: The 2010 Revision*, http://esa.un.org/unpd/wpp/index.htm (accessed June 2013).

USDA Foreign Agricultural Service: EU Protein Deficiency. Gain report no. E60050, 2001, http://gain.fas.usda.gov/Recent%20GAIN%20Publications/EU%20Protein%20Deficiency_Brussels%20USEU_EU-27_2011-08-30.pdf (accessed January 2013).

WHO: Obesity and overweight. Factsheet nr. 311. World Health Organization, Geneva, Switzerland, 2006.

WWF Denmark: Industrial biotechnology: more than green fuel in a dirty economy? Exploring the transformational potential of industrial biotechnology on the way to a green economy. 2009, http://www.bio-economy.net/reports/files/wwf_biotech.pdf (accessed January 2013).

Section 4
Access to energy

CHAPTER 15

Increasing energy access in rural areas of developing countries

Xavier Lemaire

15.1 INTRODUCTION

Increasing energy access is recognized as an absolute priority for the development of rural areas. For instance, electrification can be achieved via the extension of the grid or increasingly via decentralized generation, more and more frequently with a mix of technologies including renewable energy technologies (RETs). In most cases, public-private partnerships can greatly accelerate the rate of electrification and energy access (Kammen and Kirubi, 2008).

Rural energy services companies can maintain energy systems and guarantee the long-term delivery of energy services, which is of particular importance when up-front cost investments are high, like in the case of most RETs. However, private partners need stable and consistent policies and regulations. Rural energy service companies can deliver electricity, but also a range of energy services and complement the action of conventional utilities. They can be used to promote individual systems for rural inhabitants, but also small or medium size (mini-grid) systems suitable for agricultural exploitation.

This chapter focuses on the way to deliver energy services in rural ares of developing countries, where 0.5 billion small holdings are supporting over 2 billion people (International Fund for Agricultural Development). It deals with the policies, regulation and business models needed for the large-scale implementation of now mature small-scale electricity generating RETs; it will also very briefly introduce energy generation as by-product of farming activities; it provides cases of successful large-scale dissemination of decentralized renewable energy technologies, emphasizing notably the opportunity created by the decrease in the cost of RETs, in particular solar photovoltaic, and more generally the lessons learned in terms of (decentralized) rural electrification, but also rural energization in developing countries.

15.1.1 *The current situation of energy access in developing countries and the opportunity offered by the RETs*

15.1.1.1 *Contrasting situation across continents*
Access to electricity can vary considerably across continents. Latin American countries have reached a high level of rural electrification. Now fifty years after declarations of independence in the majority of the region's countries, grid electrification reaches less than 5% of the rural population of sub-Saharan Africa. Overall, the rate of electrification of sub-Saharan Africa is the lowest of the world. A few countries in Africa like northern African countries and South Africa have achieved high electrification rates (70–100%), while the majority still have rates of electrification below 50%, if not 20% (REEGLE/SERN, 2013). Most sub-Saharan African countries have a very low population density and to electrify the majority of the African rural population requires dramatic increases in the use of stand-alone and mini-grid systems for the most scattered inhabitants. In some South East Asian countries the particulars of geography or the failure of policy-makers to provide reliable on-grid electricity also make decentralized generation desirable. Over 1.3 billion people around the world are still without electricity. Furthermore, the number of people without access to modern sources of fuel for cooking – more than three billion – is also increasing (UNDP – WHO, 2009).

15.1.1.2   *The rationale for decentralized generation with RETs*

Energy needs in rural areas of developing countries are often smaller than in developed countries. Giving farmers in developing countries access even to small supplies of electricity to charge a mobile phone (which is now a strategic tool for them to get information and sell their crops) can make a huge difference. Larger supplies can power devices related to farming activity, such as water pumping; other RETs like the use of passive solar thermal products can be used also for cold/dry chain for preservation/processing of crops.

Inhabitants of rural areas expect more and more to get a minimum level of comfort and basic amenities in their house, like lighting without having to use candles and paraffin, or power for a TV or radio (Gustavsson and Ellegård, 2004). An increasing number of rural inhabitants with an income are not willing to wait anymore for the grid to be extended, and will purchase these devices with (or sometimes without) credit.

The belief that grid extension could reach progressively rural areas may have, for a long time, impeded the development of distributed generation. However, this belief has in a lot of countries been dispelled. The absence of electrical infrastructure outside urban areas and the fast decreasing cost of now mature RETs on the market could create an opportunity for countries with low electrification rates to leapfrog to extensive distributed generation with RETs (Moner-Girona *et al.*, 2006). Distributed generation in developing countries is often quite different from that in European countries. In the latter, connected local systems are complementary to the grid; in developing countries, with for some of them, low rural population density, stand-alone systems and mini-grids are, and probably will remain for a long time, the only source of electricity for isolated inhabitants.

The cost of RETs may still be perceived as high, but costs have been decreasing recently making them in more and more instances competitive with stand-alone conventional generation; when a socio-economic lifecycle comparison is made, costs of RETs-generated energy services can be now in rural areas lower than with traditional or conventional energies. This is even more the case when the local environmental and social benefits of RETs are taken into account, as RETs generate less pollution and contribute to the creation of more local jobs than conventional energies. Actually, in some remote areas, life-cycle costs of even the most expensive RETs, like photovoltaics (PV), have been lower than stand-alone diesel generation, for instance for water pumping for a number of years now (SELF, 2008). Furthermore, RETs allow for more flexibility than the extension of the grid, as systems can be installed and upgraded according to the needs of users. RETs are attractive because their output matches the low electricity demand levels in rural areas, and enable the avoidance of costly distribution and transmission networks. They also avoid disruptions linked to the unreliable transport of fossil fuel and access to parts in remote areas.

Overall, in spite of high initial investment costs, RETs help to reduce dependency on costly fuel imports. Each country can specialize and develop its energy mix according to its available renewable energy sources. Even if not every country is equally provided with renewable energy sources in terms of wind or hydropower, the solar productivity of photovoltaics is often high in developing countries.

15.1.1.3   *How to deliver energy services to remote places, and what services to deliver?*

Several known financial delivery mechanisms like micro-credit or fee-for service can be set up to increase energy access (see Section 15.2.3.1). They rely mainly on financing systems to cover up-front costs and local networks of retailers to guarantee the maintenance of the energy systems. The same mechanisms can also help to deliver other services like water, heat or refrigeration to remote places.

In rural electrification strategies, a distinction is generally made between productive uses of electricity and non-productive uses of electricity. Productive uses of electricity are directly linked to the growth of economic activities; some RETs appear to be ideal for this, like pumping water for agriculture with solar or wind. Productive use is linked to the maximization of impact of providing electricity by targeting measures to increase income (De Gouvello and Durix, 2008). Another strategy can be to prioritize access for social institutions like health centers or schools in rural areas. A further option is to provide light and cover basic needs for inhabitants of rural areas. This kind of electrification is sometimes considered as only a source of comfort and therefore not a priority for decision-makers. But, it has been demonstrated that domestic lighting is in itself increasing productivity, extending opening hours for small shops, enabling people to work at home later and children to study (Gustavsson and Ellegård, 2004; van Campen *et al.*, 2000). Rural energy service companies need a customer base as large as possible in an area, and should include both kinds of productive and less productive use in their services.

Levels of services can be adjusted to the needs and the financial capacity of end-users. Each level of service – whatever the power generated – has a fixed cost and an operational cost. The increased marginal cost linked to better services with bigger systems can be small, as fixed installation cost is high anyway for a system, whatever its size; the cost of maintenance can also be the same per unit of system whatever the size of the system (mainly labor and transport costs). But the larger the individual system, the larger the needs covered and the fewer complaints from end-users or necessity for later upgrading. A basic system can also be provided (and partly subsidized), leaving it to the end-users to opt for bigger systems if/when they have enough funding for it.

Any rural electrification strategy should also include the possibilities offered today by pico-photovoltaic (one to 10 Watts peak). With the emergence of light emitting diodes (LEDs), providing lighting can be done now in a cost-effective way with very small portable systems, like solar lanterns, which can be sometimes also used to charge a mobile phone. This market is quite different from the fixed solar home system one: as up-front cost can be very low, rural electrification with pico-photovoltaic could happen as "spontaneously" and quickly as for the diffusion of mobile phones in Africa. This spontaneity needs nevertheless to be supported.

Any rural energization strategy should include all the variety of technologies available, combining renewables and conventional sources. As long as financing and maintenance are guaranteed by the existence of local rural energy service companies with sustainable business plans, RETs can often deliver for a lower cost an energy service which is in demand from end-users. Success of rural electrification or rural energization therefore lies more in creating the conditions of emergence of a market, rather than just giving subsidies to disseminate technologies or extend the grid.

15.2   POLICY AND INSTITUTIONS FOR ENERGY ACCESS

Power reforms in their current state are not conducive to rural electrification or to the uptake of RETs, unless they are accompanied by ad hoc regulation for RETs and rural electrification (Bacon and Besant-Jones, 2002; Kozloff, 1998; Moonga Haanyika, 2006). Renewable energy laws, by clarifying the role of each stakeholder and guaranteeing a stable framework for private investors, reduce the risk of investing in rural areas.

15.2.1   *The role of energy regulators and rural electrification agencies*

The function of regulation of decentralized RETs used to be largely ignored by energy regulators, but an increasing number of developing countries are implementing rural electrification agencies and putting RETs higher in their political agenda, and there is a growing interest in off-grid regulation (Lemaire and Kerr, 2010; Mostert, 2008). The lack of a proper regulatory framework deters investors from investing in rural areas (Monroy and Hernandez, 2008). Regulation for rural electrification may be delegated to rural electrification agencies. The latter can cover both rural electrification with conventional generation and rural electrification with RETs.

Regulators and policy-makers need to ensure that all consumers have access to energy services at a reasonable cost and within a reasonable timescale (REEEP-UNIDO, 2008). The decision-making process within utilities and regulatory bodies should therefore include staff with a broad range of technological knowledge and allow a comparison between all sources of energy available in a given location, taking a life-cycle perspective in appraising and comparing projects.

Software toolkits can facilitate the process of cost-comparison of various energy sources and implementation of RETs[1].

Private energy companies need appropriate incentives to be encouraged to provide the poorest households and the most remote locations with access to energy services. Regulators and policy-makers can alter the natural bias of utilities in favor of electrifying dense urban areas by facilitating the entry of new competitors to off-grid areas, and maximizing access to alternatives to conventional energy sources. The implementation of small RET operators can reduce transaction costs, accelerate rural electrification and offer a better quality of service, while minimizing the negative environmental impact of electricity generation. For instance, cooperatives are traditionally used to reduce cost and accelerate the rate of rural electrification (Barnes, 2007; Zomer, 2001).

15.2.1.1 *Light-handed regulation*
In some developing countries, one of the major barriers to the use of decentralized generation is simply the requirement of full licenses for generating electricity and the same level of technical standards for all operators, without differentiating requirements according to the size of the operator. Small systems for self-generation should not need a license, which should instead be given to the installer of the system for satisfying good installation practices, and not on a case by case basis for every system. Small-scale systems run by mini-grid operators should need just a declaration (Reiche *et al.*, 2006). Adapted regulation should minimize the amount of information that is required, minimize the number of separate regulatory requirements and decisions, use standardized documents, and make use of documents used by other agencies, to the maximum extent possible, reducing the need for case-by-case negotiation.

In Kenya, RETs incorporated in a hybrid system not exceeding 1 MW at medium transmission voltage are not required to go through the rigorous licensing procedure. According to the Kenyan Electricity Act, an "authorization" from the Minister of Energy is sufficient. In Uganda, electricity generation plants not exceeding 0.5 MW only require registration with the Electricity Regulatory Authority while in Namibia, no generation license is required for electricity generation equipment below 500 kVA for own use. In Tanzania, small power producers up to 1 MW do not apply for a license, but instead just need to register.

An important amendment to the electricity act of a country is also the setting up of a standard power purchase agreement (PPA). Instituting a "standard PPA" – a "standard offer" from the buyer (single buyer or distribution company for a mini-grid) to purchase all energy produced by specific renewable energy based IPP at a pre-announced price-limits market uncertainty, which stands in the way of substantial investment in renewable energy-based electricity generation. In Tanzania for instance, such guidelines for small power producers either using a RETs or cogeneration of up to 10 MW have been established in 2009, as a guide to the steps necessary to acquire permits and clearances to develop and operate as a small power producer[2], with a standard power purchase agreement and a tariff methodology. The absence of such a "standard offer" implies lengthy processes and high transaction costs between private developers and utilities, which inhibits the scaling up of small (renewable) energy investments.

15.2.1.2 *Standards and codes of practices*
Standards and labels are essential tools to raise the quality of products and services. Without these, consumers may only have access to sub-standard products at low prices. After a while the growth of the market is affected. The role of regulators (with the bureau of standards and customs) in defining and monitoring imports is to prevent developing countries being the "dumping ground" of sub-standard quality products. This is particularly true for the regulation, for instance, of pico-photovoltaic and small stand-alone systems: quality standards are essential due to the often low

---

[1] RET screen: http://www.retscreen.net/; Homer/NREL: http://www.homerenergy.com/
[2] http://www.ewura.go.tz

quality of some imported products. Once these products have entered into a country it becomes difficult to monitor their quality.

Regulated codes of practices for suppliers and installers of goods and services are an essential tool to ensure high standards of service. In the past, poor performance from local or international companies has been a major reason for the failure of numerous renewable energy projects (Lemaire, 2009a). There is a need for environmental standards, even for so-called environmentally-friendly renewable energies (e.g. standards requiring recycling of the batteries from solar systems).

It is recommended that standards, labels and codes of practices need not to be too strict, in order to let small businesses flourish, but strict enough to ensure the quality of the energy service provided; they need to be effectively monitored. Many countries have already established standards and codes of practices for various RETs, and these can often be used without too many modifications. International standards have been established notably by PV GAP and the World Bank[3].

15.2.1.3   *Planning*

When rural electrification is planned in an area, all options (extension of the grid or distributed generation, conventional energies or RETs) are to be evaluated (ESMAP, 2008); a clear delineation of the future extension of the grid enables private investors to invest in small stand-alone systems or distributed generation systems. An analysis combining levels of population density, the geography of dispersion of the population, local financial resources, and renewable energy sources potential in an area, can help to decide when to support grid extension and when to give priorities to other forms of electrification (De Gouvello and Maigne, 2002).

Planning processes should take into account the environmental and social benefits of RETs (Silva and Nakata, 2009). RET projects should involve more efficient appliances than conventional ones: small photovoltaic systems imply the connection of 12/24 Volt DC appliances that are highly efficient. Evaluations should rely on services provided, and not just on the increase in energy consumption. Therefore, comparisons should not be between kilowatt-hours (kWh) delivered by a conventional system and a RET system, but on a given number of lighting points and appliances connected. Least cost planning should also integrate all the components of the systems (like internal wiring being always included in the cost of PV systems) for a complete comparison. Integrated resource planning (IRP) should be actively promoted. As noted by Lindlein and Mostert (2009):

> *"Traditional methods for least-cost power planning [...] do not quantify the price of fuel price uncertainty, including it as a cost component in the levelized cost of production per kWh of generators. Conventional project analysis for least cost planning compares alternatives on a plant to plant comparison using a fixed forecast fuel price; with the sensitivity of results to the fuel price assumption being shown in a separate risk analysis. By omitting a cost component of conventional power altogether, this approach has an inherent bias against RET."*

The right combination of grid extension, mini-grid and individual system diffusion can accelerate the rate of electrification in a country by giving the role of installing, maintaining, training, and monitoring to the most appropriate institutions at every level, so as to reduce transaction costs. However, subcontracting and delegating require a clear definition of the role of each operator. The effective participation of end-users and communities can also help to minimize costs. Decentralized energy planning can facilitate rational decision making by aggregating data at a micro-level (Urmee *et al.*, 2009).

When electricity is provided by a mini-grid, towns or villages with high population density are likely to follow a conventional scheme of grid connection to a small central organization acting like a local utility with metering; for a lower population density, small-scale RETs providing

---

[3] http://www.pvgap.org

electricity directly to the end-users without costly infrastructure to supply the energy imply a different set of institutions and organizations.

Rural electrification strategies rely, for each location, on the right combination of institutions, the population's needs to be covered and appropriate technologies. For mini-grids, this could be done while keeping in mind that at some point the grid can be extended to a location, and encompass areas connected to mini-grids while keeping in place the RETs that were used to feed the mini-grid. Any long-term plans for grid extension that can affect the economics of small private electricity companies need to be taken into account. Compensation needs to be paid by utilities to small solar companies who may have to transfer their installations to unconnected customers, and bear the financial cost of this transfer. When the connection of a mini-grid to the main grid is possible, then the private company should be able to sell its kWh to the main utility. For small-scale systems, some may also be connected in to the grid in the long-term (e.g. micro-hydro), but this is unlikely to happen for small stand-alone solar home systems, which will have to be moved or kept as alternative sources of energy when the grid reaches an area.

Solar home systems have been presented sometimes as pre-electrification systems, encouraging people to get familiar with electricity, before moving to "real electricity". But it is also the case that there may never be any grid connection in a remote location for several decades: one of the interesting features of small systems like solar systems is that they are modular and can be upgraded. This flexibility needs also to be taken into account in any long-term rural electrification strategies. In the past, programs tended more often to propose only a unique standard system for everyone; people should have more choice and benefit from the modularity of solar systems to tailor to their needs (Nieuwenhout et al., 2001). This should be part of any long-term strategy: instead of relying on a totally hypothetical grid connection, RET-based electrification could be conceived as the final source of energy for remote locations.

15.2.1.4   *Who should be regulating off-grid electricity services, and why?*

Off-grid electricity services need to be regulated in a non-traditional way. The national regulator is not necessarily the entity that can best perform this task. It can be delegated to a rural electrification agency that has a better knowledge of off-grid electrification. In any case, dedicated trained staff with a good knowledge of RETs is required (Reiche et al., 2006).

Off-grid regulation is different from on-grid regulation in the sense that renewable operators are located in remote places that are sometimes difficult to access, there is often no metering system (too costly, notably for small systems), and operators are very small and cannot mobilize resources to comply with all the requirements and answer all the queries of the regulator. Therefore the rationale for off-grid regulation is first and foremost to make sure that operators do not abuse their monopoly position with high tariffs, and deliver effectively the basic energy services they are supposed to provide (number of connections, number of hours of electricity).

End-users need to be able to raise complaints with regulatory bodies when in dispute with the energy service provider. Involvement of local communities in the assessment of the quality of service delivered can reduce the cost of regulation, and improve the quality of services through direct feedback to operators and regulators. This channel of information needs to be organized. Cooperatives are a way to organize this communication as end-users are also shareholders of the company, as long as their customer base is not too large (large cooperatives often do not function effectively and end-users remain passive). Representatives/mediators between local inhabitants and the rural agency can also fulfill this function sometimes against a small compensation claim (Barnes, 2007).

15.2.2   *Funding and the question of subsidies*

The financial system can be reluctant to finance projects using technologies that can be still perceived to be unproven, or more risky and more difficult to implement than conventional centralized energy projects. The latter have the advantage in that development banks can consume large sums of credit at once using well identified procedures, while RET-based projects imply the need to put in place new standards with new institutions. Subsidies were and remain important in the energy sector and especially for rural electrification. Immediate social objectives that are behind energy subsidies can contradict long-term environmental objectives; as it seems no rural electrification program can be engaged without a well-designed set of subsidies, they should give priority to sustainable energy sources and could be funded by removing subsidies to fossil fuels.

15.2.2.1 *Targeted subsidies*

Subsidies have been used in most national rural electrification program since the US rural electrification program in the 1930s. If they are to be part of any program with RETs or extension of conventional grid, they need to be targeted and properly managed (Barnes and Halpern, 2001; ESMAP, 2000; Nexant, 2004; Mostert n.d.; World Bank, 2008; Zomer, 2001).

Ideally, subsidies should be at least decreased, or preferably removed from the established conventional generating technologies and, if appropriate, their supply chains. This is because there is no point in subsidizing conventional energies that are polluting (e.g. the case of heavily subsidized – and even in some states provided free of charge – electricity for farmers in India; or subsidies to gas which incentivized farmers to run motor pumps with cylinders of butane gas in Morocco). Price distortions on the bulk power market are caused by subsidized prices for the fossil fuel consumption of thermal power plants, and by import duties and VAT on RE components.

A widespread factor in energy price distortion is that components and fuels for thermal power production are exempt from import duty and VAT, whilst investments in RETs (in particular photovoltaic systems, which are treated like consumer goods) are not offered the same privilege. In fossil fuel-exporting developing countries, fuels consumed at power plants are typically priced below their net-back value as an export product.

During the "pilot project years" of collaboration programs for RETs, donors accepted price distortions in favor of conventional energies. Since donor policy has shifted to fund now only "RE-mainstreaming" collaboration programs, donors increasingly refuse to finance RE-collaboration programs unless steps are taken to eliminate the pricing bias in favor of conventional energies. It has become common practice for donors to make assistance to a RE-program conditional on a government commitment to the phasing out of import duties and VAT on RETs (Lindlein and Mostert, 2005).

Otherwise, subsidies can be given through lower electricity tariffs for end-users that do not reflect market prices. Initially conceived for social objectives, electricity tariffs are fixed at a far lower rate than the marginal costs for utilities; in that case, they benefit the urban middle class who consume the most electricity, and not the poorest who cannot afford white appliances. To minimize the free-rider effects, subsidies need to be redesigned in a way that will prioritize the benefit effectively to the targeted group (Lindlein and Mostert, 2005), for instance by being restricted to a lifeline of up to 50 kWh per month (e.g. the case of free basic electricity in South Africa).

As it is politically difficult to completely remove at once all subsidies to conventional generation and move toward cost-reflective tariffs, too-broad subsidies can be progressively reduced and energy tariffs increased, while reducing the burden for the poorest by implementing energy efficiency programs, on top of providing a lifeline or cash exclusively for very low income households. This has been done relatively successfully, notably in Indonesia and Ghana (World Bank, 2009).

Cross-subsidization to finance electrification in rural areas should be transparent. It seems practically unwise for low-income countries to unbundle the network from generation due to the small size of utilities, but it is possible at least to implement separate accountancy, and to be able to determine the true cost of generation and distribution.

Output-based subsidies given to companies rather to end users for consumption of electricity are to be privileged. Financial support can be effective in giving indirect subsidies to rural energy service companies (RESCOs), with financing assistance for business planning, capacity building or market development. For instance, when it comes to rural electrification, comprehensive market surveys can be funded and made available for free to potential operators. A complete mapping of the renewable energy sources of a country is also extremely valuable for potential operators.

15.2.2.2 *Subsidies for mini-grid technologies*

A quite simple and transparent financial support scheme could be a renewable energy premium tariff (RPT). Electricity bought by local utilities receives a premium tariff paid by the rural electrification agency or any qualified body, while the end-users pay only the common tariff (Moner-Girona, 2009; Solano-Peralta *et al.*, 2009). The main idea is to have a tariff guaranteed by law for a long period. By guaranteeing investors that they will be able to connect their system and sell their electricity at a given tariff for a certain period, this RPT reduces the risk in investing in RETs and in mini-grids. This tariff is then a subsidy to generation (and not to consumption), where the small local energy provider is rewarded for the proper delivery of electricity. It can be funded by a small levy on electricity, or by the increase in income generated by the reduction of subsidies for the consumption of fossil fuels. The RPT scheme could work for large mini-grids with a metering system at the level of generation. For smaller grids (or individual systems) where metering costs would be too high or inappropriate, a higher capital subsidy per functional system seems preferable. This scheme is being tested in several countries, notably in Tanzania and Argentina.

Capital costs of small mini-grid operators (either using conventional, RET or hybrid generation) also need to be heavily subsidized. There are multiple ways of combining subsidies according to the policy of the country. Mini-grid rural electrification programs generally include a combination of grants and soft loans with long repayment periods of 10–20 years, or even longer, provided by the State and/or development agencies to cooperatives (e.g. China with small hydropower), or to mini-grid private operators who invest in generation, but also in grid extension. A subsidy linked to the number of connections is often adopted. Even when the operators are heavily subsidized, end-users will have to pay an initial connection fee and then a bill. In the poorest areas, to increase the number of connections, end-users need to be given the option of spreading the remaining part of the connection cost over a number of years. It is also sometimes worthwhile being flexible in the collection of the bill, and adapting to the sources of income of some categories of rural inhabitants, especially farmers whose income can fluctuate according to the seasons and yields.

15.2.2.3 *Subsidies for decentralized stand-alone systems*

The capital cost for rural electricity, as with conventional electricity, may need to be subsidized because the cash purchasing power of inhabitants in the most remote areas remains low, and there is limited capacity in local financial institutions to offer loans to small rural companies or end-users. Reasons of social equity are also put forward in some countries (e.g. Chile, Costa Rica): rural inhabitants are not asked to pay more for a decentralized system that they would have been asked to for connection to the grid, which means they pay the same amount in connection fee and for the wiring and appliances connected to the system, but not the generating system itself.

Even when credit can be accessed, it does not seem that rural inhabitants – except a minority – could cover the full capital cost for instance of a complete solar home system on top of operating costs. It is only when end-users already have a relatively high level of income, and that a network of rural banks exists, that systems of larger sizes can be funded without proper financial mechanisms (and a capital subsidy), which is rare in developing countries. People in rural areas can have an income and be willing to pay a significant part of their income for electricity, but often they cannot afford up-front costs linked to small RET systems. Furthermore, as with connection to the grid, even when a capital grant subsidy is offered, it may be worth spreading the remaining capital cost over several years; facilities for payment need to be given to those who have irregular sources of incomes, like farmers.

15.2.3 *The role of rural energy service companies (RESCOs)*

Regional or local networks of operators and chains of supply can be created according to the size of the country. The mix of imported technologies and technologies manufactured locally will depend of the RET and the industrial policy of the country; tax duties on RETs should be removed, and custom duties should be kept if products can be manufactured locally or otherwise removed.

In any case, networks of local retailers and operators are needed. Rural energy services companies are pivotal in the dissemination of new technologies; they can be cooperatives, associations or private entrepreneurs (Sanchez, 2006). NGOs specializing in RETs do have a role to play, but they need either to focus on capacity building or evolve into commercial entities.

Rural energy support schemes, carefully designed according to the geography of the country, can accelerate the implementation of such networks. Large-scale diffusion of small and medium off-grid RET systems in developing countries implies adherence to a certain number of rules. The main hindrance remains a general bias toward short-term investments in conventional energies and often a lack of interest – till recently – from national stakeholders towards innovative rural electrification schemes. Relations with existing utilities need be clarified and public authorities, once connection targets are assigned, should avoid any interference in the management of energy providers. Confidence of potential investors in the stability of the regulatory framework, creation of a value chain and involvement of utilities can only be created through long-term energy and industrial policies and real commitment at the highest political level.

15.2.3.1   *Different business models for increasing energy access in rural areas with small decentralized RET systems*

There is a wide consensus to say there is no "one size fits all" solution in terms of rural electrification schemes: institutions and programs should be designed according to the particularities of the country. Lessons have been drawn from the dissemination of RETs, notably solar home systems, which can be applied to a number of RETs not necessarily linked to electricity generation. This section develops notably the example of commercialization of solar home systems combined with disseminating Liquefied Petroleum Gas, biogas or efficient cook-stoves. Rural energy service companies and retailers have already been successful in disseminating a number of RETs, hybrid systems and conventional decentralized energy systems.

Conventional utilities have rarely the capacity to intervene on small RET systems; a network of small enterprises specialized in the installation and maintenance of RET systems can better intervene in rural areas. There are mainly three schemes for the large-scale dissemination of small-scale decentralized energy systems like solar home systems: cash purchase, the micro-credit scheme and the fee-for-service concessions scheme (EDRC, 2003; Krause and Nordström, 2004; Scheutzlich *et al.*, 2002; Schultem *et al.*, 2003).

15.2.3.2   *Cash purchase and micro-credit models*

15.2.3.2.1   Cash purchase and network of retailers

The first option is the modular cash purchase of the different parts of solar systems bit by bit, which enables customers to overcome the up-front cost barrier. Kenya is a well-known example of a country with widespread diffusion of solar home systems (more than 200,000 solar home systems) relying on cash purchase (Jacobson, 2006). Nevertheless, it is common that systems are very small (often in the range of 12–20 Watt peak in Kenya) due to limited personal funding, which permits only very low consumption, often leading to the disappointment of end-users. Furthermore, in developing countries, this rarely allows access to bigger/more costly systems like solar photovoltaic pumps useful for agricultural exploitation.

15.2.3.2.2   Micro-credit models

The second delivery mechanism is the micro-credit scheme. End-users buy the system and get a loan to make the purchase that they will reimburse progressively. Adaptation of the micro-credit offer to specificities of the local market is important (GVEP, 2004). In the micro-credit scheme, rural energy service companies (RESCOs) can focus on the maintenance of the systems and leave the financing aspect to specialized institutions. Micro-finance institutions (MFIs) need to provide funding at two levels: funding for RESCOs and consumer loans for end-users.

There is a need to find reliable and well-implemented MFIs that can reach end-users and provide them with consumer loans. They should be able to provide credit to several thousand households and manage to get reimbursed. Formal commercial banks can provide funds to smaller MFIs which are better implemented in rural areas and provide large lines of credit for RESCOs.

The success of micro-credit in Asia, notably in Bangladesh (see case study below) has been well described, but it remains to be proven how easily it can be replicated in other continents notably in Africa, as MFIs are weak and the population density in some rural areas of African countries can be low. Furthermore micro-finance schemes, even when successful, are not without income-generation difficulties, as micro-credit schemes can often be financially sustainable only through encouraging high level of micro-debts (Dichter and Harper, 2008). It appears than in African countries, unlike in Asia, micro-credit institutions are very small and that existing micro-credit schemes most of the time cover only 10–100 US$ loans, a level that is not sufficient to cover the cost of a solar home system. Informal African institutions like tontines cannot provide funding for a conventional solar home system in the range of 500–1000 US$ (Basu *et al.*, 2004). For other sources of energy, like micro-hydro, up-front costs seem to make them out of reach of most micro-credit schemes.

Some authors (Miller, 2009) tend to favor then the reduction of the size of solar home systems (10–20 Watt peak instead of 50 Watt peak) to make them more affordable by micro-credit (or even cash payment). Nevertheless, as a number of sub-standard photovoltaic products exist, the quality of the systems disseminated needs to be closely monitored (Duke *et al.*, 2002; Otieno, 2003); with the advent of LEDs for lighting/pico-photovoltaic products, the dissemination of photovoltaic systems could then be limited to the purchase of solar lamps or small systems to charge mobile phones.

15.2.3.2.3   Case study: SHS, biogas and improved cook stove diffusion by Grameen Shakti in Bangladesh

One of the most successful providers in South Asia is Grameen Shakti. Grameen Shakti is a renewable energy company established in 1996; this company belongs to the family of companies set up around the Grameen Bank. It has benefited from the successful experience of the latter in providing micro-loans in Bangladesh since 1983, and could rely from its inception on an already extensive network of micro-lenders and a good customer base. Nevertheless, the move to expand its solar business – one of its three renewable energy programs with wind and biomass – was quite slow. Up to July 1999, Grameen Shakti sold only 1147 systems, with 200 technicians trained (Barnes, 2007).

At the very beginning, with limited funds Grameen Shakti could offer only one year finance (Miller, 2009). But from 1998, by securing lines of credit from the International Finance Corporation/World Bank, Grameen Shakti has been able to provide a more diversified offer of loans up to three years. The choice given to consumers is now between three options (as shown in Table 15.1).

Systems proposed today are ranging from 10 Watt peak to 130 Watt peak, with 50 Watt peak accounting for most of sales; the cost of a 50 Watt peak system with a battery imported from Japan is around 380 US$. As of September 2010, Grameen Shakti has installed more than 464,000 solar home systems, and the program is growing at a rate of 20,000 new clients per month[4].

Table 15.1.   Options for purchase of a solar home system with Grameen Shakti (Source: Miller, 2009).

Options	Cash basis	Loan 2 years	Loan 3 years
Initial payment	100% initial payment with 4% discount	25% initial payment	15% initial payment
Interest rate		8% interest rate	12% interest rate

---

[4]http://www.gshakti.org/index.php

The particular success of Grameen Shakti needs to be compared to other similar institutions in Bangladesh, as the closest installs only 1000 solar home systems per month. It seems to be linked to the existence of an extensive network of now more than 950 decentralized branch offices, but also to an adapted door-to-door marketing strategy.

The beneficial impact of solar home systems for rural households appears to be substantial in Bangladesh, in terms of access to a modern source of energy and reducing energy costs; potential demand could increase with further reduction in costs of photovoltaic modules (Komatsu *et al.*, 2011). Payback periods for solar home systems replacing kerosene lamps are typically less than 3 years (Mondal, 2010). Nevertheless, the quality of installation and performance of components of solar home systems in Bangladesh can fluctuate, and specifications seem too flexible to safeguard end-users interests.

Systematic training of the heads of households in the proper use of solar panels is provided by 45 Grameen technology centers. Technicians from the company do a monthly visit during the financing period and propose an annual maintenance contract for a few US dollars per month afterwards. As in rural communities in Bangladesh, it seems inconceivable for a man to enter a house and interact with women during the absence of male member of the family who work outside during the day, 6700 women have been trained to be technicians; they often follow a four-year technical degree paid by the Grameen Technology Centers before working back in rural areas (Sovacool and Drupady, 2011). Grameen Shakti leads also a program of dissemination of biogas and cook stoves; the company employs more than 8,400 people.

15.2.3.3   *Fee-for-service models*
15.2.3.3.1   The principles
Another option is the fee-for-service scheme. In many locations, inhabitants do not have any kind of access to credit or external funding. RESCOs using a fee-for-service scheme then handle maintenance and have to collect fees for the system they install. Small companies often benefit from a long-term concession, and they can obtain a loan from the government to buy the systems; they then act as small utilities that charge a low cost for the installation of systems, and receive a monthly payment from their customers for the delivery of an energy service.

Local authorities can trust RESCOs to recover the cost of the systems, unlike if the systems were given directly to individuals. For commercial banks this scheme is simpler, as they need only to provide larger commercial loans to RESCOs. Levels of subsidy need just to be adjusted to the customer base that public authorities want the RESCOs to reach. End-users do not own the system, which remains the property of RESCOs, so they do not need to provide collateral: RESCOs get a guarantee from the government to buy systems.

This scheme seems simpler to put in place when there is no existing sound MFIs to rely on, with one single organization to put in place, which serves as unique partner for banks and the government, reducing transaction costs. Nevertheless the management of this kind of RESCO is more complex, even if synergies can be created between maintaining RET systems and collecting fees. Systems belong to the RESCOs and not to the end-users, who may be less motivated to properly maintain the systems.

15.2.3.3.2   The case of solar and LPG concessions in South Africa
The rural concessions scheme for solar home systems in South Africa represents one of the most ambitious projects of rural electrification using fee-for-service solar energy in Africa. This case study describes briefly one of the five concessions launched after the end of apartheid in Kwazulu-Natal (Lemaire, 2011).

The solar company in Kwazulu-Natal employs more than 70 people. Eight energy stores are located in the concession that covers 10,000 km^2, and more than 13,000 solar home systems have been installed. Energy stores stock parts and sell not only small photovoltaic components, but also LPG. The sale of LPG enables the stores to increase their turnover by supplementing their provision of energy services to rural households ("energization" approach).

The company gets a capital grant from the government, which represents the major part of the total cost of the solar home system. To get connected, customers need to pay just a small installation fee. The governmental subsidy for solar energy is roughly equal to the one given to the main electric utility to connect households to the grid. The customers have to pre-pay a monthly fee, which may (or may not) partially be paid by local municipalities. Some municipalities agree to pay half of the monthly fee, other municipalities do not. This creates considerable distortion between clients.

Technicians visit the installations only when there is a problem or during the planned routine visits which take place every six months. The energy stores are central to the process. People come to the energy stores mainly to charge a token which gives them a credit for electricity. The token also contains data on the functioning of the system, which can be transferred to a computer. All the data can be manipulated at the energy store, but are also immediately centralised at the headquarters. Even with this system of reporting, the process of resolving a failure can take several weeks, while the contract requires people to pay even when the system is not functioning. A point of dissatisfaction for clients is the limited amount of electricity provided by the system. Furthermore, during a survey, 2% of the systems were non-operational due to theft of components of the system (Lemaire, 2011).

This concession has not been able to extend its customer base as fast as initially scheduled due to lack of funding from the government after 2006. But, in 2011, the government and international donors began funding the installation of more systems again, and today the total number of installed systems in the solar fee-for service concessions in South Africa is increasing.

15.2.3.4   *Fee-for-service versus micro-credit models*

The fee-for service model has been used notably in Zambia (Lemaire, 2009), in Morocco (De Gouvello and Maigne, 2002), in Argentina (Alazraki and Haselip, 2007; Best, 2011), in Honduras and the Dominican Republic (Rogers *et al.*, 2006) and with mitigated success in the Pacific Islands (Dornan, 2011; Mala *et al.*, 2009). The micro-credit model can be found more often in South Asian countries like Bangladesh or India (Rogers *et al.*, 2006).

From an organisational point of view, energy companies using a fee-for service model have to collect fees themselves; in the micro-credit scheme, they leave that task to MFIs and focus on installation and maintenance; close mutual cooperation between MFIs and energy companies is required; selling can be done by an entity directly related to a MFI.

A fee-for service scheme may be more appropriate in areas where there are no already established MFIs; but, it implies a long-term commitment from the government to subsidize a number of solar home systems and protect operator's concession for a a long period. As fee-for-service companies are owners of their solar systems, they need to generate enough income to reimburse their loan, while the micro-credit scheme gives more flexibility for operators, who do not have to invest in RET systems and can adjust to the demand of end-users.

The main criteria of sustainability of RET systems relies on the capacity of rural inhabitants being targeted to cover operating costs only. For instance for solar home systems, whether rural inhabitants pay a monthly fee (fee-for-service scheme) or reimburse a credit (micro-credit scheme), this will represent a significant part of their income (up to 25%). They will pay only if the system works properly. As experience shows, if systems malfunction, then they will immediately stop their payment whatever the scheme adopted: it does not seem there is any difference in attitude between micro-credit or fee-for-service scheme beneficiaries (Nieuwenhout *et al.*, 2001).

So, operating costs of RET systems are fixed at a high level to cover the costs of an appropriate maintenance network, and it is already an accomplishment for small operators to run their business with no operating subsidies. This implies a good design of the concession/area covered and good management by operators. Concessions can rely on different mechanisms which can be used mainly for the dissemination of solar home systems, but often other products for delivering energy services in rural places as well (e.g. cook-stoves, LPG), which helps the rural company to generate income.

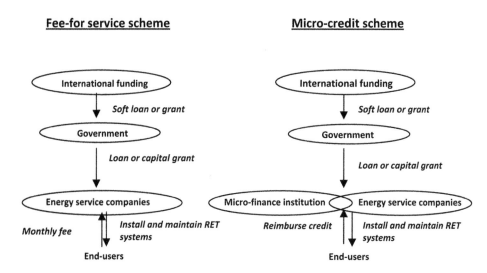

Figure 15.1. Comparison of the fee for service scheme and the micro-credit scheme (drawn by Lemaire).

15.2.3.5 *Increasing energy access by using by-product of agriculture*

There are numerous RETs which can be used by farmers either to generate electricity or to produce heat/cold. Of particular interest for farmers are the RETs which use by-product of agriculture. This section will introduce briefly the case of cogeneration in Mauritius, where synergies between farming activity and generation of electricity have been developed.

15.2.3.5.1 Biomass, biogas and cogeneration

In some cases it is possible to recycle waste or use by-products from agriculture to generate energy and reduce expenses for a farm, or even increase income by selling energy while reducing environmental impact.

Biogas is an example of a way to use manure from cattle or pigs and produce gas for cooking. This has been widely developed in countries like China with more than 30 million biogas digesters disseminated and knowledge accumulated over 40 years (Ekins and Lemaire, 2012). The questions of maintenance and of financing are the same as for other decentralized generation systems. Not so many countries have replicated the success of China, as this implies the existence of networks of farms with cattle and pigs to justify the creation of a local market for biogas digesters and bring their cost down; otherwise these products tend to remain too expensive.

Farmers can either use waste or cultivate on a small or sometimes very large-scale crops that produce biomass or biofuels either for their own use or for exports; biofuels can increase energy access by providing fuels for farmer's machineries. Biomass can be used to produce electricity locally or to feed the grid. A good example of synergy between agriculture and energy generation is cogeneration by burning waste on large estates to produce electricity to feed the grid; the example of Mauritius detailed below is quite well-documented.

15.2.3.5.2 Case of cogeneration in Mauritius

Sugar is one of the main industries of Mauritius. It generates a considerable amount of waste bagasse which can be used to produce power. Boilers in sugar factories used to produce steam for the sugar processing and electricity for their own use. In 1957, a sugar factory sold electricity to the grid for the first time. The supply of power remained linked to the crop season, until 1982 when the first agreement between the grid company and a sugar plant was signed for the supply of electricity to the network all year round, using coal during the inter-crop season.

In 1985, the Sugar Sector Package Deal Act aimed at increasing the production of bagasse for the generation of electricity. In 1988, the electricity generated by the sugar industry counted for 19% of the total requirement of the country. That same year, the Sugar Industry Efficiency Act gave tax incentives towards investments in the generation of electricity and encouraged small plantations to provide bagasse for this purpose. In 1991, to optimize the use of bagasse, a Bagasse Energy Development Program was launched with the World Bank.

Since then, the modernization of the sugar sector has led to the centralization of milling activities in fewer factories, which means more possibilities of cogeneration with increased efficiency. The price of the electricity sold by sugar factories has been determined in reference to the avoided cost of the implementation of conventional diesel production. Equivalent bagasse used for electricity generation increased from around 150,000 tons in the late eighties to more than 500,000 tons in the late nineties. In 2002, the electricity generated by the sugar industry counted for 40% of the total demand of the country, and generated a gross revenue equivalent to 90% of that earned by the miller for cane processing (Deepchand, 2001; 2005).

This use of a renewable source of energy in a country with no fossil fuel resources proves to be particularly interesting. It could be replicated in other countries with a significant sugar industry, as it generates important income for this industry in a period where the price of sugar is low and helps the modernization of this sector (REEEP – Afrepren, 2007).

15.3   CONCLUSION

Energy access in developing countries can be greatly improved by implementing consistent policies in favour of rural areas (Barnes and Foley, 2004). Countries with high population density in rural areas can accelerate grid-connection by implementing simplified regulatory frameworks which reduce the cost of connection and give incentives either to the utility or alternatively – in some circumstances – to rural electrification agencies and local providers to connect low-income areas. In other countries, where extension of the grid is too costly, large-scale diffusion of decentralised generation systems is proving to be of particular importance.

With decreasing costs, RETs – notably solar photovoltaics – are now in some countries like South Africa, Kenya, Bangladesh, India and China being adopted on a very large scale. Markets for RETs need to be nurtured and this requires a long period of preparation in implementing institutions and capacity building programmes. Once this has been achieved, and tested on small-scale projects, then it is possible to launch RET programmes on a very large-scale and increase the rate of electrification or connection of new RET projects quickly.

But energy access is not just about increasing connections to the grid or increasing the number of households with individual systems; actually rural energy access is not just about electricity, but about energy services adapted to the needs of rural inhabitants (Zerriffi, 2011). The delivery of an energy service relying on electricity cannot be measured in a number of kWh, but by the number of devices connected like charging mobile phones, TVs, radios, light bulbs, pumps, and fridges. Furthermore, a number of energy services like heat, hot water, cooking, water pumping, cold chain, and dryers can also be delivered without electricity generation, but by using passive solar technologies (solar stoves, solar dryers, solar water heaters) or other RETs (water mills). All these products are now mature and can be installed and maintained by rural energy service companies, which can specialise on a product or market several products (Restio, 2008).

Rural inhabitants in developing countries are often poor and electrification often needs to be subsidized. Nevertheless the wealthiest rural inhabitants, notably farmers, do have an income and are more than ready to pay for energy services as long as the quality of service is good (Ellegård *et al.*, 2004) and the scheme is adapted to their situation. Irregular income according to the seasons and preference for small payments are constraints that RESCOs need to take into account.

Rural energy service companies with their network of retailers are pivotal for the installation and long-term maintenance of RETs or conventional generators in rural areas; for the management of local micro and mini-grids, and for the relations they keep with end-users (Byakola *et al.*, 2009). They can be private entrepreneurs, cooperatives or associations; the quality of their management is crucial and implies intensive training of their staff (Bates *et al.*, 2003; Jacobson and Kammen,

2007). A stable regulatory framework, and permanent institutional support, with a "champion" inside the administration for rural electrification, and well-defined and adapted energy policies determine the success or the failure of energy access programs.

ACKNOWLEDGEMENTS

The research that has led to this contribution has been funded since 2005 by the Renewable Energy and Energy Efficiency Partnership (REEEP) while coordinating the Sustainable Energy Regulation Network (SERN). I would like to thank Daniel Kerr and an anonymous reviewer for their comments on this chapter.

REFERENCES

Alazraki, R. & Haselip, J.A.: Assessing the up-take of small-scale photovoltaic electricity production in Argentina: the PERMER project. *J. Clean. Prod.* 15 (2007), pp. 131–142.
Bacon, R.W. & Besant-Jones, J.: Global electric power reform, privatization and liberalization of the electric power industry in developing countries. World Bank, ESMAP, Washington DC, 2002.
Barnes, D. (ed): The challenge of rural electrification – strategies for developing countries. Resources for the Future – ESMAP, Washington DC, 2007.
Barnes, D. & Foley, G.: Rural electrification in the developing world – a summary of lessons from successful programs. UNDP/World Bank (ESMAP), Washington DC, 2004.
Barnes, D. & Halpern, J.: Reaching the poor – designing energy subsidies to benefit those that need it. *Refocus* 2:6 (2001), pp. 32–34+37(3).
Basu, A., Blavy, R. & Yulek, M.: Microfinance in Africa: experience and lessons from selected African countries. IMF Working Paper, Washington DC, 2004.
Bates, J., Gunning, R. & Stapleton, G.: PV for rural electrification in developing countries: a guide to capacity building requirements. Report IEA-PVPS T9-03, Paris, France, 2003.
Best, S.: Remote access: expanding energy provision in rural Argentina through public-private partnerships and renewable energy – a case study of the PERMER programme. IIED, London, UK, 2011.
Byakola, T., Lema, O., Kristjansdottir, T. & Lineikro, J.: Sustainable energy solutions in East Africa. status, experiences and policy recommendations from NGOs in Tanzania, Kenya and Uganda. Norges Naturvernforbund, Oslo, Norway, 2009.
Deepchand, K.: Bagasse-based cogeneration in Mauritius: a model for eastern and southern Africa. Afrepren, Nairobi, Kenya, 2001.
Deepchand, K.: Sugar cane bagasse energy cogeneration: lessons from Mauritius. Paper presented to the Parliamentarian Forum on Energy Legislation and Sustainable Development, Cape Town, South Africa, 2005.
De Gouvello, C. & Durix. L.: Maximising the productive uses of electricity to increase the impact of rural electrification programs. World Bank ESMAP, Report 332/08, Washington DC, 2008.
De Gouvello, C. & Maigne, Y.: Decentralised rural electrification: an opportunity for mankind, techniques for the planet. Systèmes Solaires, Paris, France, 2002.
Dichter, T. & Harper, M. (ed): *What's wrong with microfinance?* Practical Action Publishing, Rugby, UK, 2007.
Dornan, M.: Solar-based rural electrification policy design: the renewable energy service company (RESCO) model in Fiji. *Renew. Energy* 36 (2011), pp. 797–803.
Douglas, B. (ed): The challenge of rural electrification – strategies for developing countries. Washington, Resources for the Future – ESMAP, Washington DC, 2007.
Duke, R.D., Jacobson, A. & Kammen, D.M.: Photovoltaic module quality in the Kenyan solar home systems market. *Energy Policy* 30 (2002), pp. 477–499.
EDRC (Energy and Development Research Centre): A review of international literature of ESCOs and fee-for-service approaches to rural electrification (solar home systems). Cape Town, South Africa, 2003.
Ellegård, A., Arvidson, A., Nordström, M., Kalumiana, O.S. & Mwanza, C.: Rural people pay for solar: experiences from the Zambia PV-ESCO project. *Renew. Energy* 29:8 (2004), pp. 1251–1263.
Ekins, P. & Lemaire, X.: Sustainable consumption and production for poverty alleviation. UNEP, Paris, France, 2012.
ESMAP: Energy services for the world poor. UNDP/World Bank, Washington DC, 2000.

ESMAP: Designing sustainable off-grid rural electrification projects: principles and practices. World Bank, Washington D.C, USA, 2008.

Gustavsson, M. & Ellegaard, A.: The impact of solar home systems on rural livelihoods: experiences from the Nyimba Energy Service Company in Zambia. *Renew. Energy* 29 (2004), pp. 1059–1072.

GVEP: *Proceedings of the Global Village Energy Partnership (GVEP) workshop on consumer lending and microfinance to expand access to energy services.* Manila, Philippines, May 19–21, 2004.

Jacobson, A.: Connective power: solar electrification and social change in Kenya. *World Develop.* 35:1 (2006), pp. 144–162.

Jacobson, A.D. & Kammen, M.: Engineering, institutions, and the public Interest: evaluating product quality in the Kenyan solar photovoltaics industry. *Energy Policy* 35 (2007), pp. 2960–2968.

Kammen, D. & Kirubi, C.: Poverty, energy, and resource use in developing countries – focus on Africa. *Annales New York Academic Science* 1136 (2008), pp. 348–357.

Komatsu, S., Kaneko, S. & Ghosh, P.: Are micro-benefits negligible? The implications of the rapid expansion of solar home systems (SHS) in rural Bangladesh for sustainable development. *Energy Policy* 39 (2011), pp. 4022–4031.

Kozloff, K.: Electricity sector reform in developing countries: implications for renewable energy. Renewable Energy Power Project, Washington DC, 1998.

Krause, M. & Nordström, S. (eds): Solar photovoltaics in Africa – experiences with financing and delivery models. UNDP and GEF, May, 2004.

Lemaire, X.: De quelques mythes et réalités de l'énergie solaire photovoltaïque en milieu rural africain. In: M.-J. Menozzi, F. Flipo & D. Pecaud (eds): *Energie et société: sciences, gouvernance et usages*. EDISUD (coll. Ecologie Humaine), 2009a, pp. 122–130.

Lemaire, X.: Fee-for-service companies for rural electrification with photovoltaic systems: the case of Zambia. *Energy Sustain. Develop.* 13 (2009b), pp. 18–23.

Lemaire, X.: Off-grid electrification with solar home systems: the experience of a fee-for-service concession in South Africa. *Energy Sustain. Develop (special issue on Rural Electrification)* 15 (2011), pp. 277–283.

Lemaire, X. & Kerr, D.: Off-grid regulation and rural electrification in developing countries. A literature review. REEEP-SERN, London, UK, 2010.

Lindlein, P. & Mostert, W.: Financing renewable energy – instruments, strategies, practice approaches. KfW Bankengruppe, Frankfurt am Main, Germany, 2005.

Mala, K., Schläpfer, A. & Pryor, T.: Better or worse? The role of solar photovoltaic (PV) systems in sustainable development: case studies of remote atoll communities in Kiribati. *Renew. Energy* 34 (2009), pp. 358–361.

Miller, D.: *Selling solar – the diffusion of renewable energy in emerging markets.* Earthscan, London, UK, 2009.

Mondal, A.M.: Economic viability of solar home system: case study of Bangladesh. *Renew. Energy* 35 (2010), pp. 1125–1129.

Moner-Girona, M.: A new tailored scheme for the support of renewable energies in developing countries. *Energy Policy* 37 (2009), pp. 2037–2041.

Moner-Girona, M., Ghanadan, R., Jacobson, A. & Kammen, D.M.: Decreasing PV costs in Africa – opportunities for rural electrification using solar PV in Sub-Saharan Africa. *Refocus*, January/February (2006), pp. 40–45.

Monroy, C.R. & Hernandez, A.S.S.: Strengthening financial innovation in energy supply projects for rural exploitations in developing countries. *Renew. Sustain. Energy Rev.* 12 (2008), pp. 1928–1943.

Moonga Haanyika, C.: Rural electrification policies and institutions in a reforming power sector. *Energy Policy* 34:17 (2006).

Mostert, W.: *Financing rural electrification and renewable energy – basic principles for subsidy policy – Selected Essays.* n.d. Site www.mostert.dk (accessed 12 September 2013).

Mostert, W.: Review of experiences with rural electrification agencies: lessons for Africa. Draft EUEI Report, Brussel, Belgium, 2008.

Nexant: Subsidizing rural electrification in South Asia: an introductory guide. USAid – Sari Energy Program, New Dehli, India, 2004.

Nieuwenhout, F.D.J., van Dijk, A., Lasschuit, P.E., van Roekel, G., van Dijk, V.A.P., Hirsch, D., Arriaza, H., Hankins, M., Sharma B.D. & Wade, H.: Experience with solar home systems in developing countries: a review. *Progr. Photovoltaics: Res. Applicat.* 9 (2001), pp. 455–474.

Otieno, D.: Solar PV in Kenya. *Refocus* Sept–Oct. 2003.

REEGLE/SERN: Country policy reviews. Site: www.reegle.org (accessed 8 March 2013).

REEEP – Afrepren: Financing cogeneration and small-hydro projects in the sugar and tea industry in east and southern Africa training, Nairobi, Kenya, 2007.

REEEP – UNIDO: Sustainable energy regulation and policy-making for Africa – training manual. Vienna, Austria, 2008.

Reiche, K., Tenenbaum, B. & Torres de Mästle, C.: Electrification and regulation: principles and a model law. World Bank, Energy and Mining Sector, Washington DC, 2006.

Restio: Integrated energy utility roadmap. REEEP Project, Cape Town, South Africa, 2008.

Rogers, J., Hansen, R. & Graham, S.: Innovation in rural energy delivery – accelerating energy access through SME's. Navigant Consulting – Soluz, Chemslford, MA, 2006.

Sanchez, T.: Electricity services in remote rural communities: the small enterprise model. ITDG, Rugby, UK, 2006.

Scheutzlich, T., Pertz, K., Klinghammer, W., Scholand, M. & Wisniwski, S.: Financing mechanisms for solar home systems in developing countries: the role of financing in the dissemination process. Report IEA PVPS T9-01, Paris, France, 2002.

Schultem, B., van Hermert, B.H. & Sluijsc, Q.: Summary of models for the implementation of solar home systems in developing countries. Report IEA PVPS T9-02, Paris, France, 2003.

SELF (Solar Electric Light Fund): A cost and reliability comparison between solar and diesel powered pumps. Washington DC, 2008.

Silva, D. & Nakata, T.: Multi-objective assessment of rural electrification in remote areas with poverty considerations. *Energy Policy* 37 (2009), pp. 3096–3108.

Solano-Peralta, M., Moner-Girona, M., Wilfried, G.J.H.M, Van Sark, X. & Vallvè, X.: "Tropicalisation" of feed-in tariffs: a custom-made support scheme for hybrid PV/diesel systems in isolated regions. *Renew. Sustain. Energy Syst. Rev.* 13 (2009), pp. 2279–2294.

Sovacool, B.K. & Drupady, I.M.: Summoning earth and fire: The energy development implications of Grameen Shakti (GS) in Bangladesh. *Energy* 36 (2011), pp. 4445–4459.

UNDP – WHO: The energy access situation in developing countries: A review focusing on the least developed countries and Sub-Saharan Africa, Geneva, Switzerland, 2009.

Urmee, T., Harries, D. & Schlapfer, A.: Issues related to rural electrification using renewable energy in developing countries of Asia and Pacific. *Renew. Energy* 34 (2009), pp. 354–357.

van Campen, B., Guidi, D. & Best, G.: The potential and impact of solar photovoltaic systems for sustainable agriculture and rural development. FAO, Rome, Italy, 2000.

Varun, R., Prakash, I. & Krishnan, B.: Energy, economics and environmental impacts of renewable energy systems. *Renew. Sustain. Energy Syst. Rev.* 13 (2009), pp. 2716–2721.

World Bank: Designing sustainable off-grid rural electrification projects: principles and practices. World Bank, Washington DC, 2008.

World Bank: Climate change and the World Bank Group. Phase 1. An evaluation of World Bank win-win energy policy reforms. Washington DC, 2009.

Zerriffi, H.: *Rural electrification: strategies for distributed generation.* Springer, Berlin, Germany, 2011.

Zomer, A.N.: *Rural electrification.* PhD Thesis, University of Twente, Twente, The Netherlands, 2001.

Subject index

2°C increase (GHG) 28
450 policy scenario (GHG) 4, 24, 27

acetic acid 337, 349, 450, 361, 362
acetotrophic pathway 337
acid
    amino 283, 284, 406, 407, 412
    fatty 283
        saturated 283, 284, 287, 288
    organic 306, 379, 406, 407
    palmitoleic 284
Acrocomia aculeate 327
active regeneration technology (diesel engine) 260,
        261
additive (fuel) 198, 284, 288, 331, 332
adsorption refrigeration cycle 114, 115
advanced digestion 346
Africa 30, 31, 135, 137, 150, 419, 421, 428,
        429
agricultural
    development 10, 124
    effluent, treatment 13
    frontier 3
    land 22, 28, 126, 163, 175, 388, 401, 402
    policy 157, 158, 386, 390, 391
    product 7, 13, 138, 200, 384, 390, 391, 398
    production 3, 12, 13, 49, 97, 99, 101, 104–106,
        110, 120, 123, 124, 126, 127, 171–173,
        175–177, 195, 176, 199, 201–203, 387,
        391, 392, 411
        scale 127–130
    productivity 22, 125, 171, 175, 273
        increasing 171
    residue/waste 155, 325, 356, 370
    sector 4, 6, 7, 34, 78, 105, 140, 143, 156,
        173–175, 189, 205, 383, 386, 391
    sustainability requirement 106, 120
agricultural-related GHG emission 27
agriculture (*see also* farm *and* farming)
    energy (*see main entry* energy)
    greenhouse gas emission (*see main entry*
        greenhouse gas emission)
    sustainable policy option 155–162
agri-food 22, 31, 123, 125, 126, 131–133, 140, 142,
        150, 163
    chain 22, 32, 124, 127, 132–134, 138, 140–142,
        145, 148, 155, 157
    industry 28, 126
    management option 33
    sector 22, 28, 125, 126, 130, 133, 145, 147, 156,
        159, 162, 163, 401
agrochemical 3, 63, 64, 172, 175, 176, 179, 181,
        183, 189, 198
agro-ecological farming practice 33

air
    circulation 44
    cooling 46, 270
    drying 109, 232
air-source heat pump 134
air-water heat pump system for drying 113
Aleurites fordii 327
algae 150, 279, 281, 282, 301, 307, 355, 400, 406,
        409
allothermal gasification 362
alternative fuel (*see also* biofuel) 146, 313, 323,
        324, 326, 332, 355
ambient air-cooling 46
amino acid 283, 284, 406, 407, 412
ammonia (NH_3) 212, 240, 268, 269, 290, 337, 338,
        345, 346, 359, 372, 376, 377, 409
Anacardium occidentale 327
anaerobic digestion (*see* biofuel, biogas, anaerobic
        digestion)
anaerobic filter 348
animal
    food manufacturing 100
    house 213–215, 218
        heating 215–218
    manure 63, 67, 92, 172, 176, 207, 389,
        408, 409
    production 19, 53, 92, 195–201, 204, 205, 215,
        219, 239, 356
        waste 356
    welfare 197, 205, 212, 213, 215, 229, 232, 389
aquaculture 13, 30, 42, 92, 125, 132–134, 136, 137,
        146, 147, 150–154, 277, 282, 284, 292,
        295, 301
arable land 8–10, 78, 87, 175, 388, 389
arctic polar ice cap 6
Ariachis hypogaea 327
Arthrospira 293, 295, 301
ash
    content 355, 357–361, 364, 366, 377
    fusion temperature 358, 359, 370
Atlantic 86–88
atmosphere 19, 20, 24, 28, 29, 72, 130, 200, 207,
        290, 294, 324, 326, 328, 384, 388, 392
atmospheric CO_2 concentration 30
Attalea funifera 327
Attheya septentrionalis 287, 300
autoflocculation 300
autonomous equipment for cropping 71
autothermal gasification 362
*Avena sati*va 327
avocado 327

B20 blend (*see also* biodiesel) 323, 326, 327, 332
B30 blend (*see also* biodiesel) 326

Sustainable Energy Developments

Series Editor: Jochen Bundschuh

ISSN: 2164-0645

Publisher: CRC Press/Balkema, Taylor & Francis Group

9. Advanced Oxidation Technologies – Sustainable Solutions for Environmental Treatments
Editors: Marta I. Litter, Roberto J. Candal & J. Martín Meichtry
2014
ISBN: 978-1-138-00127-5 (Hbk)

10. Computational Models for CO_2 Geo-sequestration & Compressed Air Energy Storage
Editors: Rafid Al-Khoury & Jochen Bundschuh
2014
ISBN: 978-1-138-01520-3 (Hbk)

Printed and bound by CPI Group (UK) Ltd, Croydon, CR0 4YY

18/10/2024

01776249-0005